Lecture Notes in Electrical Engineering

Volume 822

The book series *Lecture Notes in Electrical Engineering* (LNEE) publishes the latest developments in Electrical Engineering - quickly, informally and in high quality. While original research reported in proceedings and monographs has traditionally formed the core of LNEE, we also encourage authors to submit books devoted to supporting student education and professional training in the various fields and applications areas of electrical engineering. The series cover classical and emerging topics concerning:

- Communication Engineering, Information Theory and Networks
- Electronics Engineering and Microelectronics
- Signal, Image and Speech Processing
- Wireless and Mobile Communication
- Circuits and Systems
- Energy Systems, Power Electronics and Electrical Machines
- Electro-optical Engineering
- Instrumentation Engineering
- Avionics Engineering
- Control Systems
- Internet-of-Things and Cybersecurity
- Biomedical Devices, MEMS and NEMS

For general information about this book series, comments or suggestions, please contact leontina.dicecco@springer.com.

To submit a proposal or request further information, please contact the Publishing Editor in your country:

China

Jasmine Dou, Editor (jasmine.dou@springer.com)

India, Japan, Rest of Asia

Swati Meherishi, Editorial Director (Swati.Meherishi@springer.com)

Southeast Asia, Australia, New Zealand

Ramesh Nath Premnath, Editor (ramesh.premnath@springernature.com)

USA, Canada:

Michael Luby, Senior Editor (michael.luby@springer.com)

All other Countries:

Leontina Di Cecco, Senior Editor (leontina.dicecco@springer.com)

**** This series is indexed by EI Compendex and Scopus databases. ****

More information about this series at https://link.springer.com/bookseries/7818

Sathans Suhag · Chitralekha Mahanta ·
Sukumar Mishra
Editors

Control and Measurement Applications for Smart Grid

Select Proceedings of SGESC 2021

Springer

Editors
Sathans Suhag
Department of Electrical Engineering
National Institute of Technology Kurukshetra
Kurukshetra, Haryana, India

Chitralekha Mahanta
Department of Electronics and Electrical Engineering (EEE)
Indian Institute of Technology Guwahati
Guwahati, Assam, India

Sukumar Mishra
Department of Electrical Engineering
Indian Institute of Technology Delhi
New Delhi, India

ISSN 1876-1100 ISSN 1876-1119 (electronic)
Lecture Notes in Electrical Engineering
ISBN 978-981-16-7663-5 ISBN 978-981-16-7664-2 (eBook)
https://doi.org/10.1007/978-981-16-7664-2

This Springer imprint is published by the registered company Springer Nature Singapore Pte Ltd.
The registered company address is: 152 Beach Road, #21-01/04 Gateway East, Singapore 189721, Singapore

Preface

This book, one of the three volumes, stems from the peer-reviewed select research papers presented in the International Conference on Smart Grid Energy Systems and Control, organized during March 19–21, 2021, by the Department of Electrical Engineering, National Institute of Technology Kurukshetra.

Electric power networks across the globe are experiencing transformational changes enforced by the environmental concerns, evolving information and communication technologies, concept of Internet of Things, advanced networking, and all important needs to control this emerging power network reliably and efficiently. These aims all converge in the smart grid energy system which is resilient, has much greater penetration of renewables, and uses real-time measurement and control. Currently, smart grid is an important research area in almost all parts of the world; however, it has got its associated issues and challenges.

The aim of the book is therefore to provide a perspective on evolving research trends in smart grid control and how control enables smart grid applications to present advanced research around the use of various control concepts, machine learning tools, evolutionary optimization algorithms, and the concepts like Internet of Things in smart grid. The quality research contributions in the book will be valuable in describing the important concepts and technologies that are enabling smart grid control research and applications and as such will be a source of high educational value for the reader/researcher who is interested in the domain area. In addition, this book is also an attempt to bring home the point that emerging landscape of technologies in the smart grid research is multidisciplinary.

The editors wish to thank the research contributors, the learned keynote speakers, and the organizing team members for their value additions. The editors also wish to place on record special thanks to Springer Nature for associating with the conference and contributing in bringing out this book. In addition, the financial support extended by Technical Education Quality Improvement Programme, Phase III, National Institute of Technology Kurukshetra is gratefully acknowledged.

Wishing all well!!

Kurukshetra, India — Sathans Suhag
Guwahati, India — Chitralekha Mahanta
New Delhi, India — Sukumar Mishra

Contents

About the Editors

Prof. Sathans Suhag received his B.Tech., M.Tech., and Ph.D. degrees from NIT Kurukshetra where he is currently working as Professor in the Department of Electrical Engineering with overall teaching, research, and administrative experience of more than 26 years. He has guided and is currently guiding quite a good number of Ph.D. scholars around the areas of hybrid energy systems, microgrid and its control, evolutionary algorithms and their applications in Hybrid Power System etc. He has published more than 80 research articles in well-known international/national journals and conferences of repute. He has served as Dean (Students' Welfare) for 3 years and is currently serving as the Coordinator, Technical Education Quality Improvement Programme, Phase-III at NIT Kurukshetra. At the National level, he has served for 3 years as the Chairman of All India Board of Technician Education of AICTE, New Delhi and also as the President's Nominee for faculty selections at NITs across the country. He is serving on various high level committees/bodies/task forces as member/chairman at the national level. He has travelled to USA, Australia, France, UAE, Canada, and Japan for various academic interactions and has delivered talks in reputed international conferences.

Prof. Chitralekha Mahanta after completing her Doctoral studies (Ph.D.) from IIT Delhi in India, Dr. Chitralekha Mahanta joined as an Assistant Professor in the Department of Electronics and Communication Engineering (ECE), IIT Guwahati in the year 2000. Since then she has been involved in active research in the area of control theory and its applications. She has offered a variety of courses in undergraduate and post graduate studies in the field of control systems at IIT Guwahati. She has been a full time Professor in the Department of Electronics and Electrical Engineering (EEE), IIT Guwahati since April, 2012. Starting her research at IIT Guwahati in the field of Intelligent Control, currently she is involved in the areas of robust and adaptive control with applications in robotics and flight control. Her work has been published in well-known peer reviewed international journals and conference proceedings. Dr. Mahanta has shouldered various administrative responsibilities in the institute level. She was the Organizing Chairperson of GATE 2010 and Associate Dean of Research and Development, IIT Guwahati from 2012 to 2014. She served as the Head, Electronics and Electrical Engineering Department from 3.11.2014 till 5.11.2017. Dr. Mahanta is currently serving as the Dean of Academic Affairs of IIT Guwahati.

Prof. Dr. Sukumar Mishra received his M.Tech. and Ph.D. in Electrical Engineering from National Institute of Technology, Rourkela in 1992 and 2000 respectively. After spending 9 years as a lecturer at Sambalpur University (Orissa), Prof. Mishra joined BPUT (Orissa) as a Reader at Electrical Department and served there for a period of 2 years. Presently, Dr. Mishra is a Professor at Indian Institute of Technology Delhi and has been its part for the past 17 years. Professor Mishra has won many accolades throughout his academic tenure of 27 years. He has been a recipient of Young Scientist award (1999) by Orissa Bigyan Academy, INSA medal for young scientist (2002), INAE young engineer award (2002), INAE silver jubilee young engineer award (2012), The Samanta Chandra Shekhar award (2016), Bimal Bose award (2019) and NASI-Reliance Platinum Jubilee award (2019). He has been selected as Mission Innovation National Champion (2019) under the Mission innovation initiative to accelerate Clean energy in India.

He has been granted fellowship from academies like NASI (India), INAE (India) and professional societies like IET (UK), IETE (India), IE (India). He has also been recognized as the INAE Industry Academic Distinguish Professor. Apart from all research and academic collaborations, Prof. Mishra is very actively involved in industrial collaborations. Professor Mishra is currently acting as ABB Chair professor and has previously delegated as the NTPC, INAE and Power Grid Chair professor. He has also served as an Independent Director of the Cross Border Power Transmission Company Ltd. and the River Engineering Pvt. Ltd.

Professor Mishra has also carried out many important industrial consultations with TATA Power, Microtek and others. Professor Mishra is a great enthusiast of innovation and has supported many startups in the past and recently, has incorporated a new company named Silov Solutions Private Limited under the Companies Act, 2013. The company specifically deals in products related to renewable energy sources utilizable at household scale as well as at commercial setups for example bidirectional electric vehicle supply equipment, AC based EV charger, DC based EV charger, grid connected solar inverters, smart DC home management systems, online UPS etc.

Professor Mishra's research expertise lies in the field of Power Systems, Power Quality Studies, Renewable Energy and Smart Grid. Also, he has been functioning as Associate Dean R&D of IIT Delhi from March, 2020. Professor Mishra has so far authored more than 70 IEEE Transactions/Journals, 30 IET Journals and 30 other international journal papers. He has supervised 33 Ph.D. students (18 ongoing), 40 Masters students and four Post Doctorate Fellows (two ongoing). Professor Mishra has also authored five book chapters so far and has 16 patents to his credit.

Professor Mishra has been working in close association with IEEE Delhi Section Executive Committee for past few years and is currently serving as an Editor for the *IEEE Transactions on Smart Grid*, *IEEE Transactions on Sustainable Energy* and was an Area Editor for the *IET Generation, Transmission & Distribution* journal.

Combination BFPSO Tuned Intelligent Controller for Maximum Power Point Tracking in Solar Photovoltaic Farm Interconnected to Grid Supply

B. Jegajothi, C. Yaashuwanth, K. Prathibanandhi, and S. Sudhakar

Abstract Renewable energy resources are non-pollution resources that can meet up the electricity needs without inflicting any environmental troubles. In this research work, Maximum Power Point Tracking (MPPT) behavior is taken into account for improving the power output for the grid-integrated solar photovoltaic (PV) farm with the help of a combination BFPSO tuned intelligent ANN controller. An Artificial Intelligence (AI)-based MPPT technique is utilized in solar PV arrays to maximize the electrical power output and satisfy the power demand. The combination BFPSO algorithm is selected for optimizing the connection weights in the ANN controller, and the developed ANN controller regulates the duty cycle of the DC/DC converter by monitoring the voltage and current profile of the solar PV farm. The developed optimization algorithm is implemented to get maximum feasible power from the 400 kW PV farm. Also, the proposed combination BFPSO tuned ANN controller is evaluated through means of predictable procedures like Particle Swarm Optimization (PSO), Bacterial Foraging Optimization Algorithm (BFOA). The simulation part of the proposed work is carried out in MATLAB/SIMULINK software.

Keywords Combination BFPSO algorithm · ANN controller · MPPT method · Solar PV farm · DC/DC converter · AI technique

---

B. Jegajothi (✉) · C. Yaashuwanth
Department of Information Technology, Sri Venkateswara College of Engineering (Autonomous), Sriperumbudur, Chennai, India

K. Prathibanandhi
Department of Electrical and Electronics Engineering, Sri SaiRam Engineering college, Chennai, India

S. Sudhakar
Department of Electrical and Electronics Engineering, Bharath University, Chennai, India

S. Suhag et al. (eds.), *Control and Measurement Applications for Smart Grid*,
Lecture Notes in Electrical Engineering 822,
https://doi.org/10.1007/978-981-16-7664-2_1

1 Introduction

The electricity demand is ever-increasing day by day due to industrialization. Traditional electricity resources are depleting due to insufficient reserves of coal, natural gas, and oil. Furthermore, the majority of electricity resources are not environmentally friendly. As a result, there is a need to create a new mechanism that can be utilized as an alternative source of energy [1]. The superlative nature of the energy power system is designed effectively with exceptional qualities and functionalities irrespective of the grid-connected or standalone mode under various loads and system conditions. The PV systems which are grid-connected inevitably provide an infusion of energy into the grid. The advantages of PV systems fail due to their high cost during the initial investment for obtaining grid parity. To boost the effectiveness of the PV cells, the production and fabrication process of the inverter is initiated to neglect the potential of increasing the throughput by improving the Maximum Power Point Tracker (MPPT) [2–4]. The effective functioning of the solar power PV system includes an optimal transfer of energy generated in the array for loading produced by the deft blend of battery supplier and load. The load provided to the PV array by the battery tends to change when a fluctuation is observed in charge of the battery and the temperature. When a DC–DC convertor is introduced in between, maximum power energy transfer is achieved by the PV array [5]. The PV array power and current qualities are extremely nonlinear, which are deeply enhanced by the temperature and the irradiance deviation. MPPT has to deal with inconvenience related to power and ensures the effective functioning of the PV system at the maximum power point (MPP). In the grid-connected PV system, the production of maximum possible power is obtained by the projected techniques [6].

The rest of this paper is structured as follows. Section 2 details the literature which was based on the adaptive controller strategy for grid-connected solar PV power conditioning system is described. Sect. 3, briefs about the related challenges in MPPT control methodology. Sect. 4 elaborates on the proposed combination BFPSO tuned ANN controller for grid-integrated solar power conditioning system. In Sect. 5, the simulation results of this proposed approach are discussed. At last, this paper is concluded with Sect. 6.

2 State of Art

Sera et al. [7] have presented a Hill-Climbing (HC) based MPPT control method for the GCPV system: The Perturb-and-Observe (P&O) and Incremental Conductance (INC) methods fail to track the MPP when the Sun's radiation varies. The authors created a control technique based on the HC optimization algorithm to avoid oscillations around the Maximum Power Point in the PV curve.

Putri et al. [8] have developed a control method for MPPT in solar PV systems using the Incremental Conductance (IC) algorithm to achieve maximum efficiency.

The proposed IC-based MPPT tracker was designed to manage the switching pulses of the boost converter circuit. In this way, the MPPT tracker will extract the maximum available power whenever the irradiation and temperature changes occur.

Xiao et al. [9] have proposed a Hill-Climbing (HC) based MPPT strategy which is anticipated. The proposed HC based MPPT tracker finds the Maximum Power Point (MPP) and produces the higher electricity without oscillation phenomenon. The outcome from the investigated method shows that the HC based tracker gives better performance when compared to PO and INC methods.

Lin et al. [10] have presented a reactive power controller with Probabilistic Wavelet Fuzzy Neural Network (PWFNN) for grid-connected three-phase photovoltaic (PV) system in grid faults. The anticipated method reduces the flow of reactive power injection and balances the flow of active power into the source and grid side through low-voltage ride through (LVRT) regulation.

Several algorithms and controller strategy are applied for tracking the MPPT in the solar PV system, such as PO, PSO, CS, ANN, and adaptive controllers to refine the efficiency of solar PV system Maximum Power Point Tracking. If we apply conventional methods like the P&O method for MPPT tracking, it fails to track the MPP whenever the solar radiation changes (G) or ambient temperature variation (T) occurs. Also, the PSO based MPPT approach is not suitable for modern controllers, the adaptability of the parameters based on the PSO algorithm is not suitable for the controller. This literature review details the variety of topics, techniques, methods, and approaches related to intelligent MPPT controllers. The highlights of drawbacks in existing MPPT methods lead to developing an outline of this research toward improving the tracking presentation beneath the change in environmental conditions.

3 Challenges in MPPT Methodology

The classical Maximum Power Point Tracking (MPPT) scheme utilizes the algorithms like P&O, INC, and HC. These deterministic algorithms fail to track the MPP under varying weather circumstances. Normally, the solar PV panel parameters are highly nonlinear, so the panel output voltage and current will always be nonlinear. To handle such nonlinear parameters and to reduce the oscillations around the MPP, the bio-inspired algorithms are hybridized with controllers for increasing the tracking efficiency and reduce the oscillations due to varying weather conditions. The dissimilarity of voltage is typically confined by way of 70–80% of the open-circuit voltage beneath any situation of temperature (T) and solar radiation (G) [11–13]. Therefore, the mixture of two conventional algorithms like PSO and BFOA is adopted in ANN regulator for improving the tracking efficiency in PV farm interconnected with grid supply. The developed intelligent MPPT controller works efficiently and decreases the oscillation around the MPP when the environmental conditions varied.

4 Proposed Combination BFPSO Tuned MPPT Controller

The proposed combination BFPSO based intelligent MPPT control method is utilized to track the Maximum Power Point in the 400 kW PV farm under varying solar irradiance (G) and varying cell temperature (T) circumstances. Here, the solar PV farm consists of four PV array systems each delivering 100 kW at 1000 W/m^2. The proposed method hybridizes the BFPSO algorithm and ANN regulator for improving the tracking speed and efficiency. Figure 1 depicts the block diagram for the proposed combination BFPSO based intelligent MPPT controller for grid-incorporated PV systems. In this case, the proposed regulator detects the variations in the 400 kW solar PV farm and automatically adjusts the duty ratio of the DC/DC converter circuit. In this manner, the power output of the solar PV farm will automatically improve without distressing the system performance. That is, whatever may be the interruption occurs in solar PV farm the proposed intelligent controller response well with the BFPSO method. The AI-based ANN controller weight fine-tuning is done with the help of the proposed BFPSO algorithm. After tuning the controller weights, the ANN generates a set of data for controlling the output power of the PV farm. Once trained, the ANN controller will work offline with unseen inputs to predict the best power output. The steps are included in the optimization of the control parameters that are described in the following section.

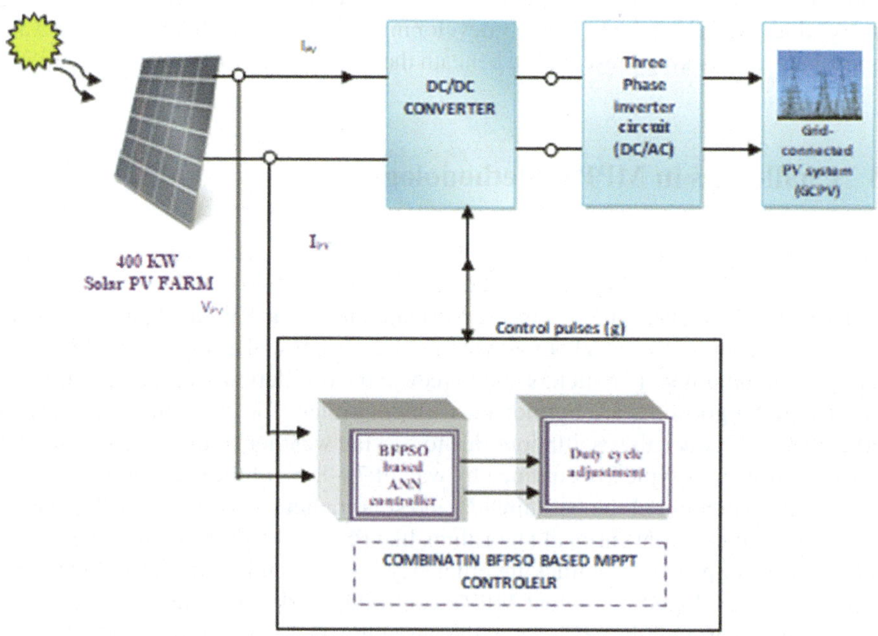

Fig. 1 Block diagram for the proposed combination BFPSO tuned ANN regulator

Combination BFPSO Algorithm: Bacterial Foraging Optimization Algorithm (BFOA) is a bio-inspired algorithm that is annoyed by the group behavior strategy of the *E. coli* bacterium. The bacterium foraging stratagem seeks to exploit the amount of food is gained by moving thru the surroundings. This foraging activity prompted recent researchers to use it in the optimization procedure for maximizing the output. The BFOA approach prevents the local minima and makes results in global maxima. However, while the tumbling process in BFOA, the unit length is randomly generated. This may lead to consuming more time for getting an optimal solution. The convergence speed of the BFOA algorithm is improved with the help of proposed intelligent controller combined with the Particle Swarm Optimization (PSO) algorithm. The PSO is a stochastic optimization that was stimulated by the actions of a group of birds swarming behavior. The proposed combination method produces the best global solutions through the grounded work of bacterial foraging (BF). The combination BFPSO algorithm combines the merits of BFOA and PSO approaches. In connection with the optimization process, the tumbling behavior of each swarm is taken by the individual best and global best positions [4, 5, 14]. The mathematical modeling of the BFPSO algorithm is as follows,

Step 1: The essential parameters of both BFOA and PSO algorithms are initiated first.

Step 2: The chemotactic movement of *E. coli* bacteria through swimming and tumbling thru flagella can be computed and modeled here,

$$\theta_i(j+1,k,l) = \theta_i(j,k,l) + C(i)\frac{\Delta(i)}{\sqrt{\Delta_T(i)\Delta(i)}} \tag{1}$$

where in Eq. (1) $\theta_i(j,k,l)$ stands for the current best position of the ith bacterium, j, k, and l indicates the numbers of chemotactic step, reproduction step, and elimination dispersal events correspondingly, $\Delta(i)$ stands for the randomly generated vector whose elements are in $[-1, 1]$ for *E. coli* bacterium movement toward favorable search, and $C(i)$ the unit step length [15–19]. The cell–cell communication between the *E. coli* swarm can be modeled as follows,

$$\begin{aligned} J_{cc}(\theta, p(j,k,l)) &= \sum_{i=1}^{s} J_{cc}(\theta, \theta_i(j,k,l)) \\ &= \sum_{i=1}^{s}\left[-d_{\text{attractant}} \exp\left(-w_{\text{attractant}} \sum_{m=1}^{p} \left(\theta_m - \theta_m^i\right)^2\right)\right] \\ &\quad + \sum_{i=1}^{s}\left[h_{\text{repellant}} \exp\left(-w_{\text{repellant}} \sum_{m=1}^{p} \left(\theta_m - \theta_m^i\right)^2\right)\right] \end{aligned} \tag{2}$$

where in Eq. (2), $J_{cc}(\theta, p(j,k,l))$ is the objective function value to be added to the actual objective function (to be minimized) to create a time-varying objective function, S is the total number of bacteria, p is the number of parameters to be optimized that are present in each bacterium, and $\theta = \left[\theta_1, \theta_2, \ldots \theta_p\right]^{\mathrm{T}}$ is a point in the p-dimensional search domain. $d_{\text{attractant}}, w_{\text{attractant}}, h_{\text{repellant}}, w_{\text{repellant}}$ are different coefficients should be selected correctly [1, 9].

Step 3: After the successful completion of each chemotactic step, the position of each particle will be changed with the velocity Eq. (4) modeled using the PSO algorithm [20].

The health status of each bacterium is calculated after each successful chemotactic process. The best fitness values of the *E. coli* bacteria are sorted in array order arrange from low to high value. Here, the worst values of the bacterial are replaced from the array group to update the positions. This makes the populace of bacteria steady. After the chemo-tactic step, the health of each bacterium during its life-time is accumulated and is designed as Eq. (3),

$$J_{\text{health}}^{i} = \sum_{j=1}^{N_c+1} J(i, j, k, l) \tag{3}$$

where N_c are the chemotactic steps and j is the value of error.

Step 4: After each chemotactic mechanism, each movement computes the best fitness value. Best fitness value J_{health} value is stored for all the iteration. The best fitness positions of bacterium values are applied in the velocity equation modeled from the PSO method. The velocity of the particle is modeled through Eq. (4).

$$V = V + C_1 * \text{rand}(P_{\text{best}} - P) + C_2 * \text{rand}(g_{\text{best}} - P) \tag{4}$$

where C_1 and C_2 are the accelerating factors.

Step 5: The position of each particle is altered with the velocity Eq. (4) computed using the PSO algorithm. With the help of the velocity Eq. (4) and repeat steps 2–5 till getting the best optimal solutions [20–23].

Step 6: Substitute this new velocity positions of the particles.

Step 7: Continue the algorithm with the significant reproduction step and elimination dispersal state.

In the proposed strategy, BFPSO algorithm is applied to optimize the DC/DC converter switching pulses through varying the duty cycle of the converter. The nonlinear solar PV array parameters are monitored automatically via the proposed intelligent controller according to the change in solar radiation (G) and cell temperature (T). The results of the proposed combination BFPSO algorithm are utilized to optimize the weights in the ANN regulator. The ANN training process involves providing input/output pairs. The main target of training ANN is to acquire a set of connection weights that gives a minimum error. Initially, feedforward and cascade forward ANN is trained using the scaled conjugate backpropagation algorithm. Accuracy is calculated and extracts the connection weight values of the network. These connection weights are then optimized by employing a combination BFPSO [21]. BFPSO calculates the objective function after each iterative step of the program as the program executes and steadily leads to an optimal solution with better fitness. Optimizing control parameters are chosen as per the fitness function and assess their ANN output equation.

Combination BFPSO Algorithm Implementation in ANN Controller: The combination BFPSO optimization algorithm is the preferred most favorable preparation dataset to lead ANN for escalating the knowledge presentation. Generally, the ANN controller contains three layers such as input layer, hidden layer, and output layer. An input layer has various nodes that are resolute by the data set. The ANN controller reads the dataset prepared from the BFPSO algorithm, once the controller trained with this data set, it is readily available to work with offline also. The entire weights are assessed to offer the abstract of this node (A), and then, the establishment task of this node is calculated. An identical function for supplementary nodes, the charge of x, Eq. (5) is calculated, and then, the establishment task of the identical node is calculated by receiving an establishment task of an output node. There are several categories of an establishment task like a step task, a sign task, and a sigmoid task. A regular establishment task is a sigmoid task which is displayed in Eq. (5)

$$f(x) = \frac{1}{1 + \mathrm{e}^{-x}} \tag{5}$$

where x is input which is specified as Eq. (5),

$$x = \sum_{b=1}^{n} w_{ab} O_b + \theta_b \tag{6}$$

The dissimilarity between objective output (O_{T}) and definite output (O_{A}) is calculated as a fault (e) that is designed in the feedforward network. Currently, the fault is stated in Eq. (6),

$$e = \frac{1}{2} \sum \sum (O_{\mathrm{T}} - O_{\mathrm{A}})^2 \tag{7}$$

Generally, the records of nodes in an input layer and an output layer are determined by the number of inputs and the number of outputs of its dataset repetitively. Presently, the Kolmogorov theorem is functioning for the number of nodes in a hidden layer which is specified in Eq. (8),

$$\text{Hidden nodes} = 2 \times \text{input} + 1 \tag{8}$$

In artificial neural networks, the artificial neuron is a simple processing component, and its interior adaptable limitations are recognized as connection weights. Artificial neuron's weight, sum, and threshold incoming signals are generating output for the proposed controller. Information is collected in the power of the interconnections or weights and the thresholds/biases. The most important intention of ANN preparation is to define a group of association weights that lessen the faulting task. The network is an available couple of input/output data and the fault (dissimilarity amongst predicted output and actual output) as a period of learning. This learning process is based on the BFOA which is an advanced count to the people of nature-inspired optimization algorithms.

5 Simulation Results and Discussion

The proposed combination BFPSO based MPPT controller is tested under diverse solar radiation patterns and cell temperature, and it is put through a wide range of application scenarios. The simulation of the proposed work is done in the MATLAB/SIMULINK 7.10.0 (R2015a) software. The proposed combination BFPSO based MPPT Simulink model is shown in Fig. 2. The developed combination model consists of 4 PV arrays each producing 100 kW at 1000 W/m^2 solar radiation (G). A single PV cluster block has 64 parallel strings, where each string consists of five modules associated in series to form the solar PV farm. In this section, discussed the output power parameters of the 400 kW PV farm are compared with the conventional methods like Particle Swarm Optimization (PSO) and BFOA. To

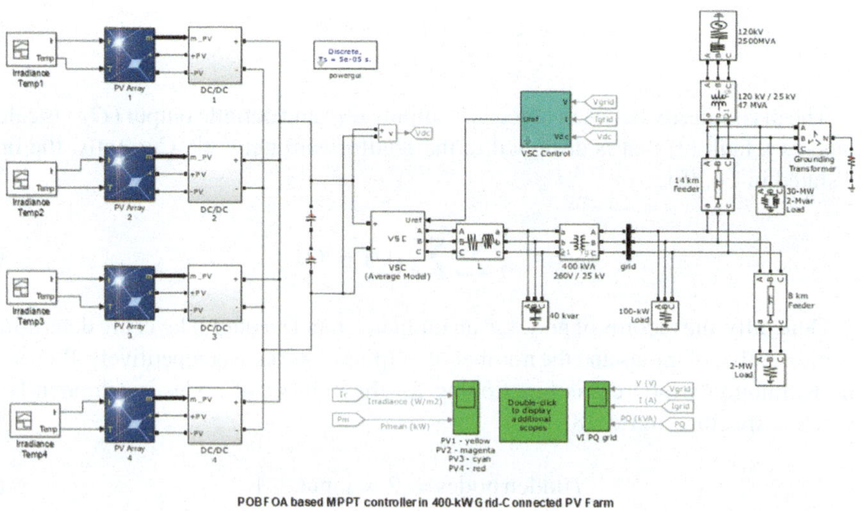

Fig. 2 Simulink diagram for the proposed combination BFPSO tuned ANN controller

Table 1 Specifications utilized in the simulated process

Parameters	Example
Model	Sun power SPR-315E-WHT-D
Peak power (W) P_{MPP}	315.072
V_{MPP} of one module	54.7 V
I_{MPP} of one module	5.76 A
No of series modules/string	5
Open-circuit voltage (V_{OC})	64.6 V
Number of parallel strings	64
Short-circuit current (I_{SC})	6.14 A
Number of series connected cells/string (N_{S})	5
Temperature coefficient of voltage (mV/°C) (K_{V})	−0.272 V/K
Temperature coefficient of current (K_{I})	0.061694 A/K

confirm the presence of the proposed method of combination BFPSO tuned ANN controller, it is compared with some traditional techniques of PSO and BFOA.

The power produced from the 400 kW PV farm is displayed for the varying solar radiation and temperature as shown in Fig. 5. The implementation parameter for the developed algorithm is given in Table 1. The performance of simulation is tested under varying solar radiation (G) and cell temperature (T). The simulations parameters are used in optimization process which are taken from solar panel side voltage and current. To examine the performances of the proposed intelligent MPPT controller during varying solar irradiance (G) and cell temperature (T) patterns. Solar radiation (G) is varied from (1000, 250, and 1000 W/m^2), and cell temperature is varied from 25 to 50 °C. During the change in environmental conditions, the developed controller adjusts the duty cycle which sets the global maximum power point in the solar PV farm. From the simulation output profile, we can see the system improved performance without oscillations and robustness under varying PV parameters. Table 1 illustrates the parameters utilized in the 400 kW PV farm while the simulation process.

Training in MATLAB **Using Neural Fitting APP**: This section discussed the training progression done in the Neural Fitting (NF) application tool in MATLAB using the proposed combination BFPSO based algorithm. Here, the feed-forward neural network is selected with three input layers, ten hidden layers, and three output processes. Neural network training is done automatically through the Levenberg training method; once trained, the ANN controller will function offline also for extracting the objective function from the PV farm. Figure 3 shows the training done in the NF tool.

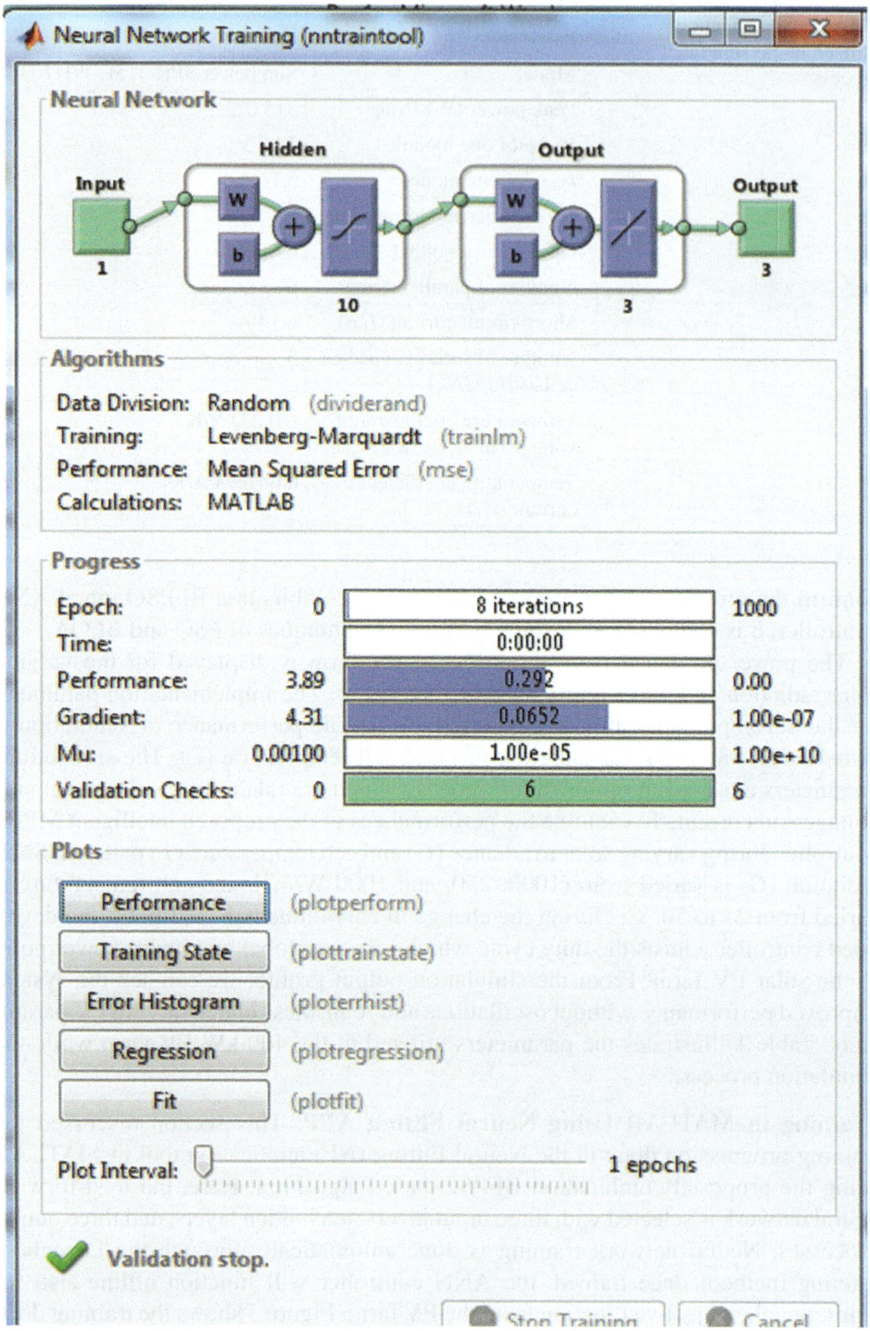

Fig. 3 Sketches about the training done in the NF tool

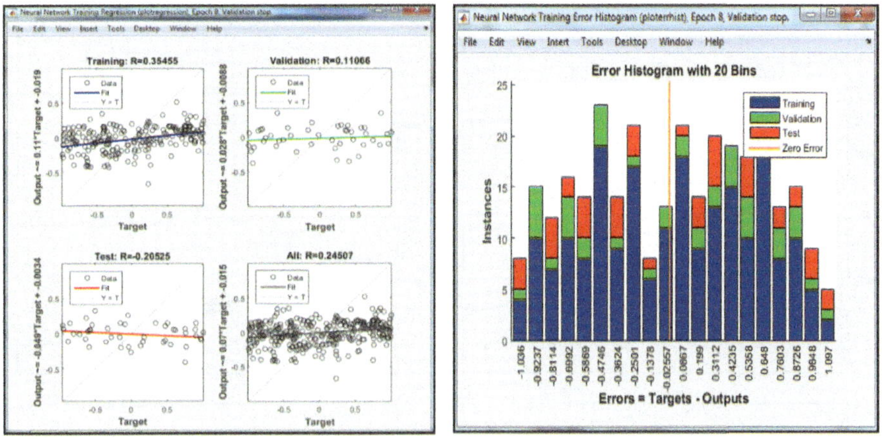

Fig. 4 Output training validation and error histogram graph for the ANN controller

The error calculation in the controller is done by the proposed algorithm of the Mean Square Error (MSE) method. According to the error value, the proposed ANN-based MPPT controller will adjust the duty ratio of the DC/DC converter circuit in this manner the output power of the solar PV farm is controlled efficiently with the proposed strategy. Each panel in a solar PV farm is trained individually with the developed ANN controller. After training in the NF tool using the optimized values, the results in the outcome graph show its effectiveness of the PV farm track. Figure 4 sketches the output training validation and error histogram plot for the developed ANN controller.

The solar PV farm parameters are examined after implementing the proposed ANN controller is interfaced with the 25 kV grid supply. Here, each PV module is controlled through an individual MPPT controller using the BFPSO algorithm. The average model of DC/DC converter is utilized in the proposed combination model, such models do not produce harmonics profile in the VSC converter interfaced with it. The output for the average model of DC/DC converter's DC link voltage of 500 V is maintained steady throughout the simulation process. The 500 V DC is converted into 260 V AC with the aid of a three-phase VSC converter connected to the 25 kVA grid supply. Figures 5 and 6a, b show the detailed simulation output of PV farm and also highlight DC link voltage profile and the grid parameters with improved performance. Figure 7 sketches about the power comparison plot for existing PSO based ANN controller and combination BFPSO based ANN controller. The efficiency comparison plot for the existing and proposed method is shown in Fig. 8.

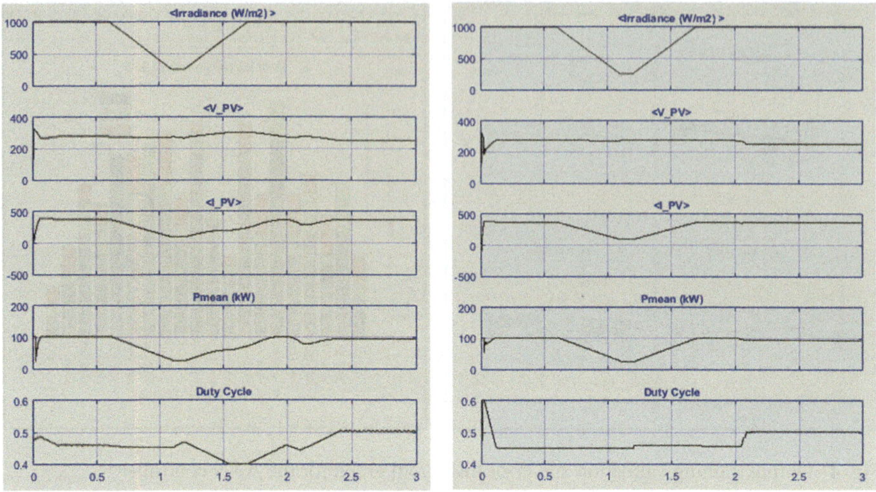

(a) (b)

Fig. 5 Detailed simulation output for the 400 kW PV farm under varying solar radiation (*G*) and cell temperature (*T*). **a** Existing PSO tuned ANN controller, **b** combination BFPSO tuned ANN controller

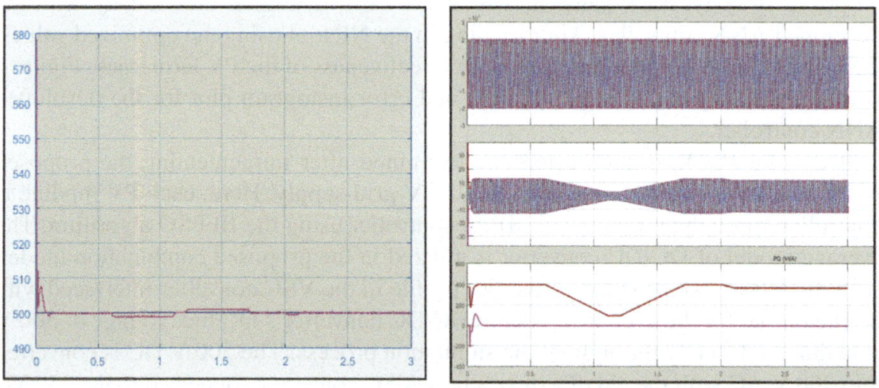

(a) (b)

Fig. 6 **a** DC link voltage profile and **b** grid output parameters for the proposed combination BFPSO tuned ANN controller

6 Conclusion

In this paper, an intelligent ANN controller-based MPPT tracking is used in a 400 kW PV farm. The suggested MPPT method is based on a combination BFPSO algorithm adapted in ANN controller algorithms. Here, the developed algorithm is implemented in four types of photovoltaic arrays independently; each has a capacity of 100 kW.

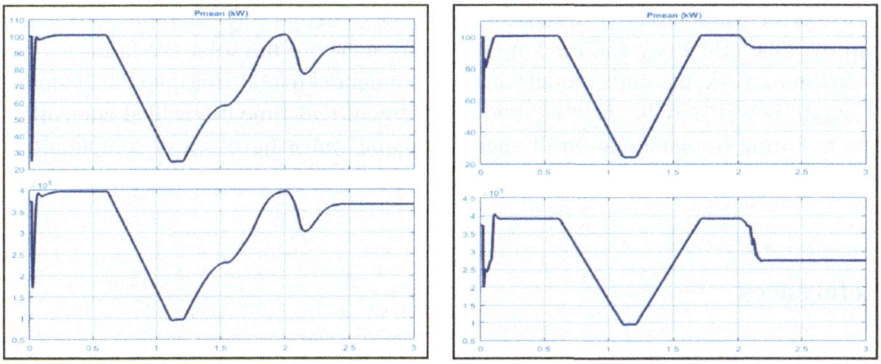

(a) (b)

Fig. 7 Output power comparison analysis for the existing and proposed models. **a** Existing PSO tuned ANN controller, **b** combination BFPSO tuned ANN controller

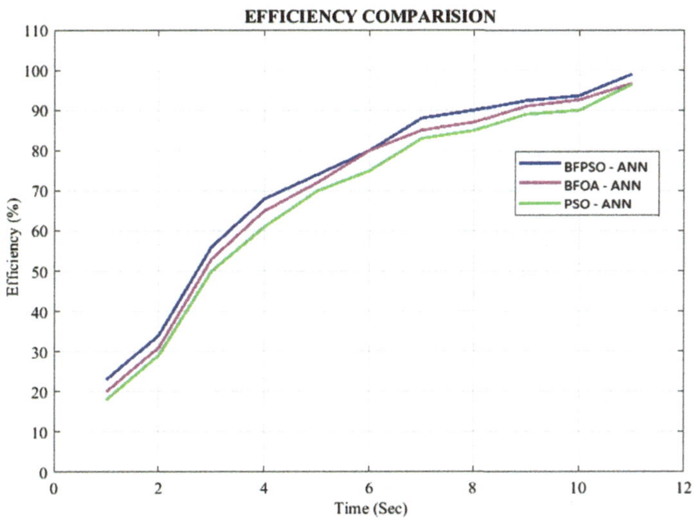

Fig. 8 Efficiency comparison graph for the existing PSO, BFOA, and proposed combination BFPSO tuned ANN controller

The motivation of the research work is to extract the maximum available power from the solar PV farm through the proposed controller is obtained. The simulation model is designed in MATLAB/SIMULINK software. The developed ANN controller continuously monitors the solar PV farm under different environmental conditions. The simulation results demonstrate the ability of the BFPSO algorithm to extract more power from the solar PV farm under the rapid change in solar radiation (G) and cell temperature (T) circumstances. The output power profile of the proposed model has less oscillation than the existing method like particle swarm optimization

(PSO) as the simulation figure shows its robustness. In this way, the proposed method improves the efficiency and tracking the performance of the solar PV farm.

In future work, the same model can be implemented in real-time and the outcomes of output power may be examined with intelligent real-time embedded controllers. The real-time implementation of such techniques must have a less computational cost.

References

1. Panwar NL, Kaushik SC, Kothari S (2011) Role of renewable energy sources in environmental protection: a review. Renew Sustain Energy Rev 15(3):1513–1524
2. Baba AO, Liu G, Chen X (2020) Classification and evaluation review of maximum power point tracking methods. Sustain Futures 2:100020
3. Pakkiraiah B, Sukumar GD (2016) Research survey on various MPPT performance issues to improve the solar PV system efficiency
4. Ji Y-H, Jung D-Y, Kim J-G, Kim J-H, Lee T-W, Won C-Y (2011) A real maximum power point tracking method for mismatching compensation in PV array under partially shaded conditions. IEEE Trans Power Electron 26(4):1001–1009
5. Fadaee M, Radzi MAM (2012) Multi-objective optimization of a stand-alone hybrid renewable energy system by using evolutionary algorithms: a review. Int J Renew Sustain Energy Rev 16(5):3364–3369
6. Kaleem Z, Yoon TM, Lee C (2016) Energy efficient outdoor light monitoring and control architecture using embedded system. IEEE Trans Embedded Syst Lett 8(1):18–21
7. Sera D, Mathe L, Kerekes T, Spataru SV, Teodorescu R (2013) On the perturb-and-observe and incremental conductance MPPT methods for PV systems. IEEE J Photovolt 3(3):1070–1078
8. Putri RI, Wibowo S, Rifa'I M (2015) Maximum power point tracking for photovoltaic using incremental conductance method. Int J Energy Procedia 68:22–30
9. Xiao X, Huang X, Kang Q (2016) A hill-climbing-method-based maximum-power-point-tracking strategy for direct-drive wave energy converters. IEEE Trans Ind Electron 63(1):257–267
10. Lin F-J, Lu K-C, Ke T-H (2016) Probabilistic wavelet fuzzy neural network based reactive power control for grid-connected three-phase PV system during grid faults. Int J Renew Energy 92:437–449
11. Youssef A, El Telbany M, Zekry A (2018) Reconfigurable generic FPGA implementation of fuzzy logic controller for MPPT of PV systems. Int J Renew Sustain Energy Rev 82:1313–1319
12. Abdel-Salam M, El-Mohandes M-T, Goda M (2018) An improved perturb-and-observe based MPPT method for PV systems undervarying irradiation levels. Int J Sol Energy 171:547–561
13. Tian H, Mancilla-David F, Ellis K, Muljadi E, Jenkins P (2012) A cell-to-module-to-array detailed model for photovoltaic panels. Int J Sol Energy 86(9):2695–2706
14. Rai AK, Kaushika ND, Singh B, Agarwal N (2011) Simulation model of ANN based maximum power point tracking controller for solar PV system. Int J Sol Energy Mater Sol Cells 95(2):773–778
15. Jegajothi B, Yaashuwanth C (2018) Embedded controller based maximum power point tracking for photovoltaic system using adaptive technique. Int J Electr Eng 18, edition-1. Print ISSN 1582-4594
16. Passino KM (2002) Biomimicry of bacterial foraging for distributed optimization and control. IEEE Trans Control Syst 22(3):52–67
17. Sudhakar Babu T, Priya K, Maheswaran D, Sathish Kumar K, Rajasekar N (2015) Selective voltage harmonic elimination in PWM inverter using bacterial foraging algorithm. Int J Swarm Evol Comput 20:74–81

18. Ishaque K, Salam Z, Amjad M, Mekhilef S (2012) An improved particle swarm optimization (PSO)–based MPPT for PV with reduced steady-state oscillation. IEEE Trans Power Electron 27(8):3627–3638
19. Babu YS, Sekhar KC (2020) Battery assisted, PSO-BFOA based single stage PV inverter fed five phase induction motor drive for green boat applications. In: Intelligent systems, technologies and applications. Springer, Singapore, pp 227–240
20. Bouselham L, Hajji M, Hajji B, Bouali H (2017) A new MPPT-based ANN for photovoltaic system under partial shading conditions. Energy Procedia 111:924–933
21. Subudhi B, Pradhan R (2017) Bacterial foraging optimization approach to parameter extraction of a photovoltaic module. IEEE Trans Sustain Energy 9(1):381–389
22. Sarvi M, Ahmadi S, Abdi S (2015) A PSO-based maximum power point tracking for photovoltaic systems under environmental and partially shaded conditions. Prog Photovolt Res Appl 23(2):201–214
23. Zhu Z, Liu G (2018) MPPT control method for photovoltaic system based on particle swarm optimization and bacterial foraging algorithm. Int J Electr Compon Energy Convers 4(1):45

Nonlinear Error Function Based Extended Kalman Filter with Improved Scaling Factor for Cancer Chemotherapy

Utkarsha L. Mohite and Hirenkumar G. Patel

Abstract The most common cause of death is cancer because the same chemical treatment is the most important and widely used method. In this article, we monitor the number of significant cell types of normal cells, body cells, as well as tumors within unpredictability in the various variables in the cancer model. The nonlinear kernel-based error function using extended Kalman filter (EKF) is designed for operation; the error is detected using standard deviation and is used for the EKF regeneration process. Here, the recovery process is based on error operation. Also, besides, a measurement is included that monitors the development of the error of the update process. The proposed control is explored with other traditional methods that will influence the injection of the drug into normal cells, body, and tumor. It has also been confirmed that the proposed NEF-EKF operates strictly with parameter uncertainty.

Keywords Drug dosage · Extended Kalman filter · Chemotherapy · Controller · Error function

1 Introduction

Cancer is one of the most significant illness in the universe for humans [1–3]. Many cancer treatments are like chemotherapy, radiotherapy, hormone therapy, immunotherapy, and surgery. In addition, it is necessary to calculate the effectiveness and effectiveness of the chemotherapy program. In reality, chemotherapy [4–6] is a conventional procedure used to treat cancer in medicine. The introduction of chemotherapy is the best treatment and has received worldwide attention these days. Currently, engineering science has greatly contributed to this analysis by developing many computational models, demonstrating the effects of chemotherapy [7–9] and its dosage. This work has been used to develop and analyze various drug control methods [10–12]. This "silico test" is costly and helps clinicians and engineers to examine the inconsistencies of new drug dosages in medical pharmacies. Currently,

U. L. Mohite (✉) · H. G. Patel
Sardar Vallabhbhai National Institute of Technology, Surat, Gujarat 395007, India

S. Suhag et al. (eds.), *Control and Measurement Applications for Smart Grid*, Lecture Notes in Electrical Engineering 822,
https://doi.org/10.1007/978-981-16-7664-2_2

the combined use of siRNAs and chemotherapy [13, 14] remains the best cancer treatment and the most sought after. Forms of siRNA and chemotherapy medicine based on the 2D [15, 16] aim to reduce the drug's side effects and reduce the normal cell damage. However, after repeated chemotherapy [17, 18], the adverse impacts caused by chemo agents weaken the result, therefore leading to clinical failure [19–21].

Furthermore, Kalman's filtering approaches play an important role in measuring dynamic countries and improving measurement precision. It also enhances temporary accuracy and precision, thereby ensuring improved assumptions in levels of drug dosage [22–24]. Besides that, the use of the Kalman filter is a significant problem that lies in the setting of covariance matrixes. Furthermore, the introduction of EKF increases the precision of dynamic state measurement (DSE). Moreover, EKF can be used to measure immune cells and may change the dosage of the drugs and, as a result, regulate immune, tumor, and normal chemotherapy [25].

2 Enhanced Kalman Filter with Nonlinear Error Function with Scaling Factor

Body cell calculation is difficult in the Web domain and requires some testing, and therefore, a non-line observer body test is required. EKF is taking advantage of the enhanced nonlinear kernel-based error function in this investigation project with a scalar function, and the following are the processing requirements.

In the first phase, an indirect schedule of time is not created as shown in Eq. (1), where x_k and y_k apply to the measurement and sound phase, similarly. There, x_k and y_k is designated as Gaussian zero-based noise, and two covariance measures are also shown as R and Q, as described in Eq. (2). b_k defines the measuring vector that requires normal cells as well as stem cells, and a_k describes a system similar to plant counts, normal cells, and antibodies. These vectors are indicated in Eq. (3).

$$\begin{aligned} b_k &= f(b_{k-1}, u_{k-1}) + y_k \\ a_k &= U b_k + x_k \end{aligned} \tag{1}$$

$$\begin{aligned} \left\{w_k w_j^{\mathrm{T}}\right\} &= Q \delta_{kj} Q > 0, \\ \left\{x_k x_j^{\mathrm{T}}\right\} &= R \delta_{kj} R > 0 \\ \left\{y_k y_j^{\mathrm{T}}\right\} &= 0 \end{aligned} \tag{2}$$

$$\begin{aligned} b_k &= [I M_k T C_k N_k]^{\mathrm{T}} \\ a_k &= [T C_k N_k]^{\mathrm{R}} \end{aligned} \tag{3}$$

The cancer model is found according to f as in Eqs. (1), (2), and (3). Equation (4) including the test parameter values is applied to the EKF monitor. Hence, f is calculated as in Eq. (4).

$$f = \begin{bmatrix} \overline{st} + \frac{\overline{\rho} I M_k T C_k}{\overline{\alpha} + T_k} - \overline{q}_1 I M_k T C_k - \overline{t}_1 I M_k - \vec{y}_1 u_{1k} I M_k \\ \overline{ra}_1 T C_k (1 - \overline{c}_1 T C_k) - \overline{q}_2 I M T C_k - \overline{q}_3 T C_k N_k - \vec{y}_2 u_{2k} T C_k \\ \overline{ra}_2 N_k (1 - \overline{c}_2 N_k) - \overline{q}_4 T C_k N_k - \vec{y}_3 u_{3k} N_k \end{bmatrix} \tag{4}$$

This EKF predicts different states by following two stages including scaling factor (i) Update, (ii) Prediction, (iii) Scaling Factor.

2.1 Prediction

Equation (5) specifies one-step estimation of $\vec{b}_{k|k-1}$ along with error covariance matrix $R_{k|k-1}$ corresponding.

2.2 Update

So here, the $\vec{b}_k$ state estimation and PR_k assessment are carried out as demonstrated in Eq. (6), in which a gain of Kalman is represented by k_k. Hence, Eqs. (5) and (6) are determined for the immune cell, respectively.

$$\begin{aligned} \vec{b}_{k|k-1} &= f\left(\vec{b}_{k-1}\right) \\ R_{k|k-1} &= E_k R_{k-1} E_k^{\mathrm{T}} + Q_k \\ E_k &= \frac{\partial f(b)}{\partial b}\Big|_{b=\vec{b}_{k-1}} \end{aligned} \tag{5}$$

$$\begin{aligned} \vec{b}_k &= \vec{b}_{k|k-1} + k_k\left(a_k - U_k \vec{b}_{k|k-1}\right) \\ k_k &= R_{k|k-1} U_k^{\mathrm{T}} \left(U_k R_{k|k-1} U_k^{\mathrm{T}} + R_k\right)^{-1} \\ R_k &= \left(R_{k|k-1}^{-1} + U_k^{\mathrm{T}} R_k^{-1} U_k\right)^{-1} \end{aligned} \tag{6}$$

2.3 Scaling Factor

The scaling component SF can be processed according to Eq. (7), where, σe signifies the standard deviation, n implies the instant number, and xe indicates the weight, where xe is measured as according to Eq. (8), here In represents the past along with

Table 1 Computation of the scaling factor

Number of instants n	Error computation Δe	Standard deviation σe	Weight factor xe
1	–	–	–
2	$e_1 - e_0$	σe_1	$\frac{1}{2} + \frac{2}{2}$
3	$e_2 - e_1$	σe_2	$\frac{1}{3} + \frac{2}{3} + \frac{3}{3}$
$\vdots$	$\vdots$	$\vdots$	$\vdots$
n	$e_{n-1} - e_{n-2}$	σe_n	$\frac{1}{n} + \frac{2}{n} + \frac{3}{n} + \cdots + \frac{n}{n}$

present instants, and CIn implies to the current instants. σe is measured according to Eq. (9), Here nu implies the number of points of data, x_i represents every data value, and $\overline{x}$ specifies the mean of x_i. Table 1 provides the model of measurement of the parameters recognized SF (Fig. 1).

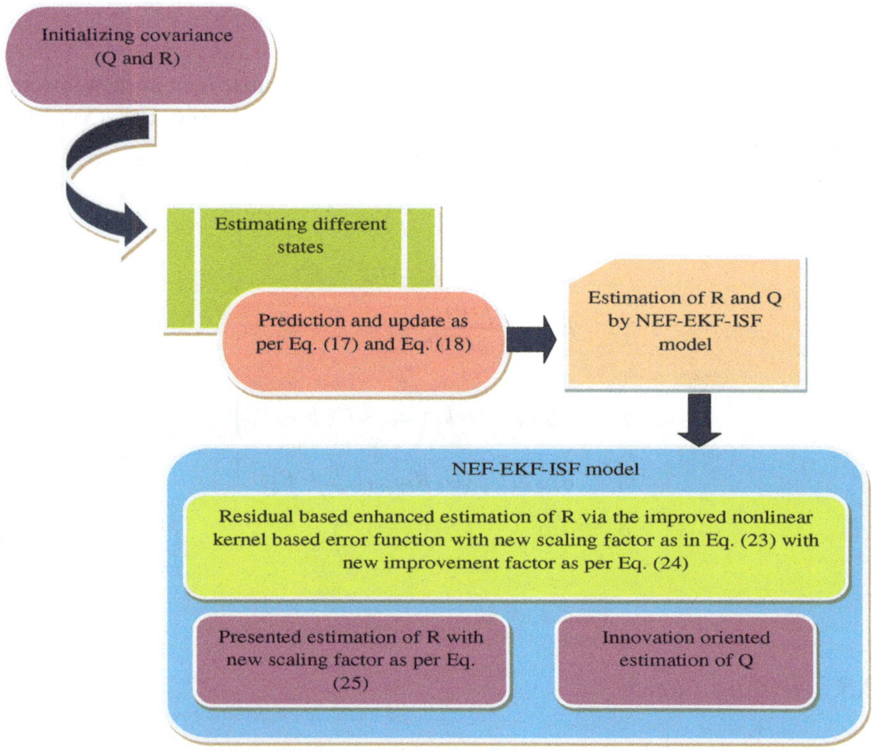

Fig. 1 Diagrammatic representation of proposed model

$$\mathrm{SF} = \frac{\sigma e}{n} \sum_{i=1}^{n} xe_i \tag{7}$$

$$xe = \frac{\mathrm{In}}{\mathrm{CIn}} \tag{8}$$

$$\sigma e = \sqrt{\sum_{i=1}^{nu} \frac{(x_i - \overline{x})^2}{nu - 1}} \tag{9}$$

3 Results and Discussion 3-D Analysis

The effect on control of three forms (immune, normal as well as tumor cells) of the introduced control theory (NEF-EKF-ISF) is shown in Fig. 2. The purpose of this

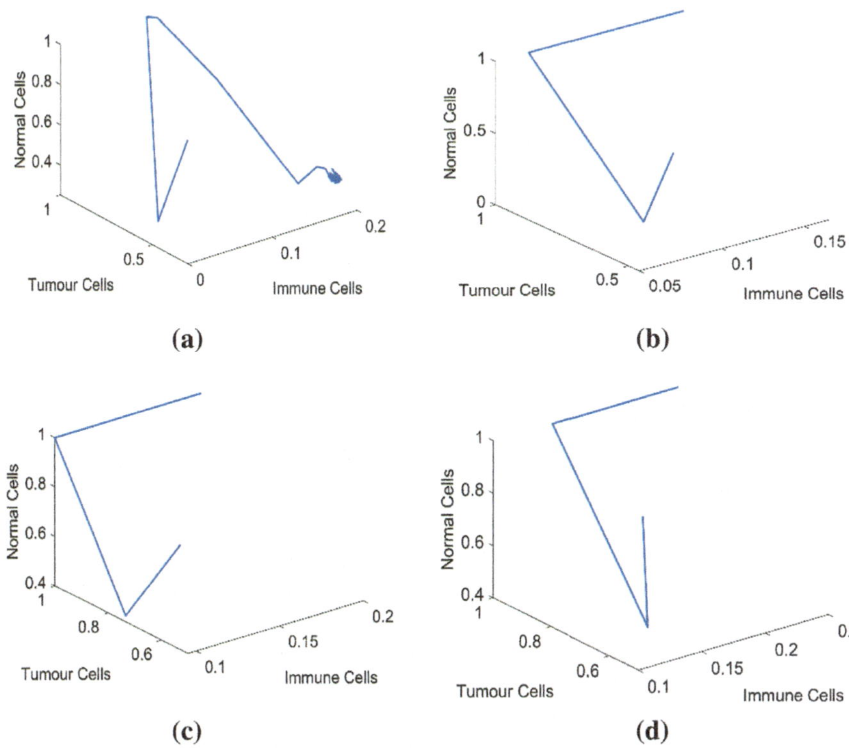

Fig. 2 3D analysis on the impact of proposed and conventional control theory **a** EKF, **b** AEKF, **c** NEF-EKF, **d** NEF-EKF-ISF

control theory is to calculate the drug dosage that can totally kill tumor cells. The obtained findings indicate the illustration of how much closer the proposed method contributes to the removal of tumor cells.

4 Overall Performance

The general exhibition of the introduced conspire over other current methodologies on following the ideal estimations of the forms (typical, immune as well as a tumor) is given in Table 2. The proposed control hypothesis from the investigation

Table 2 Overall performance of proposed control theory over other conventional theories

Cells	"EKF [10]"	"AEKF [26]"	"NEF-EKF [27]"	"NEF-EKF-ISF"
Immune cells (mol)	1.33940	1.26860	1.21190	1.04530
Tumor cells (mol)	0.024430	0.024610	0.021890	0.020430
Normal cells (mol)	0.872440	0.846260	0.846260	0.846260

is observed to have accomplished the nearest wanted qualities on the other thought about models. Moreover, the tumor cells are acquired with the least assessed values, which additionally demonstrate the non-presence of the tumor cells updated from Grey Wolf Optimization [28]. Subsequently, better boundary assessment ends up being accomplished via the introduced NEF-EKF-ISF model on the other regular methodologies.

5 Conclusion

This paper has given a hearty regulator an all-inclusive Kalman channel that impacts the medication measurements along with boundary assessment. This regulator changes drug measurements and monitors the tumor, resistant, and ordinary cells in ideal chemotherapy. There, a new NEF-EKF-ISF has been planned, and the refreshing cycle has been finished relying upon the mistake work. Further, a scaling factor was set up that considered the blunder upgrade for the refreshing cycle. The presentation of the regulator was looked at over different plans that outcomes the effect of medication measurements infusion on ordinary, insusceptible, and tumor cells; the following of wanted estimations of the tumor cells was discovered to be steady in the presence of the demonstrated spectator. Along these lines, the boundary estimations of the executed NEF-EKF-ISF method in achieving the ideal estimations of framework states have been affirmed.

References

1. Padmanabhan R, Meskin N, Haddad WM (2017) Reinforcement learning-based control of drug dosing for cancer chemotherapy treatment. Math Biosci 293:11–20
2. Lai X, Friedman A (2019) Mathematical modeling in scheduling cancer treatment with combination of VEGF inhibitor and chemotherapy drugs. J Theoret Biol 462:490–498
3. Wu X, Liu Q, Zhang K, Cheng M, Xin X (2018) Optimal switching control for drug therapy process in cancer chemotherapy. Eur J Control 42:49–58
4. Liang L, Luo H, He Q, You Y, Liang J (2018) Investigation of cancer-associated fibroblasts and p62 expression in oral cancer before and after chemotherapy. J Cranio-Maxillofac Surg 46(4):605–610
5. Khalili P, Vatankhah R (2019) Derivation of an optimal trajectory and nonlinear adaptive controller design for drug delivery in cancerous tumor chemotherapy. Comput Biol Med 109:195–206
6. Wang P, Liu R, Jiang Z, Yao Y, Shen Z (2019) The optimization of combination chemotherapy schedules in the presence of drug resistance. IEEE Trans Autom Sci Eng 16(1):165–179
7. Bazrafshan N, Lotfi MM (2016) A multi-objective multi-drug model for cancer chemotherapy treatment planning: a cost-effective approach to designing clinical trials. Comput Chem Eng 87:226–233
8. Khadraoui S, Harrou F, Nounou HN, Nounou MN, Bhattacharyya SP (2016) A measurement-based control design approach for efficient cancer chemotherapy. Inf Sci 333:108–125
9. Batmani Y, Khaloozadeh H (2013) Optimal drug regimens in cancer chemotherapy: a multi-objective approach. Comput Biol Med 43(12):2089–2095
10. Rokhforoz P, Jamshidi AA, NamaziSarvestani N (2017) Adaptive robust control of cancer chemotherapy with extended Kalman filter observer. Inform Med Unlocked 8:1–7
11. Chen T, Kirkby NF, Jena R (2012) Optimal dosing of cancer chemotherapy using model predictive control and moving horizon state/parameter estimation. Comput Methods Prog Biomed 108(3):973–983
12. Paryad-Zanjani S, Mahjoob MJ, Amanpour S, Kheirbakhsh R, Haji Akhoundzadeh M (2016) A supplemental treatment for chemotherapy: control simulation using a mathematical model with estimated parameters based on in vivo experiment. IFAC-PapersOnLine 49(26):277–282
13. Pouchol C, Clairambault J, Lorz A, Trélat E (2018) Asymptotic analysis and optimal control of an integro-differential system modelling healthy and cancer cells exposed to chemotherapy. J Math Pures Appl 116:268–308
14. Matsuda C, Ishiguro M, Teramukai S, Kajiwara Y, Fujii S, Kinugasa Y, Nakamoto Y, Kotake M, Sakamoto Y, Kurachi K, Maeda A, Komori K, Tomita N, Shimada Y, Takahashi K, Kotake K, Watanabe M, Mochizuki H, Sugihara K (2018) A randomized controlled trial of 1-year adjuvant chemotherapy with oral tegafur–uracil versus surgery alone in stage II colon cancer: SACURA trial. Eur J Cancer 96:54–63
15. Toyooka S, Okumura N, Nakamura H, Nakata M, Yamashita M, Tada H, Kajiwara S, Watanabe N, Okada M, Sakamoto J, Aoe M, Soh J, Miyoshi S, Hotta K, Matsuo K, Date H (2018) A multicenter randomized controlled study of paclitaxel plus carboplatin versus oral uracil-tegafur as the adjuvant chemotherapy in resected non-small cell lung cancer. J Thorac Oncol 13(5):699–706
16. Wu H, Hu H, Wan J, Li Y, Wu Y, Tang Y, Xiao C, Xu H, Yang X, Li Z (2018) Hydroxyethyl starch stabilized polydopamine nanoparticles for cancer chemotherapy. Chem Eng J 349:129–145
17. Gibbons A, Groarke AM (2018) Coping with chemotherapy for breast cancer: asking women what works. Eur J Oncol Nurs 35:85–91
18. Jiang S, Liu Y, Huang L, Zhang F, Kang R (2018) Effects of propofol on cancer development and chemotherapy: potential mechanisms. Eur J Pharmacol 831:46–51
19. Sun B, Luo C, Cui W, Sun J, He Z (2017) Chemotherapy agent-unsaturated fatty acid prodrugs and prodrug-nanoplatforms for cancer chemotherapy. J Controlled Release 264:145–159
20. Wang F, Porter M, Konstantopoulos A, Zhang P, Cui H (2017) Preclinical development of drug delivery systems for paclitaxel-based cancer chemotherapy. J Controlled Release 267:100–118

21. Abbasian M, Roudi MM, Mahmoodzadeh F, Eskandani M, Jaymand M (2018) Chitosan-grafted-poly(methacrylic acid)/graphene oxide nanocomposite as a pH-responsive de novo cancer chemotherapy nanosystem. Int J Biol Macromol
22. Bao T, Seidman AD, Piulson L, Vertosick E, Chen X, Vickers AJ, Blinder VS, Zhi WI, Li Q, Vahdat LT, Dickler MN, Robson ME, Mao JJ (2018) A phase IIA trial of acupuncture to reduce chemotherapy-induced peripheral neuropathy severity during neoadjuvant or adjuvant weekly paclitaxel chemotherapy in breast cancer patients. Eur J Cancer 101:12–19
23. Kimmick GG, Li X, Fleming ST, Sabatino SA, Wilson JF, Lipscomb J, Cress R, Bergom C, Anderson RT, Wu X-C (2018) Risk of cancer death by comorbidity severity and use of adjuvant chemotherapy among women with locoregional breast cancer. J Geriatr Oncol 9(3):214–220
24. Kurt B, Kapucu S (2018) The effect of relaxation exercises on symptom severity in patients with breast cancer undergoing adjuvant chemotherapy: an open-label non-randomized controlled clinical trial. Eur J Integr Med
25. Sun X, Zhang M, Du R, Zheng X, Tang C, Wu Y, He J, Huang W, Wang Y, Zhang Z, Han X, Qian J, Zhong K, Tian X, Wu L, Zhang G, Wu Z, Zou D (2018) A polyethyleneimine-driven self-assembled nanoplatforms for fluorescence and MR dual-mode imaging-guided cancer chemotherapy. Chem Eng J 350:69–78
26. Akhlaghi S, Zhou N, Huang Z (2017) Adaptive adjustment of noise covariance in Kalman filter for dynamic state estimation. Syst Control
27. Mohite UL, Patel HG (2020) Robust controller for cancer chemotherapy dosage using non-linear kernel-based error function. BioAlgorithms MedSyst 16(4). https://doi.org/10.1515/bams-2019-0056
28. Mohite UL, Patel HG (2020) Regularized error function-based extended Kalman filter for estimating the cancer chemotherapy dosage: impact of improved grey wolf optimization. Bio-Algorithms Med-Syst 18(1). https://doi.org/10.1515/bams-2020-0048

Artificial Neural Network-Based Automatic Generation Control for Two-Area Nuclear-Thermal System

Mugdha Mishra and Nitin Kumar Saxena

Abstract In this paper, artificial neural network (ANN)-based controller for two-area nuclear-thermal system has been proposed for frequency control. A nuclear-thermal interconnected two-area system is an emerging area because of their characteristics of supplying power in different demand modes such as base or peak load demand. Frequency control studies become more important especially in interconnected areas due to high complexity of operations especially at the time of system or load disturbances. In available papers, automatic generation control (AGC) has been implemented using proportional integral derivative (PID) controller. Conventional methods like Ziegler–Nichols (ZN) and nonlinear control design (NCD) have been applied for tuning the gain constants of PID controller. To overcome the limitations of conventional tuning method, ANN based robust control is proposed in this paper. The results show that the ANN based AGC provides better performance than conventional methods for two-area nuclear-thermal power system. The control method ensures that the value of error in system frequency and tie-lines power flow is maintained within the limits. These controllers are simulated in MATLAB/SIMULINK package.

Keywords Artificial neural network (ANN) · Automatic generation control (AGC) · Two-area nuclear-thermal power system · Tie-line model for nuclear · Thermal power generation

Nomenclature

f	System frequency
Ptie(max)	Tie-line power
ΔP_L	Disturbance in load demand

M. Mishra
Government Polytechnic, Bareilly, India

N. K. Saxena (✉)
KIET Group of Institutions, Delhi-NCR, Ghaziabad, UP, India

S. Suhag et al. (eds.), *Control and Measurement Applications for Smart Grid*, Lecture Notes in Electrical Engineering 822,
https://doi.org/10.1007/978-981-16-7664-2_3

$H1$	Inertia constant of thermal plant
$H2$	Inertia constant of nuclear plant
$D1$	Load frequency constant of area-1
$D2$	Load frequency constant of area-2
$\Delta\delta = \delta_1 - \delta_2$	Load angle, δ_1 = Sending end angle and δ_2 = Receiving end angle
Tt, Tg, THR1, THR2	Time constant of thermal unit
$b1$, $b2$	Bias constant
KLP, KHP	Gain of nuclear unit
Tg, THR3, THR34, T1, TRH1, TRH2	Time constant of nuclear unit

1 Introduction

The structure of power utilities has been changed in recent years. As per report published by Central Electricity Authority, India, in October 2016, total installed capacity of thermal plant is 212.4689 GW and capacity of nuclear plant is 5.78 GW. A comparative study for thermal and nuclear power plant with others is shown in Fig. 1. The pie chart shows the percentage contribution of different energy sources in India. The GW installed capacity of different power plants of India is indicated in Table 1 according to data taken from Central Electricity Authority, Ministry of Power, Government of India, 2016 [1]. In India, every year, the demand of electricity is significantly rising due to population and industrialization is climbing. The

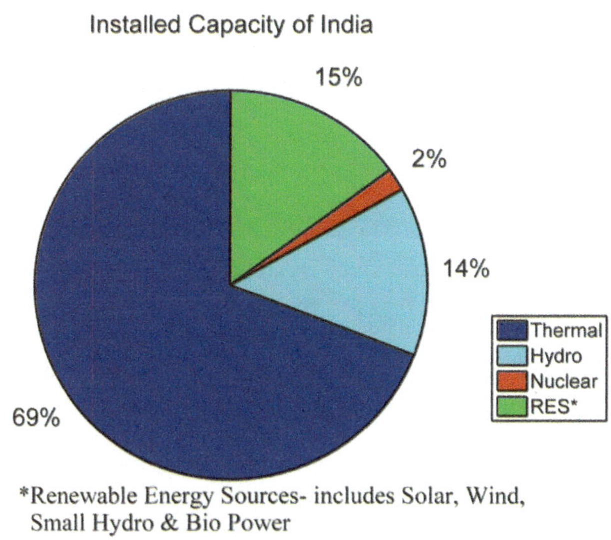

Fig. 1 % installed capacity of different plants in India

Table 1 GW installed capacity of different power plants of India

Power plant	India's installed capacity (GW)
Thermal	212.4689
Hydro	43.11243
Nuclear	5.78
RES*	45.91695
Total installed capacity	307.27828

important fact of power generation is as per law of conservation of energy, continuously fulfilling demand of power. The power production should be load following. It is proposed in literature that the nuclear power plant is suggested to operate for base load, and the thermal power plant can be operated for either base or peak load demand [2]. Thermal and nuclear plants generate and dispatch power output as per demand; nuclear plant runs at constant power because they are less flexible to follow load changes. Power generation mechanism of both thermal and nuclear unit is quite similar the only difference is source of heat generation [3]. Many research papers already published in this field in past years in which conventional techniques were applied on thermal-thermal and thermal-hydro system [4–6]. A hybrid interconnected power system participation having both the power plants together is suggested and dealt in many studies [5–9]. India's Northern region is rich in hydropower, Southern region has rich renewable power, and Western regions are leading in thermal and nuclear power. Major contributor of electricity is thermal power plant in India [10].

Following are the important features for interconnected power system:

1. Power can be transferred bidirectional from one area to another area according to the system requirement in different zones.
2. Each area should be capable to regulate the system frequency in its prescribed limit in case of any static change in load demand, and therefore, tie-line power should be zero for static change in load demand for particular area.
3. Time errors and frequency transients should be small. Steady-state frequency error generated due to step load change should die-out.

Due to the participation of two different power generating methods, frequency regulation problem in such hybrid generation method is still a big challenge, and therefore, two-area nuclear-thermal power system has been studied in this paper for advanced regulation techniques over conventional used techniques for the same [11]. Load frequency control (LFC) concept was first introduced by Prof. N. Cohn in 1971. This LFC is an integral controller which controls the power flow between areas by maintaining the system frequency within the prescribed limits. As per Indian Electricity Rules 1956, only $\pm 3\%$ frequency variation is allowed in the electricity supplied by the supplier to the consumer, while this range has been further reduced to $\pm 2\%$ in year 2008 by the Bureau of Indian Standards ET31 {6128}. Frequency stability defined by IEEE/CIGRE task force is the ability of the power system to maintain steady-state frequency, following a severe system upset, resulting in a significant imbalance between generation and load [12]. Fundamentally, LFC produce zero

error of frequency as well as tie-line power by reducing the value of area control error (ACE) to zero.

Therefore, LFC must operate for any change in power demand. For power utility, it is really a big challenge to maintain the equilibrium between power demand and power supply because of the intermittent demand profile in distribution system. This change in load demand may result in frequency variation and in absence of proper control mechanism; system frequency may go to beyond acceptable range. ACE is the function of tie-line power flow and frequency of the system.

$$\text{ACE} = f(\text{Power, frequency}) \tag{1}$$

It also shows the misbalance between the power interchange agreement between the demand and supply to the load.

Automatic generation control (AGC) is one of the most important and essential issues in power system designing and restructuring. It helps to maintain the tie-line power flow and frequency within the scheduled limits [13]. Hence, keeping the system frequency within its permissible range is the main objective of AGC. Therefore, a proper control mechanism is required for fast frequency control by satisfying all transient requirements of the utilities. The operation of AGC depends upon two main factors: gain of controller and proper design of speed control parameters. Area control error (ACE) is the control factor that decides the action of AGC.

Due to the fluctuations in load, it is very difficult to keep the tie-line power flow and system frequency to their nominal value. ACE can be applied as feedback signal to the power system network. By this, AGC system will takes action for controlling the generator prime mover. This is the problem of load frequency control (LFC), controls the ACE, and brings it to the zero value [14]. AGC consists of P, PI, or PID controller that can be tuned conventionally by Zeigler–Nichols and NCD techniques. Advanced tuning methods are required for fast and accurate frequency regulation. Fuzzy logic controller can also be used, but it has limitations that for optimum result the membership function has to be properly selected. ANN is an adaptive design technique and fast responding [15]. Therefore, ANN based PID controller has been used as AGC for nuclear-thermal power system which is given in Fig. 2. In this paper, a power system has been modeled which is categorized into different regions with interconnected load. Two-area power system consists of thermal plant in area-1 and nuclear plant in area-2. These regions are connected with the other areas. Tie-line is defined as the transmission lines which interconnects one power region with the other neighboring power regions. LFC controls the power flow between these tie-lines. The

Fig. 2 Two-area power system

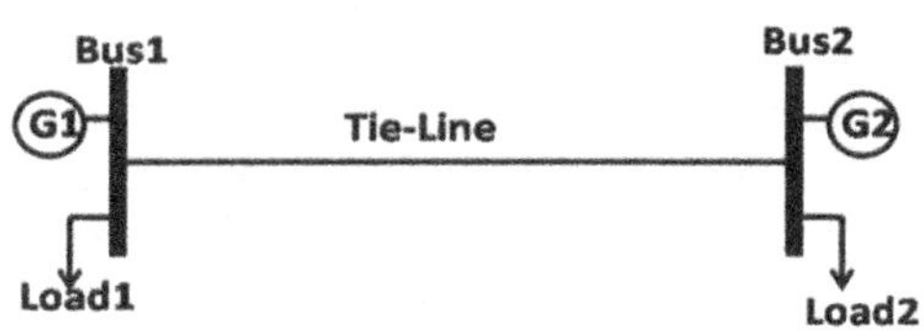

purpose of this paper is to describe the concept of stabilization of tie-line power and frequency for two-area power system using conventional and advance technique.

This paper is organized in following sections: first section comprises of introduction that is explained above; the mathematical modeling of thermal plant, nuclear plant, and the complete model of two-area nuclear-thermal power system has been described in next section. AGC tuning techniques are given and the obtained results, their discussion is presented in fourth section, and finally, conclusion is presented.

2 Modeling Methodology

2.1 Thermal Power Plant

In thermal plant, initially steam is produced at high pressure and in the boiler which is then super-heated. The steam at high pressure and high temperature strikes over turbine sections that convert it into mechanical power. Turbine power is supplied to alternator to generate electrical power. Main components of thermal unit are turbine (prime mover), alternator, and governor [16, 17].

Single thermal unit is shown in Fig. 3 that consists of all its main components that are governor, generator, prime mover, and electrical load [18]. Power system load is mainly motors load, and they are frequency sensitive load.

If H is the inertia of machine, $\Delta\delta$ is the load angle, ω is the angular speed, ΔP_{m} is the mechanical energy, and ΔP_{e} is the electrical energy, then the Swing equation for synchronous generator is given by,

$$\frac{2H1}{\omega}\frac{\mathrm{d}^2\Delta\delta}{\mathrm{d}t^2} = \Delta P_{\mathrm{m}} - \Delta P_{\mathrm{e}} \tag{2}$$

Deviation in speed can be expressed as,

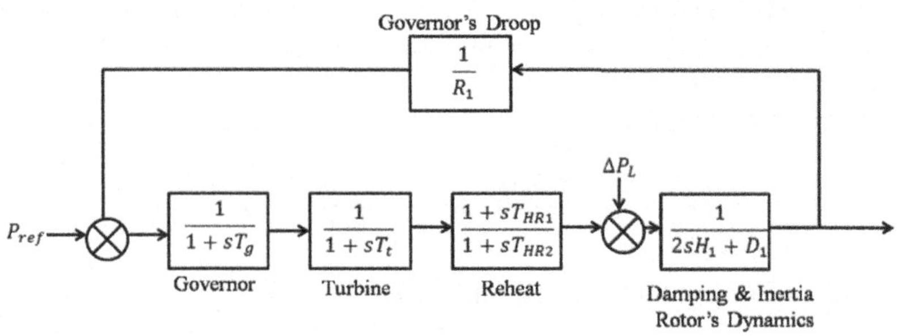

Fig. 3 Mathematical model of thermal unit

$$\frac{\mathrm{d}\Delta\frac{\omega}{\omega_s}}{\mathrm{d}t} = \frac{1}{2H1}(\Delta P_m - \Delta P_e) \tag{3}$$

The transfer function of turbine is given by,

$$\Delta P_t(s) = \Delta Y_E(s)\frac{1}{1 + sT_t} \tag{4}$$

The transfer function of governor unit is,

$$\Delta Y_E(s) = \left[\Delta P_{ref}(s) - \frac{1}{R1}\Delta F(s)\right]\left[\frac{1}{1 + sT_g}\right] \tag{5}$$

Governor modulates the speed of prime mover; it prevents from over speed and keeps the speed around prime mover rated speed. On the value speed of prime mover, the frequency of alternator depends [19].

The mathematical model of thermal plant as given in Fig. 3 has been formulated by combining mathematical Eqs. 2–5 [20].

2.2 *Nuclear Plant*

Fission and fusion are the terms associated with nuclear power. Fission-based reactors are used for generation of electricity, while fusion power technology is costly and not adopted. Gas cooled and boiling water reactors and number of different technologies have been invented. Within controlled surrounding, fission chain reaction sustain in this plant. In shielded pressure vessel, reactor core is placed where fission takes place. Fissile material fuel rods are present in the core. Graphite or water is used as moderator that varies the neutron speed to control the chain reaction. Control rods made up of boron are inserted in the reactor core to absorb the neutron, so nuclear reaction can be controlled. Water or gas is passed in the reactor acts as coolant. Then, it passed to the boiler for steam generation [3]. Nuclear reactor is the main component of nuclear plant. If $\Delta P_m(s)$ is the input power to the governor unit, $\Delta P_v(s)$ is the control signal of governor unit.

Transfer function of its governor is,

$$\frac{\Delta P_v(s)}{\Delta P_m(s)} = \frac{1}{sT_{gn} + 1} \tag{6}$$

Transfer function of high-pressure turbine is,

$$\frac{\Delta f_2(s)}{\Delta P_{v(s)}} = \frac{K_{HP}}{T_1 s + 1} \tag{7}$$

Transfer function of low-pressure turbine is,

$$T_{\mathrm{LP}}(s) = \frac{K_{\mathrm{LP}}}{sT_2 + 1} \tag{8}$$

Transfer function for reheat for LP turbine,

$$R_{\mathrm{LP}}(s) = \frac{1}{sT_{\mathrm{RH2}} + 1} \tag{9}$$

Transfer function of nuclear low-pressure turbine 1 is,

$$T_{\mathrm{LP1}}(s) = \frac{1 + sT_{\mathrm{HR2}}}{1 + sT_{\mathrm{HR3}}} \tag{10}$$

Transfer function for reheat for LP turbine 1,

$$R_{\mathrm{LP1}}(s) = \frac{1}{sT_{\mathrm{RH1}} + 1} \tag{11}$$

By combining Eqs. 6–11, the model for nuclear power plant has been formed as indicated in Fig. 4 [21].

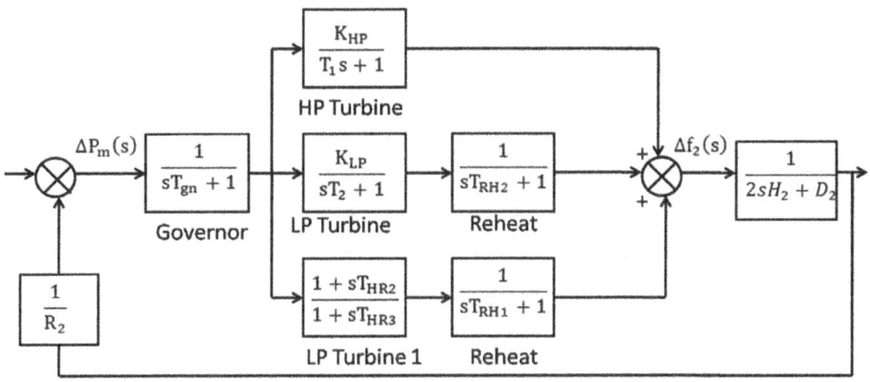

Fig. 4 Nuclear plant model

2.3 Complete Nuclear-Thermal Two-Area Power System Model

Now both generating units are coupled, and a complete model for nuclear-thermal plant is formulated as per the rating and variable values given in Appendix. MATLAB/SIMULINK designed two-area system is shown in Fig. 5.

3 Controller Tuning Techniques

Controller is the device that regulates the system properties as per the requirement. It keeps the system in stable condition. For power system stable operation, automatic generation control is used. Mainly, proportional plus integral plus derivative (PID) controller is used for this purpose of generation control. Controller evaluates the difference in reference set point and obtained output. On the basis of this calculated error signal, controller take actions [22]. The main purpose of controller is to keep this error as minimum as possible. There are three parameters involved in PID mechanism K_p, K_i, and K_d [23].

$$u(t) = K_p e(t) + K_i \int e(t)\mathrm{d}t + K_d \frac{\mathrm{d}e(t)}{\mathrm{d}t} \tag{12}$$

Controller tuning means setting up the values of controller parameters K_p, K_i, and K_d. The values of these parameters are selected such that optimum response of

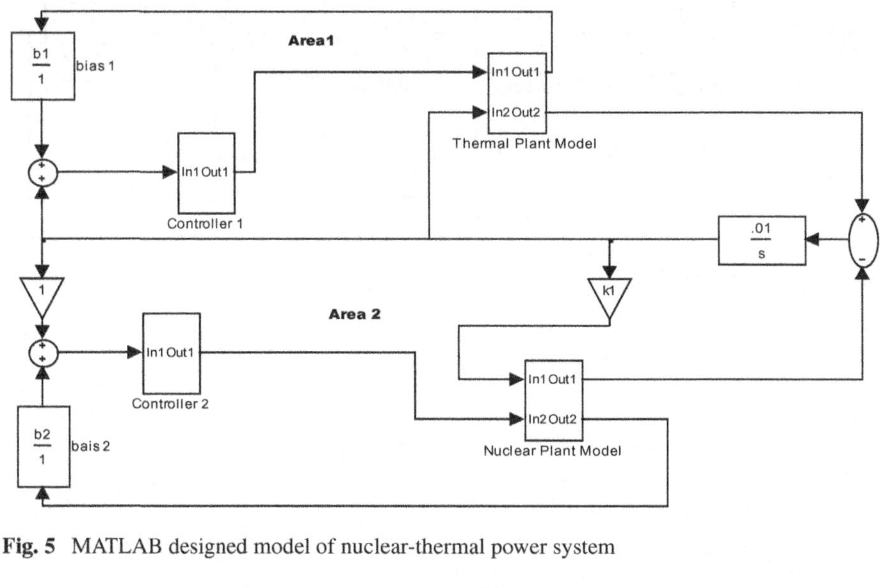

Fig. 5 MATLAB designed model of nuclear-thermal power system

the system will be achieved. Tuning is very tough task for stabilizing the system response. There are various methods of its tuning, and the traditional method is manual tuning that is very tough. Following tuning methods for controllers can be employed: Ziegler–Nichols, VRFT, genetic algorithm, nonlinear control design, block set method, etc. In this paper, Ziegler–Nichols, nonlinear control design, and artificial neural network techniques are applied for tuning of AGC controller [24].

3.1 Ziegler–Nichols Method

Ziegler–Nichols method is a tuning method which is given by John G. Ziegler and Nathaniel B. Nichols in the year of 1940 [25]. Ziegler–Nichols method is a direct method. Initially, system in closed loop mode only with proportional controller integral and derivative controller modes is not connected at this stage, then this system is tested. Gain K_p of P controller is initialized to zero and then varied from zero toward high values, till stability margin is reached. Reference/set point is adjusted slightly for initiating any oscillation. Controller K_p is also varied, so that oscillations continue at same magnitude, i.e., oscillations sustained. Gain increases slightly when amplitude of oscillations is decreased and vice versa [21].

Algorithm Ziegler–Nicholas Technique

Step 1: Initialize the value $K_p = K_d = K_i = 0$.

Step 2: Increase the value of K_p until the oscillation occur. Controller K_p is varied such that oscillations continue at same magnitude, i.e., sustained oscillations.

Step 3: If constant period and magnitude oscillations are established, then the value of oscillations period Pcr (critical) and controller gain Kcr (critical) for which oscillations were established can be calculated.

Step 4: Evaluate the parameters of PID controller as per formulas given in Table 2.

Table 2 ZN technique optimum controller parameters value

Controller type	Controller gain		
	K_p	K_i	K_d
P	0.50Kcr	–	–
PI	0.45Kcr	1.2K_p/Pcr	–
PID	0.60Kcr	2K_p/Pcr	K_p Pcr/8

3.2 NCD Optimization Method

NCD stands for nonlinear control design. Nonlinear control design technique tunable variable information and constraint bound data are automatically converted into a constrained optimization problem. Actually, this nonlinear control design block set used to reduce the highest error of constraint. At equal time points, NCD block sets generate error of constraint. There is a signal constraint block available in MATLAB Simulink; in that, the boundaries of controller parameters have been initialized, then it will automatically optimize the values of parameters. The results obtained by NCD technique and ZN technique are similar.

3.3 Artificial Neural Network

A neural network has various architectures for its class. MLP is multilayer perception, and RBF is radial basis function networks. Figure 5 shows the architecture of network for multilayer perceptron with single junction [26, 27].

Three layers are used in MLP: input layers, hidden layers, and output layers. As shown in Fig. 6, there are neurons and node junction; and g is the activation function [28–30].

$$y_i = g_i = g\left(\sum_{j=1}^{k} w_{ji}x_j + \theta_i\right) \tag{13}$$

Equation 13 represents the output of junction is given by inputs are represented by x_i where $i = 1, 2, 3 \ldots$, they are then weighted. Inputs to the neurons are firstly multiplied by the weights w_i, and all input signals are added along with a constant term θ_i which is the bias constant. Then, the signal resulted from here is applied to g activation function.

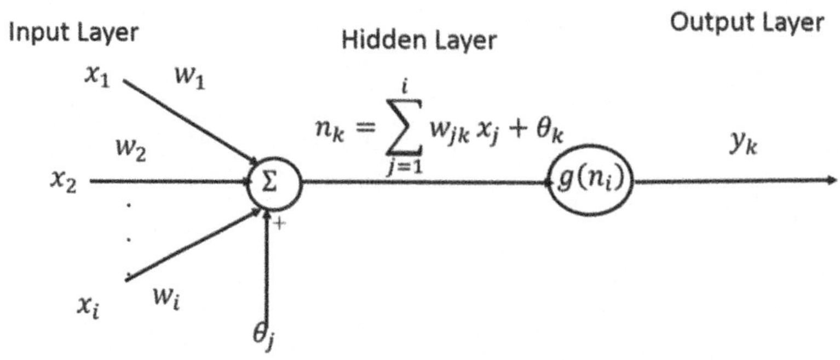

Fig. 6 Network of multilayer perception with single junction

Multilayer perceptron (MLP), the interconnected network with hidden layer, can also be formulated. First step is to fix the architecture of MLP by setting nodes (Neurons) and hidden layers. Activation function is selected for all layers; the unknown weights and bias constants can be estimated.

In this paper, multilayer perceptron function network has been used for controlling the designed thermal-nuclear power system.

4 Results and Discussions

Nuclear-thermal two-area power system has been modeled as discussed in previous section. MATLAB/Simulink designed model is formed as presented in above sections its steps are given in Fig. 7. One percent step perturbation, i.e., disturbance in load demand ΔP_L, has been implied in the system load as given in Fig. 8. This causes disturbances in the frequency of area-1 ($\Delta f1$) and area-2 ($\Delta f2$) which is shown in Fig. 9. AGC for the two-area system has been formulated as discussed in Sect. 3 that controls the system frequency and tie-line power. Firstly, for controlling tie-line power (ΔPtie), PID controller is tuned by conventional ZN technique and also controller is tuned by ANN technique. Their obtained result is given in Fig. 10 that indicates the deviation in tie-line power and its transient parameters are given in Table 3. For controlling frequency, PID controller is tuned by conventional method—Ziegler–Nichols as well as by artificial neural network technique result is shown in Fig. 11. It indicates the deviation in frequency, and its transient parameters are given

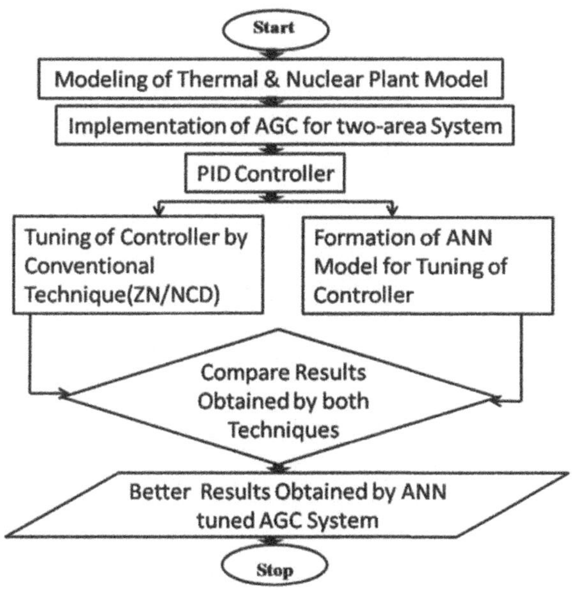

Fig. 7 Flowchart of designed system

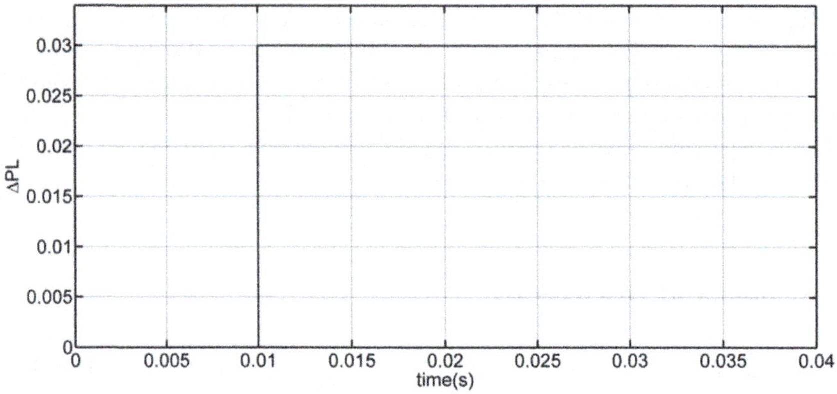

Fig. 8 Disturbance in load demand ΔP_L for system

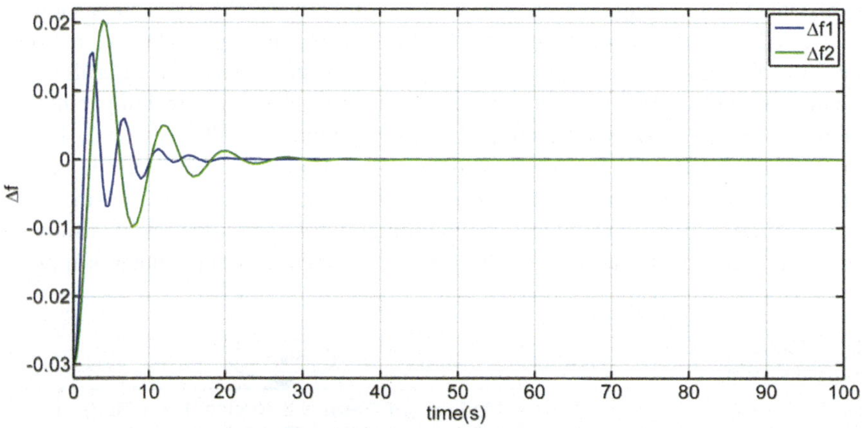

Fig. 9 Disturbance in frequency $\Delta f1$ of area-1 and $\Delta f2$ of area-2

in Table 4. It is seen from response of both techniques that the ANN tuned AGC result consists of less oscillations, and it is more stable as compared to the ZN tuned AGC. All the parameters in their respective figures are taken in pu.

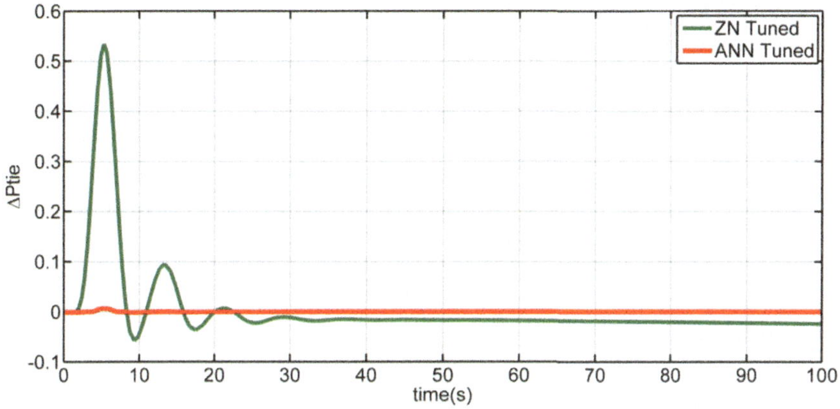

Fig. 10 Zeigler–Nichols and artificial neural network tuned tie-line power ΔPtie for 1% load disturbance

Table 3 Performance comparison of tie-line power for both tuning techniques

Transient parameter	ZN tuned	ANN tuned
ess	0.0245	$1.9836 * 10^{-4}$
Tr	0.3775	0.3869
ts	65.7674	56.1771
tp	28	28
Peak	0.5345	0.0072
Overshoot	136.3393	120.4561

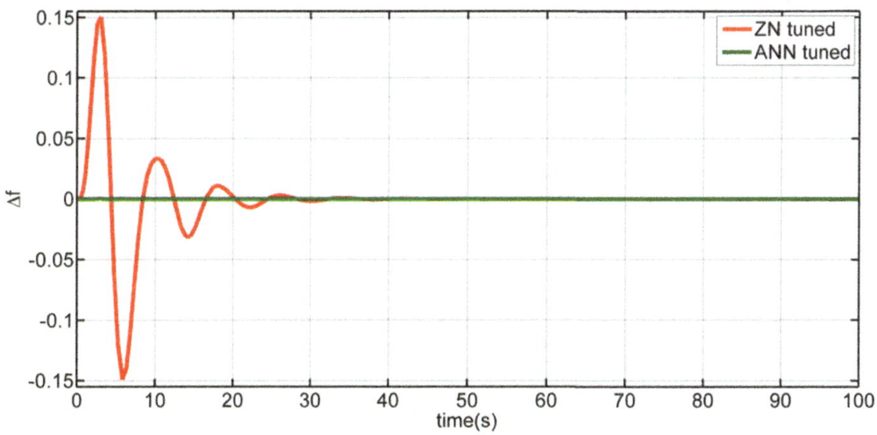

Fig. 11 Zeigler–Nichols and artificial neural network tuned frequency deviation Δf for 1% load disturbance

Table 4 Zeigler–Nichols and artificial neural network tuned frequency deviation Δf for 1% load disturbance

Transient parameter	ZN tuned	ANN tuned
ess	0.00011968	0.000040405
tr	0.0015	0.00077397
ts	53.2598	26.2376
tp	17	17
Peak	0.1507	0.00034800
Overshoot	125,940	2.5053

5 Conclusions

In this paper, the active power and frequency regulator is designed using PID controller for nuclear-thermal two-area power system. Controller is tuned by conventional Ziegler–Nicholas and ANN techniques. Investigations of obtained time-response results clearly show the superiority of ANN based controller over conventional controller. The overshoot, oscillations, and settling time are quite better in ANN response. Hence, it concluded that ANN based AGC is a better technique.

Appendix

System parameters: $f = 50$ Hz, $\text{Ptie}_{(\text{max})} = 200$ MW. $H1 = H2 = 5$ s, $D1 = D2 = 8.33 * 10^{-3}$ pu MW/Hz, $\Delta\delta = 30°$.

See Tables 5 and 6.

Table 5 Thermal plant parameter values

Thermal plant parameter	Value
$R1$	2.4
$b1$	0.425
$H1$	0.0833
$D1$	0.00833
Tt	0.425
Tg	0.8
THR1	5
THR2	10

Table 6 Nuclear plant parameter values

Nuclear plant parameters	Value
K_{LP}	0.3
$T2$	0.5
$H2$	0.0833
$D2$	0.00833
K_{HP}	0.2
Tgn	0.8
THR3	5
THR4	10
$T1$	0.5
TRH1	9
TRH2	7
$R2$	2.5
$k1$	−1

References

1. Report by Central Electricity Authority, Ministry of Power, Government of India, Oct 2016
2. Shankar R, Chatterjee K, Bhushan R (2016) Impact of energy storage system on load frequency control for diverse sources of interconnected power system in deregulated power environment. Int J Electr Power Energy Syst. https://doi.org/10.1016/j.ijepes.2015.12.029
3. Freris L. Renewable energy in power systems. Wiley
4. Singh G, Bala R (2011) Automatic generation & voltage control of interconnected thermal power system including load scheduling strategy. Int J Eng Adv Technol (IJEAT) 1(2):1–7
5. Abdulraheem BS, Gan CK (2016) Power system frequency stability and control: survey. Int J Appl Eng Res 11:5688–5695
6. Naghizadeh RA, Jazebi S, Vahidi B (2012) Modeling hydro power plants and tuning hydro governors as an educational guideline. Int Rev Model Simul 5(4)
7. Mathur HD, Manjunath HV (2007) Frequency stabilization using fuzzy logic based controller for multi-area power system. S Pac J Nat Sci 4:22–30. https://doi.org/10.1071/SP07004
8. Ramakrishna KSS, Sharma P, Bhatti TS (2010) Automatic generation control of interconnected power system with diverse sources of power generation. Int J Eng Sci Technol 2(5):51–65. https://doi.org/10.4314/ijest.v2i5.60102
9. Nasiruddin I, Bhatti TS, Hakimuddin N (2015) Automatic generation control in an interconnected power system incorporating diverse power plants using bacteria forging optimization technique. Electr Power Compon Syst 189–199
10. Mishra M, Khan MT, Kumar N (2019) Green energy: a building block for smart India. In: 1st international conference on future learning aspects of mechanical engineering (FLAME-2018), India
11. Rogers K, Ragheb M (2010) Symbiotic coupling of wind power and nuclear power generation. In: International nuclear and renewable energy conference (INREC10)
12. Kundur P, Paserba J, Ajjarapu V, Anderson G, Bose A, Canizares C, Hatzizrgyriou N, Hill D, Sankovic A, Taylor C, Cutsem T, Vittal D (2004) Definition and classification of power system stability. IEEE Trans Power Syst 19(3). https://doi.org/10.1109/TPWRS.2004.825981
13. Hakimuddin N, Nasiruddin I, Bhatti TS, Arya Y (2020) Optimal automatic generation control with hydro, thermal, gas, and wind power plants in 2-area interconnected power system. Electr Power Compon Syst 48(6–7):558–571. https://doi.org/10.1080/15325008.2020.1793829

14. Sahu RK, Gorripotu TS, Panda S (2016) Automatic generation control of multi-area power systems with diverse energy sources using teaching learning based optimization algorithm. Int J Eng Sci Technol 19(1). https://doi.org/10.1016/j.jestch.2015.07.011
15. Demiroren A, Zeynelgil HL, Sengor NS (2001) The application of ANN technique to load-frequency control for three-area power system. In: IEEE Porto power tech conference. IEEE Xplore, Porto. https://doi.org/10.1109/PTC.2001.964793
16. Elgard OI. Electric energy system theory, 2nd edn. McGraw-Hill, New York
17. Elgerd OI. Electric energy system theory: an introduction. McGraw Hill, New York
18. Mishra M, Saxena N (2016) ANN based automatic generation control for two-area power system. MIT Int J Electr Instrum Eng 40–44
19. Kundur P. Power system stability and control. McGraw-Hill Inc., New York
20. Mishra P, Mishra M, Saxena N (2015) Tuning of generation participation depending upon load demand. In: IEEE international conference on communication, control and intelligent systems (CCIS). IEEE Xplore, CCIS, India, pp 459–463. https://doi.org/10.1109/CCIntelS.2015.7437960
21. Mishra M, Saxena NK, Mishra P (2016) ANN based AGC for hybrid nuclear-wind power system. In: IEEE international conference on micro-electronics and telecommunication engineering ICMETE. IEEE Xplore, India, pp 410–415. https://doi.org/10.1109/ICMETE.2016.116
22. Nanda J, Mangla A (2004) Automatic generation control of an interconnected hydro-thermal system using conventional integral & fuzzy logic controller. In: IEEE international conference on electric utility deregulation reconstructing & power technology. IEEE Xplore
23. Tammam MA, Moustafa MA, Abo Ela MAES, Seif AEA (2010) Load frequency control using genetic algorithm based PID controller for single area power system. In: International conference on renewable energies and power quality, ICREPQ'11, Las Palmas de Gran Canaria
24. Donde V, Pai MA, Hiskens IA (2001) Simulation and optimization in an AGC system after deregulation. IEEE Trans Power Syst 16(3):481–488
25. Matušů R, Prokop R (2010) Experimental verification of design methods for conventional PI/PID controllers. WSEAS Trans Syst Control Arch 5:269–280
26. Ansarian M, Shakouri (2006) A novel neural optimal approach for decentralized load frequency control design in multi area power system. In: Proceedings of power and energy conference, p 172
27. Jaleel JA, Ahammed TPI (2008) Simulation of ANN controller for automatic generation control of hydroelectric power system. In: Conference proceedings. TENCON-2008, pp 1–4
28. Demiroren A, Zeynelgil HL, Sengor NS (2001) The application of ANN technique to load-frequency control for three-area power system. In: IEEE Porto power tech conference, Portugal
29. Bhongade S, Gupta HO (2010) Artificial neural network based automatic generation control scheme for deregulated electricity market. In: Proceedings of IPEC 2010, pp 1158–1163
30. Padhy BP, Tyagi B (2009) Artificial based multi area automatic generation control scheme for a competitive electricity market environment. In: International conference on power systems, ICPS 09, pp 1–6

Gas Leakage Alert System

Sudhanshu Sharma, Sumit Kumar, Abhishek Chawla, Paras Darmal, and Bharat Singh

Abstract The most common type of energy source which is utilized by the domestic people is propane gas which contains liquefied gas called liquefied petroleum gas (LPG). As leakage of gas is becoming common accidents nowadays which cause damages to human lives as well as public property. Hence, it becomes a major security concern. This paper presents a low-cost, power-efficient device called the gas leakage alert system. This system includes main devices such as MQ-6 sensor a gas detector device, Arduino microcontroller, and an alarm unit. Here Arduino microcontroller controls all the operations. This model also contains the GSM module which is used for sending an alarming message to the personal mobile phone or any other government authorities. In the alarm unit, we have used Piezo buzzer which is used to produce a beep sound as an alert.

Keywords Liquefied petroleum gas (LPG) · MQ-6 sensor · Alarm unit · Arduino microcontroller · GSM module

1 Introduction

Liquefied petroleum gas is a flammable gas which has a mixture of hydrocarbons gases which mostly used as fuels for domestic purposes such as in heating appliances. The mixture includes primarily propane gas (C_3H_8) and primarily butane gas (C_4H_{10}), and most commonly, it includes both propane and butane gases [1, 2].

Unlike natural gases, LPG gas is much heavier than air and because of this the gas flows through floors and they tend to settle at low spots. The main things that can happen due to this are as follows:

- Possibility of explosion, i.e. if there is an ignition source present and the mixture of LPG gas and the air is in the right amount, it could lead to an explosion.

S. Sharma · S. Kumar · A. Chawla (✉) · P. Darmal · B. Singh
Department of Electrical and Electronics Engineering, Bharati Vidyapeeth's College of Engineering (Approved by AICTE, New Delhi-Affiliated to GGSIP University, New Delhi), New Delhi, India

S. Suhag et al. (eds.), *Control and Measurement Applications for Smart Grid*, Lecture Notes in Electrical Engineering 822,
https://doi.org/10.1007/978-981-16-7664-2_4

Table 1 Properties of LPG

Gas	Formula	%LEL	%UEL	Ignition temp
Propane	C_3H_8	2.2	9.5	470
Butane	C_4H_{10}	1.8	8.4	365

- Because of a decrease in the concentration of oxygen, it could lead us to suffocation [2].

Gas leakage security system becomes essential when people who have a low sense of smell, and they may or might not respond to concentration of gas leakage. This helps us to prevent gas leakage accident. The circuit which is used for the detection of harmful gas and alerting unit has also been proposed in which alarm is activated immediately after the concentration of gas exceeds a certain level [3].

The biggest example of a gas leakage accident in India was the Bhopal gas tragedy which happened in 1984. This was one of the world's worst gas leakage industrial accident which kills many people at that time. A gas leakage alert system isn't only important to detect the leakage of gas but also to prevent it is equally essential. That is why the system not only detects the leakage of gas but also produces an alarming sound with the help of a buzzer and sends an alarming message to the user as well as government authority, i.e. firehouse [2, 3].

2 Uses of LPF and Properties

According to reports of the Census of India in 2011, 28.5% of households in India use the LPG gas for cooking fuel, heating appliances, etc., which is transferred to their houses either by using the pressurized cylinders or pipes [1].

LPG is the main source for the combined heat as well as power technologies which are known as combined heat power (CHP). The CHP is the method that uses bot electric power and single heat source. Due to this, LPG gas is not only used for households purposes but also for the decentralized generation of energy. LPG gas has greater reliability, and also, it can be stored in different ways; as LPG is a fossil fuel, and it could combine with other renewable power sources [1, 3].

The properties of LPG gas are shown in Table 1.

3 Methodology

The method of this system is divided into three steps which are shown in Fig. 1.

In the first phase, the gas which is leaking is sensed by the gas sensor called MQ-6 sensor then the gas which is detected passes the signal to the Arduino microcontroller.

Fig. 1 Flow diagram

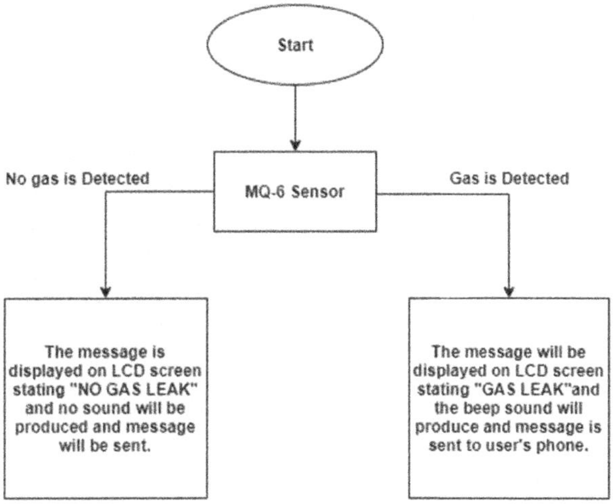

Then in the second phase, the signal which is received by the microcontroller from the gas detector, the microcontroller passes the signal to other external devices which are attached to it [3].

In the third step which is the last step, many tasks are performed such as buzzer starts working which produces a beep sound to alarm people about leaking also the warning message is sent to user's mobile phone through SMS, the message was displayed on the LCD screen [3, 4].

4 Component and Technology Used

4.1 Arduino Microcontroller

Arduino (see Fig. 2) is an open-source platform that has both software and hardware property. It is a microcontroller-based circuit board that can be programmed, and it also has ready-made software known as Arduino IDE where IDE stands for integrated development environment. Arduino is equipped with a variety of microprocessors and controllers. The board is equipped with digital pins from (D0–D13) and analog pins from (A0–A6). The board also has a feature of serial communication interfaces, which includes Universal Serial Bus which is used for uploading programs from computers. The programming which is used in Arduino microcontroller is C and C++ [4].

Fig. 2 Arduino Uno board

4.2 GSM Module

GSM module (see Fig. 3) is a device that is used to send an SMS to an individual's mobile phone. In this project, the GSM module is the main device which an issue to sense the leakage of LPG gas. Whenever the gas which is leaking is detected by the sensor, it sends a signal to the microcontroller connected to it and due to this one of the tasks of sending a message is done. The requirement of a GSM module SIM card number because to send an SMS to a particular number. The module works on +12 V DC Supply. The message which is sent is saved in microcontroller memory [5].

Fig. 3 GSM module

Fig. 4 MQ-6 sensor

4.3 *MQ-6 Sensor*

MQ-6 Sensor (see Fig. 4) is the device that is to detect the leakage of gas within a specific range. It is one of the important devices used in this project. It has a fast reaction time and has a long-lifetime. MQ-6 is highly sensitive toward LPG gas. The sensor can be operated at +5 V. The range at which it can detect the concentration of gas is 200–1000 ppm [5].

4.4 *LCD Display*

Liquid crystal display (LCD) (see Fig. 5) is a flat display screen that is mostly used in cameras, computers, watches, etc. Liquid crystal display consists of two sheets which are made up of polarizing material which consists of a liquid crystal solution. The device works when the current passes through liquid, and it causes crystals to align and does not allow the light to pass through it. Each of the crystals present in

Fig. 5 LCD display (16 × 2)

Fig. 6 Piezo buzzer

LCD acts as a shutter, i.e. either they allowed the light to pass through it or blocks them [5, 6].

4.5 *Buzzer*

Piezo buzzer (see Fig. 6) is a basic device that is used to generate a beep sound or tone. Piezo buzzer uses piezo crystals which is a special material. Its main job is to change the shape when the voltage is applied to it. When the crystals push against diaphragm, it starts generating a pressure wave which human pick this up as a sound [5, 6].

5 Working of Proposed Model

In the MQ-MQ-6 sensor, the most widely sensing material used is SnO_2, a metal oxide. When the metal oxide (SnO_2) is heated up to a certain temperature, then the potential is formed over the surface which prevents the flow of electrons. Current flows through the SnO_2 microcrystals, but the grain boundaries of the crystal absorb the oxygen which results in the formation of potential, preventing the flow of free charge carriers. All this operation takes place in the presence of deoxidizing gas which also alters the surface density of the negatively charged oxygen resulting in the fall of resistance of the sensor [7, 8].

Power supply (5 V) is given to the microcontroller which is the brain of the system. This 5 V DC supply can be achieved using step-down transformer and rectification circuit, which makes use of rectifiers and capacitors. GSM module along with other components of the system like LCD display, gas sensor, buzzer, LEDs is connected to the microcontroller. GSM is also fed from the power supply [7, 8].

If any gas leakage occurs in the surrounding, the MQ-6 sensor's output becomes high and sends an electrical signal to the microcontroller, which then processes it and turns on the red LED and the buzzer. LCD display also starts displaying the message of "GAS LEAKING." After certain seconds, an SMS is sent to the predefined mobile number having information "GAS LEAKING" which is done with the help of the GSM Module [8]. The setup and the result of the project are shown in Figs. 7 and 8.

Fig. 7 Setup

Fig. 8 Result

6 Conclusion

Many features can be added to this system, and it can be modified according to future needs. The GSM module which is used in this system is to send a message to the phone of a user when there is a leakage of gas in the house of the owner. Hence, GSM module is an important part of the system. Another modification that we can use is that we can add a circuit that turns off the main supply when the gas leakage is sensed because switching on any electrical appliances at the time of gas leakage is risky. Also, the gas regulator can be turned off. So, there is no further gas leakage. A robotic system can also be utilized in place of human beings to control the hazardous condition. So, this method will increase the reliability as well as the efficiency of the overall system [9, 10].

In recent years, it had been seen that various deaths are being caused by the explosion of gas cylinders due to leakage. So, keeping this in mind we propose an Arduino-based system which helps us to detect the leakage of gas within the specific range and sends an instant SMS to government authority on their mobile numbers. The main motive of this project is to provide security in homes. Also, when gas leakage occurs LED will turn into red color and the buzzer will also get on. This system is very useful in giving a warning to the house owner about LPG leakage. The advantage of this system is that it is user-friendly and made of low cost. Another good point of using this system is when there is nobody in the house, it will alert the owner by sending a message through the GSM module. Thus, lower the probability of accidents. Therefore, we can say that the proposed system is a very secure, reliable, cost-effective, and user-friendly way to prevent domestic fire accidents [4].

References

1. Wickramasinghe MGDD, Abhayasinghe N (2013) LPG gas leakage alarm. Electr Comput Eng 2(5):1850–1869
2. Shrivastava A, Prabhaker R, Kumar R, Verma R (2013) GSM Based gas leakage detection system. Int J Emerging Trends Electr Electron 3(2):45–48
3. Anusha M, Nagesh V, Venkata Sai B, Srikanth K, Nanda R (2020) IoT based LPG leakage detection and booking system with customer SMS alerts. Int J Mod Trends Sci Technol 6(5):1–5
4. Chaudhary J, Mishra A (2019) Detection of gas leakage and automatic alert system using Arduino. In: Proceedings of 2nd international conference on advanced computing and software engineering (ICACSE), pp 262–264
5. Abhishek BN, Bharath, Gunasheel B, Vinodh Kumar G, Veeresh H (2016) Automation of LPG cylinder booking and leakage monitoring system. Int J Combined Res Dev (IJCRD)
6. Jebamalar Leavline E, Asir Antony Gnana Singh D, Deepika H, Abinaya B (2017) LPG gas leakage detection and alert system. Int J Electron Eng Res 9(7):1095–1097
7. Siddika A, Hossain I (2018) LPG gas leakage monitoring and alert system using Arduino. Int J Sci Res (IJSR) 9(1):1734–1737
8. Jebamalar Leavline E, Asir Antony Gnana Singh D, Abinaya B, Deepika H (2017) LPG gas leakage detection and alert system. Int J Electron Eng Res 9(7):1095–1097

9. Kulothungan S, Gukan A, Arunprabu KB (2019) Automatic gas leakage detection and prevention system. Int J Eng Dev Res (IJEDR) 7(2):10–12
10. Yadav V, Shukla A, Bandra S, Kumar V, Ansari U, Khanna S (2016) Review on microcontroller based LPG gas leakage detector. J VLSI Des Signal Process 2(3):1–10

Design of Hybrid Fuzzy-PID Power Management Unit for Control of Battery–Supercapacitor HEV Using Unified LA-92 Drive Cycle

Bibaswan Bose, Vijay Kumar Tayal, and Bedatri Moulik

Abstract In hybrid electric vehicles (HEV), the power management unit helps to control the distribution of power between various power sources with the aim of meeting a set of predefined objectives. Often, the minimization of difference between power demand and supplied power is a prime objective. However, further challenges need to be addressed are the selection of suitable set of rules and adapting them to the vehicle drive cycle. In this paper, a battery–supercapacitor hybrid vehicle control with two schemes, viz. fuzzy rule-based (FRB) and hybrid fuzzy-PID controllers are proposed. The MATLAB simulation results indicate that the application of hybrid fuzzy-PID control scheme results in meeting the power demand with slower rate of discharge. This ensures improved efficacy of the proposed scheme.

Keywords Fuzzy rule-based (FRB) · Power management unit · CuK converter · Supercapacitor · PID control · Unified LA-92 · Hybrid electric vehicle (HEV) · Battery electric vehicle (BEV) · State of charge (SoC)

1 Introduction

The conventional vehicles run by fossil fuel are available since the year 1886. In order to make the vehicles safer, faster and comfortable, the vehicle market has seen numerous changes. However, pollution and greenhouse gasses emitted from these vehicles due to fossil fuel have created serious health and environmental issues. This is affecting sustainability of human life. The introduction of battery electric vehicles (BEV) in the year 1996 has been seen as a solution for these rising concerns. Even after 24 years and surpassing the sale of 5.3 million in 2018, consumers are still doubtful to buy these vehicles due to their low range, low speed, and uncertainty in availability of charging facilities. The hybrid electric vehicles (HEV) operating on conventional fuel as well as battery are being preferred over BEV due to their longer range, low rate of battery discharge, and improved drivability [1].

B. Bose (✉) · V. K. Tayal · B. Moulik
Electrical and Electronics Engineering Department, Amity University Uttar Pradesh, Noida, India

S. Suhag et al. (eds.), *Control and Measurement Applications for Smart Grid*,
Lecture Notes in Electrical Engineering 822,
https://doi.org/10.1007/978-981-16-7664-2_5

The HEVs are of various types [2, 3] depending on the power sources being used. The HEV can be classified on the basis of type of power train available into, parallel hybrid, series hybrid, and power split hybrid [4, 5]. The power split hybrid configuration has been considered as the most efficient as compared to other vehicle configurations due to its longer range and higher speed. In order to make the electric vehicle architecture more reliable for the consumers, many research studies are presented in the literature. The charge conservation techniques [6, 7] for an HEV are helpful for a vehicle to drive for long ranges. Hui and Yunbo [8] study the simulation of different power management strategies for plug-in hybrid electric vehicle (PHEV) and [9] propose the use of rule-based algorithm for electric vehicles. The conservation model of battery state of charge for HEV has been presented by [10] using different power management strategies. The drive train efficiency has been analyzed and compared by [11] for various control strategies and the use of fuzzy logic controller for parallel-type HEV as proposed in [12, 13]. The advantages of using fuzzy for a predictive energy management strategy are explained in [14, 15] which uses fuzzy for uniaxial power train. The power management unit is one of the most important components of the HEV as the power distribution from the source to the vehicle motor is decided by this unit. The power control flow is to be checked by the power management controller. Based on need and availability, the power management strategies can be classified by various algorithms [16, 17]. In order to control the power split in hybrid electric vehicle, this work provides a novel approach toward the application of fuzzy rule-based power management unit. The advantages of fuzzy control are: sturdiness to modeling errors and faulty measurement, tuning of fuzzy rules with ease and the adaptive output with variations in operating condition. The input is fuzzified using fuzzy rules, and output is defuzzied to yield control signals [18]. An intelligent situation awareness agent [19] has been introduced to fuzzy formed torque distribution algorithm. This methodology requires classification based on the type of the roads and behavior of the driver. Thus, learning vector quantization [20] is being used to determine the drive conditions which ensures much better performance improvement [20].

The aim of this work is to minimize the power difference between the demanded power and the supplied power. In order to make the system operability close to real-time characteristics experienced by the vehicle during on-road operations, dynamic models of system have been developed. In this paper, comprehensive mathematical model of DC/DC converter has been derived [21] with battery [22, 23] and supercapacitor [24] for the entire HEV powertrain. To minimize the difference between power demand and supplied power by vehicle with least rate of discharge, a fuzzy power management unit and a hybrid fuzzy-PID controller have been designed [25–27]. The comparison of simulation results obtained with fuzzy power management unit and the hybrid fuzzy-PID scheme shows that the hybrid fuzzy-PID controller yields in far better charge sustainment and drivability characteristics [28]. This organization of paper is done as follows—Sect. 2 includes introduction to all the HEV components followed by the fuzzy rules used in the work in Sect. 3. The system architecture, simulation results, and finally, the outcome of this work has been explained in Sect. 4.

2 Hybrid Electric Vehicle Components

The major components used in the modeling of hybrid electric vehicle (HEV) are battery, supercapacitor, and DC-DC converter. The dynamic modeling helps to bridge the gap between simulation model results and real-time performance.

2.1 Battery

The battery considered in this current work is a lithium–ion (Li-ion) battery. This battery has higher voltage range, low self-discharge, and longer life cycle. The battery equivalent circuit is shown in Fig. 1. The mathematical modeling of the battery is done in two parts [23]. The first part includes the electrical modeling by considering Thevenin's equivalent circuit and its mathematical Eqs. (4) and (5).

$$\dot{V}_{\mathrm{SOC}} = -\frac{I_{\mathrm{bat}}}{C_{\mathrm{cap}}} \tag{1}$$

$$\dot{V}_{\mathrm{ts}} = -\frac{I_{\mathrm{bat}}}{C_{\mathrm{ts}}} - \frac{V_{\mathrm{ts}}}{R_{\mathrm{ts}} * C_{\mathrm{ts}}} \tag{2}$$

$$\dot{V}_{\mathrm{tl}} = -\frac{I_{\mathrm{bat}}}{C_{\mathrm{tl}}} - \frac{V_{\mathrm{tl}}}{R_{\mathrm{tl}} * C_{\mathrm{tl}}} \tag{3}$$

From Eqs. (1)–(3), one can get

$$\dot{x} = \begin{bmatrix} 0 & 0 & 0 \\ 0 & -\left(\frac{1}{R_{\mathrm{ts}} * C_{\mathrm{ts}}}\right) & 0 \\ 0 & 0 & -\left(\frac{1}{R_{\mathrm{tl}} * C_{\mathrm{tl}}}\right) \end{bmatrix} x - \begin{bmatrix} \frac{1}{C_{\mathrm{cap}}} \\ \frac{1}{C_{\mathrm{ts}}} \\ \frac{1}{C_{\mathrm{tl}}} \end{bmatrix} u \tag{4}$$

Fig. 1 Battery equivalent circuit

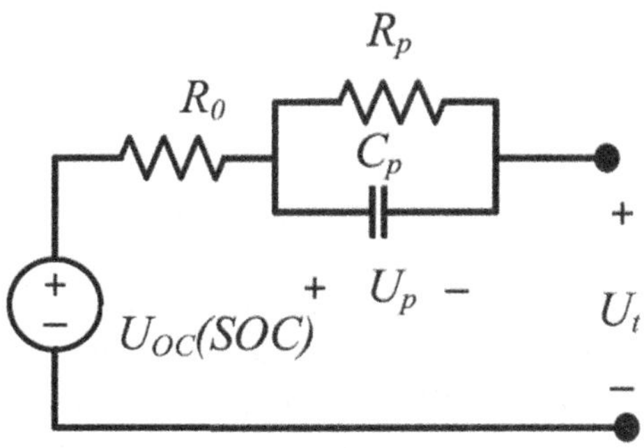

where $x = \begin{bmatrix} V_{\mathrm{SOC}} \\ V_{\mathrm{ts}} \\ V_{\mathrm{tl}} \end{bmatrix}$, $u = I_{\mathrm{bat}}$.

From Eq. (2),

$$y = V_{\mathrm{bat}} = g(V_{\mathrm{SOC}}) + V_{\mathrm{ts}} + V_{\mathrm{tl}} - R_{\mathrm{s}} * I_{\mathrm{bat}} \tag{5}$$

The second part of the battery model has been derived by expressing the voltage in terms of state of charge (SOC) and mathematically expressed by Eq. (6) as follows:

$$\begin{aligned} V_{\mathrm{OC}}(\mathrm{SOC}) = {} & (3.083) + (4.859 * \mathrm{SOC}) + \left(18.21 * \mathrm{SOC}^2\right) + \left(38.56 * \mathrm{SOC}^3\right) \\ & - \left(38.64 * \mathrm{SOC}^4\right) + \left(14.58 * \mathrm{SOC}^5\right) \end{aligned} \tag{6}$$

2.2 *Supercapacitor*

A supercapacitor is created by placing a dielectric separator between two parallel plates. This makes the system more robust, increase power density and life of the system. The supercapacitor is modeled in 2 parts. Equations (7) and (8) represent the dynamic equation of the system, and Eq. (9) represents the relation between state of charge of the system with respect to current [24].

$$\frac{D^n X}{DT^n} = A * X + B * I \tag{7}$$

$$y = C * X + D * I \tag{8}$$

where

$$X = \begin{bmatrix} U_0 \\ U_1 \\ U_2 \end{bmatrix}, n = \begin{bmatrix} \alpha \\ \beta \\ \gamma \end{bmatrix}, A = \begin{bmatrix} 0 & 0 & 0 \\ 0 & -\frac{1}{R_1 C_1} & 0 \\ 0 & 0 & -\frac{1}{R_2 C_2} \end{bmatrix}, B = \begin{bmatrix} \frac{1}{C_0} \\ \frac{1}{C_1} \\ \frac{1}{C_2} \end{bmatrix}, C = \begin{bmatrix} 1 & 1 & 1 \end{bmatrix},$$

$D = R_{\mathrm{S}}$, $y = U_0$.

$$S(t) = S(0) - \frac{1}{C_n} \int_0^t I(t)\mathrm{d}t \tag{9}$$

where $S(t)$ is the state of charge of supercapacitor as a function of time and C_{n} is the rated capacitance of the supercapacitor. The supercapacitor modeling is done based on Eqs. (7)–(9) in MATLAB/Simulink.

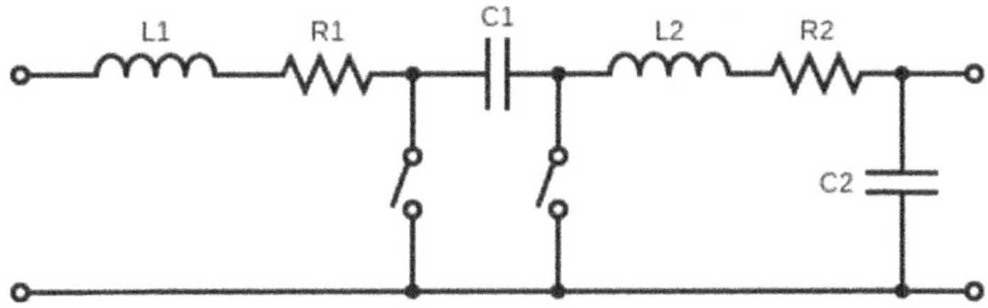

Fig. 2 CuK converter

2.3 DC-DC Converter

In this work, CuK converter, a dual of buck–boost DC-DC converter is used. One of the main assets of using the CuK converter is that it has continuous input and output currents. The CuK converter circuit used for dynamic modeling is shown in Fig. 2.

In Mode 1 of operation, S1 switch is ON and S2 switch is OFF. Upon applying of Kirchhoff law, Eq. (10) can be constructed.

$$X' = \begin{bmatrix} (-R_1/L_1) & 0 & 0 & 0 \\ 0 & (-R_2/L_2) & (-1/L_2) & (-1/L_2) \\ 0 & (1/C_1) & 0 & 0 \\ 0 & (1/C_2) & 0 & 0 \end{bmatrix} X + \begin{bmatrix} 1/L_1 & 0 \\ 0 & 0 \\ 0 & 0 \\ 0 & -1/C_2 \end{bmatrix} U \tag{10}$$

Or the generalized expression can be written as

$$X' = A_1 X + BV \tag{11}$$

where $X = \begin{bmatrix} I_1 \\ I_2 \\ U_{C1} \\ U_{C2} \end{bmatrix}$ and $V = \begin{bmatrix} U_{\text{in}} \\ I_{\text{out}} \end{bmatrix}$.

In Mode 2 of operation, S1 switch is OFF and S2 switch is ON. Upon applying of Kirchhoff law, Eq. (12) can be constructed.

$$X' = \begin{bmatrix} (-R_1/L_1) & 0 & (-1/L_1) & 0 \\ 0 & (-R_2/L_2) & 0 & (-1/L_2) \\ (1/C_1) & 0 & 0 & 0 \\ 0 & (1/C_2) & 0 & 0 \end{bmatrix} X + \begin{bmatrix} 1/L_1 & 0 \\ 0 & 0 \\ 0 & 0 \\ 0 & -1/C_2 \end{bmatrix} V \tag{12}$$

Or the generalized expression can be written as

$$X' = A_2X + BV \tag{13}$$

where $X = \begin{bmatrix} I_1 \\ I_2 \\ U_{C1} \\ U_{C2} \end{bmatrix}$ and $V = \begin{bmatrix} U_{in} \\ I_{out} \end{bmatrix}$.

The nonlinearity is introduced in the system model. This can be obtained by merging the two state-space equations, i.e., Eqs. (10) and (12).

$$A_{Tot} = A_2 + D(A_1 - A_2) \tag{14}$$

Here, the first matrix represents the A_{Linear} and the later part represents the $A_{Non\text{-}linear}$ [29, 30].

$$X' = \begin{bmatrix} (-R_1/L_1) & 0 & (-1/L_1) & 0 \\ 0 & (-R_2/L_2) & 0 & (-1/L_2) \\ (1/C_1) & 0 & 0 & 0 \\ 0 & (1/C_2) & 0 & 0 \end{bmatrix} X$$
$$+ \begin{bmatrix} 0 & 0 & (D/L_1) & 0 \\ 0 & 0 & (-1/L_2) & 0 \\ (-D/C_1) & (D/C_1) & 0 & 0 \\ 0 & 0 & 0 & 0 \end{bmatrix} X + \begin{bmatrix} 1/L_1 & 0 \\ 0 & 0 \\ 0 & 0 \\ 0 & -1/C_2 \end{bmatrix} u \tag{15}$$

3 Fuzzy Rules Used and System Architecture

This section gives a brief overview of set of rules that have been used in the present work. Figure 3 gives the various fuzzy rules where the power demanded by the vehicle has been categorized into three parts, and the three fuzzy rules used and sets have been expressed in Fig. 4:

- Low Power Motoring Mode: In this mode, all the demanded power is supplied by the battery in a power range from 0 to 2000 W. The supercapacitor is kept off during this time.
- High Power Motoring Mode: This mode ranges from 2000 W and above. Due to the high robustness of supercapacitors, all the transients occurring over 2000 W are taken up by the supercapacitor.
- Regenerative Braking Mode: During deceleration, there is a generation of power which helps in charging the battery and supercapacitor.

The proposed system has been dynamically modeled in MATLAB/Simulink. The battery and supercapacitor have been connected to a common bus bar through their

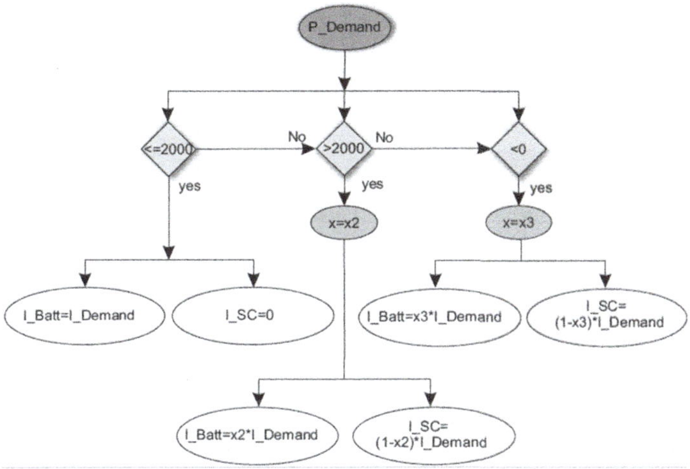

Fig. 3 Fuzzy power management unit rules

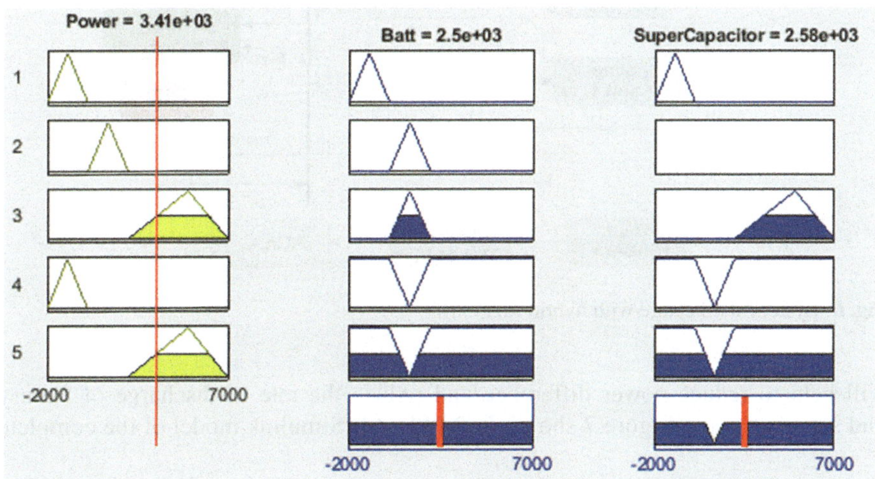

Fig. 4 Fuzzy sets and rules considered

respective DC-DC converters. The converter charges the battery and supercapacitor during regenerative braking. The power management unit gets the information about power demand based on drive cycle. Thus, duty cycle of the converter is altered to meet the power demand.

Figure 5 shows the combination of fuzzy rule-based algorithm with a PID controller that has been connected in order to tune the current exiting from the power management unit (Fig. 6). These values are tuned using PID tuning method [31]. The aim of the use of PID is to tune the duty cycle of the converter that

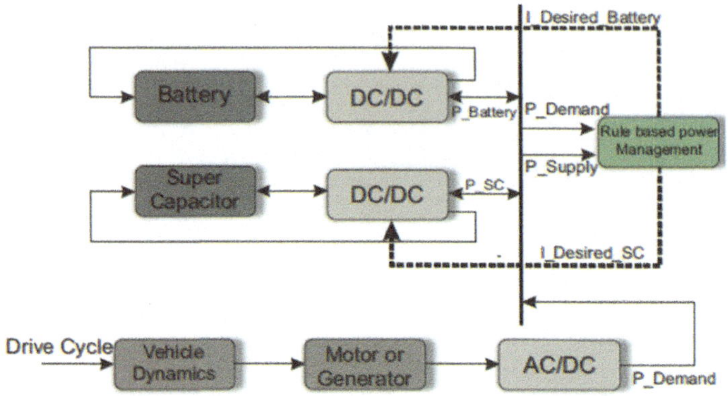

Fig. 5 Fuzzy rule-based system architecture

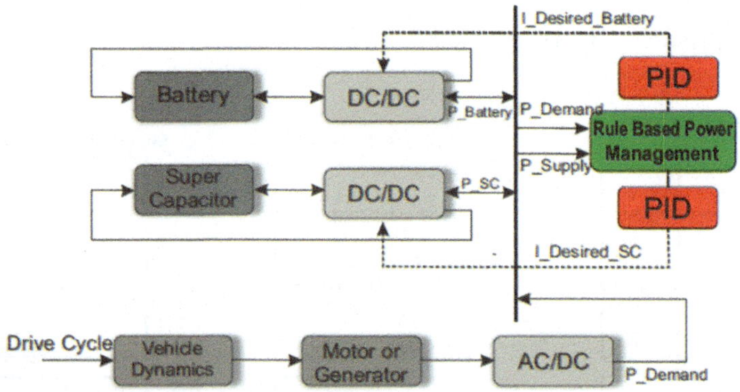

Fig. 6 System architecture with hybrid fuzzy-PID

will help to reduce power difference and reduce the rate of discharge of battery and supercapacitor. Figure 7 shows the MATLAB/Simulink model of the complete

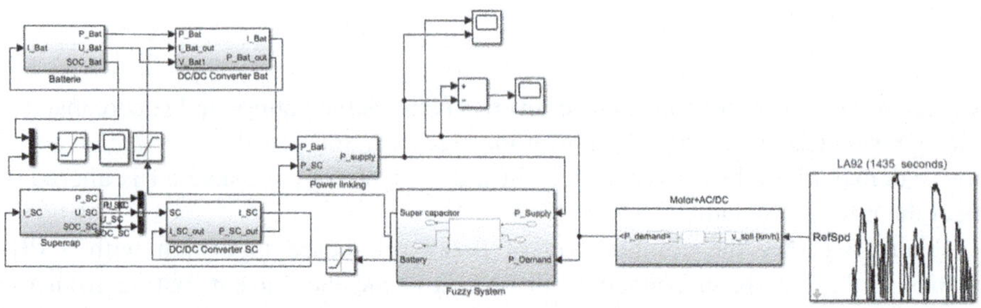

Fig. 7 Simulink model of system architecture

Table 1 PID controller gains

HEV configuration	PID controller gains		
	Kp	Ki	Kd
Battery	1.1578	0.2634	2.1247
Supercapacitor	0.7532	0.0945	0.4756

system architecture. Table 1 shows the values of PID controller gains obtained from equation 40 using PID tuner method.

In this paper, MATLAB simulation of a hybrid electric vehicle subjected to varied drive cycle with two control schemes, viz. Fuzzy Rule Base (FRB) and Hybrid Fuzzy-PID, has been carried out. In order to make the proposed system to be more robust, the dynamically modeled system is used. The fuzzy controller has been used in the power management unit to control the power split between battery and supercapacitor. Figure 8 displays the input current to the converter used to control the duty cycle, for the operation of battery and supercapacitor. This has been observed that the battery takes up only stable operating region and the supercapacitor will take up the transients thus preventing excess load on the battery. From Fig. 9, it has been seen that the variation of voltage with respect to time remains constant near to 90 V.

Magnified images of a short segment show that transients are produced due to frequent switching of converters. Figure 10 shows power supplied and the power demanded by the vehicle. It has been observed that the difference between the demanded power and the supplied power is negligible. Figure 11 shows the comparison of power difference for fuzzy rule-based (FRB) system and the fuzzy-PID system. It can be interpreted that the power difference with fuzzy and hybrid fuzzy-PID control scheme is 100 W and 10 W, respectively. Figures 12 and 13 show the comparison of state of charge of the supercapacitor and battery. This shows that rate

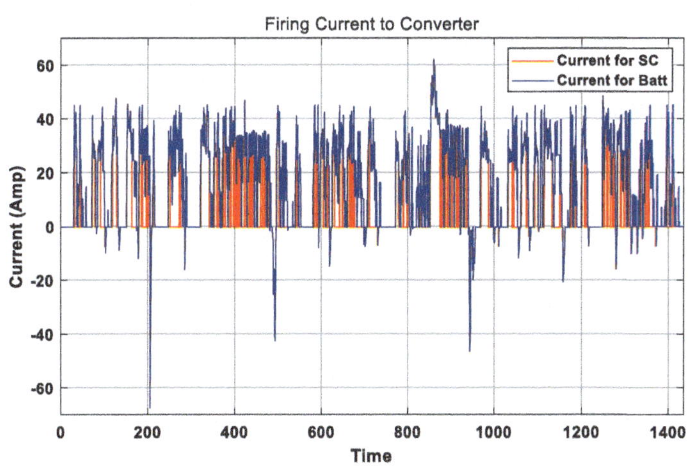

Fig. 8 Output current from converter

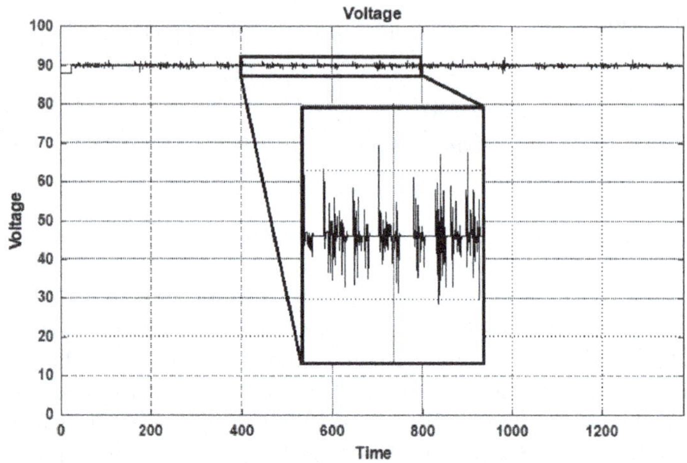

Fig. 9 Output voltage from converter

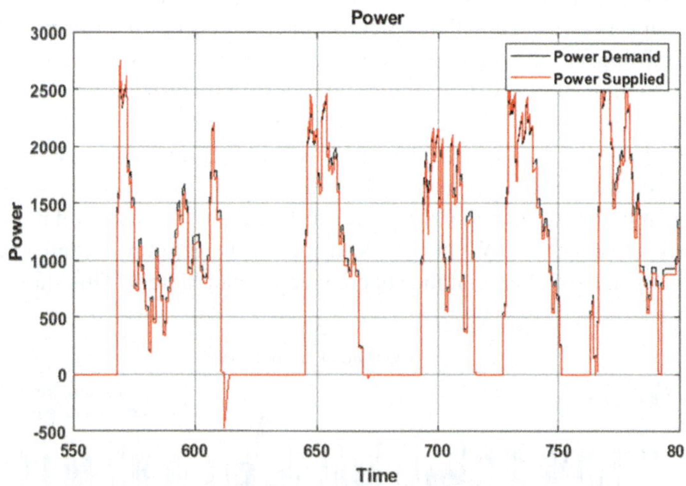

Fig. 10 Power demanded versus power supplied

of discharge with hybrid fuzzy-PID controller is also much slower as compared to fuzzy rule-based (FRB) control scheme for both battery and supercapacitor.

4 Conclusion

This work presents the design of a hybrid electric vehicle (HEV) using architecture of fuzzy rule-based power management system. The battery and supercapacitor are

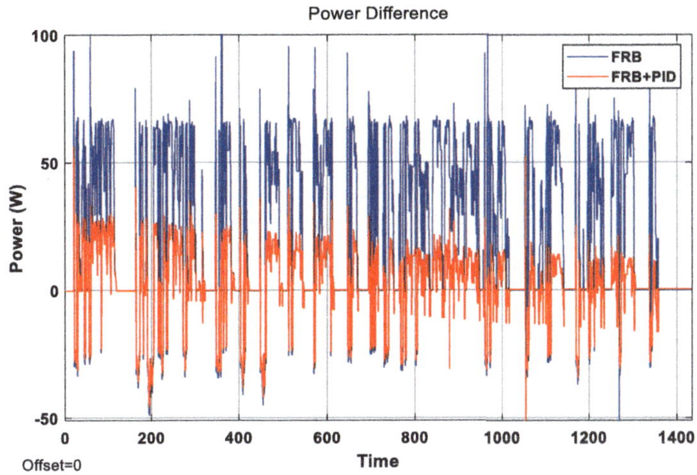

Fig. 11 Power difference

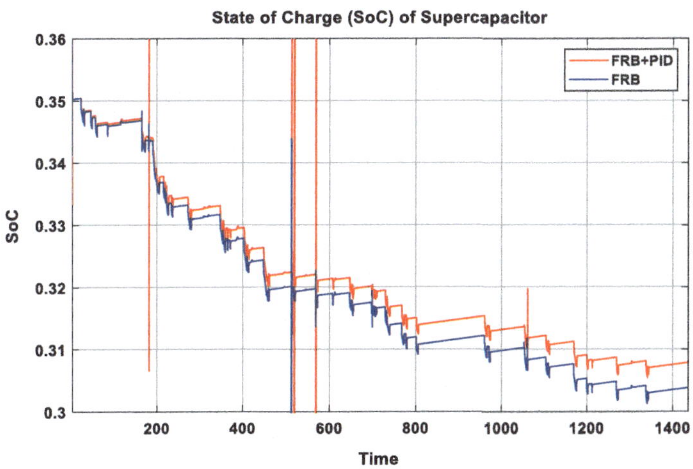

Fig. 12 SOC of supercapacitor comparison

working as power sources. The developed model is simulated in MATLAB/Simulink environment. The primary aim of maintaining the bus voltage constant and the power supply to meet the demand have been successfully attained. Further, the reduction of power difference and state of discharge of battery and supercapacitor have been successfully achieved by application of fuzzy rule base (FRB) and PID control schemes. The hybrid fuzzy-PID controller yield in much better charge sustainment and drivability characteristics. This ensures far more superior performance of the proposed scheme.

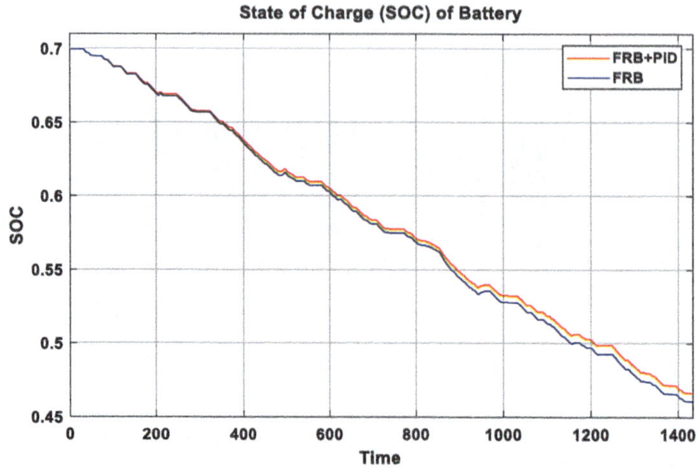

Fig. 13 SOC of battery comparison

References

1. Chen Y, Ding Z, Ma J, Liu J (2018) A comparison of the life cycle energy consumption and emissions of BEV and HEV. IOP Conf Ser Earth Environ Sci 170(4):042070. IOP Publishing
2. Singh KV, Bansal HO, Singh D (2019) A comprehensive review on hybrid electric vehicles: architectures and components. J Mod Transp 27(2):77–107
3. Fathabadi H (2018) Novel fuel cell/battery/supercapacitor hybrid power source for fuel cell hybrid electric vehicles. Energy 143:467–477
4. Rousseau A, Pagerit S, Gao DW (2008) Plug-in hybrid electric vehicle control strategy parameter optimization. J Asian Electr Veh 6(2):1125–1133
5. Dextreit C, McGeoch D, Andrews M, Harrison K, Jaguar Land Rover Ltd (2020) Control strategy for plug-in hybrid electric vehicle. U.S. patent 10,543,830
6. Boyd SJ (2006) Hybrid electric vehicle control strategy based on power loss calculations. Doctoral dissertation, Virginia Tech
7. Sharer PB, Rousseau A, Karbowski D, Pagerit S (2008) Plug-in hybrid electric vehicle control strategy: comparison between EV and charge-depleting options (No. 2008-01-0460). SAE technical paper
8. Hui X, Yunbo D (2008) The study of plug-in hybrid electric vehicle power management strategy simulation. In: 2008 IEEE vehicle power and propulsion conference, Sept 2008. IEEE, pp 1–3
9. Banvait H, Anwar S, Chen Y (2009) A rule-based energy management strategy for plug-in hybrid electric vehicle (PHEV). In: 2009 American control conference, June 2009. IEEE, pp 3938–3943
10. Lee H, Jeong J, Park YI, Cha SW (2017) Energy management strategy of hybrid electric vehicle using battery state of charge trajectory information. Int J Precis Eng Manuf-Green Technol 4(1):79–86
11. Desai C, Williamson SS (2009) Comparative study of hybrid electric vehicle control strategies for improved drive train efficiency analysis. In: 2009 IEEE electrical power & energy conference (EPEC), Oct 2009. IEEE, pp 1–6
12. Lee HD, Sul SK (1998) Fuzzy-logic-based torque control strategy for parallel-type hybrid electric vehicle. IEEE Trans Ind Electron 45(4):625–632
13. Qian LJ, Xi ZY, Zhao H (2006) Simulation of hybrid electric vehicle control strategy based on fuzzy neural network. J Syst Simul 18(5):1384–1387

14. Montazeri-Gh M, Mahmoodi-K M (2016) Optimized predictive energy management of plug-in hybrid electric vehicle based on traffic condition. J Clean Prod 139:935–948
15. Dawei M, Yu Z, Meilan Z, Risha N (2017) Intelligent fuzzy energy management research for a uniaxial parallel hybrid electric vehicle. Comput Electr Eng 58:447–464
16. Huang Y, Wang H, Khajepour A, Li B, Ji J, Zhao K, Hu C (2018) A review of power management strategies and component sizing methods for hybrid vehicles. Renew Sustain Energy Rev 96:132–144
17. Ali AM, Söffker D (2018) Towards optimal power management of hybrid electric vehicles in real-time: a review on methods, challenges, and state-of-the-art solutions. Energies 11(3):476
18. Jin F, Wang M, Hu C (2016) A fuzzy logic based power management strategy for hybrid energy storage system in hybrid electric vehicles considering battery degradation. In: 2016 IEEE transportation electrification conference and expo (ITEC), June 2016. IEEE, pp 1–7
19. So R, Sonenberg L (2004) Situation awareness in intelligent agents: foundations for a theory of proactive agent behavior. In: Proceedings. IEEE/WIC/ACM international conference on intelligent agent technology, 2004. (IAT 2004), Sept 2004. IEEE, pp 86–92
20. Yin H, Zhou W, Li M, Ma C, Zhao C (2016) An adaptive fuzzy logic-based energy management strategy on battery/ultracapacitor hybrid electric vehicles. IEEE Trans Transp Electrific 2(3):300–311
21. Rayeen Z, Bose S, Dwivedi P (2018) Study of closed loop Cuk converter controlled by loop shaping method. In: 2018 IEEE 13th international conference on industrial and information systems (ICIIS), Dec 2018. IEEE, pp 442–447
22. Thirugnanam K, Ezhil Reena Joy TP, Singh M, Kumar P (2014) Mathematical modeling of Li-ion battery using genetic algorithm approach for V2G applications. IEEE Trans Energy Convers 29(2):332–343
23. How DN, Hannan MA, Lipu MH, Ker PJ (2019) State of charge estimation for lithium-ion batteries using model-based and data-driven methods: a review. IEEE Access 7:136116–136136
24. Wang J, Zhang L, Mao J, Zhou J, Xu D (2019) Fractional order equivalent circuit model and SoC estimation of supercapacitors for use in HESS. IEEE Access 7:52565–52572
25. Bose B (2020) Modelling of microinverter and pushpull flyback converter for SPV application. In: 2020 8th international conference on reliability, infocom technologies and optimization (trends and future directions) (ICRITO), June 2020. IEEE, pp 458–462
26. Sahu AR, Bose B, Kumar S, Tayal VK (2020) A review of various power management schemes in HEV. In: 2020 8th international conference on reliability, infocom technologies and optimization (trends and future directions) (ICRITO), June 2020. IEEE, pp 1296–1300
27. Bose B, Kumar S (2020) Design of push-pull flyback converter interfaced with solar PV system. In: 2020 first international conference on power, control and computing technologies (ICPC2T), Jan 2020. IEEE, pp 117–121
28. Pratumsuwan P, Thongchai S, Tansriwong S (2010) A hybrid of fuzzy and proportional-integral-derivative controller for electro-hydraulic position servo system. Energy Res J 1(2):62–67
29. Smedley KM, Cuk S (1995) Dynamics of one-cycle controlled Cuk converters. IEEE Trans Power Electron 10(6):634–639
30. Brkovic M, Cuk S (1992) Input current shaper using Cuk converter. In: Proceedings of fourteenth international telecommunications energy conference-INTELEC'92. IEEE, pp 532–539
31. Bose B, Suryawanshi R, Shah J (2018) Simulation based V2H super capacitor charging using bidirectional DC-DC converter. In: 2018 international conference on smart city and emerging technology (ICSCET), Jan 2018. IEEE, pp 1–5

Event-Triggered Sliding Mode Controller Design for Interconnected Power System

Aradhna Patel and Shubhi Purwar

Abstract The load frequency control problem is addressed in this paper and the event-triggered sliding-mode control (SMC) technique is suggested in multi-area interrelated power frameworks. To begin with, we have considered a three-power area interconnected control system and developed the numerical model associated with it. The recommended $H\infty$ execution is utilized to quantify the sufficiency of the diminishing of the adjusting unsettling influence. For each subsystem, an appropriate sliding surface is designed, and the intermittent controller is intended to ensure that the subsequent subsystem state is asymptotically stable and reliable. The corresponding adequate conditions are therefore obtained and the law is triggered by movement events to confirm the accessibility of the sliding surface in a limited time. In the final analysis, case studies and simulations are presented to help understand the probability and feasibility of the event-initiated SMC scheme.

Keywords SMC · Load frequency control · Multi-region power systems · ET-SMC

1 Introduction

Digitally controlled systems, easy maintenance, installation, and low cost work together, resulting in increased attention in recent years [1, 2]. Due to collaboration of communication with control systems, increase in jamming of signal and data loss is observed, which may decrease system performance, increase oscillation, and instability in system. Recently, the event techniques for the possibility of reducing executive numbers administration tasks, while maintaining a promising performance of the system, provoked wide spread concern. A strategic impulse caused by events transmission of state signals only when the present trigger condition is satisfied (or

A. Patel (✉) · S. Purwar
Electrical Engineering Department, MNNIT Allahabad, Prayagraj, India

S. Purwar
e-mail: shubhi@mnnit.ac.in

S. Suhag et al. (eds.), *Control and Measurement Applications for Smart Grid*, Lecture Notes in Electrical Engineering 822,
https://doi.org/10.1007/978-981-16-7664-2_6

violated, depending on the accepted logic). This technique basically reduces the communication burden.

Consider the issue of energy supply and demand for the multi-zone power system, load frequency control (LFC) to maintain frequency deviations to some extent, in case of load fluctuations. It is achieved by controlling the reference load points of the selected group of generators. Attention to LFC studies on the last decades is significant, see, for example, there are various control schemes to accomplish the preferred dynamic performance, such as PI and PID controllers are used in LFC for multi-area power systems [3], the optimal control approach [4], reliable decentralized theory of control [5], and adaptive control schemes [6]. In power systems, LFC is efficiently used to preserve frequency and power in scheduled values with adjacent regions.

On the other hand, SMC is considered as a powerful and reliable management strategy [7–9]. Simply, SMC technique is issued to intend a switching surfaces, and it forces the closed loop state trajectories of any system towards the sliding surface in finite time. This technique is used in LFC for maintaining the stability limit of frequency deviation and tie line power flow of multi-area power system.

As of late, the sliding mode load frequency controller (SMLFC) was utilized to tackle power frameworks issues with uncertainty [10–16]. Utilizing the shaft task procedure, in [14], they demonstrated a SMLFC where the parametric vulnerabilities were considered under the connection that fulfilled the coordinating condition. Nonetheless, vulnerabilities in force frameworks were not generally fulfilled by the coordinating condition. [11] proposed a discrete variable structure regulator for the LFC of multi-zone interconnected force frameworks, where the framework nonlinearities were remembered for the re-enactment considered. Be that as it may, boundaries vulnerability because of variable activity point were not examined.

Note that the control of simulated continuous time systems. Finally, using a computer to transfer the state of the system implements only in certain temporary instances, and updating the information of the control inputs. Therefore, the sliding mode (DSMC) discrete time control occurs and becomes a study subject. A new technique for handling LFC multi-zone power systems is implemented in this paper, i.e., event-activated sliding mode load frequency control (ETSMLFC). Compared to periodic sampling, event-based sampling will save several parameters, such as network energy, power usage, and NCS computational load controller [8, 17, 18]. An intelligent fuzzy coordination control is designed in [19], and unknown input observer used in LFC problem is reported in [20].

This article suggests an event-driven SMC approach for the LFC issue in multi-region energy control systems. In particular, based on the DSMC technique, the solution for the LFC issue is given by integrating time interval analysis approaches with the event-triggered process. The associated appropriate states of the event-triggered SMC shall be established to make it easier for each subsystem to be asymptotically stable with the specified $H\infty$ performance. The significant part of our work has been outlined as follows:

(1) A new SMC approach is being developed using the event-driven technique to confirm that the carefully collaborated dynamic framework has the recommended energy-to-energy output.
(2) By addressing the sequential minimization issue with a typical MATLAB toolbox, an event-triggered SMC algorithm is suggested for the LFC method.
(3) The effectiveness of the suggested ET-SMLFC is tested using MATLAB simulation findings on a three-area power execution system.

2 Problem Formulation and Preliminaries

The LFC or AGC as an auxiliary assistance in the force framework has a significant and essential job of keeping up the dependability at a suitable level. The LFC has two principle control circles. Initial one is the essential speed control circle [21], which controls the lead representative valve position as per recurrence deviation in single-zone power framework. The auxiliary circle keeps up the harmony among age and burden considering power misfortunes in an interconnected multi-region power framework utilizing region control blunder (ACE). The enormous interconnected force frameworks are normally deteriorated into control territories or zones concurring to different models, for example, authoritative, hierarchical, specialized, recorded, topographical, and so on. These interconnected control territories are alluded as multi-zone power frameworks. In multi-area interconnected executing power system, each area has a single machine, and each machine designates power system with a load frequency control circuit as shown. The power network whereas being a complex nonlinear framework can be linearized for the examination on LFC issue. The linearized model [22] of the *i*th area power framework is displayed in Fig. 1. The dynamic model of the multi-zone power system is presented as follows:

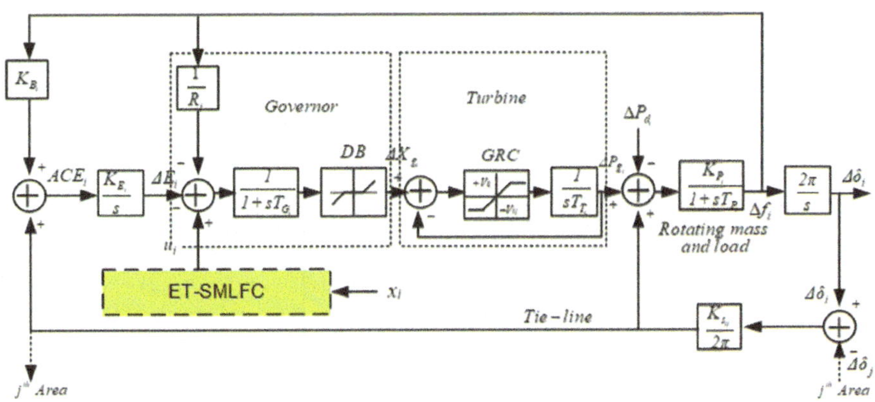

Fig. 1 Block diagram of ETSMLFC for multi-region energy system

$$\Delta \dot{f}_i(t) = -\frac{1}{T_{P_i}}\Delta f_i(t) + \frac{K_{P_i}}{T_{P_i}}\Delta P_{g_i}(t) - \frac{K_{P_i}}{T_{P_i}}\Delta P_{d_i}(t) - \frac{K_{P_i}}{2\pi T_{P_i}} \sum_{\substack{j \in N \\ j \neq i}} \{\Delta\partial_i(t) - \Delta\partial_j(t)\} \tag{1}$$

$$\Delta \dot{P}_{g_i}(t) = -\frac{1}{T_{T_i}}\Delta P_{g_i}(t) + \frac{1}{T_{T_i}}\Delta X_{g_i}(t) \tag{2}$$

$$\Delta \dot{X}_{g_i}(t) = -\frac{1}{T_{G_i} R_i}\Delta f_i(t) - \frac{1}{T_{G_i}}\Delta X_{g_i}(t) - \frac{1}{T_{G_i}}\Delta E_i(t) + \frac{1}{T_{G_i}} u_i(t) \tag{3}$$

$$\Delta \dot{E}_i(t) = K_{E_i}\left[K_{B_i}\Delta f_i(t) + \frac{1}{2\pi} \sum_{\substack{j \in N \\ j \neq i}} K_{sij}\{\Delta\partial_i(t) - \Delta\partial_j(t)\} \right] \tag{4}$$

$$\Delta \dot{\partial}_i(t) = 2\pi \Delta f_i(t) \tag{5}$$

where $i = 1, \ldots, N$ and N is number of areas.

The state space representation of above dynamics Eq. (1) to (6) can be written as:

$$\dot{x}_i(t) = A_i x_i(t) + B_i u_i + \sum_{\substack{j \in N \\ j \neq i}} E_{ij} x_j(t) + F_i \Delta P_{d_i}(t) \tag{6}$$

where

$$A_i = \begin{bmatrix} \frac{1}{-T_{P_i}} & \frac{K_{P_i}}{T_{P_i}} & 0 & 0 & -\frac{K_{P_i}}{2\pi T_{P_i}} \sum_{\substack{j \in N \\ j \neq i}} K_{sij} \\ 0 & \frac{-1}{T_{T_i}} & \frac{1}{T_{T_i}} & 0 & 0 \\ \frac{-1}{T_{G_i} R_i} & 0 & \frac{-1}{T_{G_i}} & \frac{-1}{T_{G_i}} & 0 \\ K_{B_i} K_{E_i} & 0 & 0 & 0 & \frac{K_{E_i}}{2\pi} \sum_{\substack{j \in N \\ j \neq i}} K_{sij} \\ 2\pi & 0 & 0 & 0 & 0 \end{bmatrix}; \quad B_i = \begin{bmatrix} 0 & 0 & \frac{1}{T_{G_i}} & 0 & 0 \end{bmatrix}^{\mathrm{T}};$$

$$E_{ij} = \begin{bmatrix} 0\,0\,0\,0 & \frac{K_{P_i}}{2\pi T_{P_i}} K_{sij} \\ 0\,0\,0\,0 & 0 \\ 0\,0\,0\,0 & 0 \\ 0\,0\,0\,0 & \frac{-K_{E_i}}{2\pi} K_{sij} \\ 0\,0\,0\,0 & 0 \end{bmatrix}; \; F_i = \begin{bmatrix} \frac{-K_{P_i}}{T_{P_i}} & 0 & 0 & 0 & 0 & 0 \end{bmatrix}^{\mathrm{T}};$$

$$\text{and } x_i(t) = \begin{bmatrix} \Delta f_i(t) \\ \Delta P_{g_i}(t) \\ \Delta X_{g_i}(t) \\ \Delta E_i(t) \\ \Delta \partial_i(t) \end{bmatrix} \tag{7}$$

The dimensions of system matrices are $A_i \in R^{n\times n}$, $B_i \in R^{n\times m}$, $F_i \in R^{n\times k}$, $E_{ij} \in R^{n\times n}$, $P_{di} \in R^{k\times 1}$.

According to the LFC system, nominal values must be stored for both the frequency of the control zone and the power exchanged via the connection line. SMLFC and switching surface for each region ensure the asymptotic stability of the system as a whole and are developed on the basis of the following assumptions:

Assumption 1 A_i and B_i are fully controllable.

Assumption 2 The load disturbance is bounded, and meet $\|P\|_d \leq d$, where d is a known constant.

2.1 Design of Switching Surface

The SMC control hypothesis is one of the most impressive control procedures to accomplish desired execution when managing uncertain frameworks. The details of the control structure and the stability investigation are represented in the accompanying subsections. The sampling duration is selected as T, the discrete time approximation model of the framework (6) can be gained depending on the Runge–Kutta fourth-order estimate [22].

The switching surface is designed as,

$$S_i(t) = C_i x_i(t) \tag{8}$$

where C_i is the surface gain matrix. Its dimension is $C_i \in \Re^{1\times n}$. The gain of sliding surface is calculated using pole placement methodology.

Where

$x_i(t)$	State vector
$x_j(t)$	Neighbouring state vector of x_i(t)
u_i(t)	Control vector
$\Delta P_{d_i}(t)$	Load disturbance for the ith subsystem
$\Delta f_i(t)$	Increment frequency deviation
$\Delta P_{g_i}(t)$	Incremental changes in ith subsystem's output
$\Delta X_{g_i}(t)$	Incremental changes in ith governor value position
$\Delta E_i(t)$	Incremental changes in integral control
$\Delta\partial_i(t)$	Incremental changes in rotor angle deviation
T_{G_i}	ith governor time constant
T_{t_i}	ith turbine tune constant
T_{p_i}	ith subsystem model time constant
K_{p_i}	ith power system gain
R_i	ith speed regulation coefficient
K_{E_i}	ith integral control gain
K_{B_i}	ith frequency bias factor
K_{sij}	ith reconnection gain between area i and j ($i \neq j$)

The condition of inequality only means that in all conditions, the space trajectories of the device state reaches the sliding surface for a stated amount of time, without any constraint to the trajectories of the scene's motion.

Here, to develop a variable structure controller, we adopt the following constant achievement rule:

$$u_{\text{smc}} = -\varepsilon \operatorname{sgn}(S) \tag{9}$$

where $\varepsilon > 0$, the constant ε specifies the movement speed of the system's point of motion to the switching surface $S = 0$, the corresponding SMC for ith the power system is:

$$u_i(t) = -(C_i B_i)^{-1}\left(C_i A_i x_i(t) + \varepsilon \operatorname{sgn}(S) + C_i \sum_{\substack{j \in N \\ j \neq i}} E_{ij} x_i(t) \right) \tag{10}$$

2.2 Lyapunov Stability

A Lyapunov function candidate has been taken as:

$$V(t) = \frac{1}{2}s^T(t)s(t) \tag{11}$$

Taken derivative of Lyapunov function:

$$\begin{aligned}\dot{V}(t) &= s^T(t)\dot{s}(t) \\ &= s^T(t)(C_i\dot{x}_i(t)) \\ &= s^T(t)\left(C_iA_ix_i(t) + C_iB_iu_i(t) + \sum_{\substack{j \in N \\ j \neq i}} C_iE_{ij}x_i(t) + C_iF_i\Delta P_{di}(t) \right)\end{aligned} \tag{12}$$

For stability, $\dot{V}(t) \leq 0$. Then, we have chosen control for LFC problem that is given below.

For equivalent control;

$$C_iA_ix_i(t) + C_iB_iu_i(t) + \sum_{\substack{j \in N \\ j \neq i}} C_iE_{ij}x_i(t) + C_iF_i\Delta P_{di}(t) = 0$$

$$u_{\text{eq}}(t) = -(C_iB_i)^{-1}\left(C_iA_ix_i(t) + C_i\sum_{\substack{j \in N \\ j \neq i}} E_{ij}x_i(t) \right)$$

For bounded load disturbance $\|\Delta P_{di}(t)\| \leq \gamma$

$$u_i(t) = u_{\text{eq}}(t) + u_{\text{smc}}(t)$$

$$u_i(t) = -(C_iB_i)^{-1}\left(C_iA_ix_i(t) + C_i\sum_{\substack{j \in N \\ j \neq i}} E_{ij}x_i(t) \right) - \varepsilon\text{sgn}(S)$$

$$u_i(t) = -(C_i B_i)^{-1} \left(C_i A_i x_i(t) + \varepsilon \mathrm{sgn}(S) + C_i \sum_{\substack{j \in N \\ j \neq i}} E_{ij} x_i(t) \right) \tag{13}$$

Using Lyapunov Stability, we have successfully developed control law $u_i(t)$ for LFC system.

2.3 Event-Triggered LFC Design

The event-triggered scheme is integrated with control law to minimize the updating of control law. The utilization of event trigger-based SMLFC in multi-area executing power system reduces the load on communication channel. The developed ET-SMLFC control law excluded the zeno behaviour. In this construction, we continuously monitor the state $x_i(t)$. The dynamic model, according to the scheme initiated by the event, is transmitted as:

$$\dot{x}_i(t_r) = A_i x_i(t) + B_i u_i(t_r) + \sum_{\substack{j \in N \\ j \neq i}} E_{ij} x_i(t) + F_i \Delta P_{di}(t) \tag{14}$$

With initial condition of $x_i(t) = x_0$. The error presented in the LFC system due to a control action triggered by the event is defined as $e_{x_i}(t)$. The error $e_{x_i}(t)$ is taken as:

$$e_{x_i}(t) = x_i(t_r) - x_i(t) \tag{15}$$

Theorem 1 *For system* (6), *with control law* (10) *and sequence of load frequency control update step given by* $\{t_r\}_{r \in Z \geq 0}$, *the system will be in practical state if for given* $\alpha > 0$,

$$\|C_i\| \|A_i\| \|e_{xi}\| \geq \alpha \quad \text{for all } \{t_r\}_{r \in Z \geq 0} \tag{16}$$

The state $x_i(t)$ will be updated and sent to SMC if the error fails to comply with the following, with $\sigma \in [0, 1)$ and alpha > 0.

$$\|C_i\| \|A_i\| \|e_{xi}\| \geq \sigma\alpha \tag{17}$$

Equation (17) is called event triggering rule for multi-area power system. At $t = t_r$, the error reaches zero; however, it grows from zero till it is equal to $\frac{\sigma \times \alpha}{\|A_i\| \|C_i\|}$ as per. Then, the next triggering instant can be generated as,

$$t_{r+1} = \inf\{t : t \succ t_r \text{ and } \|A_i\|\|C_i\|\|e_{xi}\| \geq \sigma\alpha\} \tag{18}$$

3 Simulation Result

The following simulations were done in MATLAB to validate the design of the sliding mode dependent control for load frequency problems (SMLFC). As fluctuations in the load of the power system cause changes in the operating point of the governor, this results in uncertainties in the parameters of the power system. The proposed controller is simulated against step load disturbance 0.06, −0.04 and −0.05 pu applied in each area at $t = 0$ s. Due to continuous change in power demand, the load uncertainty is continuously changing system frequency. Here, we develop ET-SMLFC controller to stabilize the system frequency and improve the system reliability of multi-region energy executing system. Three area power executing system is considered for simulation study. The data for the three areas are as follows:

Simulation data is presented for Area 1:

$$A_1 = \begin{bmatrix} -0.05 & 6 & 0 & 0 & -0.955 \\ 0 & -4.445 & 4.445 & 0 & 0 \\ -11.574 & 0 & -16.667 & 0 & 0 \\ 4 & 0 & 0 & 0 & 1.592 \\ 6.283 & 0 & 0 & 0 & 0 \end{bmatrix} F_1 = \begin{bmatrix} -6 \\ 0 \\ 0 \\ 0 \\ 0 \end{bmatrix} B_1 = \begin{bmatrix} 0 \\ 0 \\ 16.667 \\ 0 \\ 0 \end{bmatrix}$$

$$E_{13} = \begin{bmatrix} 0 & 0 & 0 & 0 & 0.478 \\ 0 & 0 & 0 & 0 & 0 \\ 0 & 0 & 0 & 0 & 0 \\ 0 & 0 & 0 & 0 & -0.796 \\ 0 & 0 & 0 & 0 & 0 \end{bmatrix} E_{12} = \begin{bmatrix} 0 & 0 & 0 & 0 & 0.478 \\ 0 & 0 & 0 & 0 & 0 \\ 0 & 0 & 0 & 0 & 0 \\ 0 & 0 & 0 & 0 & -0.796 \\ 0 & 0 & 0 & 0 & 0 \end{bmatrix}$$

Simulation data is presented for Area 2:

$$A_2 = \begin{bmatrix} -0.04 & 4.5 & 0 & 0 & -0.716 \\ 0 & -4.041 & 4.041 & 0 & 0 \\ -11.431 & 0 & -18.519 & 0 & 0 \\ 4 & 0 & 0 & 0 & 1.592 \\ 6.283 & 0 & 0 & 0 & 0 \end{bmatrix} F_2 = \begin{bmatrix} -4.5 \\ 0 \\ 0 \\ 0 \\ 0 \end{bmatrix} B_2 = \begin{bmatrix} 0 \\ 0 \\ 18.519 \\ 0 \\ 0 \end{bmatrix}$$

$$E_{21} = \begin{bmatrix} 0\,0\,0\,0 & 0.358 \\ 0\,0\,0\,0 & 0 \\ 0\,0\,0\,0 & 0 \\ 0\,0\,0\,0 & -0.796 \\ 0\,0\,0\,0 & 0 \end{bmatrix} \quad E_{23} = \begin{bmatrix} 0\,0\,0\,0 & 0.358 \\ 0\,0\,0\,0 & 0 \\ 0\,0\,0\,0 & 0 \\ 0\,0\,0\,0 & -0.796 \\ 0\,0\,0\,0 & 0 \end{bmatrix}$$

Simulation data is presented for Area 3:

$$A_3 = \begin{bmatrix} -0.05 & 5.75 & 0 & 0 & -0.915 \\ 0 & -3.809 & 3.809 & 0 & 0 \\ -12.698 & 0 & -19.047 & 0 & 0 \\ 4 & 0 & 0 & 0 & 1.592 \\ 6.283 & 0 & 0 & 0 & 0 \end{bmatrix} \quad F_3 = \begin{bmatrix} -5.75 \\ 0 \\ 0 \\ 0 \\ 0 \end{bmatrix} \quad B_3 = \begin{bmatrix} 0 \\ 0 \\ 19.047 \\ 0 \\ 0 \end{bmatrix}$$

$$E_{32} = \begin{bmatrix} 0\,0\,0\,0 & 0.458 \\ 0\,0\,0\,0 & 0 \\ 0\,0\,0\,0 & 0 \\ 0\,0\,0\,0 & -0.796 \\ 0\,0\,0\,0 & 0 \end{bmatrix} \quad E_{31} = \begin{bmatrix} 0\,0\,0\,0 & 0.458 \\ 0\,0\,0\,0 & 0 \\ 0\,0\,0\,0 & 0 \\ 0\,0\,0\,0 & -0.796 \\ 0\,0\,0\,0 & 0 \end{bmatrix}$$

The switching surfaces are chosen for all three area of power executing system. Using the pole placement stability theory, the sliding surface gain is selected. The sliding vector for all three areas is given below:

$$\begin{aligned} &C_1 = [3.6\ 8.2\ 5.5\ 2.7\ 6.1];\ C_2 = [9.2\ 2.5\ 7.1\ 1.8\ 11.9]; \\ &C_3 = [3.1\ 6.6\ 1.2\ 8.3\ 12.6]; \end{aligned} \tag{19}$$

where C_i is the sliding surface gain.

In the case of the load aggravation applied, the frequency deviation in the three region is given in Fig. 2a. The frequency deviation of the three regions is transient and shifts under load disturbance towards steady state (zero) esteem. On comparable lines, the error apparent from tie-line power deviations given by Fig. 2b. From this figure it is evident, that all areas frequency deviation and tie line powers using ET-SMCLFC approaches are settled to zero within 10 s. For all three regions, the area control error (ACE) is shown in Fig. 2c. The ACE decreases continuously over time. Control effort for all three-area executing power system are shown in Fig. 2d.

In addition, under the event triggered condition, Fig. 3a–c shows the release instants and release intervals for the three-area method. It is obvious, from Table 1, that ET-SMLFC control uses less sample for updating the control law as compare to normal SMLFC.

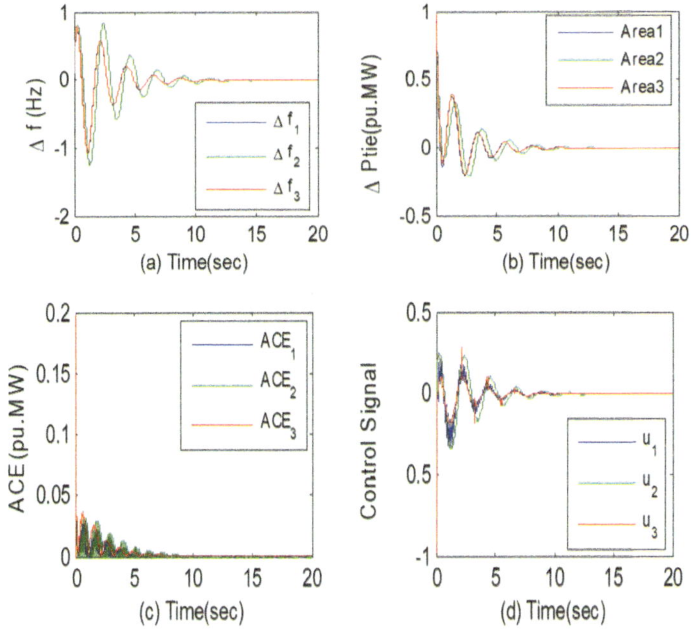

Fig. 2 **a** Frequency deviation, **b** tie line power deviation, **c** ACE response, **d** control effort

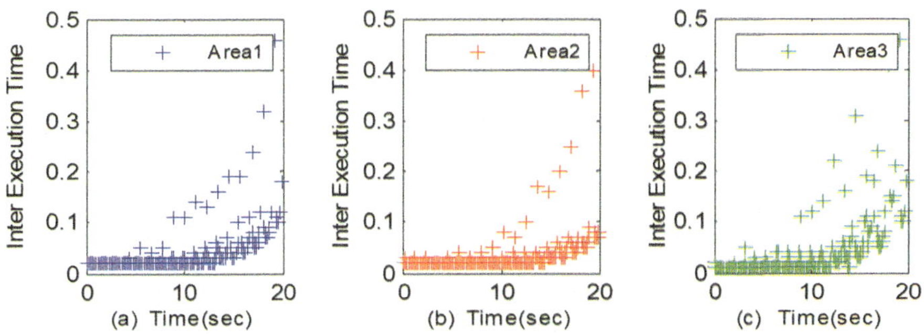

Fig. 3 Inter execution time, **a** for Area 1, **b** for Area 2, **c** for Area 3

Table 1 Evaluation of ET-SMLFC and constant sampling SMLFC for three area power executing system

System parameters	Total no. of sample for $T = 0.01$ s	
Control effort for	ET-SMLFC	SMLFC
Area 1	1545	2001
Area 2	1365	2001
Area 3	1473	2001

4 Conclusion

Based on the event-based SMC methodology, the LFC issue in multi-zone energy systems is solved in this article. In particular, a new discrete sliding surface triggered by the event function was proposed for each subsystem within the energy of several device regions. The sliding mode regulator is intended to guarantee that every subsystem in the multi-zone power framework is steady and impervious to outer aggravations, including load varieties and recurrence varieties. Finally, with the assistance of discrete and on-to-the-point models and reproductions viability and adequacy of the proposed new plan techniques have been delineated.

Further improvement of the proposed algorithm can be achieved via additional research in this area; for example, extending existing methods for controlling the frequency of charging of power systems against a delayed entry cyber-attack and monitoring the situation when certain components of the LFC system states are unavailable.

References

1. Tan C, Li L, Zhang H (2015) Stabilization of networked control systems with both network-induced delay and packet dropout. Automatica 194–199
2. Wu Z, Shi P, Su H, Chu J (2013) Network-based robust passive control for fuzzy systems with randomly occurring uncertainties. IEEE Trans Fuzzy Syst 966–971
3. Tan W (2010) Unified tuning of PID load frequency controller for power systems via IMC. IEEE Trans Power Syst 341–350
4. Rerkpreedapong D, Hasanovic A, Feliachi A (2003) Robust load frequency control using genetic algorithms and linear matrix inequalities. IEEE Trans Power Syst 855–861
5. Alrifai M, Hassan F, Zribi M (2011) Decentralized load frequency controller for a multi-area interconnected power system. Int J Electr Power Energy Syst 198–209
6. Pan CT, Liaw CM (1989) An adaptive controller for power system load frequency control. IEEE Trans Power Syst 122–128
7. Niu Y, Ho DWC, Wang X (2007) Sliding mode control for Itô stochastic systems with Markovian switching. Automatica 43(10):1784–1790
8. Shi P, Xia Y, Liu GP, Rees D (2006) On designing of sliding-mode control for stochastic jump systems. IEEE Trans Autom Control 97–103
9. Xia Y, Jia Y (2003) Robust sliding-mode control for uncertain time-delay systems: an LMI approach. IEEE Trans Autom Control 1086–1092
10. Goro S (1979) Load frequency control using Lyapunov's second method: bang-bang control of speed changer position. Proc IEEE 67(10):1458–1459
11. Ai-Hamouz ZM, Abdel-Magid YL (1993) Variable structure load frequency controllers for multi area power systems. Electr Power Energy Syst 15(5):293–300
12. Kumar A, Malik OP, Hope GS (1987) Discrete variable structure controller for load frequency control of multi area interconnected power systems. Proc Inst Electr Eng Gener Transm Distrib C 134(2):116–122
13. Goshaidas R, Sitansu D, Bhattacharyyam TK (2004) Multi-area load frequency control of power systems: a decentralized variable structure approach. Electr Power Compon Syst 33(3):315–331
14. Siaramakrishna AY, Hariharam MV, Srisailam MC (1984) Design of variable structure load-frequency controller using pole assignment technique. Int J Control 40(3):487–498

15. Vrdoljak K, Perić N, Petrović I (2010) Sliding mode based load-frequency control in power systems. Electr Power Syst Res 80(5):514–527
16. Vrdoljak K, Tezak V, Peric N (2007) A sliding surface design for robust load-frequency control in power systems. In: Proceedings of IEEE Lausanne power tech, July 2007, pp 279–284
17. Behera AK, Bandyopadhyay B (2015) Decentralized event-triggered sliding mode control. In: 2015 10th Asian control conference (ASCC), pp 1–5
18. Liu Y, Yang M (2015) The study of sliding mode load frequency control for single area time delay power system. In: 27th Chinese control and decision conference (CCDC), pp 602–607
19. Oshnoei A, Kheradmandi M, Muyeen SM (2020) Robust control scheme for distributed battery energy storage systems in load frequency control. IEEE Trans Power Syst 35(6):4781–4791
20. Haes Alhelou H, Hamedani Golshan ME, Hatziargyriou ND (2020) Deterministic dynamic state estimation-based optimal LFC for interconnected power systems using unknown input observer. IEEE Trans Smart Grid 11(2):1582–1592
21. Kothari DP, Nagrath IJ (2011) Modern power system analysis, 4th edn. McGraw-Hill
22. Yang M, Yang F, Wang C (2013) Decentralized sliding mode load frequency control for multi-area power systems. IEEE Trans Power Syst 4301–4309

PSO Tuned PID Controller for DC Motor Speed Control

Aaditya Sharma, Veena Sharma, and O. P. Rahi

Abstract The purpose of this paper is to plan a PSO algorithm application to tune the parameters of the PID regulator. This paper employs the model of a DC motor as a plant. As the conventional tuning of PID regulator using Ziegler–Nichols (Z-N) technique delivers a major overshoot, the present-day heuristics approach named particle swarm optimization (PSO) has been utilized here to upgrade the proficiency of old conventional technique. Four different performance indices (IAE, ISE, ITAE, and ITSE) are used while comparing PSO-based PID and ZN-PID in this paper. The results have shown the better performance of the PID tuning utilizing the PSO-based optimization approach.

Keywords IAE · ISE · ITAE · ITSE · PID · PSO · Z-N

List of Symbols

b	Motor Viscous Friction Constant
$C(s)$	Controller Transfer Function
e	Control Error Signal
e_V	Back EMF
$G(s)$	Plant Transfer Function
i	Armature Current
IAE	Integral Absolute Error
ISE	Integral Square Error

A. Sharma (✉) · V. Sharma · O. P. Rahi
Department of Electrical Engineering, National Institute of Technology Hamirpur, Hamirpur, HP 177005, India

V. Sharma
e-mail: veena@nith.ac.in

O. P. Rahi
e-mail: oprahi@nith.ac.in

S. Suhag et al. (eds.), *Control and Measurement Applications for Smart Grid*, Lecture Notes in Electrical Engineering 822,
https://doi.org/10.1007/978-981-16-7664-2_7

ITAE	Integral Time Absolute Error
ITSE	Integral Time Square Error
J	Moment of Inertia of Rotor
K	Motor Torque Constant
Kt	EMF Constant
Kw	EMF Constant
K_{p}	Proportional Gain
K_{i}	Integral Gain
K_{d}	Derivative Gain
L	Inductance
OF	Objective Function
PID	Proportional Integral Derivative
PSO	Particle Swarm Optimization
PV	Process Variable
r	Input Signal
R	Armature Resistance
T	Torque
y	Output Signal
Z-N	Ziegler–Nichols

1 Introduction

Though the control theory has touched new heights, even today the most inescapable form of feedback compensation comes from the PID controller. PID controllers have become the backbone of the motion control system in the industry [1]. Self-adjusting PID has transformed the shape of the industrial world and made it suitable for engineers to bring the finest control of a plant. The output generated by the PID controller is the aggregation of the outputs of proportional, integral, and derivative controllers. As per the literature concerned, above 95% of industrial control is done by the PID controllers. A PID regulator ceaselessly figures an error $e(t)$ as the distinction amid the process variable (PV) and set point (SP). PID's accuracy and optimized automatic control make it a problem solver. As it has all the necessary dynamics like reduced rise time, steady-state error and improves the transient response which makes the system stable. Ziegler–Nichols developed the first tuning rules with two methods [2]. The first one is for an open-loop system, and the second one is for the closed-loop system. The Z-N adjusting in feedback loop requires the critical gain and critical period. In this technique, the controller is put on automatic mode while the integral and derivative actions are shut off. The regulator gain is raised until an interruption which causes continuous oscillations in the process variable. As the computational approaches are modernized in recent times for the sake of desirable results in industrial process control via tuning of the regulator's optimization, algorithms came into role [3–8]. The principal proponents of the PSO algorithm were Kennedy and Eberhart in 1995 [9]. In order to get the finest results, different agents are employed. These agents move in a group and every agent tries to provide the finest outcome. In PSO individual particle's latest location is decided by a velocity term which redirects the attraction of global best and its own best throughout the history of the particle and random coefficients.

2 Problem Formulation

The block diagram of a feedback control system of the closed-loop type is shown in Fig. 1. PID regulator comprises of following gains (a) proportional, (b) integral, and (c) derivative. The scheme of feedback type is reflected in Fig. 1 where reference input, control error, and measured output are denoted as r, e, y correspondingly.

In this control scheme shown in Fig. 1, the target system (plant) is represented as $G(s)$ and regulator as $C(s)$, which is specified by Eq. (1) as given below:

$$C(s) = K_p + K_i/s + K_d s \quad (1)$$

where K_p, K_i, K_d are separately recognized as proportional, integral, and derivative coefficients of the PID regulator that are heading for adjustment.

DC Motor Modeling

See Fig. 2.

System equations are represented as follows:

Torque equation:

$$T = K * i \quad (2)$$

Back EMF equation:

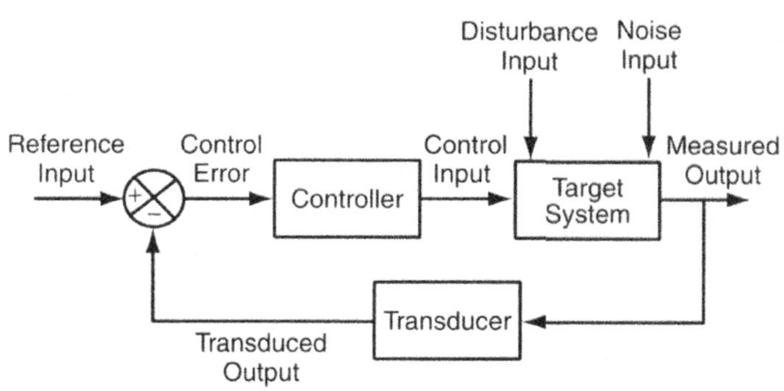

Fig. 1 Block diagram representation of a feedback control system [10]

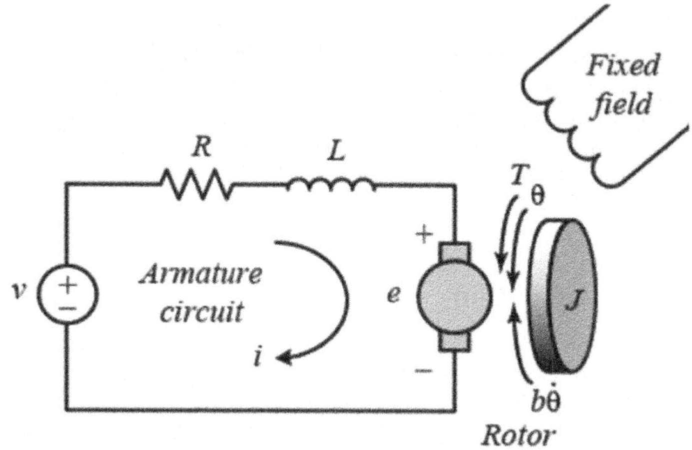

Fig. 2 Electrical equivalent circuit of DC motor [11]

$$e_V = K_W * (\theta') \tag{3}$$

Torque equation using Newton's second law:

$$T = J * (\theta'') + b * (\theta') \tag{4}$$

Terminal voltage equation using Kirchhoff's voltage law:

$$V = L\frac{di}{dt} + R * I + K_W * (\theta') \tag{5}$$

Using Eqs. (2)–(5) in Laplace domain, we get

$$P(s) = \frac{(\theta'(s))}{V(s)} = \frac{K}{(Js + b)(Ls + R) + K * Kw} \tag{6}$$

Using the block diagram, the obtained transfer function between angular position and voltage is given in Fig. 3:

$$G(s) = \frac{\theta(s)}{V(s)} = \frac{K}{JLs^3 + (BL + JR)s^2 + (BR + K * Kw)s + K * Kt} \tag{7}$$

The physical parameters of the DC motor are given below:

$J = 1$ kg m^2, $B = 8$ N m s, $R = 1\ \Omega$, $L = 1$ H, $K = 1$ N m/A, $K_W = 15$ V/Rad/s and Kt = 15 V/Rad.

On substituting the above parameters in Eq. (7), the transfer function model of the DC motor is given by Eq. (8):

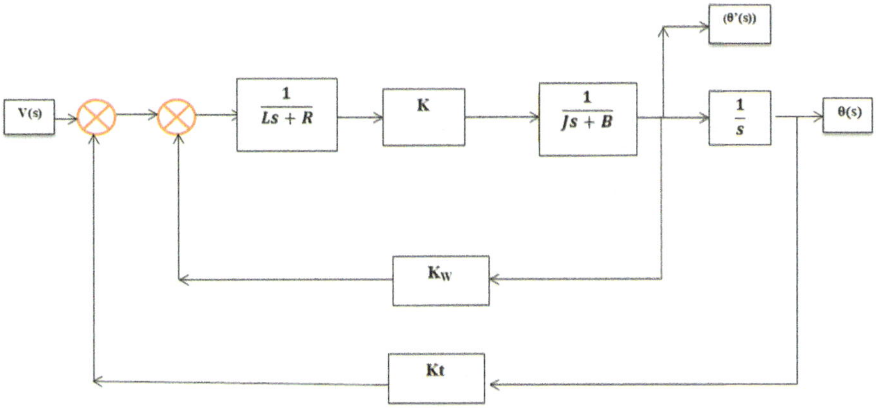

Fig. 3 Block diagram of DC motor [12]

$$G(s) = 1/\left(s^3 + 9s^2 + 23s + 15\right) \tag{8}$$

Moreover, error-based performance index to design an optimum PID regulator is being used in this work. Error-based performance criteria are utilized as a measurable tool to rate the functioning of the PID regulator system which has been designed in this paper. Utilizing the instant strategy an 'ideal framework' can be frequently planned, and a bunch of coefficients of PID regulator in the framework can be changed accordingly to achieve the necessary conditions. Framework execution of PID regulator is portrayed in ISE, IAE, ITAE, and ITSE. They are characterized as follows:

$$\text{ISE} = \int_0^\infty e^2(t)\mathrm{d}t \tag{9}$$

$$\text{IAE} = \int_0^\infty |e(t)|\mathrm{d}t \tag{10}$$

$$\text{ITAE} = \int_0^\infty t|e(t)|\mathrm{d}t \tag{11}$$

$$\text{ITSE} = \int_0^\infty te^2(t)\mathrm{d}t \tag{12}$$

3 Solution Methodology

A. Two solution methodologies have been used in this research work. First one is tuning of PID using conventional Ziegler–Nichols (ZN) technique, and the second one is PID tuning using particle swarm optimization.

B. Z-N METHOD USED FOR TUNING OF PID REGULATOR

We have used the second technique of ZN METHOD, i.e., tuning of PID with closed-loop method. This approach demands that the critical gain and critical time to be calculated. The derivative time constant (T_d) is kept at zero while integral time constant (T_i) is retained at infinity. All this can be accomplished by altering the regulator gain (K_u) until continuous oscillations are met by the system (Table 1).

C. PSO-BASED PID REGULATOR TUNING

The transformative computational procedure is the base of PSO optimization. Collective ventures in a school of fish and group of birds were the reason and inspiration which brought out a new technique of optimization, and the proponents of this technique were Kennedy and Eberhart in 1995. A very few numbers of parameters are assigned to the PSO algorithm in contrast to other metaheuristic algorithms. Initialization of a cluster of simulated birds is done with random locations *Xi* and velocities *Vi*. Every bird in the swarm is scattered arbitrarily in the first stage throughout the *D* dimensional search space. Every particle within the cluster starts adjusting its velocity and location under the inspection of the objective function, companion's experiences and their own experiences. Every particle remembers its best position attained by it along with the best global place attained by any other particle in the swarm throughout the exploration of an optimal solution. $X_I = (x_{i1}, x_{i2}, \ldots, x_{iD})$ is the representation of the *i*th particle. Gbest denoted by symbol g is a representative of the finest particle amid all particles in the population. $P_I = (p_{i1}, p_{i2}, \ldots, p_{iD})$ is represented as pbest and $V_I = (v_{i1}, v_{i2}, \ldots, v_{iD})$ is represented as velocity of the particle.

The particles are modified corresponding to the equations below:

$$V_{id}{}^{n+1} = w * V_{id}{}^{n} + c_1 * \text{rand}() * \left(p_{id}{}^{n} - X_{id}{}^{n}\right) + c_2 * \text{rand}() * \left(p_{gd}{}^{n} - X_{id}{}^{n}\right) \quad (13)$$

$$X_{id}{}^{n+1} = X_{id}{}^{n} + V_{id}{}^{n+1} \quad (14)$$

In this, two positive acceleration coefficients c_1 and c_2 are used. A random number between 0 and 1 is produced by rand () operator, and n represents iteration. The new

Table 1 Parameters used for Ziegler–Nichols closed-loop tuning [13]

Controller type	KP	Ti	Td
P	0.5 Kcr	∞	0
PI	0.45 Kcr	1/1.2 Pcr	0
PID	0.6 Kcr	0.5 Pcr	0.125 Pcr

results of the speed and location in comparison with the earlier results from its own top experience are evaluated by Eq. (7). According to Eq. (14), the particle moves toward a novel location. Shi and Eberhart [14] proposed the concept of inertial weight which was missing in the primary algorithm suggested by Kennedy and Eberhart. The work reports that the global exploration is facilitated by large inertia weight factor while local exploration is facilitated by a small weight factor. Moreover, particle swarm optimization exploration potential can be enhanced for multi-dimensional issues by regulating the weight of inertia which was proposed by various scientists. Later in 1999 Clerc [15] presented his own version of PSO which resulted in guaranteed convergence of algorithm for inertial weight $w = 0.729$.

D. EXECUTION OF PSO TUNED PID CONTROLLER

The PSO-based technique has been used using MATLAB/Simulink model as shown in Figs. 4 and 5.

Metaheuristic optimization can be implemented to alter PID regulator gains to guarantee the performance of the regulator at minimal functioning states. In Eq. (8), using the plant transfer functions K_p, $K_{i,}$ and K_d are altered using PSO in offline mode. It primarily creates an initial horde of particles in a domain characterized by a matrix. Every particle denotes a unique result for PID coefficients. The values of these coefficients are varied on a scale between 0 and 100. As there are primarily three coefficients, we have to solve for three-dimensional space with velocity and position characterized by matrices of dimension 3× swarm size. For PSO algorithm, these values such as swarm size of 40, acceleration coefficients $c_1 = c_2 = 1.494$, and weight of inertia (w) which is reducing linearly from 0.9 to 0.4 as the number of iterations reach its maximum value are referred from the existing literature [17].

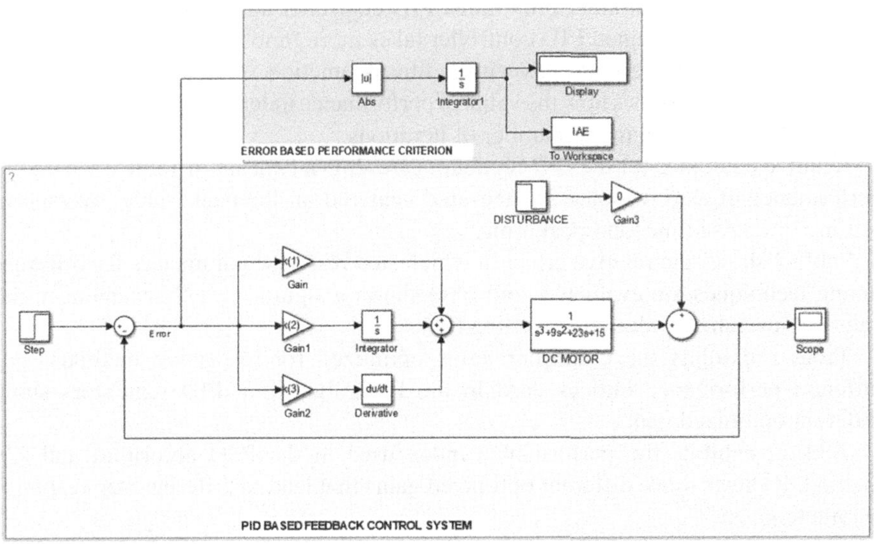

Fig. 4 Simulink diagram of the PID-based control system [16]

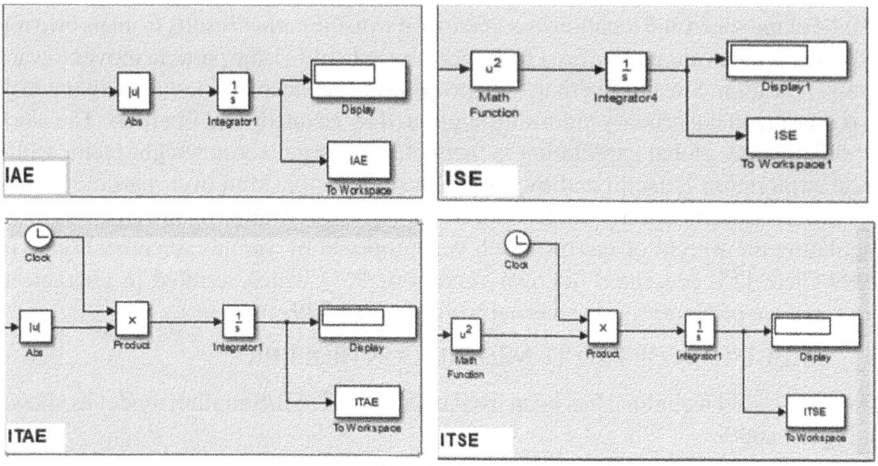

Fig. 5 Simulink diagram of four performance indices

4 Results and Discussion

Using the traditional Ziegler–Nichols (Z-N) technique, the step response of the system creates large overshoot, instead superior performance is found with the execution of the PSO tuned PID controller. As compared to the Z-N technique, PSO has less overshoot and less settling time. The PSO process is simulated for 50 iterations. Collective step response is plotted for the Z-N method and PSO tuned controller with different performance index shown in Fig. 6. In Fig. 7 on applying step disturbance signal on the transducer side, PSO tuned PID controller attenuates the disturbance successfully and Z-N tuned PID controller takes more time to settle down.

Figure 8 denotes the plot between the fitness function value and the number of iterations. This plot shows how the value of performance index converges to minimum value with the increase in the number of iterations.

Relative outcomes for the PID regulators are shown beneath in Table 2 where the performance of step response is measured centered on the peak value, overshoot, settling time, rise time, and peak time.

Table 2 shows the relative effect in which step response parameters for different tuning techniques are evaluated, and PSO shows a significant enhancement in the values of overshoot values and settling time.

Table 3 exhibits the controller gains optimized for respective methods. For different performance indices used in the PSO algorithm, PID regulators show different optimized gains.

Table 3 exhibits the performance index used in the PSO algorithm and ZN method. It shows those different optimized gains that lead to different step response parameters.

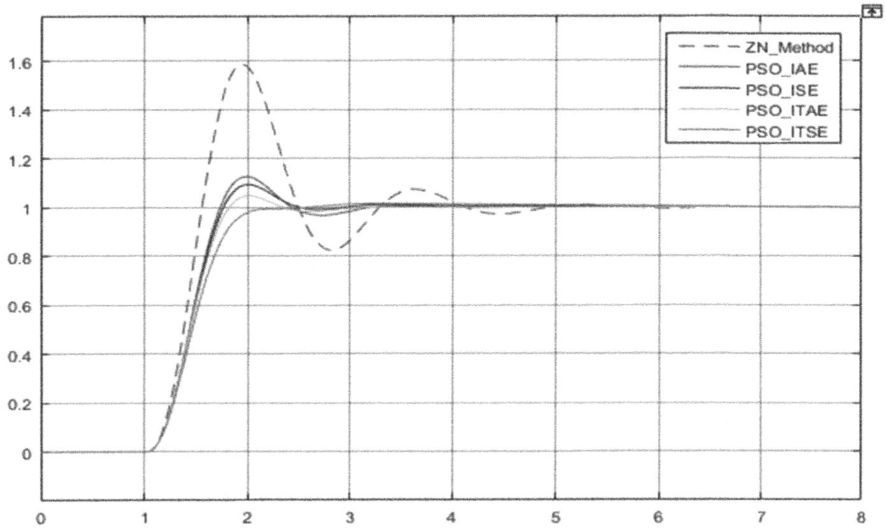

Fig. 6 Step response plot for Z-N and PSO tuned PID controller

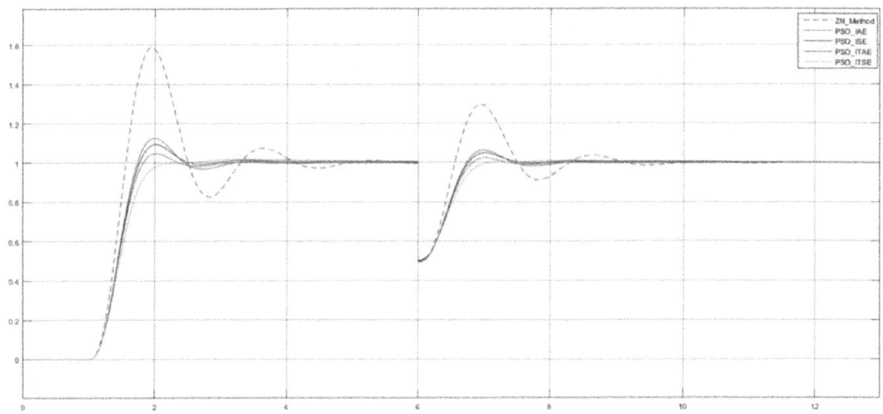

Fig. 7 Step response plot for Z-N and PSO tuned PID controller when a disturbance is introduced at the transducer side

5 Conclusion and Future Scope

As inferred from the results that PSO works well as compared to that of traditional Ziegler–Nichols method by attenuating the disturbance produced at the transducer side. PSO in comparison with Ziegler–Nichols method has less overshoot and less settling time, and because of this, it becomes a more efficient computational method. While a few parameters are to be adjusted in PSO, it has an edge over Ziegler–Nichols

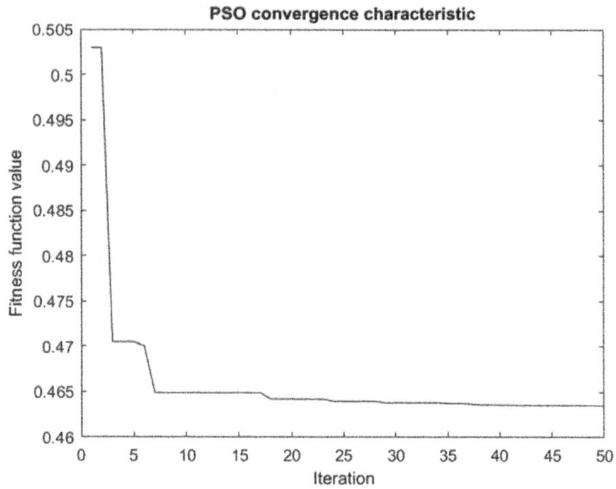

Fig. 8 Performance index plot of PSO algorithm of IAE type

Table 2 Parameters of step response for proportional integral derivative controllers

Technique used for tuning	Peak	Overshoot	Peak time	Rise time	Settling time
ZN METHOD	1.5876	58.8061	1.9496	0.3319	4.6684
PSO_IAE	1.1267	12.6882	1.9949	0.4506	2.9590
PSO_ISE	1.0943	9.2795	2.0083	0.4696	2.3884
PSO_ITAE	1.0486	4.8792	2.0151	0.4949	2.2451
PSO_ITSE	1.0165	1.3478	3.2557	0.5905	2.0216

Table 3 Controller gains optimized for respective methods

Technique used for tuning	K_p	K_i	K_d
ZN METHOD	115.2	175.9	18.9
PSO_IAE	100.0	38.1933	19.8595
PSO_ISE	100.0	40.2469	22.6765
PSO_ITAE	100.0	34.0138	24.5222
PSO_ITSE	93.4686	34.5348	28.7630

method and successfully gives optimal solutions. Moreover, indistinguishable performances are observed when PSO tuned PID regulators are optimized with various error-based performance indices. In relation to future scope other metaheuristic optimization techniques such as moth flame optimization (MFO) can be carried out for tuning of PID regulator as it has achieved a faster rate of convergence and less peak overshoot value. Here in this paper, third-order DC motor is used as a plant; however, higher-order DC motor model can also be considered for further research.

References

1. Tavakoli S, Sadeghi J, Griffin I, Fleming PJ (2016) PI controller tuning for load disturbance rejection using constrained optimization. Int J Dyn Control 6(1):188–199
2. Ogata K (1987) Modern control systems. Prentice Hall, Minneapolis, MN
3. Gaing Z-L (2004) A particle swarm optimization approach for optimum design of PID controller in AVR system. IEEE Trans Energy Convers 19(2):384–391
4. Zhao J, Li T, Qian J (2005) Application of particle swarm optimization algorithm on robust PID controller tuning. In: Advances in natural computation. Springer, Berlin/Heidelberg, pp 948–957
5. Mahony TO, Downing CJ, Fatla K (2000) Genetic algorithm for PID parameter optimization: minimizing error criteria. In: Process control and instrumentation. University of Stracthclyde, pp 148–153
6. Wang Q, Spronck P, Tracht R (2003) An overview of genetic algorithms applied to control engineering problems. In: Proceedings of the international conference on machine learning and cybernetics (IEEE Cat. No.03EX693)
7. Latha K, Rajinikanth V, Surekha PM (2013) PSO-based PID controller design for a class of stable and unstable systems. ISRN Artif Intell 2013:1–11
8. Vincent A, Nersisson R (2017) Particle swarm optimization based PID controller tuning for level control of two tank system. IOP Conf Ser Mater Sci Eng 263:052001
9. Kennedy J, Eberhart RC (1995) Particle swarm optimization. In: Proceedings of IEEE international conference on neural networks (Perth, Australia). IEEE Service Center, Piscataway, NJ, pp IV: 1942–1948
10. Hellerstein J (2004) Introduction and overview, Ch 1. In: Feedback control of computing systems. Wiley, New York City, NY, pp 3–28
11. Messner B, Tilbury D (2017) DC motor speed: system modeling. Accessed 01 Feb 2021. [Online]. Available: https://ctms.engin.umich.edu/CTMS/index.php?example=MotorSpeed§ion=SystemModeling
12. Zaccarian L (2012) Lecture notes, topic: "Dc motors: dynamic model and control techniques". University of Rome, Rome
13. Ziegler JG, Nichols NB (1993) Optimum settings for automatic controllers. ASME J Dyn Syst Meas Control 115(2B):220–222
14. Shi Y, Eberhart R (1998) A modified particle swarm optimizer. In: Proceedings of the IEEE world congress on computational intelligence, pp 69–73
15. Clerc M (1999) The swarm and the queen: towards a deterministic and adaptive particle swarm optimization. In: Proceedings of the conference on evolutionary computation, pp 1951–1957
16. Bequette BW (2003) Process control: modeling, design, and simulation. Prentice Hall PTR, India
17. Solihin MI, Tack LF, Kean ML (2011) Tuning of PID controller using particle swarm optimization (PSO). Int J Adv Sci Eng Inf Technol 1(4):458

A Real-Time Implementation of Performance Monitoring in Solar Photovoltaics Using Internet of Things

Dishore Shunmugham Vanaja and Albert Alexander Stonier

Abstract The most feasible source of power output is from solar power-based photovoltaic systems. Due to the penetration of solar photovoltaic system, the demand in electrical energy is satisfied. This paper explores to track solar photovoltaic systems via the Internet of things (IoT) in real time. For monitoring the photovoltaic's and converting it to the AC to meet the need for AC load a positive output DC-to-DC conversion with the ability to track maximum power is proposed. The ThinkSpeak open-source IoT cloud platform is used to track the PV parameters such as voltage, current, and temperature in real time. The device's prototype and experimental setup are also done with Arduino circuit, voltage, current, and sensors.

Keywords IoT · MPPT · Photovoltaic · Power management · Real-time monitoring

1 Introduction

Solar photovoltaic (SPV) systems are most popular since it is available abundantly in nature, and during the conversion process, no harmful gasses are evolved. It has a small influence on the environment. Furthermore, solar power is growing to a greater extent in the production of energy in the world. The goal is to provide useful energy by using methods for photovoltaics by a photovoltaic structure, also called PV system. It contains several components of solar panels, converters, inverters, batteries, and so forth. Because it does not work with moving elements or environmental pollutants, solar energy has been developed over the last few years. Photovoltaic systems have

D. S. Vanaja
Department of Electrical and Electronics Engineering, Rajadhani Institute of Engineering and Technology, Thiruvananthapuram 695102, India
e-mail: dishoresv@rietedu.in

A. A. Stonier (✉)
Department of Electrical and Electronics Engineering, Kongu Engineering College, Perundurai 638060, India
e-mail: albert@kongu.ac.in

S. Suhag et al. (eds.), *Control and Measurement Applications for Smart Grid*,
Lecture Notes in Electrical Engineering 822,
https://doi.org/10.1007/978-981-16-7664-2_8

become a promising technology to produce energy in a manner that enables them to be in niche markets. While the output of solar power is rising, its power is not compelling. The identification of distributed generation in the global and regional situation is extremely important to improve its efficiency [1].

Almost every month, a photovoltaic system can supply a certain quantity of electricity. However, due to their increased erection price and poor power these devices have a significant drawback [2]. In addition, changes in the environment would be affected, resulting in bad efficiency, variability in output, and a reduction of the photovoltaic system's dependability [3]. The power production of photovoltaic panels must be increased. Maximum powerpoint tracking is indeed a method for obtaining full power under such circumstances from the photovoltaic panel. In this photovoltaic panel, the voltage is at the peak and is referred to as maximum powerpoint limit. The maximum possible power obtained from PV changes with the change in temperature and radiation [4]. The authors in [5] implemented a fault diagnosis system for solar photovoltaic system using IoT. The implementation of IoT determines the faults in PV system and autocorrects itself from the data available in the cloud. In [6], the authors have discussed the hot-spotting issue which could affect the efficiency of PV module. Tests are conducted and compared with seven different progressive MPPT techniques. Using beta methods, the tracking accuracy ranges from 92 to 94%. In [7], the authors discussed a modified Perturb and Observe method. The maximum power is predicted using intelligent control to ensure the result. The efficiency of the system is obtained as 90%. A MPPT and limited power point tracking (LPPT) were implemented for a grid-connected system in [8].

The authors in [9] proposed an adaptive P&O method. The working methodologies for the novel P&O method rely on the following three consecutive points of action with power–voltage functionality. This approach gives a steady-state output free of oscillation by reducing the current iteration by 90% of the previous one. This approach benefits from low complexity and less numerical burden than other approaches. In [10], the authors proposed a grid integrated system in which the PV panels are interfaced with DC-DC boost converters. The structure is implemented with P&O method since it has superior performance when compared with other MPPT techniques. In [11], a solar updraft tower is proposed. The primary drawback of this system is its performance and cost. The authors in [12] proposed an incremental–proportional–integral derivative (IPID) controller to track maximum power. On comparing the suggested controller with P and O, the IPID performs better. In [13], the authors proposed an MPPT method with geometric control. A whale optimized differential evolution technique is introduced in [14]. This strategy has a tolerable constant and complex output under rapidly changing irradiance and calefaction levels. Moreover, in comparison the proposed method has supremacy over standard methods. In [15], the authors proposed the overall distribution particle swarm optimization (ODPSO) MPPT approach to enhance the certainty of MPPT.

Fuzzy logic-based MPPTs are proposed in [16]. The working point changes are an extensive assortment with the aid of surrounding environmental conditions. To boost working point, the proposed MPPT multifunctional FL solution is simple, definitive, and less convergence. With a nice process, fewer ripples and rapid dynamism, the suggested methodology completes the real power extraction and arrives on the stationary stage within a minimum of time. The authors in [17] stated that the inter-harmonics due to maximum power point monitoring, emissions occur on multilevel inverter. The inter-harmonic emissions would also increase as the sampling rate of the MPPT increases. The inter-harmonics in the output current can be effectively minimized by altering the MPPT measurement to select the moderate inspection value. At the fast-sampling frequency, comparable efficiency is retained. For this process, the single-phase grid was experimented. The authors in [18] proposed an improved MPPT for PV applications. The authors in [19] presented a grid integrated system using Harris hawks optimization. The effective implementation of P&O method improves the system efficiency.

Latest methods of monitoring maximum power and increasing maximum power through the converter, based on the comprehensive literature analysis cited above, have the following disadvantages:

- Conventional methods are not applicable for high-rated applications.
- The device for remote control by standard approaches such as data logging is complicated relative to the IoT.
- Poor scalability because of large variations found in the characteristics of solar cells.

The authors in [20, 21] implemented the soft computing approach to extract maximum power from solar PV system. Implementing the optimization techniques improved the system efficiency. The primary objective of the proposed method is to monitor the developed solar PV conversion system and observe its performance from any location using IoT. The secondary objective is to implement positive output cascaded step-up chopper to increase output power under any environmental conditions by giving switching signals to the MPPT controller. The manuscript is structured as: Sect. 2 provides the drawbacks of the existing methods, Sect. 2 details the anticipated system, Sect. 3 put forth the methodology of the proposed system, Sect. 4 mulls the hardware implementation, and lastly the article is concluded in Sect. 5.

2 Anticipated System

The method employs the controller to trail the highest power and then raise the system power and manage IoT parameters. The controller helps to track the maximum power and makes the converter to operate at that point by giving converter pulses. The suggested converter increases the output power and supplies DC input to the DC–AC conversion.

The layers of Internet of things (IoT) are obtained from output by sensors to capture different parameters intending perception, data dispensation to the network forms the network layer, and finally, the display is captured in application layer. The perception layer uses a large number of sensors to derive different parameters from the solar PV system, which interacts with the Internet communication. This equipment allows users to remotely monitor through a cloud interface. Using the current time data from the specifications the output of solar PV can be identified right away. The solution suggested would boost solar photovoltaic performance. This way the optimal and constant energy from the photovoltaic panels is collected. Apart from being selective and receptive, the proposed solution is quick, fast, effective, and consistent. The system works in different environmental circumstances and not only in normal conditions.

There are a large number of renewable sources of energy available in India. The PV of solar energy is also the best alternative for electricity. By carbon footprints, it will comfort the buyer's existing usage at the perfect rate. The foremost purpose of the strategy provided is to evolve a maximum power consumption of the solar photovoltaic energy, the real-time monitoring mechanism which tracks, exposes, and designs various grid arrangement requirements in different regions using an internet connection (i.e., IoT). Sensors on both sides of the sensors attached to the Arduino are solar PV-connected. Arduino Nano converts analogue signals into digital signals. The digitals are transmitted to the ES P 8266, and the information are shown in the application server.

Finally, the structure as an innovation (IoT) must be built into a distance or rustic configuration to confirm the continuity and optimum potential of the solar photovoltaic accumulation. Viable sources of energy as production sources are needed for grid-tied, solar-oriented Photovoltaic systems as a provider to achieve ecological objectives. In comparison, nations such as the USA have maintained standards such as the Clean Power Plan which decide that carbon dioxide emissions are declining.

3 Methodology

In preparation for remote control, especially used for safety, the approaching practice is to track the maximized power and scrutinize the output of the solar photovoltaic through an IoT-based web portal. Specifications such as current, voltage, strength, and temperature are tested in the atmosphere using a variety of sensors. The data was gathered and sent via a controller to a web server. Figure 1 displays the machine functional diagram. Using MPPT controller, Perturb and Observe algorithms constantly track the power from solar PV for the highest significance. Pulses given to the suggested boost converter increase the DC voltage.

IoT panel etiquette design begins with the lowest insight panel, which involves different types of sensors. The details of the collected parameters are eventually entered into a forum where a smart phone application/web server may also be used, which represents the final device framework represented as an application layer (Layer 3). It is possible to save the data and monitor it each month.

The role of P&O Algorithm is to increment or decrement the switching sequences of the converter based on the operating point. Increase in DC voltage is put forth by the suggested chopper circuit. The importance of inverter is to supply the AC loads with the AC power. Viable use of electronic sensors senses the current, voltage and

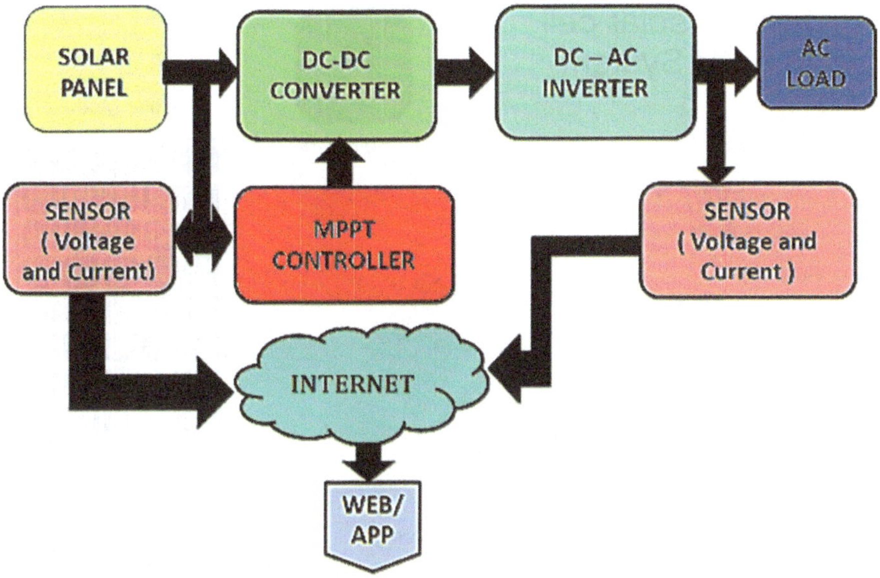

Fig. 1 Functional block diagram

power of the system in real time and processes the data to an application through the Internet of things.

Solar photovoltaic arrays consume daylight to produce power. The photovoltaic systems indicate the solar photovoltaic set of a photovoltaic structure which reproduces and generates power from solar. Each unit is defined under customary conditions by its DC energy derived and typically limited by 100–365 W. Figure 2 shows the configuration of the solar panel.

3.1 Positive Output Boost Converter

The suggested circuit of DC–DC converter is a two-stage boosting multi-circuit, which allows the creation of increasing voltage with a simple structure in the geometrical series. The circuit configuration is seen in Fig. 3.

A DC-AC converter normally named as an inverter has the capability of converting the direct current (DC) to alternating current (AC). The inductor ripple current $iL2$ is given in Eq. 1.

$$\Delta i_{L2} = \frac{3V_1}{L_2}kT = \frac{V_2 - 3V_1}{L_2}(1-k)T \tag{1}$$

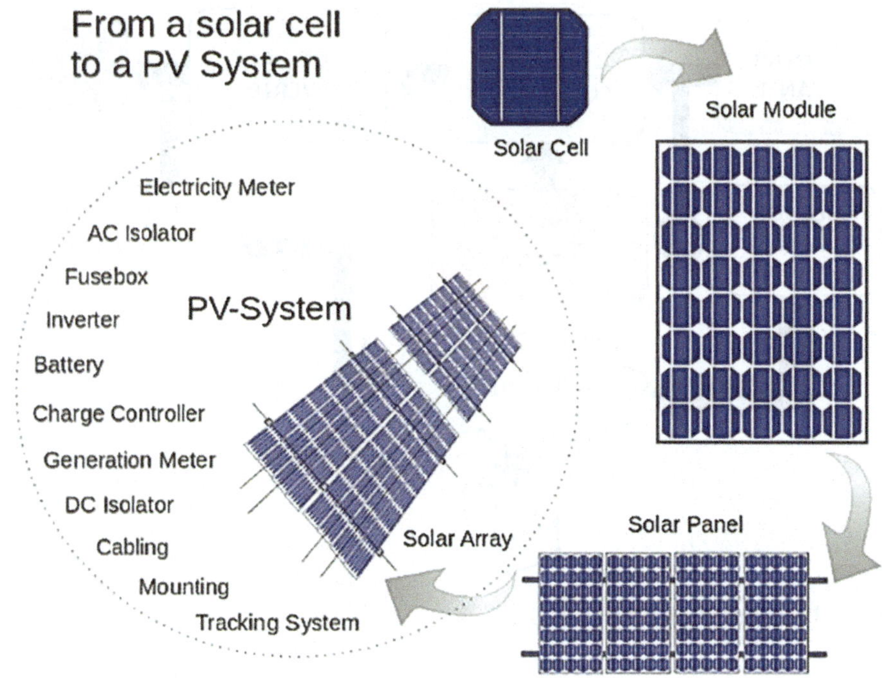

Fig. 2 Transformation of cell to a panel

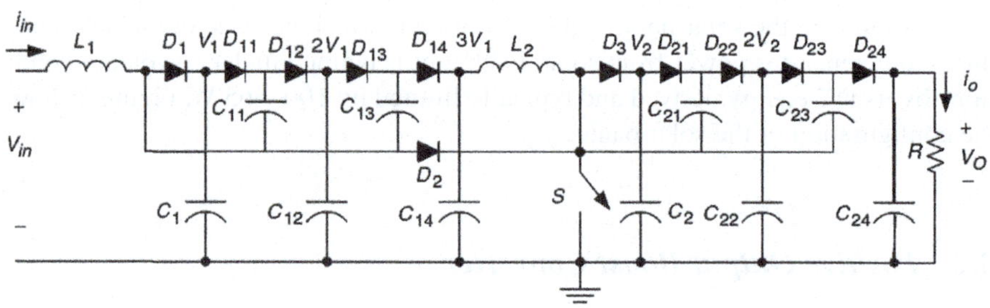

Fig. 3 Positive output cascaded boost converter

$$V_2 = \frac{3}{1-k}V_1 = 3\left(\frac{1}{1-k}\right)^2 V_{\text{in}} \tag{2}$$

$$V_{\text{O}} = V_{C2} + V_{C23} = 3V_2 = \left(\frac{3}{1-k}\right)^2 V_{\text{in}} \tag{3}$$

$$G = \frac{V_{\text{O}}}{V_{\text{in}}} = \left(\frac{3}{1-k}\right)^2 \tag{4}$$

V_o is the output voltage and G is the transfer gain, respectively.
Mechanistically,

$$\Delta i_{L1} = \frac{V_{\text{in}}}{L_1} kTI_{L1} = \left(\frac{2}{1-k}\right)^2 I_O \tag{5}$$

$$\Delta i_{L2} = \frac{V_1}{L_2} kTI_{L2} = \frac{2I_O}{1-k} \tag{6}$$

Current through the $L1$ is,

$$\xi_1 = \frac{\Delta i_{L1}/2}{I_{L1}} = \frac{k(1-k)^2 T V_{\text{in}}}{8L_1 I_o} = \frac{k(1-k)^4}{16} \frac{R}{fL_1} \tag{7}$$

Current through the $L2$ is,

$$\xi_2 = \frac{\Delta i_{L2}/2}{I_{L2}} = \frac{k(1-k)TV_1}{4L_2 I_o} = \frac{k(1-k)^2}{8} \frac{R}{fL_2} \tag{8}$$

Usually ξ_1 and ξ_2 are small.
The ripple VO is given by,

$$\Delta v_O = \frac{\Delta Q}{C_{22}} = \frac{I_O(1-k)^T}{C_{22}} = \frac{1-k}{fC_{22}} \frac{V_O}{R} \tag{9}$$

The variation ratio of VO is

$$\varepsilon = \frac{\Delta v_O/2}{V_O} = \frac{1-k}{2RfC_{22}} \tag{10}$$

4 Hardware Implementation

Figure 4 displays the full hardware implementation in real time. The photovoltaic module is coupled to voltage, current, and power measurement sensors and the MPPT controller that controls the full power and transmits pulses to the converter. The sensor possesses an input DC voltage in the range of 0–25 V. It has an 0.004849 V analogue resolution. The voltage indicator has the VCC that is the positive termination of the voltage that is to be measured. ACS712 is a linear current sensing system based on Hall effect with voltage insulation.

Arduino Nano features the ATMEGA 328 microcontroller which comes with the IDE platform, the A1–A7 analogue pins and the D2–D12 optical pins. The ESP8266 is the on-board Wi-Fi module connecting to the Internet through Arduino. This

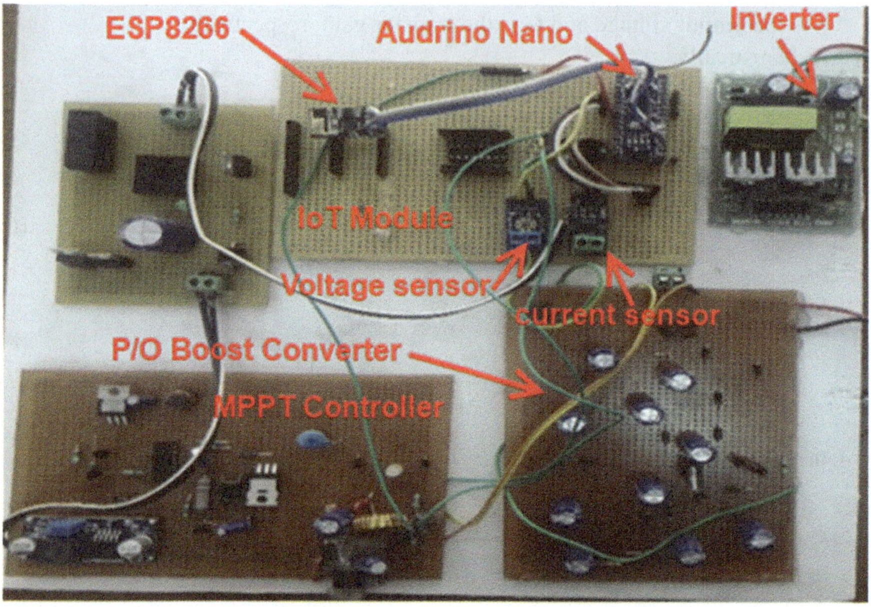

Fig. 4 Hardware implementation

small board will work with a Wi-Fi connection and two GPIO pins as a standalone microcontroller. The Arduino Nano is supplied with the sensor output. Arduino is fed to the ESP8266 output signal. The Arduino Nano is supplied with 5 V and the ESP8266 with 3.5 V.

It retains the full power and also improves the power of the chopper circuit. Online tracking increases the effectiveness of the solar panel and its output. The average photovoltaic device monitoring IoT parameter outcome is shown in Table 1. IoT controls the voltage, current, and power (results are shown in Fig. 5).

Table 1 Parameters and values of PV system monitoring

Parameters	Value (bytes)
Total bytes uploaded to IED	1328.00
Total bytes downloaded from IED	3033.73
Total bytes upload + download from/to IED	4361.73
Upload speed to IED	3746.02 bytes/s
Download speed from IED from IED	7328.35 bytes/s
Data transfer speed (upload + download) from/to IED	11,230.93 bytes/s
Total execution of PV system monitoring-IoT to IED	1.04 ± 0.66 s
Total execution time of PV system monitoring-IoT to cloud database	8.71 ± 12.12 s
Total of all data acquisition program execution time	19.54 ± 18.00 s

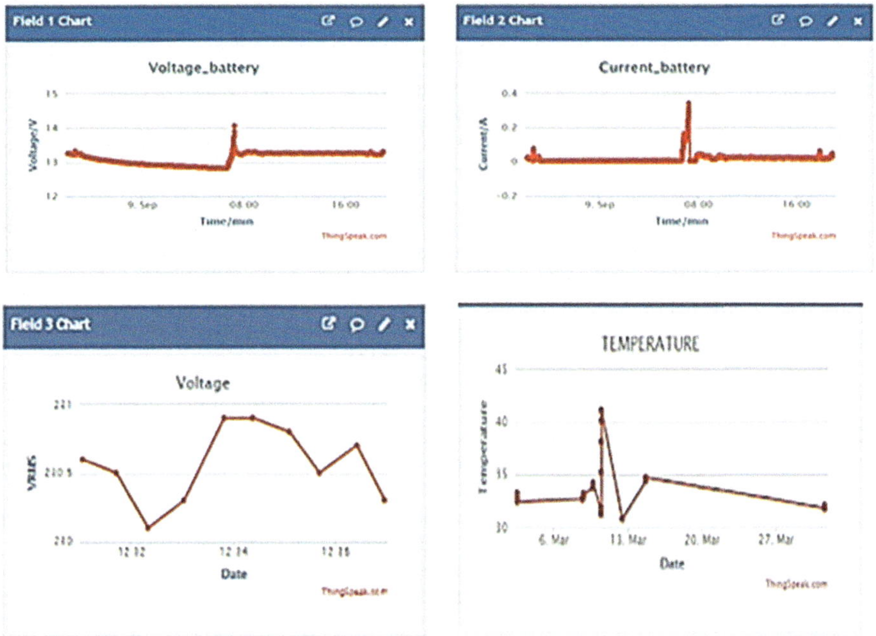

Fig. 5 Remote monitoring of real-time results in ThingSpeak IoT cloud

Victimizing advanced MPPT techniques could expand the study in the future. The progress can also be made in converter controllers. The networked solar photovoltaic systems can also be built with advanced technologies to develop a power delivery system. Other systems are highly complex, cost-effective, and sizable relative to the mostly based IoT system. A system for remote advance care of PV solar boards, such as remote shutdown, would later be fused with this technique.

5 Conclusion

The integration of IoT for solar photovoltaic identification eliminates the drawbacks and perks up the effectiveness of PV system. This work suggested the use of a suitable power converter with the controllers for the simple photovoltaic system connected to the AC load. DC-DC converter, which was fed into AC load and MPPT interface, and DC-AC converter. The outcome is replicated by MATLAB/Simulink, and the results are acquired. The test is also performed on the hardware setup, and the results are reviewed. In the future, sophisticated MPPT methods are used to expand the function. The progress can also be made in converter controllers.

References

1. Kumar AL, Alexander SA, Rajendran M (2020) Power electronic converters for solar photovoltaic systems, 1st edn. Academic Press, USA
2. Mutlu ÖÇ, Büchner D, Theurich S, Zeng T (2021) Combined use of solar and biomass energy for sustainable and cost-effective low-temperature drying of food processing residues on industrial-scale. Energies 14(3):561
3. Shanmuga Aravind P, Albert Alexander S (2013) Harmonic minimization of a solar fed cascaded H bridge inverter using artificial neural network. In: International conference on energy efficient technologies for sustainability. IEEE, Nagerkoil, pp 163–167
4. Stonier A, Murugesan S, Samikannu R, Venkatachary SK, Senthil Kumar S, Arumugam P (2020) Power quality improvement in solar fed cascaded multilevel inverter with output voltage regulation techniques. IEEE Access 8:178360–178371
5. Albert Alexander S, Lehman B (2018) An intelligent based fault tolerant system for solar fed cascaded multilevel inverters. IEEE Trans Energy Conversion 33(3):1047–1057
6. Dhimish M (2019) Assessing MPPT techniques on hot-spotted and partially shaded photovoltaic modules: comprehensive review based on experimental data. IEEE Trans Electron Devices 66(3):1132–1144
7. Ahmed J, Salam Z (2018) An enhanced adaptive P&O MPPT for fast and efficient tracking under varying environmental conditions. IEEE Trans Sustain Energy 9(3):1487–1496
8. Premkumar M, Kumar C, Sowmya R, Pradeep J (2021) A novel salp swarm assisted hybrid maximum power point tracking algorithm for the solar photovoltaic power generation systems. Automatika 62(1):1–20
9. Remoaldo D, Jesus I (2021) Analysis of a traditional and a fuzzy logic enhanced perturb and observe algorithm for the MPPT of a photovoltaic system. Algorithms 14(1):24
10. Shunmugham Vanaja D, Stonier A (2020) A novel PV fed asymmetric multilevel inverter with reduced THD for a grid-connected system. Int Trans Electr Energy Syst 30(4):1–25
11. Ashour AI, Almitani KH, Abu-Hamdeh NH (2021) Developing and improving a prototype scale concentrating solar power tower-system. Sustain Energy Technol Assess 45:101105
12. Subudhi B, Pradhan R (2019) A new adaptive maximum power point controller for a photovoltaic system. IEEE Trans Sustain Energy 10(4):1625–1632
13. Dang VT, Yang MG, Shim Y, Lee W, Baek KH (2021) An accurate time-based MPPT circuit with two-period tracking algorithm and convergence range averaging technique for IoT applications. IEEE Access 9:31401–31410
14. Windarko NA, Nizar Habibi M, Sumantri B, Prasetyono E, Efendi M (2021) A new MPPT algorithm for photovoltaic power generation under uniform and partial shading conditions. Energies 14(2):483
15. Charin C, Ishak D, Zainuri MA, Ismail B, Jamil MK (2021) A hybrid of bio-inspired algorithm based on levy flight and particle swarm optimizations for photovoltaic system under partial shading conditions. Solar Energy 15, 17(2):1–4
16. Rezk H, Aly M, Al-Dhaifallah M, Shoyama M (2019) Design and hardware implementation of new adaptive fuzzy logic-based MPPT control method for photovoltaic applications. IEEE Access 7:106427–106438
17. Dishore SV, Johny Renoald A, Albert Alexander S. Experimental investigations on a PV fed Modular STATCOM in WECS using intelligent controller. Int Trans Electr Energy Syst. https://doi.org/10.1002/2050-7038.12845
18. Vanaja DS, Stonier AA, Mani G, Murugesan S (2021) Investigation and validation of solar photovoltaic-fed modular multilevel inverter for marine water-pumping applications. Elec Eng 1–16. https://doi.org/10.1007/s00202-021-01370-x
19. Dishore SV, Stonier AA (2021) Grid integration of modular multilevel inverter with improved performance parameters. Int Trans Electr Energy Syst 31(1):1–34

20. Albert Alexander S, Manigandan T (2014) Power quality improvement in solar photovoltaic system to reduce harmonic distortions using intelligent techniques J Renew Sustain Energy 6(4):043127 (1)–(19)
21. Shunmugham Vanaja D, Stonier AA, Moghassemi A, A novel control topology for grid-integration with modular multilevel inverter. Int Trans Elec Energy Syst e13135

One Hour Ahead Solar Irradiation Forecast by Deep Learning Network Using Meteorological Variables

Pardeep Singla, Manoj Duhan, and Sumit Saroha

Abstract With the exponential growth in electricity demand, a renewable energy resource comes with the best alternate especially solar cells. The penetration of these resources, interconnection with the grid, the accurate and precise forecasting of next hour electricity generation is the most important factor to manage the grid. This paper discusses the comparative study of two deep learning models: long short term memory (LSTM) network and gated recurrent unit (GRU) network for the forecasting of global solar irradiance (GHI). The meteorological parameters: wind direction, dew point, pressure, temperature, solar zenith angle, relative humidity, wind speed and precipitation are considered to train these deep learning networks. The study used the clear sky index (CSI) calculation to stationarize the data. The models are trained using the one year of hourly datasets, while the testing is performed for one hour ahead monthly solar irradiation forecast. The experimental results of GRU and LSTM networks are compared to the naïve model (benchmark model) in terms of root mean square error (RMSE), mean absolute percentage error (MAPE) and coefficient of determination (R^2). The result shows that the GRU network outperform LSTM network and benchmark model with annual average RMSE of 69.8117 (w/m^2), annual average MAPE of 71.777(w/m^2) and R^2 of 0.86.

Keywords Long short term memory · Gated recurrent unit · Deep learning · Solar global horizontal irradiance · Solar forecasting · Clear sky Index

1 Introduction

Owing to the setting up of large numbers of new industrial and residential sectors, electricity demand is rising day by day [1]. In 2019, Marsal-Pomianowska et al. reported 40–50% electricity consumption only from the buildings which were only

P. Singla (✉) · M. Duhan
Deenbandhu Chhotu Ram University of Science and Technology, Murthal, Haryana, India

S. Saroha
Guru Jambheshwar University of Science and Technology, Hisar, Haryana, India

S. Suhag et al. (eds.), *Control and Measurement Applications for Smart Grid*,
Lecture Notes in Electrical Engineering 822,
https://doi.org/10.1007/978-981-16-7664-2_9

16% in 2013 [2]. However, the use of renewable energy sources will meet this rising demand, which is becoming a promising solution to this issue [3]. In addition, solar energy is one of the popular and eco-friendly solutions among all renewable sources which generate electricity by converting solar radiations received from sun [4, 5]. Nowadays, the solar plants are used as a power generation source not only in standalone mode but also in grid interconnected mode [6]. In fact, the scale and numbers of these solar plants are also growing at an exponential rate [7]. Despite this growth, the uncertainty and randomness of power generation by PV panels is still a major challenge. Therefore, the forecasting of the next step electricity generation is necessary to manage the grid planning, maintenance and operation in case of grid interconnected plants [8]. A precise and accurate forecast of solar irradiation helps not only in grid management but also prevents penalty.

Generally, there are two methods to forecast the solar GHI: direct and indirect methods [9]. Direct methods are the methods which only consider the historical variables as input features whereas the meteorological variables are considered in indirect method of forecasting [3]. Numerous studies have been published in the literature to forecast the solar GHI using artificial neural network (ANN), support vector machine (SVM), regression method, etc. K. Mohammadi et al. developed a forecasting model based on SVM. This study utilized the wavelet transform (WT) decomposition to make the model hybrid and showed WT + SVM performed better than single SVM [10]. Likewise, SVM along with k-means clustering was used by the Bae et al. This combination again outperforms to backpropagation neural network (BPNN) [11]. In addition to this, wavelet combination with the neural network was also performed by Sharma et al. [12]. The comparative study of regression-based model and SVM was conducted by Sharika et al. in their study [13]. The ability of deep learning to manage large data has made it popular in recent years in the field of forecasting. But very few studies are available for forecasting solar GHI using deep learning networks. Sharadga H. et al. predicted the PV power output using the bidirectional long short term memory (Bi-LSTM) network [14]. Whereas Srivastava S. et al. forecasted the solar GHI using the LSTM model [1]. In addition to this, Gao M. et al. proposed the LSTM network to forecast the PV power of a plant. However, for a smooth dataset, this study used the meteorological variables as inputs, whereas the time series data has been considered for non-ideal weather condition [3]. Moreover, solar GHI has also been forecasted by Aslam M et al. In their paper, GRU deep learning network was used to forecast hourly solar GHI for a year. The developed model was also compared with the benchmark model: SVM feed-forward neural network (FFNN) [15].

Therefore, motivated from the power of deep learning, the paper developed two deep learning networks to forecast one hour ahead solar GHI. The research contributions to the paper are:

(i) A brief literature of the deep leaning for solar GHI forecast.
(ii) A theoretical overview of LSTM network and GRU network.
(iii) Calculation of CSI to stationarize the meteorological data.

(iv) Simulation of naïve model (benchmark model), LSTM network and GRU network for solar GHI forecast.
(v) Performance evaluation of developed model using RMSE, MAPE and R^2.

The paper organization is as follows: Section 2 describes the theoretical background of LSTM and GRU deep learning techniques. Section 3 discusses the experimental setup of this paper. This section discusses the data description along with the developed forecasting model and error metrics. The result and analysis are provided in Sect. 4. Finally, the study is concluded in Sect. 5.

2 Theoretical Background

This section provides the theoretical background of LSTM and GRU networks in brief.

2.1 Long-Short Term Memory Network

Generally, the vanishing gradient problem is generated in the simple recurrent neural network (RNN) due to its limited memory [16]. This problem is solved in case of LSTM network. In LSTM, the memory blocks are presents instead of summation unit like in RNN [1]. The gate 'g_i' collects the previous time step output say '$h_{t-1.}$' However, the data in context of present time is also input to the gate of the cell. The outputs of four gates are depicts as input '$i_{t,}$' output '$O_{t,}$' update 'g_i' and forget '$f_{t.}$' The information passed to the cell is decided by the input gate as:

$$i_t = \text{sigm}(\theta^i x_t + \theta^{hi} h_{t-1}) \tag{1}$$

The amount of previous state information that has to pass is decided by forget gate and expressed as:

$$f_t = \text{sigm}(\theta^f x_t + \theta^{hf} h_{t-1}) \tag{2}$$

whereas the amount of information of internal state that has to pass is decided by output gate.

$$O_t = \text{sigm}(\theta^o x_t + \theta^{ho} h_{t-1}) \tag{3}$$

So, the internal memory state 'C_t' will be updated as:

$$C_t = \text{sigm}(f_t * C_{t-1} + i_t * \tilde{C}_t) \tag{4}$$

$$\tilde{C}_t = \tanh(\theta^g x_t + \theta^{hg} h_{t-1}) \tag{5}$$

$$h_t = \tanh(C_t) * O_t \tag{6}$$

2.2 Gated Recurrent Unit (GRU)

Similar to LSTM, the GRU has reset gate and update gate as a gating unit. The purpose of gating unit is to modulate the information passage in the unit [17]. The reset gate for a GRU cell is expressed as:

$$r_t = \text{sigm}(\theta^r x_t + \theta^{hr} h_{t-1}) \tag{7}$$

Similarly, update gate,

$$z_t = \text{sigm}(\theta^z x_t + \theta^{hz} h_{t-1}) \tag{8}$$

The current hidden state of the cell can be finally expressed as (Fig. 1):

$$h_t = (1 - Z_t) * h_{t-1} + Z_t * \tilde{h}_t) \tag{9}$$

$$\tilde{h}_t = \tanh(\theta^h x_t + (r_t * h_{t-1})\theta^h) \tag{10}$$

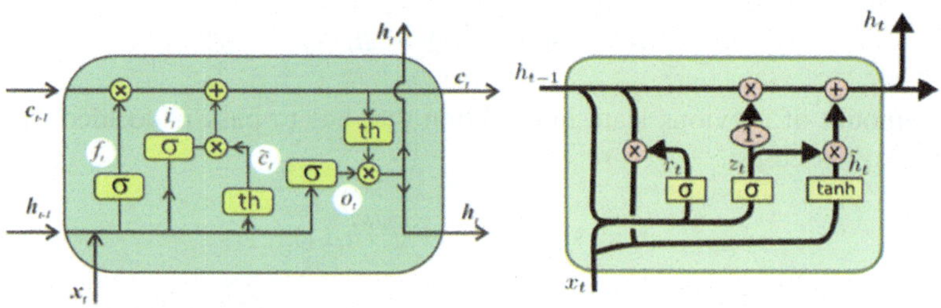

(a) LSTM architecture (b) GRU architecture

Fig. 1 **a**, **b** Architecture of LSTM and GRU

3 Materials and Methodology

3.1 Data Description

In order to compare LSTM and GRU network performance, the dataset of an 'Ahmadabad, Gujarat' has been considered. The location is a city of Gujarat state of India located at latitude longitude of 23° 0.05′/72° 0.35′ having climatic condition of extreme type. The entire year is divided into three different climatic seasons: summer, winter and monsoon [18]. The dataset for the study was collected from the database of National energy renewable laboratory on hourly basis. Eight different meteorological variables have been collected as: wind direction, dew point, pressure, temperature, solar zenith angle, relative humidity, wind speed and precipitation. The model was trained using the data of one year and one step ahead forecasting is performed for month basis.

3.2 Forecasting Process

Figure 2 represents the flow graph of process used to perform the forecasting using the LSTM and GRU.

At stage-I, the meteorological data was collected for the targeted site. The data collected is often in raw form as well as some missing and incorrect vales are included. So, the data quality check was performed to remove the night hours as well as missing and false observations. The night hours were removed due to the non-availability of GHI in the night. Once the quality of the data has been checked, the clear sky index was calculated to make the data stationary. The clear sky index can be calculated as:

$$K_t = \frac{Y}{Y_{CS}} \tag{11}$$

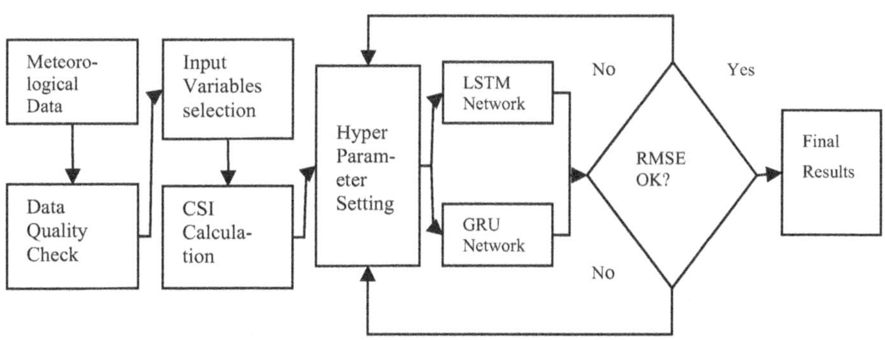

Fig. 2 Flow Process of forecasting using LSTM and GRU

where, $Y_{CS} = E_o \exp^{-\frac{\tau}{\text{Sin}^{a}(\mu(t))}} \text{Sin}(\mu(t))$.

where, Y is the solar irradiation, $\mu(t)$ is the height of panel in degree, E_o is the extraterrestrial irradiation, and a is the fitting parameter.

However, the normalization techniques can be used in place of finding the clear sky index. The correlation coefficient of each variable was performed in the next step with the target variable. The variable with a strong correlation has been selected, while the weaker correlated variable has been removed. The next and important step is the selection of the hyperparameters for deep learning networks. There is, however, no particular rule provided in the literature for selecting these hyperparameters. For the same method, only the error and trail method must be practiced. This study conducted various experiments to select proper hypermeter. The hidden units were varied from '20–100' with a learning rate of '0.2. 0.02, 0.002, 0.0007.' But, hidden units of '70' with '500' epochs, performed best for both of the networks with an initial learning rate of '0.02.' The '*Adam*' function was used as an optimizer as prescribed in the literature also. The learning rate drop factor was '0.2' with a drop period of '125.' Both the models were trained with one year of meteorological data, while one step ahead monthly GHI was forecasted. After completing the training and testing, the error metrics were calculated. If the results are satisfactory then finalize the model otherwise reselect the hyperparameters and repeat the process.

3.3 Evaluation Metrics

$$\text{RMSE} = \sqrt{\frac{1}{n}\sum_{i=1}^{n}\left(\text{GHI}_{p,i} - \text{GHI}_{a,i}\right)^2} \tag{12}$$

$$\text{MAPE} = \frac{1}{n}\sum_{i=1}^{n}\left|\frac{\text{GHI}_{p,i} - \text{GHI}_{a,i}}{\text{GHI}_{a,i}}\right| \tag{13}$$

$$R^2 = 1 - \frac{\text{var}(\text{GHI}_{a,i} - \text{GHI}_{p,i})}{\text{var}(\text{GHI}_{p,i})} \tag{14}$$

where, '$\text{GHI}_{p,I}$' is predicted/forecasted irradiation, and '$\text{GHI}_{a,I}$' is real/actual irradiation.

4 Results and Analysis

Table 1 presents the observations of RMSE (w/m^2), MAPE (%) and R^2 for the benchmark model, LSTM and GRU network for different months, respectively.

Table 1 Results for naïve, LSTM and GRU

Month	RMSE (w/m^2)			MAPE (%)			R^2		
	Naïve	LSTM	GRU	Naïve	LSTM	GRU	Naïve	LSTM	GRU
1	145.98	42.27	42.29	37.09	5.89	6.05	0.52	0.95	0.95
2	160.61	51.71	51.66	33.84	6.10	5.73	0.50	0.94	0.94
3	157.30	35.27	34.59	25.43	3.88	3.59	0.52	0.97	0.97
4	146.43	28.77	52.95	20.77	3.09	5.71	0.53	0.98	0.96
5	144.82	44.50	42.06	21.07	5.39	4.80	0.53	0.95	0.96
6	145.12	86.90	63.68	22.29	10.75	8.62	0.48	0.78	0.80
7	150.54	123.14	121.98	28.69	24.13	23.17	0.56	0.68	0.70
8	169.30	155.66	158.83	30.74	27.10	27.32	0.43	0.54	0.54
9	149.01	151.88	126.64	29.23	35.88	29.97	0.62	0.65	0.70
10	149.02	65.75	73.23	28.63	8.52	11.64	0.52	0.89	0.89
11	135.18	39.30	39.87	33.97	6.88	7.11	0.54	0.96	0.96
12	134.01	38.13	37.89	31.99	4.56	4.47	0.54	0.96	0.96
Avg.	148.94	71.77	69.81	28.64	11.85	11.52	0.52	0.85	0.86

According to Table 1, the RMSE for GRU network is better than LSTM network except for three months: April, august and October. The minimum RMSE obtained from GRU network is 34.59 (w/m^2) in month of March, whereas maximum RMSE is 158.83 (w/m^2) in the month of September. On the other hand, the minimum RMSE obtained from LSTM network is 28.77 (w/m^2) in month of April, whereas it is at maximum in the month of September with 155.66 (w/m^2). In addition, the benchmark mode obtained minimum RMSE [134.01 (w/m^2)] in month of December and maximum RMSE [169 (w/m^2)] in month of August. As far as MAPE is concern, the minimum MAPE obtained from LSTM network is 3.09% in month of April, whereas it is maximum in month of September with value of 35.88%. The GRU network achieved the minimum MAPE in month of March with 3.59% and maximum in month of September with value of 29.97%. Whereas the benchmark model obtained minimum MAPE (20.77%) in month of April and maximum MAPE (37.09%) in the month of January. However, the overall performance on the basis of annual average RMSE and MAPE, GRU still a good choice over the LSTM for forecasting using climatic variables. The higher RMSE and MAPE in the months June, July, August, September and October are due to the presence of uncertainty in the data due to rainy and cloudy days.

The accuracy parameter R^2 statistics is also observed for these models to evaluate the curve fitting progress of the model. From the results, it also shows that the overall maximum R^2 (0.86) is obtained by model GRU model, while the benchmark model and LSTM network obtained R^2 equal to 0.52 and 0.85, respectively. For more analysis, Fig. 3a, b shows the actual GHI tracing by the GRU model for one week data of 'May' and 'September' month, respectively. The figure plot clearly shows that the month 'May' have smooth data set which can be traced by LSTM and GRU properly.

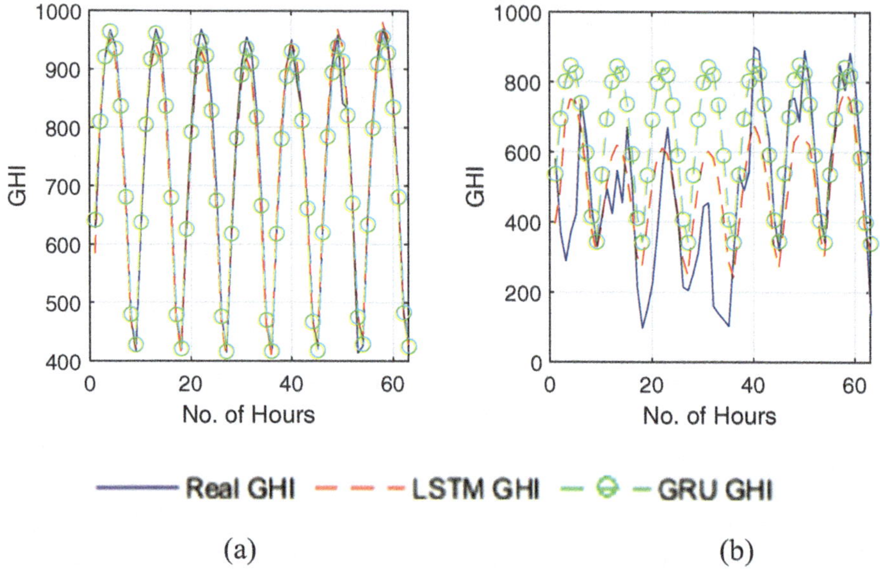

Fig. 3 **a**, **b** Performance of LSTM and GRU on one week of May and September

But the month 'September' has larger variations or randomness and the uncertainty in the data which is not traced by LSTM model and GRU model properly. Only GRU model traced this GHI much precisely than LSTM model. In addition, Fig. 4a, b represents the annual GHI plot for LSTM model and GRU model, respectively.

Therefore, the GRU model achieved annual average RMSE of 69.8117 (w/m^2), whereas it was 148.94 (w/m^2) and 71.7772 (w/m^2) for benchmark model and LSTM network, respectively. Moreover, the MAPE obtained from GRU, LSTM network and benchmark model was 11.5205%, 11.8521% and 28.64%, respectively. In addition to this, the LSTM, GRU and benchmark model obtained annual R^2 is equal to 0.85, 0.86 and 0.52, respectively. Table 2 shows the comparative results of the performed experiments with the other study also.

This study used the clear sky index as a calculation to make the data stationary instead of simple normalization. The results mentioned in Table 2 show that both of the networks performed well with the clear sky index calculation. Among both of the networks, GRU is better in terms of RMSE from LSTM for meteorological data. This study achieved RMSE of 69.8117 (w/m^2) which was 122.45 (w/m^2) in Ref. [17] and 127.3 (w/m^2) in Ref. [19].

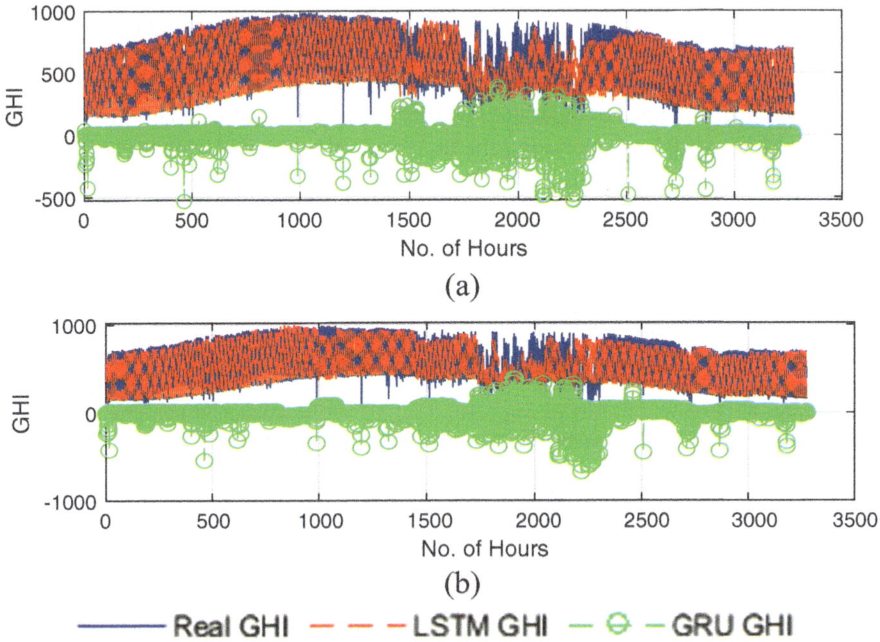

Fig. 4 **a**, **b** Performance of LSTM and GRU for annual dataset

Table 2 Comparative results of the performed experiments with the other study

Refs.	Study area	Model	RMSE (w/m^2)
[17]	Denver, USA	GRU + weather forecast	122.45
[19]	Qingdao, China	DFT + PCA + Elman	127.3
This Study	Ahmadabad, India	LSTM, GRU	69.81

5 Conclusion

This paper studies the LSTM and GRU network using meteorological variables to forecast the GHI for 'Ahmadabad, Gujarat' area. On the basis of the intensity of their correlation with the target variable, the metrological variables are chosen. The research used the clear sky index calculation to stationarize the data to enhance the model's efficiency. In the analysis, one step ahead monthly forecast was carried out to compare the performance of LSTM and GRU network with benchmark model using RMSE, MAPE and R^2 statistics. The model configuration was selected from the variations of different hyperparameters where an initial learning rate of 0.02 with

70 hidden layers performed best. The GRU and LSTM network obtained 69.8117 (w/m^2) and 71.777(w/m^2) of RMSE, whereas 11.5205 and 11.8521% of MAPE. Moreover, the LSTM and GRU network obtained R^2 statistics equal to the 0.85 and 0.86, respectively. These comparative results show that the GRU network is better than LSTM network by considering the meteorological variables as input features.

References

1. Srivastava S, Lessmann S (2018) A comparative study of LSTM neural networks in forecasting day-ahead global horizontal irradiance with satellite data. Sol Energy 162:232–247. https://doi.org/10.1016/j.solener.2018.01.005
2. Radwan A, Ookawara S, Ahmed M (2016) Analysis and simulation of concentrating photovoltaic systems with a microchannel heat sink. Sol Energy 136:35–48. https://doi.org/10.1016/j.solener.2016.06.070
3. Gao M, Li J, Hong F, Long D (2019) Day-ahead power forecasting in a large-scale photovoltaic plant based on weather classification using LSTM. Energy 187:115838. https://doi.org/10.1016/j.energy.2019.07.168
4. Lai JP, Chang YM, Chen CH, Pai PF (2020) A survey of machine learning models in renewable energy predictions. Appl Sci 10. https://doi.org/10.3390/app10175975
5. Wan C, Zhao J, Song Y, Xu Z, Lin J, Hu Z (2016) Photovoltaic and solar power forecasting for smart grid energy management. CSEE J Power Energy Syst 1:38–46. https://doi.org/10.17775/cseejpes.2015.00046
6. Ismail AM, Ramirez-Iniguez R, Asif M, Munir AB, Muhammad-Sukki F (2015) Progress of solar photovoltaic in ASEAN countries: a review.https://doi.org/10.1016/j.rser.2015.04.010
7. Sobri S, Koohi-Kamali S, Rahim NA (2018) Solar photovoltaic generation forecasting methods: a review. https://doi.org/10.1016/j.enconman.2017.11.019
8. Singla P, Duhan M, Saroha S (2021) A comprehensive review and analysis of solar forecasting techniques. Front Energy:1–37. https://doi.org/10.1007/s11708-021-0722-7
9. Paulescu M, Brabec M, Boata R, Badescu V (2017) Structured, physically inspired (gray box) models versus black box modeling for forecasting the output power of photovoltaic plants. Energy 121:792–802. https://doi.org/10.1016/j.energy.2017.01.015
10. Mohammadi K, Shamshirband S, Tong CW, Arif M, Petković D, Sudheer C (2015) A new hybrid support vector machine-wavelet transform approach for estimation of horizontal global solar radiation. Energy Convers Manage 92:162–171. https://doi.org/10.1016/j.enconman.2014.12.050
11. Bae KY, Jang HS, Sung DK (2017) Hourly solar irradiance prediction based on support vector machine and its error analysis. IEEE Trans Power Syst 32:935–945. https://doi.org/10.1109/TPWRS.2016.2569608
12. Sharma V, Yang D, Walsh W, Reindl T (2016) Short term solar irradiance forecasting using a mixed wavelet neural network. Renew Energy 90:481–492. https://doi.org/10.1016/j.renene.2016.01.020
13. Sharika W, Fernando L, Kanagasundaram A, Valluvan R, Kaneswaran A (2018) Long-term solar irradiance forecasting approaches—a comparative study. In: 2018 IEEE 9th international conference on information and automation for sustainability, ICIAfS 2018. Institute of Electrical and Electronics Engineers Inc. https://doi.org/10.1109/ICIAFS.2018.8913381
14. Sharadga H, Hajimirza S, Balog RS (2020) Time series forecasting of solar power generation for large-scale photovoltaic plants. Renew Energy 150:797–807. https://doi.org/10.1016/j.renene.2019.12.131
15. Aslam M, Lee JM, Kim HS, Lee SJ, Hong S (2019) Deep learning models for long-term solar radiation forecasting considering microgrid installation: a comparative study. Energies 13. https://doi.org/10.3390/en13010147

16. Heidari A, Khovalyg D (2020) Short-term energy use prediction of solar-assisted water heating system: application case of combined attention-based LSTM and time-series decomposition. Sol Energy 207:626–639. https://doi.org/10.1016/j.solener.2020.07.008
17. Gao B, Huang X, Shi J, Tai Y, Xiao R (2019) Predicting day-ahead solar irradiance through gated recurrent unit using weather forecasting data. J Renew Sustain Energy. 11:043705. https://doi.org/10.1063/1.5110223
18. Ahmedabad climate: average temperature, weather by month, Ahmedabad weather averages. Climate-Data.org. https://en.climate-data.org/asia/india/gujarat/ahmedabad-2828/. Last accessed 2020/04/24
19. Lan H, Zhang C, Hong YY, He Y, Wen S (2019) Day-ahead spatiotemporal solar irradiation forecasting using frequency-based hybrid principal component analysis and neural network. Appl Energy 247:389–402. https://doi.org/10.1016/j.apenergy.2019.04.056

Heat Transfer Analysis Through Ducts of Different Geometries with Extended Surfaces

M. Bardalai, B. K. Das, I. A. Chaudhury, S. Kumar, and P. P. Dutta

Abstract The present study discusses about heat transfer in fluids passing through different ducts with rectangular and triangular extended surfaces. The entire analysis in this work is performed with the help of ANSYS software. The ducts of three types of geometries are considered for the heat transfer analysis namely rectangular, triangular and circular. The parameters considered for the analysis are pressure drop ($\triangle P$), Nusselt number (Nu), heat transfer coefficient (h_f) and friction factor (f). The results obtained from the computational analysis show that Nu in triangular duct is highest, whilst fluid temperature is lowest because of higher rate of heat transfer. Again, the total pressure and (P_t) in triangular duct is higher in comparison with rectangular and circular. Based on these observations, the triangular duct is chosen for further analysis. Rectangular and triangular shaped fins are applied on the outer surface of the triangular duct in order to investigate the rate of heat transfer in both the cases. It is observed that using triangular fins, Nu and pressure are found to be higher as compared to rectangular fins. The analysis shows that difference of Nu between the ducts with and without the fins gradually increases with Re. Twelve number of fins were tested and found that highest rate of heat transfer is achieved by using only four number of fins in the given geometry. The efficiencies of the triangular and rectangular fins were estimated as 80.65% and 58.38%, respectively. Similarly, the effectiveness of the triangular and rectangular fins is found to be 1.186 and 1.054, respectively. From these results obtained in this analysis, triangular duct with triangular fins are found to be best with respect to heat transfer, efficiency and effectiveness.

Keywords Rectangular duct · Triangular duct · Circular duct · Nusselt number · Total pressure

M. Bardalai (✉) · B. K. Das · I. A. Chaudhury · S. Kumar · P. P. Dutta
Tezpur University, Tezpur, Assam, India

S. Suhag et al. (eds.), *Control and Measurement Applications for Smart Grid*,
Lecture Notes in Electrical Engineering 822,
https://doi.org/10.1007/978-981-16-7664-2_10

1 Introduction

Heat exchangers are the devices that can be used for various industrial and domestic applications for exchanging thermal energy between two fluids. Ducts are the important parts of heat exchangers through which fluids flow and heat transfer takes place. In order to increase the rate of heat transfer fins are applied on the surface of the ducts. Optimization of fin number helps in selecting an alternative way which is cost effective and efficient under the given conditions. Analysis of heat transfer and pressure drop are two deciding factors while designing a heat exchanger. The dimensionless numbers such as Reynold Number (Re), Prandtl Number (Pr), Nusselt number, etc., are found to be frequently used in the analysis of heat transfer. These analyses can be efficiently performed by commercial software based on Computational Fluid Dynamics (CFD) such as ANSYS Fluent, COMSOL Multiphysics etc.

O'Brien et al. [1] carried out the study in order to estimate the forced convection heat-transfer coefficients and friction factors for the flow through a corrugated duct. The range of Re used in their study was considered from 1500 to 25000 and the Pr ranged from 4 to 8. They found the improvement of heat transfer by a factor of 2.5 when compared with conventional parallel-plate channel. Olek et al. [2] studied the conjugated transient heat transfer in pipe for fully developed flow. The study was based on constant temperature and heat flux at the wall. This study revealed that the degree of conjugation and viscous dissipation play significant role on the distribution of temperature in the fluid. The analysis on entropy generation in fluid flow through a circular duct was carried out by Dagtekin et al. [3]. They used number of different shaped longitudinal fins in the study to investigate the effects. Chandratilleke et al. [4] studied convective heat transfer in airflow through a duct with wall thermal radiation. Their study included investigation of thermal instability and effects of wall thermal radiation on the rate of overall heat transfer.

Yang et al. [5] performed the analysis of finned heat exchanger. The number of fins optimization was done in this study along with the fluid flow. Sahu and Singh [6] investigated heat transfer and flow due to natural convection in a heated triangular cylinders of different sizes inside a square enclosure. The computation was carried out with the help of Ansys fluent software. Another study of fluid flow was done by Karupaaraj et al. [7] on artificially roughened flat surface using Ansys Fluent k-ε model. Charles et al. [8] performed a comparative heat transfer study using different types of fins. The results of the study revealed that heat transfer rate is the highest in inverted trapezoidal fins as compared to trapezoidal and rectangular fins. Mirapalli et al. [9] also investigated the effect of rectangular and triangular fins on the periphery of the engine cylinder. The CFD study was done by Kumar et al. [10] to analyse the rate of heat transfer in semi-circular rib-roughened equilateral triangular duct. The analysis was performed with the help of Ansys Fluent. Shrirao et al. [11] studied the air flow analysis in a duct with and without internal threads using Ansys Fluent software. They compared the experimental results with the analytical calculations. Bhargava and Arya [12] developed a simplified MATLAB solution scheme for heat transfer equations. This type of scheme can be used for heat transfer and fluid flow problems in different situations. Sharma et al. [13] studied an experimental investigation in rectangular duct with some pentagonal ribs. Their study showed that pentagonal ribs can improve the heat transfer with minimum pressure drop. A parametric study by using triangular fins was carried out by Yadav and Pandey [14] to improve the heat transfer in forced convection. Gurumurthy et al. [15] performed the CFD analysis to compare the heat transfer coefficient for circular and noncircular duct. Their study concluded that elliptical cross section shows better heat transfer coefficient than circular cross section.

Over these extensive literatures as stated above, it is found that the comparative study of heat transfer in circular, rectangular and triangular duct with fin has not been done so far. Therefore, the present study aims to carry out the heat transfer analysis using rectangular, circular and triangular ducts. The study also covers the comparison between triangular and rectangular fins and determination of optimum number of fins.

2 Experimental and Methodology

2.1 Problem Definition

The conjugate heat transfer analysis is carried out in circular, rectangular and triangular ducts with fins using ANSYS software. In order to perform the analysis, the necessary assumptions are taken as follows [9]:
The flow is laminar.
Inlet velocity of the fluid = 0.5 m/s
Inlet temperature of fluid = 343 K
Constant wall temperature = 300 K.
Thermal conductivity of the tube material (aluminium), $k_t = 202.4$ W/m K
Thermal conductivity of the fluid (air), $k_f = 0.0242$ W/m K
Density of the fluid (air), $\rho = 1.225$ kg/m^3
Height of fins = 23 mm
Gap between the fins = 2 mm
The model of circular, triangular and rectangular ducts are shown in Fig. 1. The diameter of the circular duct is 50 mm side of the equilateral triangular duct is 68 mm and length and breadth of the rectangular duct are 45 and 40 mm, respectively.

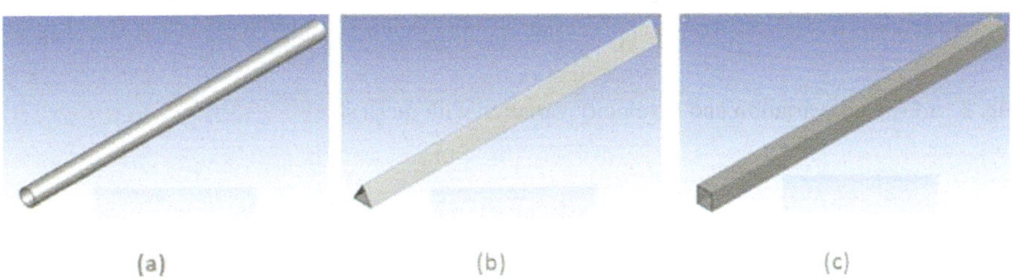

Fig. 1 Model of **a** circular, **b** triangular and **c** rectangular duct

2.2 Methodology

2.2.1 Design Modeling

The geometry of the ducts were modeled in the Design Modeler of ANSYS Workbench. The required mesh generation was done with the help of ICEM CFD 16.0.

2.2.2 Mesh Independence Test

Mesh or grid independence study is performed to estimate the optimum number of grids where the solutions are found to be relatively accurate with the expense of minimum computational resources. Figure 2a, b are showing the variation and pressure and velocity with the mesh size. In both the cases, the variation is found to be negligible after the mesh size of 2 mm. Therefore, the mesh size of 2 mm is taken for the analysis.

2.2.3 Solution Scheme

For convergence of the results, Semi Implicit Method for Pressure Linked Equations Consistent (SIMPLEC) for setting the correlation of pressure velocity coupling. The constant wall temperature boundary condition is used by maintaining the wall temperature at 323 K.

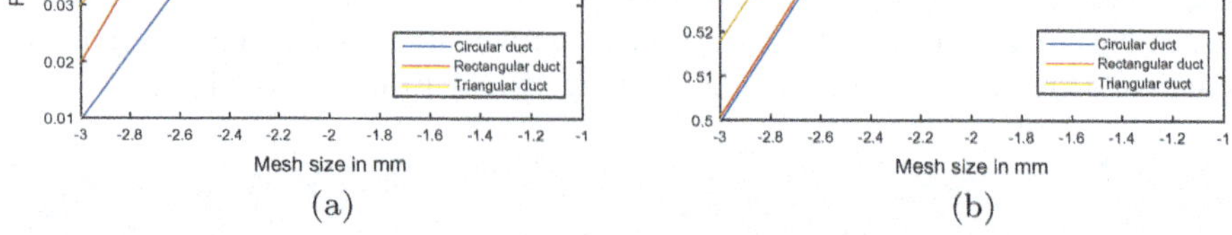

Fig. 2 **a** Pressure variation and **b** velocity variation with mesh size

2.2.4 Mathematical Equations

The mathematical equations for compressible fluid used in the computation model are continuity, momentum and energy equations as shown in Eqs. 1–3.

$$\frac{\partial \rho}{\partial t} + \frac{\partial}{\partial x_i}(\rho u_i) = 0 \tag{1}$$

$$\frac{\partial}{\partial t}(\rho u_j) + \frac{\partial}{\partial x_i}(\rho u_i u_j - P_{ij}) = 0 \tag{2}$$

$$\frac{\partial E}{\partial t} + \frac{\partial}{\partial x_i}(u_i E - u_j P_{ij} + q_i) = 0 \tag{3}$$

where ρ is the density of fluid, u is the velocity, P_{ij} is the stress tensor, E is the energy and q is the heat flux.
The efficiency of any fin is given by the Eq. 4.

$$\eta = \frac{Q_{\mathrm{act}}}{Q_{\mathrm{max}}} \tag{4}$$

$$Q_{\mathrm{max}} = hA_f(T_b - T_\infty) \tag{5}$$

The Q_{act} for rectangular fin can be expressed as Eq. 6.

$$Q_{\mathrm{act}} = \sqrt{hpkA_c}(T_b - T_\infty)\frac{\sinh(ml) + \left(\frac{h}{mk}\right)\cosh(ml)}{\cosh(ml) + \left(\frac{h}{mk}\right)\sinh(ml)} \tag{6}$$

where

$$m = \sqrt{\frac{hp}{kA_c}} \tag{7}$$

P is perimeter of tip, h is heat transfer coefficient, A_f is heat transfer area of fin and A_f is the cross sectional area of the fin.
The Q_{act} of the triangular fin is represented by the Eq. 8.

$$Q_{\mathrm{act}} = L\sqrt{2hbk\theta}\frac{I_1(2B\sqrt{l})}{I_0(2B\sqrt{l})} \tag{8}$$

$$B = \sqrt{\frac{2lh}{bk}} \tag{9}$$

where L is length of the fin, l is height of the triangular fin, b is base of the fin, $I_0(x)$ is the first kind Bessel function of zeroth order and $I_1(x)$ is the first kind Bessel function of first order.
The effectiveness of the fins are calculated using the Eq. 10.

$$\epsilon = \frac{Q_{\text{with fin}}}{Q_{\text{without fin}}} \tag{10}$$

3 Results and Discussion

3.1 Temperature Variation of Fluid

The contours obtained by computation in Ansys software are presented in Figs. 3, 4 and 5. In the contours, the red, orange, yellow, purple and blue colours are indicating the highest to lowest temperature region, respectively. The temperature contours in circular duct at inlet and outlet are shown by Fig. 3a, b, respectively. At the inlet, the temperature of the fluid is higher, and therefore, the colour of the contour is seen to be red. However, as the fluid moves from inlet to outlet of the duct, heat is transferred to the wall and thus at the outlet, temperature has decreased from the centre to the wall. This variation of temperature can be seen by the colours of the contour from red (at centre) to blue (at the wall) through yellow, green, purple, etc. Similarly, for triangular and rectangular ducts, the temperature variation at inlet and outlet are shown by Figs. 4 and 5, respectively. The contours of temperature variation represent the heat transfer from the fluid to the duct walls. The temperature variation along

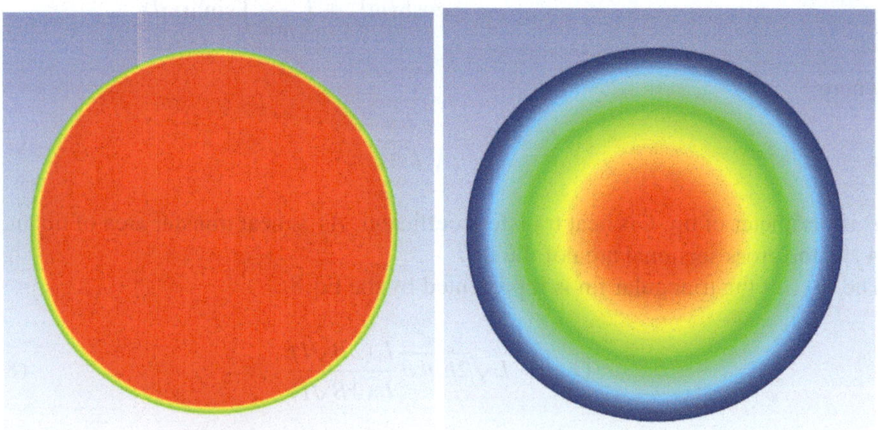

(a) at inlet (b) at outlet

Fig. 3 Temperature contour in circular duct

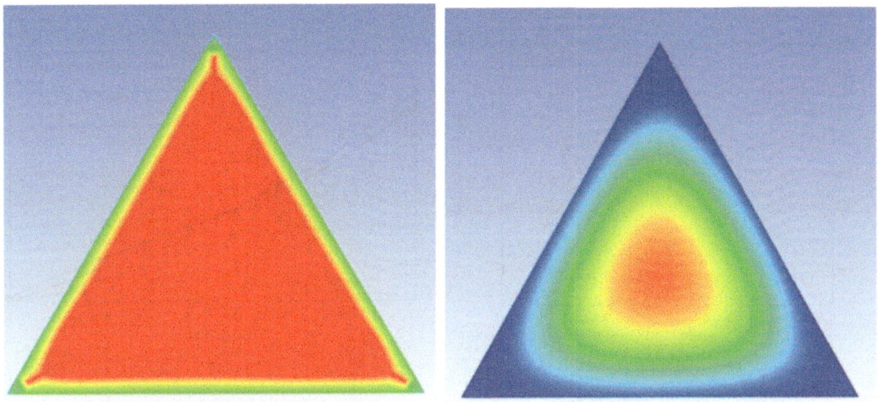

(a) at inlet (b) at outlet

Fig. 4 Temperature contour in triangular duct

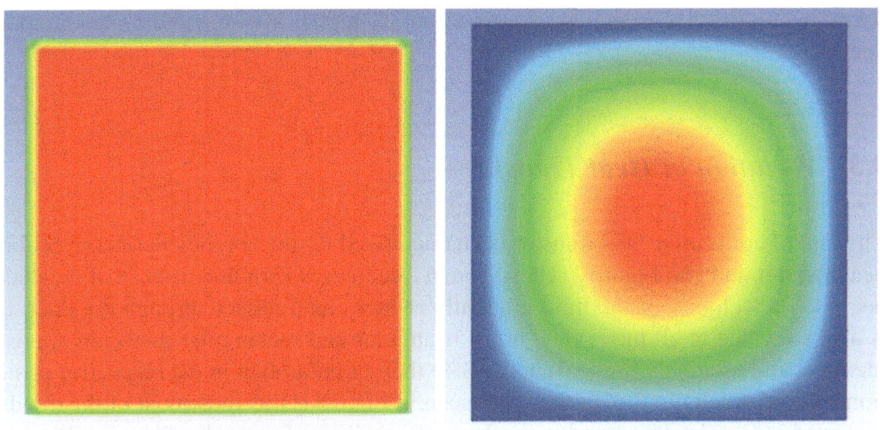

(a) at inlet (b) at outlet

Fig. 5 Temperature contour in rectangular duct

the space of the ducts are shown in Fig. 6a. The figure reveals the highest amount of temperature reduction (about 2.4%) in the triangular duct.

3.2 *Pressure Variation*

The total pressure of the fluid decreases as it passes through the ducts. The variation of the pressure in the different ducts are seen in Fig. 6b. Although, the trends of variation of pressure in all the ducts are identical, the total pressure in the triangular duct is

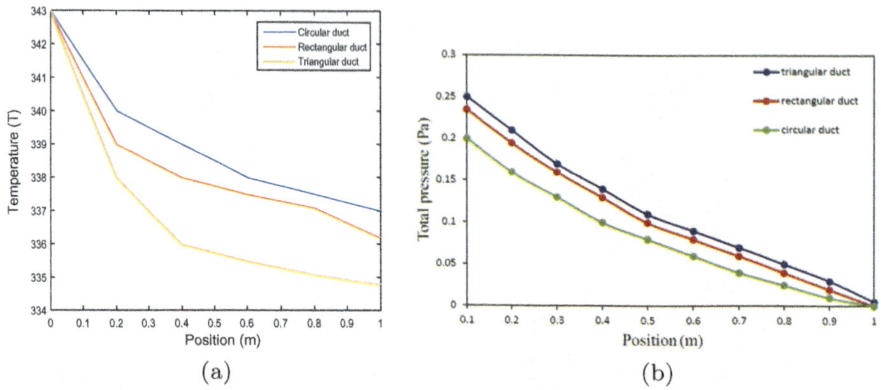

Fig. 6 Temperature and pressure variation along the position of the ducts

found to be highest in all the positions. It has been observed that at the position of 0.5 m, the pressure in triangular duct is about 3% higher as compared to the circular duct.

3.3 Variation of Heat Transfer

The rate of convective heat transfer is characterized by the coefficient of convection heat transfer and Nu. Figure 7a shows that h_f decreases with the distance of the duct because the velocity of the fluid gradually reduces as it passes through the duct. It is also seen that at all the points, h_f of triangular and rectangular ducts are almost identical and relatively higher (about 33%) than circular duct in the respective positions. The Nusselt number variation as seen in Fig. 7b is the agreement of the trends observed in Fig. 7a. This trend is similar to the study performed by Chandratilleke et

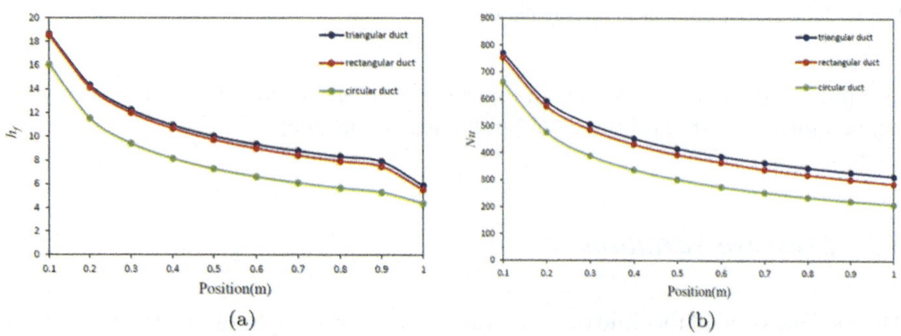

Fig. 7 Heat transfer coefficient and Nu variation along the position of the duct

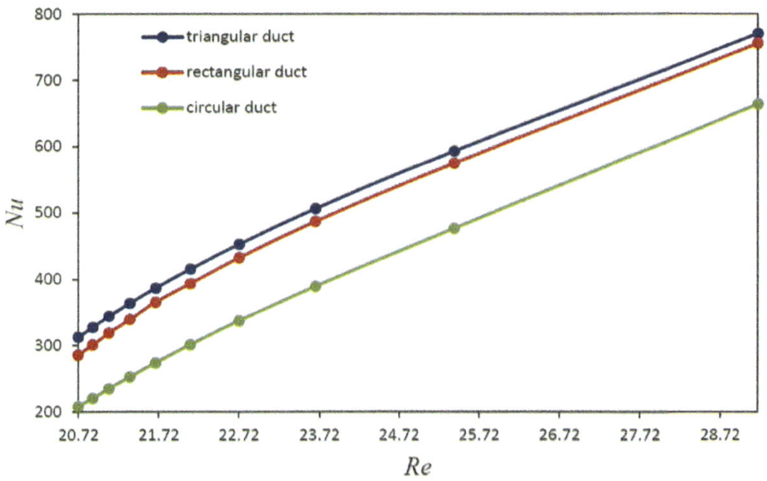

Fig. 8 Nusselt number variation along the position of the ducts

al. [4]. In triangular and rectangular ducts, the Nusselt number is about 50% higher than circular duct at a given position. The mixing of fluids in triangular and rectangular ducts is higher and therefore, the heat transfer rate is relatively higher in comparison with circular duct. Figure 8 reveals that for all Re, the Nu of triangular and rectangular ducts are higher than circular duct. However, Nu in triangular duct is slightly higher than rectangular duct.

3.4 Analysis of Heat Transfer with Fins

3.4.1 Variation of Temperature

Triangular duct has shown the maximum heat transfer as compared to the rectangular and circular ducts. Therefore, triangular duct has been selected to study the heat transfer by introducing fins over it. Longitudinal rectangular and triangular fins are attached on the triangular ducts as seen in Fig. 9a, b. The dimensions of rectangular and triangular fins are 15 mm × 5 mm and 30 mm × 30 mm × 5 mm, respectively. Temperature contours of the triangular duct with fins are shown in Figs. 10 and 11. At the inlet of the ducts, temperature of the air in the entire region is equal to the inlet temperature of air and therefore, the entire region appears as red (Figs. 10a and 11a). Two significant points can be observed from the contours seen in Figs. 10b and 11b. Firstly, the blue coloured portions, i.e. low temperature regions are comparatively larger than temperature contours of ducts without fins as seen in Figs. 4b and 5b. This indicates that more heat has transfered from the fluid to the wall as compared to without fin condition. Secondly, the hot core portion in Fig. 11b is smaller than

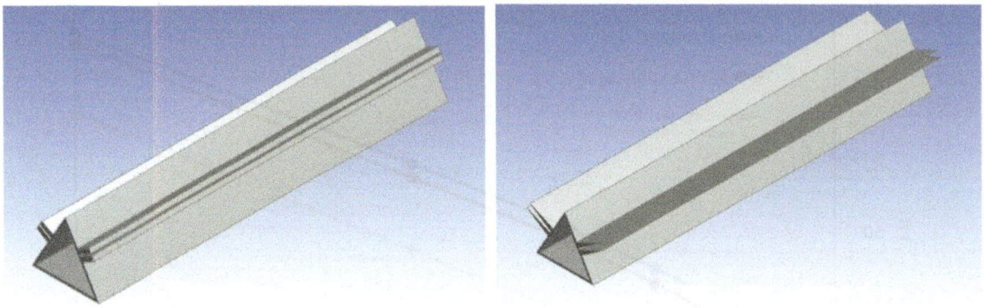

(a) Rectangular fin (b) Triangular fin

Fig. 9 Rectangular and triangular fins on triangular duct

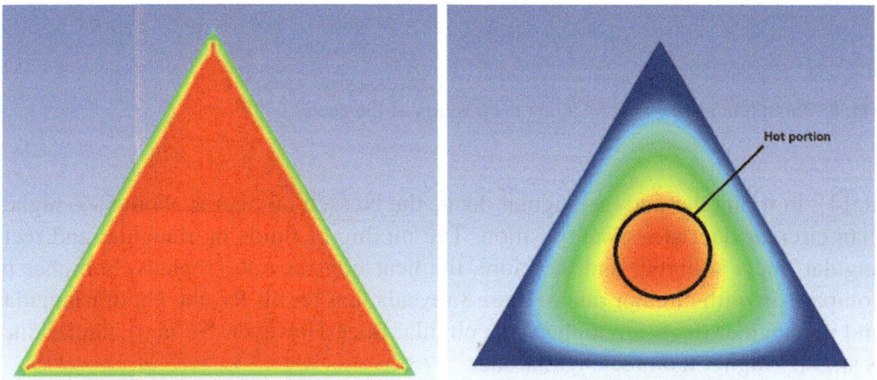

(a) at inlet (b) at outlet

Fig. 10 Temperature contour in triangular duct with rectangular fins

in Fig. 10b, which reveals that rate of heat transfer is higher when triangular fins are used than rectangular fins.

3.4.2 Pressure and Heat Transfer Analysis

Total pressure variation in the triangular ducts with rectangular and triangular fins are shown in Fig. 12. At the beginning, the pressure with triangular fins is higher than rectangular fins and this difference gradually decreases to the outlet of the pipe. At inlet, the pressure with triangular fins is about 20% higher than with rectangular fins and becomes negligible near the outlet of the duct.

The rate of decreasing of Nusselt number in the ducts with both triangular and rectangular fins are higher at the beginning and gradually decreases towards the end of the duct as seen in Fig. 13a. Nusselt number in both the cases are almost identical, however, at the beginning, it is slightly higher in case of triangular fins. The variation

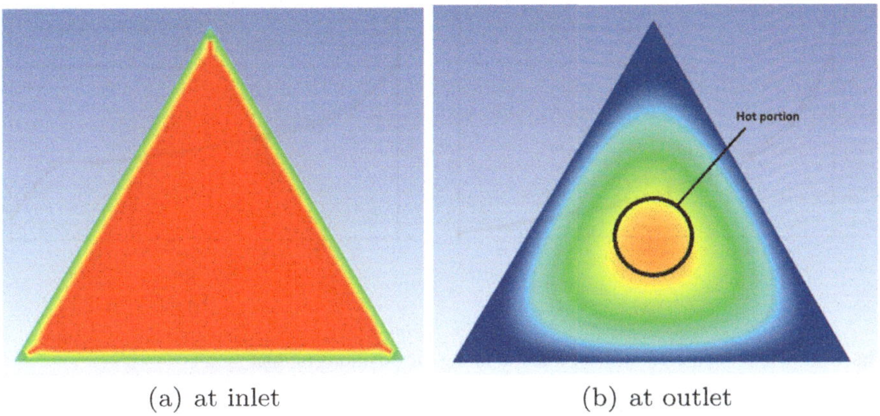

Fig. 11 Temperature contour in triangular duct with triangular fins

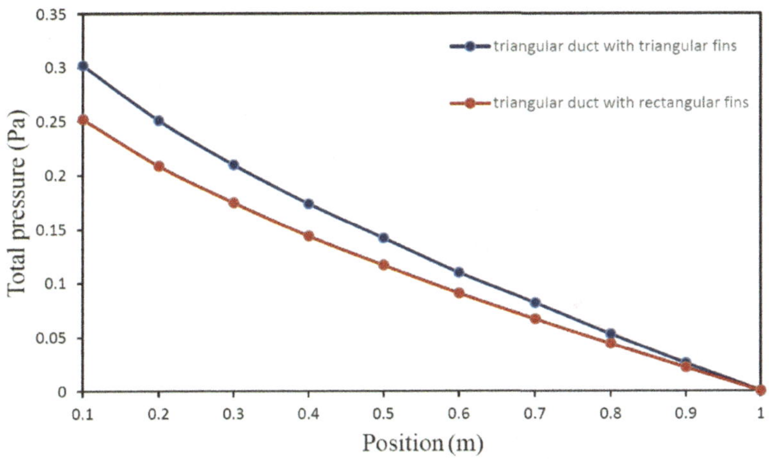

Fig. 12 Pressure variation in the triangular duct with fins

of Nu in the triangular duct can be compared with results obtained by Talukdar and Shah [16]. Similar to Nusselt number, h_f is also found to be identical in both the cases which gradually decreases with the positions (Fig. 13b). However, in case of triangular fins, h_f is slightly higher than rectangular fins. Since Nu varies with Re, by increasing Re, Nu is found to be increased in the finned duct and at higher Re (e.g., Re>25). The Nu in triangular duct with triangular fins is about 3% higher than rectangular fins as shown in Fig. 14. Therefore, the triangular fins are relatively beneficial as compared to rectangular fins.

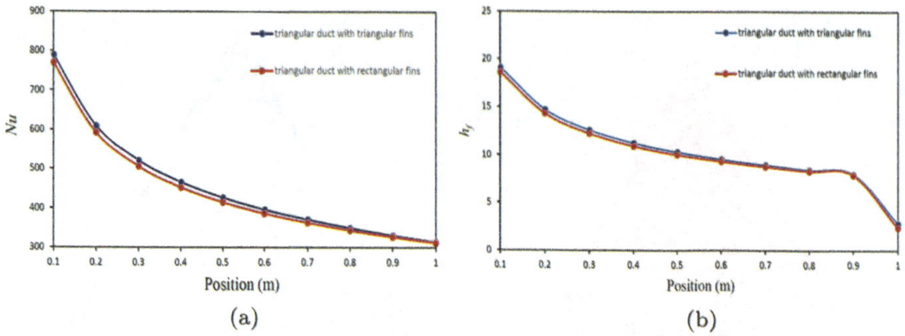

Fig. 13 Nu and coefficient of heat transfer along the positions of duct

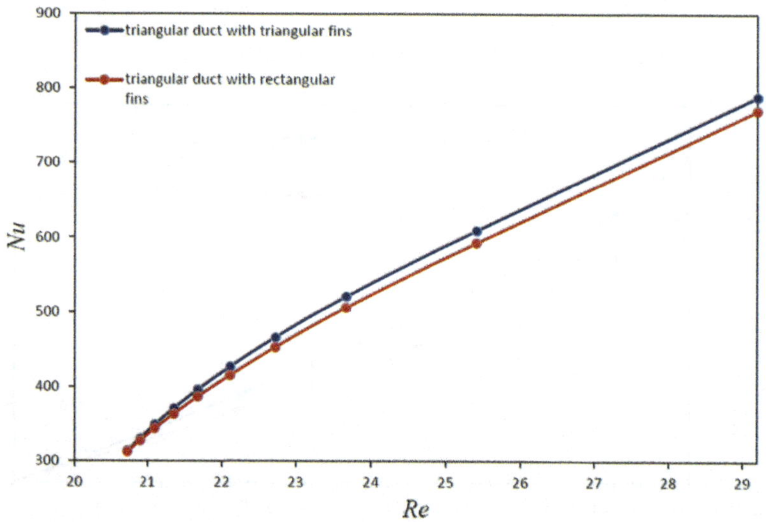

Fig. 14 Nusselt number variation in finned ducts

3.5 *Influence of Fin Number on Heat Transfer*

Few number of fins were applied on the triangular duct in order to investigate the heat transfer. It is found that, using four number of fins, the Nu is highest beyond which it firstly decreases and then increases with a very small rate. Figure 15 shows that when the number of fin is 4, the highest Nu reaches the maximum value, i.e. 427.

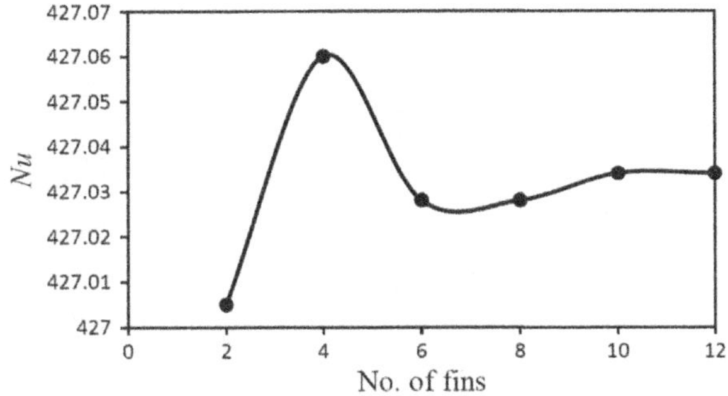

Fig. 15 Nusselt number variation with number of fins

Table 1 Efficiency and effectiveness of fin

	Triangular fin (%)	Rectangular fin (%)
Efficiency (η)	80.65	58.38
Effectiveness (ϵ)	1.186	1.054

3.6 Efficiency and Effectiveness of Fins

Based on the geometry under consideration, the efficiency and effectiveness of both rectangular and triangular fins were calculated using Eqs. 4–10. The results tabulated in Table 1 indicates that efficiency of triangular fin is about 22% higher than rectangular fin. On the other hand, effectiveness of both type of fins are almost same. Because of the geometric profile, triangular fins require less volume and material for similar heat transfer area as compared to the rectangular fins, and thus, they are more efficient and economic.

4 Conclusions

Heat transfer analysis of air flowing through the rectangular, circular and triangular ducts were carried out. The analysis shows that rate of heat transfer is highest in triangular duct as compared to rectangular and circular duct. Therefore, two types of fins such as rectangular and triangular fins were applied on the triangular duct in order to observe the heat transfer. More heat transfer was found when triangular fins are used in comparison with rectangular fins. The influence of number of fins on heat transfer was studied. Based on the given geometry, maximum heat transfer was found when four number of fins were used. Further, both efficiency and effectiveness of triangular fins were found to be higher as compared to rectangular fins.

References

1. O'Brien JE, Sparrow EM (1982) Corrugated-duct heat transfer, pressure drop, and flow visualization. J Heat Transfer 104(3)
2. Olek S, Elias E, Wacholder E, Kaizerman S (1991) Unsteady conjugated heat transfers in laminar pipe flow. Int J Heat Mass Transf 34(6)
3. Dağtekin İ, Öztop HF, Şahin AZ (2005) An analysis of entropy generation through a circular duct with different shaped longitudinal fins for laminar flow. Int J Heat Mass Transfer 48(1)
4. Chandratilleke TT, Narayanaswamy R, Wangdhamkoom P (2010) Convective heat transfer in airflow through a duct with wall thermal radiation. IOP Conf Ser Mater Sci Eng 10:012026
5. Yang MH, Yeh RH, Hwang JJ (2012) Forced convection in a channel with transverse fins. Int J Numerical Methods Heat Fluid Flow 22(3)
6. Sahu KB, Singh RK (2014) Analysis of heat transfer and flow due to natural convection in air around heated triangular cylinders of different sizes inside a square enclosure. Proc Eng 90
7. Karupaaraj RT, Sekhar YR (2014) Heat transfer and fluid flow analysis of flat surface with artificially roughened square transverse wire rib for application in solar air heater. ARPN J Eng Appl Sci 9(5)
8. Charles R, Wang CC (2015) Novel heat dissipation fin design applicable for natural convection augmentation. Int Commun Heat Mass Transf 59
9. Mirapalli S, Kishore PS (2015) Heat transfer analysis on a triangular fin. Int J Eng Trends Technol 19(5)
10. Kumar R, Kumar A, Varun (2017) Computational fluid dynamics based study for analyzing heat transfer and friction factor in semi-circular rib-roughened equilateral triangular duct. Int J Numer Methods Heat Fluid Flow 27(4)
11. Shrirao NP, Pente SS, Gaddamwar SS (2017) Fluid flow analysis in air duct flow with and without internal threads using computational fluid dynamics (CFD). Int J Mech Eng Technol 8(3)
12. Bhargava CK, Arya RK (2017) Simplified MATLAB solution schemes of heat transfer equations. Chem Eng Technol 40(1)
13. Sharma N, Tariq A, Mishra M (2019) Experimental investigation of heat transfer enhancement in rectangular duct with pentagonal ribs. Heat Transf Eng 40
14. Yadav S, Pandey KM (2018) A parametric thermal analysis of triangular fins for improved heat transfer in forced convection. J Mech Eng 64
15. Gurumurthy KM, Manjunatha TG, Eswaramoorthy M (2018) Comparative CFD analysis on heat transfer coefficient of circular and non circular duct. AIP Conf Proc 2039
16. Talukdar P, Shah M (2008) Analysis of laminar mixed convective heat transfer in horizontal triangular ducts. Numer Heat TR A-Appl 54

Frequency Regulation of Interconnected Power Generating System Using Ant Colony Optimization Technique Tuned PID Controller

V. Kumarakrishnan, G. Vijayakumar, D. Boopathi, K. Jagatheesan, S. Saravanan, and B. Anand

Abstract Ant colony optimization technique (ACO) is proposed for frequency regulation of single area power generating network. The proposed system consists of thermal power generating system with reheater, turbine, governor, hydrogen aqua electrolyzer (HAE), and fuel cell (FC). Proportional–integral–derivative (PID) regulator acts as an auxiliary regulator to maintain the frequency deviation during unexpected load demand. The proposed PID controller gain values are tuned by utilizing ACO tuned PID regulator with ITAE objective function. To show the superiority of the ACO technique and the performance, responses were equated with conventional method PID controller's results for the identical power network. The dynamic performance of the suggested ACO-PID regulator gives improved response over conventional PID in terms of quick settling time.

Keywords Ant colony optimization · Proportional–integral–derivative controller · Fuel cell · Hydrogen aqua electrolyzer · Frequency regulation

1 Introduction

The requirement of electrical power is always in the upward direction, because of industrials and domestic development in the world. Balancing the power demand is not an easy task in the power network. While the system size increases, complication in the maintenance of power quality also raises. In order to maintain power quality, load frequency control (LFC) scheme is employed. Many of the researchers are taken these issues to solve by applying various optimization techniques and controller [1,

V. Kumarakrishnan (✉) · D. Boopathi · K. Jagatheesan
Department of EEE, Paavai Engineering College, Namakkal, Tamil Nadu, India

G. Vijayakumar · S. Saravanan
Department of EEE, Muthyammal Engineering College, Rasipuram, Tamil Nadu, India

B. Anand
Department of EIE, Hindusthan College of Engineering and Technology, Coimbatore, Tamil Nadu, India

S. Suhag et al. (eds.), *Control and Measurement Applications for Smart Grid*, Lecture Notes in Electrical Engineering 822,
https://doi.org/10.1007/978-981-16-7664-2_11

2]. Sine cosine optimization (SCO) tuned PD-PID cascade controller implemented for two-area thermal power plant with regenerated energy sources (RESs) and energy storage system [3]. In [4], firefly algorithm (FA)-based PD-PID cascade controller employed for interconnected power grid. Author in [5] moth flame-optimization (MFO) optimized PI/PID/PIDD controller to multi-sources interconnected power network.

ACO optimized PID regulator has been implemented by [6] for three-area thermal power grid AGC, and in [7], the same controller is utilized for single-area nuclear energy system; the authors [8, 9] are used ACO-PID for two-area thermal power system. ANN tuned PID regulator implemented in [10] for thermal power plant with distributed generation system. Quantum-inspired evolutionary algorithm tuned PID regulator employed by [11], glow swarm optimization (GSO) technique-based PI controller implemented in [12] AGC of thermal power unit. Firefly algorithm is implanted [13] to harmony PID gain values for reheated thermal power network. Artificial bee colony (ABC) technique optimized PID was inspected multi-area thermal units for LFC [14], and in [15], fuzzy gain scheduling controller is tuned by genetic algorithm (GA) for AGC of grid-connected thermal power network.

Model predictive control (MPC) employed in hydro-thermal power system AGC [16], tilt integral controller proposed in [17] for thermal, gas, and hydro interconnected power grid. Hybrid fuzzy logic PID regulator is implemented in [18] for thermal power network with GRC and boiler AGC, 2DoF-PID controller optimized by TLBO designed in [19] for two-area non-reheated thermal power network, bat algorithm (BAT) optimized PID controller implemented in [20] for thermal power network. ACO technique is also utilized to solve various computational issues like mobile edge display [21, 22], traveling salesman problem [23], path planning problem [24], in [25] drilling of curved and tilt bones. Artificial bee colony technique implemented for load balancing [26], and in [27], Hungarian method is involved for load balancing issues, MPC used in [28] for power generation issues. Author in [29] handled LFC of single-area multi-source power system consists of (thermal–hydro–gas–nuclear) with PV by adopting PSO-PID as auxiliary controller. Author in [30, 31] discussed many computational algorithms for various application. Literature review gives more information about the LFC/AGC issues in power network related to the thermal power plant (PP). Novelty of the research work is most of the researchers used thermal PP as one of the sources for the suggested power network, In this proposed research work, thermal power plant separately investigated with small micro-grid (HAE and FC).

Manuscript organization

Frequency regulation of the suggested power network is carryout with 6 separate chapters. Section 1 provides detailed introduction about the need of frequency regulation and literature review for the above-mentioned issues, and Sect. 2 gives the modeling and mathematical expression of power network. Section 3 is about the design stagey of suggested regulator, and Sect. 4, discusses the tuning methods to obtain optimized controller parameters. Section 5 discusses the performance analysis

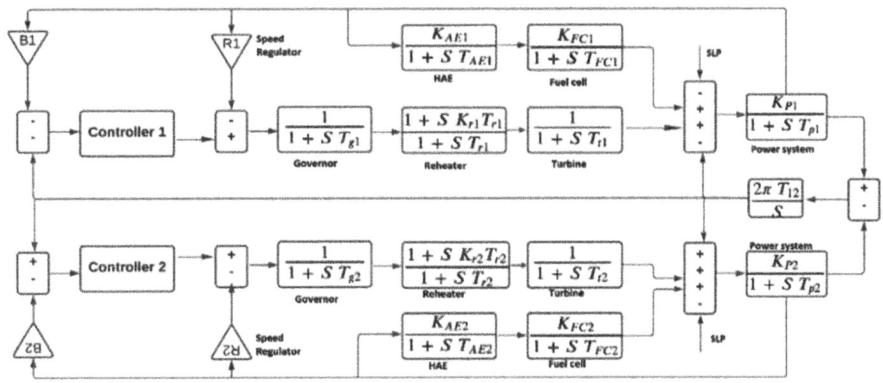

Fig. 1 Model of proposed power system

of the proposed tuning method over conventional tuning method. Section 6 gives a conclusion about the research work.

2 System Modeling

The proposed investigated thermal power system incorporated with distributed generation units such as (HAE and FC). Thermal power unit comprises of speed governor, reheated turbine where K_r, K_{AE}, $K_{P,}$ and K_{FC} are the gain parameters and T_g, T_t, T_{AE}, T_{FC}, and T_P are the time constants. Each area is controlled by separate auxiliary controller. The mathematical model is shown in Fig. 1 [3]. Nominal parameters of power system in the block diagram are reported in annexure 1.

3 Controller Design and Objective Function

PID controller involved in this research work for frequency regulation of power network, because it is most common industrial controller. Its performances well for various controlling and operating situations, efficient in all application and also easy to implement and cost-efficient. PID is the compensation of three different controllers [Integral (I), Proportional (P), and Derivative (D)]. Mathematical expression of PID controller is given in Eq. 1. Structure of PID controller is given in Fig. 2 [26].

where K_P, K_I, K_D are P, I, and D controller gain values, respectively.

$$u(t) = K_P e(t) + K_I \int_0^t e(t)\mathrm{d}t + K_D \frac{\mathrm{d}e(t)}{\mathrm{d}t} \tag{1}$$

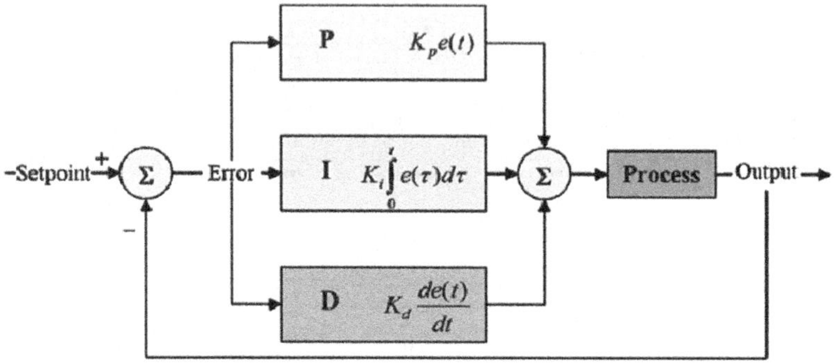

Fig. 2 Structure of PID controller

Proper selection of the objective function is essential one to achieve expected output response. Integral time absolute error (ITAE) is used to obtain optimal gain values of controller in this research work. Mathematical expression of ITAE cost function is given in Eq. 2 [6]. The selection of the cost function is the most important task to obtain optimal gain parameters for the designed controller. In the literature review, many of the researchers [3, 5–8, 10, 11, 14, 15, 17, 18, 20, 29] are used ITAE cost function to get optimal controller gain parameters for various structures of power plants, and also they achieve better result than other cost functions.

$$J = \text{ITAE} = \int_0^{t\text{sim}} t.|e(t)|\text{d}t \tag{2}$$

4 Investigation of Power System

The proposed power network is investigated by considering PID controller as an auxiliary controller. Controller gain values are tuned by two different tuning methods such as conventional method and ant colony optimization techniques. The details of two techniques are given in the following sections.

4.1 Conventional Method-Based PID Controller

The conventional tuning method is the most classical method of tuning. It is a trial and error pattern-based tuning method for PID regulator. The tuning procedure is as follows. The performance indices curve for ITAE cost function is shown in Fig. 3. The tuned gain values are reported in Table 1 [7].

Conventional method (trial and error) gain value tuning procedure:

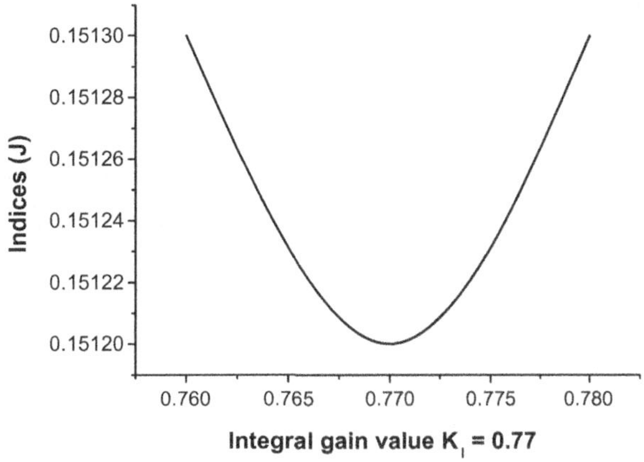

Fig. 3 Performance indices curve of ITAE

Table 1 Conventional PID controller gain values

Gain value/tuning method	K_{P1}	K_{I1}	K_{D1}	K_{P2}	K_{I2}	K_{D2}
Conventional	4.1	0.77	0.12	4.1	0.77	0.12

Step 1 Tune K_I by kept K_P and K_D is zero.
Step 2 Tune K_P by fix optimal K_I and kept K_D is zero.
Step 3 Tune K_D by fix optimal K_I and K_P.

4.2 ACO Technique-Based PID Controller

ACO was first implemented by Marco Dorigo in the 1990s. This process is applied based on the foraging behavior of an ant for discovering a path in between their colony and food source. It was originally used to address the well-known dilemma of traveling salesmen. Later on, it is used to solve various difficult problems with optimization. Functional flowchart is given in Fig. 4.

After completing the tuning process gain values were tuned by ACO technique are reported in Table 2.

5 Result and Discussion

The Simulink model of the suggested network is simulated for 60 s time slot in MATLAB 2014a version platform by applying both conventional and ACO technique tuned gain values in PID controller with 1% SLP. Frequency deviation response comparison of conventional and ACO-PID of area 1 and area 2 are shown in Figs. 5

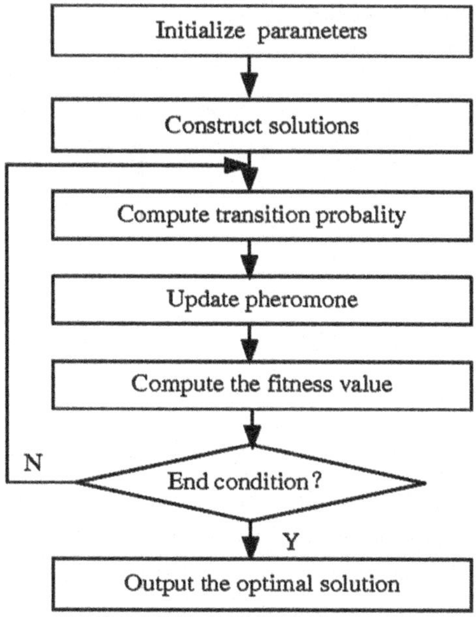

Fig. 4 ACO technique flowchart

Table 2 ACO tuned PID controller gain values

Gain value/tuning method	K_{P1}	K_{I1}	K_{D1}	K_{P2}	K_{I2}	K_{D2}
ACO	0.98	0.99	0.25	0.62	0.9	0.36

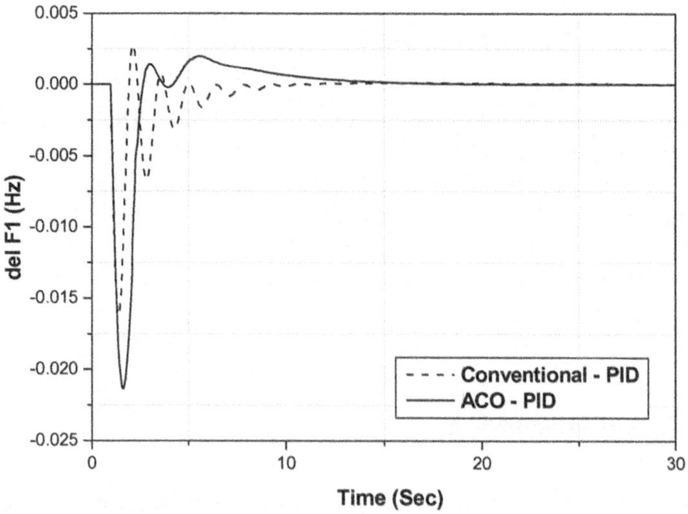

Fig. 5 Frequency deviation comparison of area 1

and 7, respectively, and area control error (ACE) of areas 1 and 2 is shown in Figs. and 8, respectively.

The time domain parameters of del F in area 1 are reported in Table 3, with the help of Fig. 5.

The time domain parameters of ACE in area 1 are reported in Table 4, from Fig. 6.

The time domain parameters of del F of area 2 are reported in Table 5, with the help of Fig. 7.

The time domain parameters of ACE in area 2 are reported in Table 6, with the help of Fig. 8

Tie line power flow of conventional, ACO tuned PID regulator performance comparison is shown in Fig. 9., and, the time domain parameters of tie line power flow are reported in Table 7.

Table 3 Time domain parameters of del F in area 1

Time domain parameters/tuning method	T_S (s)	OS (Hz)	US (Hz)
Conventional-PID	35	2.8×10^{-3}	0.0162
ACO-PID	24	2×10^{-3}	0.0215

Table 4 Time domain parameters of ACE in area 1

Time domain parameters/tuning method	T_S (s)	OS (Hz)	US (Hz)
Conventional PID	40.5	7.65×10^{-3}	1.5×10^{-4}
ACO-PID	26	0.0107	0.4×10^{-3}

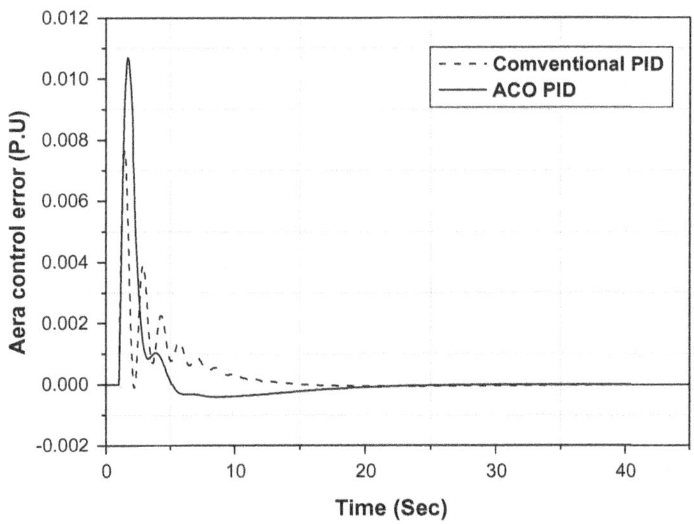

Fig. 6 Area control error of area 1

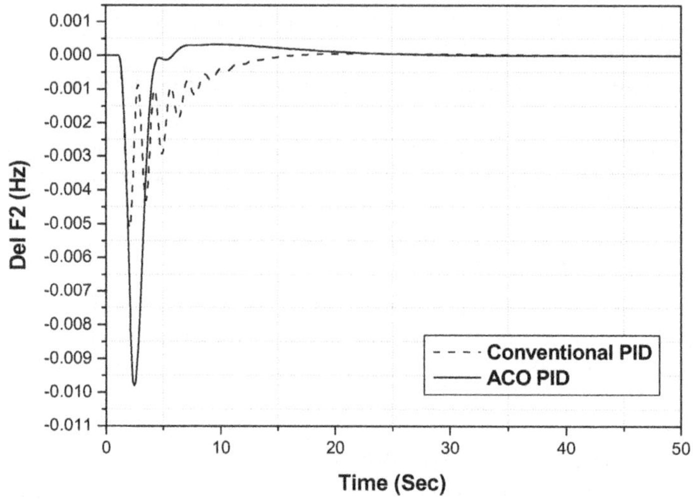

Fig. 7 Frequency deviation comparison of area 2

Table 5 Time domain parameters of frequency deviation comparison of area 2

Time domain parameters/tuning method	T_S (s)	OS (Hz)	US (Hz)
Conventional PID	48	0.5×10^{-3}	5.15×10^{-3}
ACO-PID	30	0.3×10^{-3}	9.8×10^{-3}

Table 6 Time domain parameters of ACE in area 2

Time domain specific parameters/tuning method	T_S (s)	OS (Hz)	US (Hz)
Conventional PID	46	1×10^{-3}	6×10^{-4}
ACO-PID	28	1.8×10^{-3}	8.5×10^{-4}

A bar chart comparison of settling time in del *F* of two areas 1 and 2, and del P_{tie} flow (between areas 1 and 2) is conformed to prove the superiority of the suggested ACO optimized PID controller. The bar chart clearly shows that settling time of all area is quick in ACO-PID. The bar chart comparison is shown in Fig. 10. Also, the percentage of improvement of the proposed controller over conventional PID is given in Table 8.

To find the reliability of the proposed controller performance, a robustness test is carried out with 1, 2, and 5% SLP for the designed power network.

ACO-PID controller for the suggested power network is performed well at all critical loading condition, and it is clearly evident by Figs. 11 and 12.

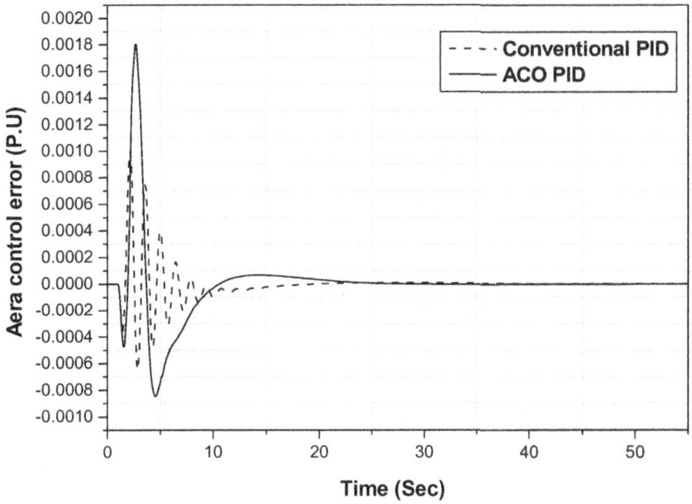

Fig. 8 Area control error of area 2

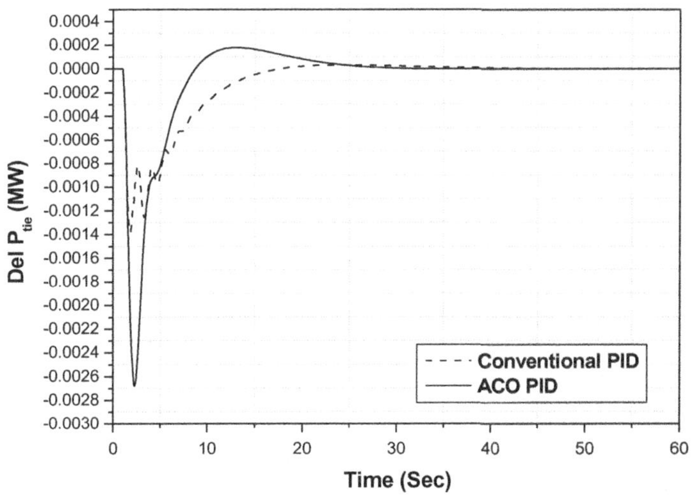

Fig. 9 Tie line power flow comparison

Table 7 Time domain parameters of tie line power flow

Time domain parameters/tuning method	T_S (s)	OS (Hz)	US (Hz)
Conventional PID	58	3×10^{-5}	1.4×10^{-3}
ACO-PID	35	2×10^{-4}	2.7×10^{-3}

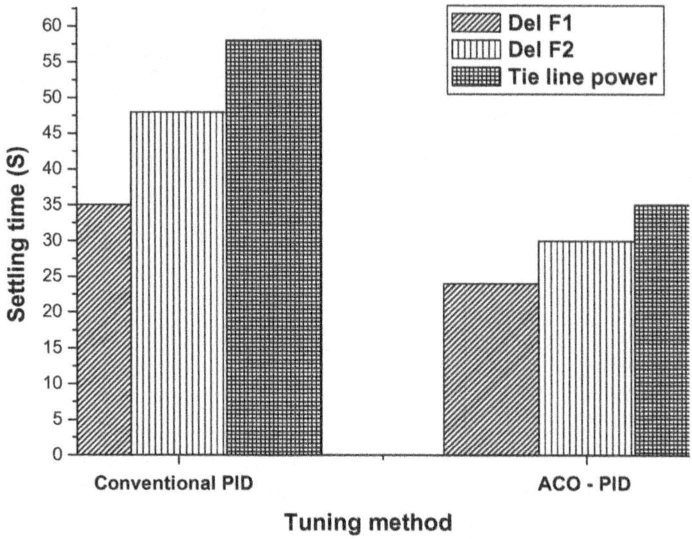

Fig. 10 Bar chart comparison of setting time

Table 8 Percentage of improvement of ACO-PID over conventional PID

Percentage of improvement over conventional PID	Settling time	% of improvement (%)
del *F* of area 1	30	37.5
del *F* of area 2	28	39.1
del tie line power flow	35	39.6

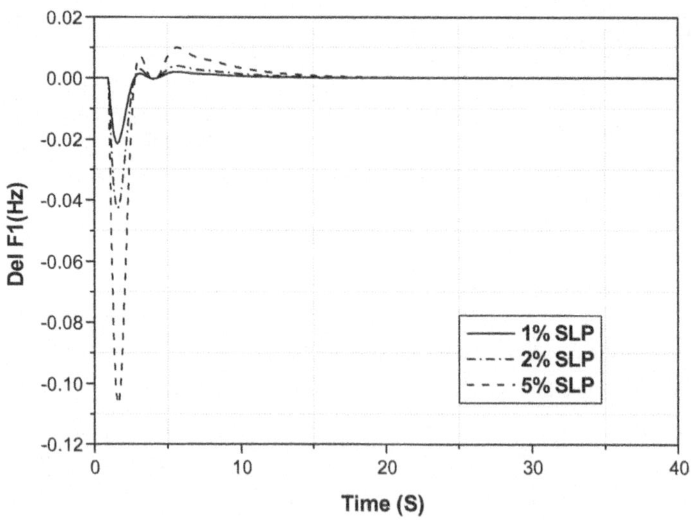

Fig. 11 Robust test performance of del *F* of area 1

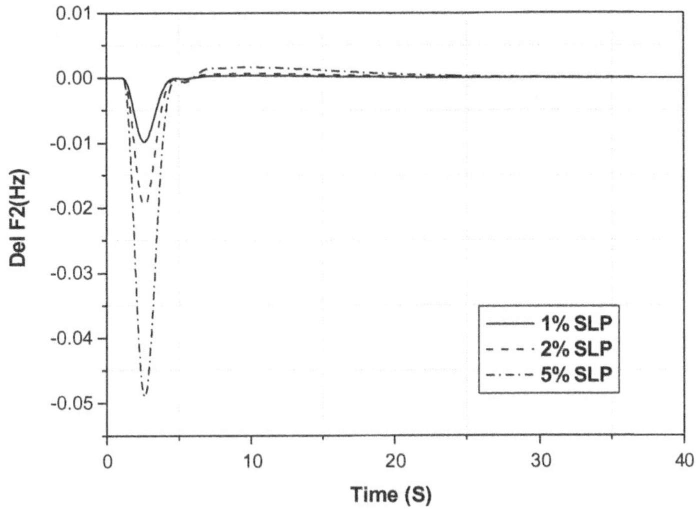

Fig. 12 Robust test performance of del *F* of area 2

6 Conclusion

Frequency regulation of the proposed (two areas interconnected) power network is investigated by implementing ACO-based PID controller. Supremacy and the improvement of the suggested regulator and tuning technique, performance response was equated with conventional method tuned PID regulator for the identical power network. At the result, ACO-PID shows its betterment in all the area's frequency and tie line power deviation over conventional method tuned controller. ACO-PID regulator gives quick settling time in del *F* in area 1 and area 2 (24 and 30 s), and del P_{tie} in between areas 1 and 2 (35 s).

Future research direction

The following points are related to the future research scope in this article:

- In this paper, reheated thermal power plant is used as major sources for the production of electrical power, due to the pollution and shortage of fossil fuel renewable energy sources may be incorporated to balance the load demand.
- PID controller is used as auxiliary controller to manage the power fluctuation; it may be change by other controller.
- PSO technique is utilized to optimize the controller gain parameters in this article, and the optimization technique may be changed in the future.

Annexure 1

$P_w = 2000$ MW, $f = 50$ Hz, $T_{g1} = T_{g2} = 0.03$ s, $T_{t1} = T_{t2} = 0.3$ s, $K_{r1} = K_{r2} = 0.5$, $K_{\mathrm{AE}} = 0.002$, $T_{\mathrm{AE}} = 0.5$ s, $K_{\mathrm{FC}} = 0.01$, $T_{\mathrm{r1}} = T_{\mathrm{r2}} = 10$ s, $T_{\mathrm{FC}} = 4$ s, $R_1 = R_2 = 4$ p.u/MW, $B_1 = B_2 = 0.425$ p.u/MW, $K_{\mathrm{p1}} = K_{\mathrm{p2}} = 120$, $T_{\mathrm{p1}} = T_{\mathrm{p2}} = 20$ s.

References

1. Elgerd OI (1982) Electric energy systems theory: an introduction
2. Nagrath J, Kothari DP (1994) Power system engineering. Tata Mc-Graw Hill Publishing Company Limited, New Delhi
3. Gorripotu TS, Ramana P, Sahu RK, Panda S (2020) Sine cosine optimization based proportional derivative-proportional integral derivative controller for frequency control of hybrid power system. In: Computational intelligence in data mining. Springer, Singapore, pp 789–797
4. Satapathy P, Debnath MK, Mohanty PK (2019) Optimal load frequency control of an unified power system with SMES and TCPS. In: International conference on intelligent computing and applications. Springer, Singapore, pp 215–225
5. Mohanty B (2019) Performance analysis of moth flame optimization algorithm for AGC system. Int J Model Simul 39(2):73–87
6. Jagatheesan K, Anand B, Dey KN, Ashour AS, Satapathy SC (2018) Performance evaluation of objective functions in automatic generation control of thermal power system using ant colony optimization technique-designed proportional–integral–derivative controller. Electr Eng 100(2):895–911
7. Dhanasekaran B, Siddhan S, Kaliannan J (2020) Ant colony optimization technique tuned controller for frequency regulation of single area nuclear power generating system. Microprocess Microsyst 73:102953
8. Nguyen GN, Jagatheesan K, Ashour AS, Anand B, Dey N (2018) Ant colony optimization based load frequency control of multi-area interconnected thermal power system with governor deadband nonlinearity. In: Smart trends in systems, security and sustainability. Springer, Singapore, pp 157–167
9. Omar M, Soliman M, Ghany AA, Bendary F (2013) Optimal tuning of PID controllers for hydrothermal load frequency control using ant colony optimization. Int J Electr Eng Inform 5(3):348
10. Debnath MK, Agrawal R, Tripathy SR, Choudhury S (2020) Artificial neural network tuned PID controller for LFC investigation including distributed generation. Int J Numer Modell Electron Netw Devices Fields 33(5):e2740
11. Jagatheesan K, Samanta S, Choudhury A, Dey N, Anand B, Ashour AS (2018) Quantum inspired evolutionary algorithm in load frequency control of multi-area interconnected thermal power system with non-linearity. In: Quantum computing: an environment for intelligent large scale real application. Springer, Cham
12. Baral KK, Barisal AK, Mohanty B (2016, Oct) Load frequency controller design via GSO algorithm for nonlinear interconnected power system. In: 2016 international conference on signal processing, communication, power and embedded system (SCOPES). IEEE, pp 662–668
13. Naidu K, Mokhlis H, Bakar AHA, Terzija V (2014) Comparative performance analysis of firefly algorithm for load frequency control in automatic generation control of interconnected reheat thermal power system
14. Naidu K, Mokhlis H, Terzija V (2017) Performance investigation of ABC algorithm in multi-area power system with multiple interconnected generators. Appl Soft Comput 57:436–451
15. Arya Y, Kumar N (2016) Fuzzy gain scheduling controllers for automatic generation control of two-area interconnected electrical power systems. Electr Power Components Syst 44(7):737–751

16. Liu X, Kong X, Lee KY (2016) Distributed model predictive control for load frequency control with dynamic fuzzy valve position modelling for hydro–thermal power system. IET Control Theory Appl 10(14):1653–1664
17. Topno PN, Chanana S (2018) Load frequency control of a two-area multi-source power system using a tilt integral derivative controller. J Vib Control 24(1):110–125
18. Haroun AG, Li YY (2017) A novel optimized hybrid fuzzy logic intelligent PID controller for an interconnected multi-area power system with physical constraints and boiler dynamics. ISA Trans 71:364–379
19. Sahu RK, Panda S, Rout UK, Sahoo DK (2016) Teaching learning based optimization algorithm for automatic generation control of power system using 2-DOF PID controller. Int J Electr Power Energy Syst
20. Kaliannan J, Baskaran A, Dey N, Ashour AS, Kumar R (2019, Jan) Bat algorithm optimized controller for automatic generation control of interconnected thermal power system. In: Information technology and intelligent transportation systems: proceedings of the 3rd international conference on information technology and intelligent transportation systems (ITITS 2018), vol 314. Xi'an, China, 15–16 Sept 2018. IOS Press, p 276
21. Vimal S, Khari M, Dey N, Crespo RG, Robinson YH (2020) Enhanced resource allocation in mobile edge computing using reinforcement learning based MOACO algorithm for IIOT. Comput Commun 151:355–364
22. Tuani AF, Keedwell E, Collett M (2020) Heterogenous adaptive ant colony optimization with 3-opt local search for the travelling salesman problem. Appl Soft Comput:106720
23. Jiang X, Lin Z, He T, Ma X, Ma S, Li S (2020) Optimal path finding with beetle antennae search algorithm by using ant colony optimization initialization and different searching strategies. IEEE Access 8:15459–15471
24. Patino-Ramirez F, Layhee C, Arson C (2020) Horizontal directional drilling (HDD) alignment optimization using ant colony optimization. Tunnell Undergr Space Technol 103:103450
25. Gopal M (2002) Control systems: principles and design. Tata McGraw-Hill Education
26. Sharma S, Saini H (2019) Efficient solution for load balancing in fog computing utilizing artificial bee colony. Int J Ambient Comput Intell (IJACI) 10(4):60–77
27. Mondal RK, Ray P, Nandi E, Biswas B, Sanyal MK, Sarddar D (2019) Load balancing of unbalanced assignment problem with hungarian method. Int J Ambient Comput Intell (IJACI) 10(1):46–60
28. Hu C, Bi L, Piao Z, Wen C, Hou L (2018) Coordinative optimization control of microgrid based on model predictive control. Int J Ambient Comput Intell (IJACI) 9(3):57–75
29. Kumarakrishnan V, Vijayakumar G, Boopathi D, Jagatheesan K, Anand B (2020) Optimized PSO technique based PID Controller for load frequency control of single area power system. Solid State Technol 63(5):7979–7990
30. Dey N (ed) (2017) Advancements in applied metaheuristic computing. IGI Global
31. Dey N, Ashour AS, Bhattacharyya S (eds) (2020) Applied nature-inspired computing: algorithms and case studies. Springer, Singapore

Duhem Hysteresis Modelling of Single Axis Piezoelectric Actuation System

D. V. Sabarianand and P. Karthikeyan

Abstract In this paper, the control of PEA regarding hysteresis impact is introduced. The hysteresis demonstrating of PEA utilizing Duhem model and Prandtl-Ishlinskii model as rate independent or utilizing backlash-like operator. The exhibition of both models is looked at and limit the nonlinearity utilizing the improved PID controller. The non-linear control configuration is combined to minimize the hysteresis effect of PEAs on the micropositioning systems.

Keywords Duhem hysteresis model · Backlash-like hysteresis model · Hysteresis nonlinearity

1 Introduction

Piezoelectric actuators (PEA) offer the benefits of high inflexibility and quick reaction. The utilizations of PEAs, for example, rapid AFM scanners [1], adaptive optics [2, 3], machine tools [4], aviation parts, micromanipulators [5–9] and synchrotron imaging systems [10]. Smart material actuators are typically used for smaller than expected/nanopositioning structures. Amongst these materials, piezoelectric is far the most notable due to their amazing objective, high return power, high response speed and unprecedented capacity to weight extent [11].

The piezoelectricity phenomenon found in the nineteenth century is a shared activity between mechanical power and electric field. In PEAs, the electric field that is brought about by input voltage gives the actuator relocation in nanometer scales with a high goal. Prior to applying a voltage to piezoelectric material heads, no relocations are noticed on account of irregular direction of dipoles. Yet, from that

D. V. Sabarianand (✉) · P. Karthikeyan
Department of Production Technology, Madras Institute of Technology, Anna University, Chennai, India
e-mail: dvsabarianandh@annauniv.edu.in

P. Karthikeyan
e-mail: pkarthikeyan@mitindia.edu

S. Suhag et al. (eds.), *Control and Measurement Applications for Smart Grid*, Lecture Notes in Electrical Engineering 822,
https://doi.org/10.1007/978-981-16-7664-2_12

point onwards, dipoles change their direction, and an anisotropic field is developed that adjustments in material length [12].

The main trouble, whilst utilizing PEA is their non-direct elements, including hysteresis, creep and straight vibrational elements [13]. This control conduct makes the control task testing and could weaken the exhibition of the entire framework. Vibration elements shows up close to the resounding recurrence of the framework. This wonder, despite the fact that could upset the presentation is testing just in the high frequencies and consequently isn't considered in this investigation.

Creep is a non-straight conduct in PEA. In the event that a consistent voltage is applied to a piezoelectric material, a large portion of the removal happens in the principal milliseconds [14, 15]. Be that as it may, in the following minutes, the actuator length keeps as creep testing in open-circle control positions. In this examination creep isn't worried, it goes about as position input. The fundamental focal point of this investigation is on hysteresis which the main non-straight elements saw in piezoelectric, ferroelectric and shape memory combination (SMA) materials [16–21]. Yield of a hysteresis component in a specific moment of time, notwithstanding framework input, is an element of framework state in the past moments. Applying a sinusoidal contribution to a PEA, the yield position in rising and falling ways doesn't coordinate. The shape and size of hysteresis and recurrence of the info. The accompanying highlights should be considered to pick suitable actuator, for example, basic plans of piezo actuators; Stiffness; Load Capacity and Force Generation; Protection from Mechanical Damage; Power Requirements; Mounting Guidelines.

In everything, except the most elevated exactness mechanical positioners, hysteresis negligibly affects situating precision and repeatability, and much of the time, the impacts of backfire significantly outperform those of hysteresis. In any case, PEAs, which depend on material strain to deliver movement, can encounter hysteresis of 10 to 15 per cent of the directed development. Working piezo actuators in a shut circle framework can decrease or wipe out hysteresis impacts [22].

2 Modelling of Piezoelectric Actuator

The mechanical device model of PEA seems in Fig. 1a, b. The numerical model for a voltage governable PEA (PA) framework are often roughly delineated by the differential condition [23].

$$m\ddot{x} + k_f\dot{x} + k_g x + k_e(u(t)) = F \tag{1}$$

where $u(t)$ is that the time-varying motor terminal voltage, $x(t)$ is that the position, K_f is that the damping coefficient made by the motor K_g is that the mechanical stiffness, K_e is that the input control coefficient, m is that the effective mass, and F is that the system nonlinear disturbance.

Maybe the most acknowledged disrupting impacts present in the components of the PEA is the hysteresis wonder. The hysteresis, all around, deters high-precision development, and hysteresis assessment is in this way a crucial development towards the affirmation of first-class PEA structures. As explained in [24], the hysteresis is a rubbing like a wonder. The hysteresis is an erosion-like marvel. In [25], a spring coupled to an unadulterated Coulomb grating component is utilized to demonstrate the piezo drive with hysteresis. In [26], the Maxwell slip model is employed to deal with physical phenomenon. In any case, these current composing use the static pounding like physical phenomenon models that get merely a small amount of the physical phenomenon hysteresis models. This is the explanation [23] introduced a dynamical model for hysteresis, which is depicted by

$$\dot{z} = \alpha\dot{x} - \beta|\dot{x}|z - \gamma\dot{x}|z| \tag{2}$$

In this model, the hysteretic relationship between a state variable z and excitation x is given within the nonlinear mechanical wave assessment. The cut-off α controls the re-establishing power ampleness, and β and γ management the state of the physical phenomenon circle, notwithstanding, the dynamical model in [23] is clear. In [15], an all the in addition unsupportive dynamical granulating model from a frictional power was introduced. The model contains a state variable tending to the common redirection of versatile strands that are a depiction of the geography of the showing up at surfaces. The ensuing model shows by a protracted shot the larger a part of the acknowledged contact immediate, like physical phenomenon pounding slack, unsteady split away powers, and stick–slip advancement. It is sufficiently thorough to get dynamical hysteresis impacts. The dynamical model of the system is given under as

$$F = \alpha_0 z + \alpha_1\dot{z} + \alpha_2\dot{x} \tag{3}$$

with

$$\dot{z} = \dot{x} - \frac{|\dot{x}|}{h(\dot{x})}z,$$

$$h(\dot{x}) = \frac{F_c + (F_s - F_c)e^{-(x/x_0)^2}}{\alpha_0}$$

where F_c, F_s, $\dot{x}_s$, σ_0, σ_1 and σ_2 are positive constants that are typically unknown.

3 System Identification and Modelling

The PEA is considered as the mass-spring-damper system as shown in Fig. 1a and forces acting on a PEA is shown in Fig. 1b. The dynamics of PEAs demonstrates the behaviour of a distributed parameter system.

In practical, the principal recurrence of the reference signal is kept beneath the main reverberation recurrence of the PEA, with the goal that the mechanical reverberation over the primary reverberation recurrence of the stage can be disregarded. Along these lines, the dispersed boundaries nature of the PEA can be securely dismissed and the PEA model can be decreased to lumped boundaries framework particularly when PEAs are coordinated with flexure stage [27–29].

This part models the elements of the PEA when input voltage V_{a} is applied to it and the comparing yield of the framework is dislodging x of the example. When all is said in done, a second-request direct model has been used for portraying the elements of the PEAs. To a first estimation, PEA can be considered as a spring-mass-damper system as appeared in Fig. 1a. The whole mass of the PEA has lumped into one mass M_{a}, B_{a} and k_{a} individually model the primary damping and flexible conduct of PZT actuators. As indicated by Newton's subsequent law, the amount of power F_i following up on the PZT actuator should be equivalent to the result of mass and speeding up for example.

$$\sum_i F_i = M_{\mathrm{a}}\ddot{X} \tag{4}$$

As shown in Fig. 1b, there are four forces acting on the PZT actuator F_{d}, F_{s} and F_{a}. The damping force F_{d} can be assumed to be proportional to the velocity of the actuator.

$$F_{\mathrm{d}} = -B_{\mathrm{a}}\dot{x} \tag{5}$$

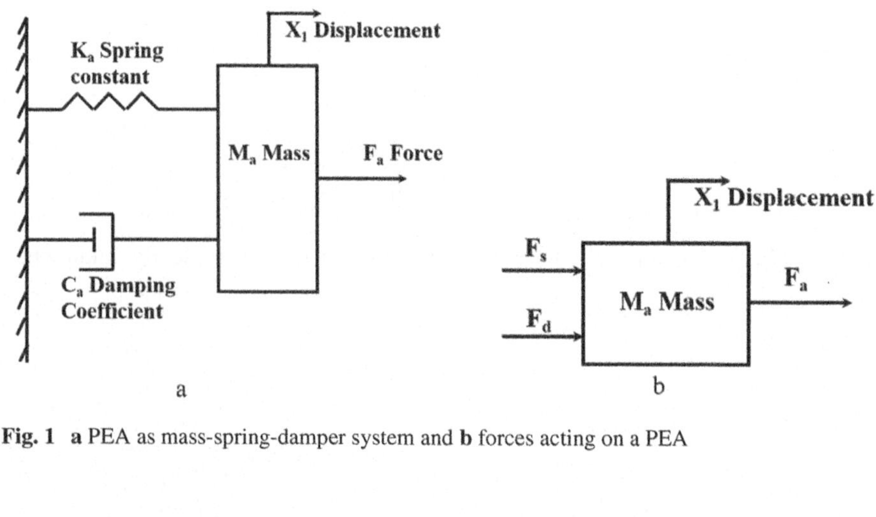

Fig. 1 **a** PEA as mass-spring-damper system and **b** forces acting on a PEA

The force F_s exerted by the spring is proportional to the displacement of the PZT actuator.

$$F_s = -K_a x \tag{6}$$

And as above mentioned, the induced or blocking force F_a is proportional to the applied voltage V_a volts.

$$F_a = N d k_a' V_a = K_A V_A \tag{7}$$

According to Newton's law, the equation of the force can be written as

$$\sum_1^3 F = M_a \ddot{x} = F_s + F_d + F_a \tag{8}$$

The relationship between applied voltage and resultant displacement of the sample is the second order differential equation of the PEA is given as

$$M_a \ddot{x} + c_a \dot{x} + k_a x = k_A V_A \tag{9}$$

$$\frac{X(s)}{F_a(s)} = \frac{1}{M_a S^2 + B_a s + k_a} \tag{10}$$

where F_a represents the force generated by the PZT actuator and x is the displacement of the stage/sample, V_a is the applied voltage, $\dot{x}$, $\ddot{x}$ are the first and second derivative of the x with respect to time (velocity and acceleration, respectively). Considering the single axis PEA as mass-spring-damper system. The nonlinear response of the PEA the hysteresis nonlinearity is modelled using backlash-like model and Duhem differential equation based model were discussing in the following chapters.

4 Hysteresis Modelling of Piezoelectric Actuators

4.1 Backlash-Like Hysteresis Model

The backlash-like model is based on the simplified Duhem model [30], which was also called SSSL model [31]. The backlash-like model is based on the differential equation is described as

$$\frac{dw}{dt} = \alpha \left| \frac{dv}{dt} \right| |cv - w| + B_1 \frac{dv}{dt} \tag{11}$$

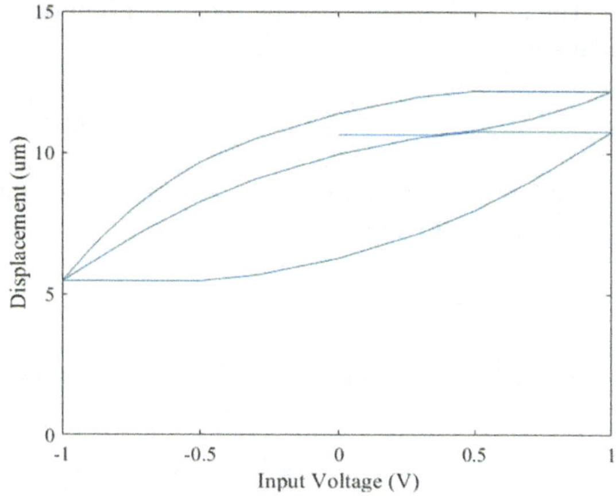

Fig. 2 Hysteresis response of PEA using backlash operator

where α, c and B_1 are constant, satisfying $c > B_1$. Following the derivation in [32], the solution of the model (11) is explicitly expressed as

$$w(t) = cv(t) + d(v(t)) \tag{12}$$

with

$$d(v(t)) := [w_0 - cv_0]e^{-\alpha(v-v_0)\operatorname{sgn}(\dot{v})} + e^{-\alpha v \operatorname{sgn}(\dot{v})} \int_{v_0}^{v} [B_1 - c]e^{\alpha\zeta v \operatorname{sgn}(\dot{v})} d\zeta \tag{13}$$

for steady, were and are the underlying condition, and sgn is the notable Sign capacity. The advantage for picking the backlash-like model is that the hysteresis nonlinearity is communicated as a straight capacity of the info signal in addition to a limited aggravation. Accordingly, the traditional strong control approaches can be used to manage the hysteresis without building the opposite of the hysteresis [33, 34]. The model of hysteresis using Backlash operator as shown in Fig. 2.

From graph, Fig. 2 shows the output of the PEA using Prandtl-Ishlinskii hysteresis model it results as the rate independent hysteresis or backlash operator based hysteresis of the PEAs.

4.2 Duhem Hysteresis Model

Duhem has an unequivocal differential articulation, through changing the boundary α, coefficients of f and g of Duhem model; diverse hysteresis qualities could be reflected; whilst recognizing the boundaries of the Duhem model precisely. The summed up Duhem model is the key for displaying hysteresis of PEAs [19]. The constituent equations of Duhem model are given by

$$\begin{aligned} \dot{x}(t) &= f[x(t), u(t)]g[\dot{u}(t)], x(0) = x_0, t \geq 0 \\ y(t) &= h[x(t), u(t)] \end{aligned} \tag{14}$$

where x is that the state, x_0 is that the initial state, u and y area unit, respectively, the input and out-put of the physical phenomenon, f, g and t are continuous functions. The parameters of the Duhem model are given as follows.

$$\begin{aligned} f[x(t), u(t) &= \left[-\gamma x(t) + \frac{\mathrm{d}\phi[u(t)]}{\mathrm{d}u(t)}, \gamma x(t) + \frac{\mathrm{d}\phi[u(t)]}{\mathrm{d}u(t)}\right]; \\ g[\dot{u}(t)] &= [\dot{u}_+(t), \dot{u}_-(t)]^T \\ y(t) &= x(t) + \mu u(t) + h_0[u(t)]; \end{aligned}$$

where $\dot{u}_+(t) \triangleq \max\{0, \dot{u}\}$ and $\dot{u}_-(t) \triangleq \max\{0, \dot{u}\}$. These parameters defines the shape of the hysteresis with respect to the input and output of the PEAs.

In order to describe the PEA, chosen the same functions $f(v)$ and $g(v)$ as those in [35].

$$f(v) = \begin{cases} a.v_s \text{ for } x > v_s \\ a.v \text{ for } |v| \leq v_s \\ -a.v_s \text{ for } v < -v_s \end{cases} \tag{15}$$

and

$$f(v) = \begin{cases} 0 \text{ for } x > v_s \\ \overline{\mathrm{b}} \text{ for } |v| \leq v_s \\ 0 \text{ for } v < -v_s \end{cases} \tag{16}$$

The yield changes its attributes when the information alters its course. Utilizing this model, rate-ward and rate autonomous hysteresis can be demonstrated. Each consistent u gives balance. The rate-subordinate implies that the hysteresis map relies upon the pace of the information and shape-subordinate implies that the hysteresis map relies upon the state of the information [36, 37].

From the graph, Fig. 3 results the input and output of the PEA. It depends on the memory of the history state. In order to determine the output of the system, both the input signal and the history state of the system should be considered.

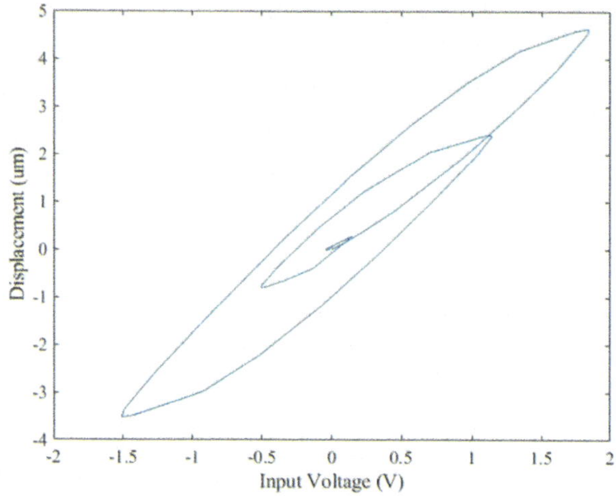

Fig. 3 Hysteresis response with respect to input and output of the PZT actuator

5 Control Algorithm

5.1 PID Controller

The PID controller is most common control algorithm using for the tracking performance of the piezoelectric actuators. The constituent equation of PID controller is given by

$$u(t) = K_{\mathrm{p}}e(t) + K_i \int e(t)\mathrm{d}t + K_{\mathrm{d}}\frac{\mathrm{d}e}{\mathrm{d}t} \tag{17}$$

where $u(t)$ is the PID control variable, K_{p} is the proportional gain, $e(t)$ is the error value, K_i is the integral gain, de is the change in error value, and dt is the change in time as shown in Fig. 4.

After controller implementation the nonlinear characteristic of model hysteresis can be minimized as shown in Fig. 5.

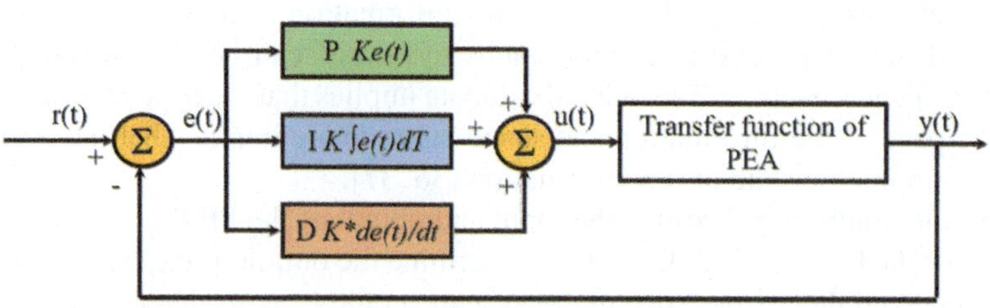

Fig. 4 PID controller with piezoelectric actuators

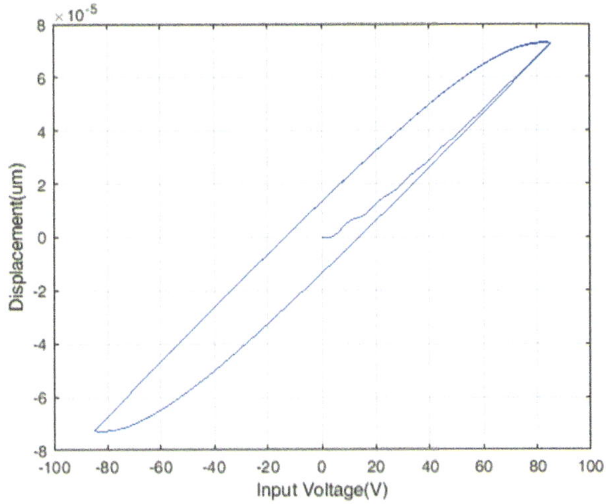

Fig. 5 Input voltage versus displacement

6 Error Analysis

The regulator plan relations, which depends on an exhibition file that thinks about the whole shut circle reaction, are created. Three execution lists will be viewed as, for example,

Integral of the absolute value of the error (IAE)

$$\text{IAE} = \int_{0}^{\infty} |e(t)|\mathrm{d}t \tag{18}$$

where the error signal $e(t)$ is the difference between the set point and the assessment. Notice that for the PID controller, where offset, the crucial given by Equation doesn't consolidate. In these cases, one can use a changed integrand, which replaces the slip-up by $y(\infty) - y(t)$, since this term approaches zero as t goes to infinity.

Integral of squared error (ISE)

$$\text{ISE} = \int_{0}^{\infty} [e(t)]^2\mathrm{d}t \tag{19}$$

Integral of time-weighted absolute error (ITAE)

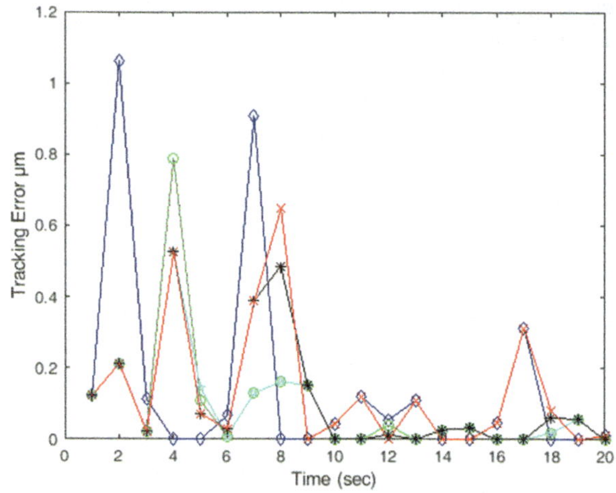

Fig. 6 Tracking error of piezoelectric actuator

$$\mathrm{ITAE} = \int_{0}^{\infty} t|e(t)|\mathrm{d}t \tag{20}$$

The ISE will rebuff the response that has colossal mix-ups, which generally speaking occur close to the beginning of a response because the bungle is squared. The ITAE will rebuff a response which has confused that persist with a long time. The IAE will treat all botches in a uniform manner; subsequently, it allows a greater deviation than ISE. When in doubt, ITAE is the supported essential slip-up standard since it achieves the most conservative controller settings. These relations rely upon the chief solicitation notwithstanding time-concede model, for instance, Eq. (20), and the ideal PID controller. Note that the ideal controller settings are different depending upon whether step responses to load or set point are considered. For load changes, the store and system move limits are relied upon to be unclear [37] as shown in Fig. 6.

7 Conclusion

The single axis PEA with its nonlinear characteristics of hysteresis is modelled using two different mathematical model such as Duhem and backlash-like. The hysteresis nonlinear characteristics is minimized using PID controller algorithm and tracking error of piezoelectric actuator were analyzed. The future work is to implement the modelled control strategy in real-time application.

References

1. Schitter G, Åström KJ, DeMartini BE, Thurner PJ, Turner KL, Hansma PK (2007) Design and modeling of a high-speed AFM-scanner. IEEE Trans Control Syst Technol 15:906–915
2. Song H, Vdovin G, Fraanje R, Schitter G, Verhaegen M (2009) Extracting hysteresis from nonlinear measurement of wavefront-sensorless adaptive optics system. Opt Lett 34:61–63
3. Wang X, Pommier-Budinger V, Reysset A, Gourinat Y (2014) Simultaneous compensation of hysteresis and creep in a single piezoelectric actuator by open-loop control for quasi-static space active optics applications. Control Eng Pract 33:48–62
4. Neugebauer R, Denkena B, Wegener K (2007) Mechatronic systems for machine tools. CIRP Ann 56:657–686
5. Li Y, Xu Q (2010) Development and assessment of a novel decoupled XY parallel micropositioning platform. IEEE/ASME Trans Mechatron 15:125–135
6. Tang H, Li Y (2015) Feedforward nonlinear PID control of a novel micromanipulator using Preisach hysteresis compensator. Robot Comput-Integr Manuf 34:124–132
7. Xu Q, Li Y, Xi N (2009) Design, fabrication, and visual servo control of an XY parallel micromanipulator with piezo-actuation. IEEE Trans Autom Sci Eng 6:710–719
8. Yang Y-L, Wei Y-D, Lou J-Q, Fu L, Fang S, Chen T-H (2018) Dynamic modeling and adaptive vibration suppression of a high-speed macro-micro manipulator. J Sound Vib 422:318–342
9. Li S, Feng Y, Cao S, Yanru J (2010) Design and analysis of a 3-DOF micromanipulator driven by piezoelectric actuators, presented at the international conference on mechanic automation and control engineering, Wuhan, China, 2010
10. Liu P, Yan P, Özbay H (2018) Design and trajectory tracking control of a piezoelectric nano-manipulator with actuator saturations. Mech Syst Signal Process 111:529–544
11. Devasia S, Eleftheriou E, Moheimani SOR (2007) A survey of control issues in nanopositioning. IEEE Trans Control Syst Technol 15:802–823
12. Choi SB, Han YM (2016) Piezoelectric actuators: control applications of smart materials. CRC Press
13. Habineza D, Rakotondrabe M, Le Gorrec Y (2014) Modeling, identification and feedforward control of multivariable hysteresis by combining Bouc-Wen equations and the inverse multiplicative structure. In: Proceedings of the American control conference, pp 4771–4777
14. Gharieb W (2009) Creep compensation in piezoelectric actuators for nanopositioning technology. Int J Nanomanuf 4:267–273
15. Yanfang Liu JS, Qi N (2013) Creep modeling and identification for piezoelectric actuators based on fractional-order system. Mechatronics 23:840–847
16. Peng JY, Chen XB (2012) Novel models for one-sided hysteresis of piezoelectric actuators. Mechatronics 22:757–765
17. Jiang H, Ji H, Qiu J, Chen Y (2010) A modified prandtl-ishlinskii model for modeling asymmetric hysteresis of piezoelectric actuators. IEEE Trans Ultrason Ferroelectr Freq Control 57:1200–1210
18. Gu GY, Zhu LM, Su CY (2014) Modeling and compensation of asymmetric Hysteresis nonlinearity for Piezoceramic actuators with a modified Prandtl-Ishlinskii model. IEEE Trans Industr Electron 61:1583–1595
19. Wang G, Chen G (2017) Identification of piezoelectric hysteresis by a novel Duhem model based neural network. Sens Actuators A 264:282–288
20. Zhu W, Rui X-T (2016) Hysteresis modeling and displacement control of piezoelectric actuators with the frequency-dependent behavior using a generalized Bouc-Wen model. Precis Eng 43:299–307
21. Rodriguez-Fortun JM, Orus J, Alfonso J, Buil F, Castellanos JA (2011) Hysteresis in piezoelectric actuators: modeling and compensation, presented at the 18th world congress the international federation of automatic control, Milano, Italy, 2011
22. Alem IISF, Sheikholeslam F (2017) Adaptive sliding mode control of hysteresis in piezoelectric actuators

23. Low TS, Guo W (1995) Modeling of a three-layer piezoelectric bimorph beam with hysteresis. J Microelectromech Syst 4:230–237
24. Damjanovic D, Setter N, Robert G (200) Separation of nonlinear and friction-like contributions to the piezoelectric hysteresis. In: Proceedings of the 12th IEEE international symposium on applications of ferroelectrics Honolulu, HI, USA, 2000, pp 699–702
25. Goldfarb M, Celanovic N (1997) For control of micromanipulation. In: IEEE control systems, pp 69–79
26. Oh J, Choi GS, Choi GH (1999) Repetitive tracking control of a coarse-fine actuator. In: IEEE/ASME international conference on advanced intelligent mechatronics, Atlanta, GA, USA, 1999, pp 335–340
27. Hwang EJ, Min KS, Song SH, Ahn IH, Choi WC (2007) Optimal design of a flexure-hinge precision stage with a lever. J Mech Sci Technol 21:616–623
28. Xu Q (2012) New flexure parallel-kinematic micropositioning system with large workspace. IEEE Trans Rob 28:478–491
29. Li L, Li C-X, Gu G, Zhu L-M (2017) Positive acceleration, velocity and position feedback based damping control approach for piezo-actuated nanopositioning stages. Mechatronics 47:97–104
30. Salapaka SM, Salapaka MV (2008) Scanning probe microscopy. IEEE Control Syst Mag 28:65–83
31. Ekanayake DBI, Iyer RV (2008) Study of a play-like operator. Phys B 403:456–459
32. Su YSC, Svoboda J, Leung TP (2000) Robust adaptive control of a class of nonlinear systems with unknown backlash-like hysteresis. IEEE Trans Autom Control 45:2427–2432
33. Gu LZG, Su C, Ding H (2013) Motion control of piezoelectric positioning stages: modeling, controller design, and experimental evaluation. IEEE/ASME Trans Mechatron 18:1459–1471
34. Wen C, Zhang Y, Zhou J (2004) Adaptive backstepping control of a class of uncertain nonlinear systems with unknown backlash-like hysteresis. IEEE Trans Autom Control 49:1751–1759
35. Banning R, Adriaens HJMTA, Koops RK (2001) State-space analysis and identification for a class of hysteretic systems. Automatica 37:1883–1892
36. Zhou M, Wang J (2013) Research on hysteresis of piezoceramic actuator based on the Duhem model. Sci World J:1–6
37. Sabarianand DV, Karthikeyan P, Muthuramalingam T (2020) A review on control strategies for compensation of hysteresis and creep on piezoelectric actuators based micro systems. Mech Syst Signal Process 140:1–17

Attitude and Vibration Control of a Solar Paneled Satellite Using Quantitative Feedback Theory Based Robust Controller

Aritra Sinha, Sandipan Prasad Chakravarty, Pratik Dutta, and Prasanta Roy

Abstract In this paper, a quantitative feedback theory based robust controller is proposed to control the vibration and attitude of a solar paneled satellite. The proposed controller provides reduced sensitivity, robust stability, robust tracking with respect to parametric uncertainty. A two-degrees-of-freedom (2-DOF) single input multiple output control structure consisting of a feedback controller and a pre-filter is used to control the system. The controller provides the required sensitivity, stability, disturbance rejection and partial tracking. The pre-filter is responsible to meet up the residual part of tracking specifications explicitly. Lastly, effectiveness of the controller is validated by time domain response in Simulink environment.

Keywords Quantitative feedback theory · Robust control · Single input Multiple output System (SIMO) · Satellite control · 2-DOF

1 Introduction

With a rapid progress in space technology, satellites are considered as an essential application due to it's wide range of applicability in telecommunication, data transmission, spatial observations, navigation, earth monitoring and so on. Accurate coordinate orientation and high-precision control in angular velocity is required to stabilize and orient the satellite in desired spatial direction to carry out the objec-

A. Sinha (✉) · S. P. Chakravarty · P. Dutta · P. Roy
Department of Electrical Engineering, National Institute of Technology Silchar, Silchar, Assam, India
e-mail: aritra_pg@ee.nits.ac.in

S. P. Chakravarty
e-mail: sandipan_rs@ee.nits.ac.in

P. Dutta
e-mail: pratik_pg@ee.nits.ac.in

P. Roy
e-mail: prasanta@ee.nits.ac.in

S. Suhag et al. (eds.), *Control and Measurement Applications for Smart Grid*, Lecture Notes in Electrical Engineering 822,
https://doi.org/10.1007/978-981-16-7664-2_13

tives under different functional conditions. Along with this, most of the high and medium attitudes satellites are carrying solar panels. The solar panels with it's large co-axial length creating a considerable vibrating zone at the time of orbital motion or maneuver of satellite. Combination of this solar panels with rigid satellite body is responsible for coupled dynamics of center body during it's re-orientation time. With this amount of complexity, high attitude, speed, disturbance torques due to inertia or deformation, instantaneous position changing action make a huge impact on the uncertainty of the system model. This huge uncertainty in system model parameters represent itself as an ideal problem to imply robust control theory to perform several control tasks simultaneously.

1.1 A Brief Introduction to QFT

Horowitz [1] first proposed a frequency domain approach as a continuation of the pioneering work done by H. W. Bode. It was further upgraded to today's form known as Quantitative Feedback Theory (QFT) during the last few decades [2, 3]. Here, frequency response of an open-loop system is represented in Nichol's Chart for achieving several performance norms simultaneously notwithstanding of parametric unpredictability. Now to achieve closed loop specifications, they are mapped into appropriate bounds on a nominal open loop transfer function. It is followed by reshaping of those to find out the optimum position of nominal open loop transfer function as per design objectives. Diminution of the expense of feedback is a major advantage of QFT as it furnish a equitable trade-off betwixt the doable performance and the loop gain (expense of feedback) in pre-defined frequency range [4]. Thus, QFT can incorporate a control scheme to stabilize the satellite in it's respective orbit with the orientation in any spatial direction irrespective of disturbances and different mode of vibrations.

1.2 Primary Objectives

- To design a QFT based robust 2-DOF control architecture to regulate the angular motion of satellite.
- To perform stability, sensitivity and tracking specification simultaneously in defined frequency range.
- To attenuate the solar panel vibration with the same controller as SIMO model.

1.3 Literature Survey

A short literature study is discussed here to point out some major reported works on application of different control propositions in satellites. A variable structure model reference control with simplified flexible satellite structure first developed by Zeng et al. [5]. Feedback linearization with observer based estimation method was employed for model uncertain parameters in same control structure like previous one by Singh et al. [6]. Application of same control structure with assumed modes method based discretized plant model was carried out by Hu et al. [7]. Sliding mode control model for flexible satellite without considering any torsional disturbance was developed by Bang et al.[8]. Adaptive sliding mode controller with piezoelectric actuator was used to attenuate the vibration with attitude control simultaneously in [9]. Azadi et al. [10] applied same control mechanism on Lagrange-Rayleigh-Ritz theorem based discretized flexible plant model. L_1 adaptive controller with feedback linearization was employed in same structure like [6] in [11]. Assumed mode technique based discretized satellite plant has controlled with a nonlinear PD controller in [12]. Two parallel PID controller was employed to control vibration and attitude separately with finite time element based discrete plant model in [13]. An adaptive prediction based back-stepping controller was developed for satellite vibration control by Huo et al. [14]. Event triggered strategy based distributed adaptive flexible satellite control topology was proposed in [15]. A linear time varying model predictive control algorithm for flexible satellite attitude control has proposed in [16]. Fractional order PD sliding mode controller with it's integer order counter part has developed for a nonlinear flexible satellite model by Chakrabarty et al. [17]. A constrained magnetic linear quadratic regulator based control scheme has developed by converting time-variant representation of satellite system to a time in-variant structure in [18].

2 Design of the QFT Based Controller

2.1 Mathematical Modelling

Aerodynamics, gravitational and propulsive forces as well as different angular motions play the most influential role on performance of attitude control system of satellite having solar panels. A typical single input multiple output closed loop 2-DOF control architecture along with corresponding model description is proposed here in Fig. 1 to control the angular motion of central body ($\theta_C(t)$) and deformation of solar panel ($y_P(t)$) at the time of rotation.

The parameters description of satellite system model are given below.

- J_C, the moment of inertia (kg m^2) of satellite central body.
- T_M, torque (Nm) about the axis of rotation.
- M, mass (kg) of the solar panel consider as point mass at the end of each panel.

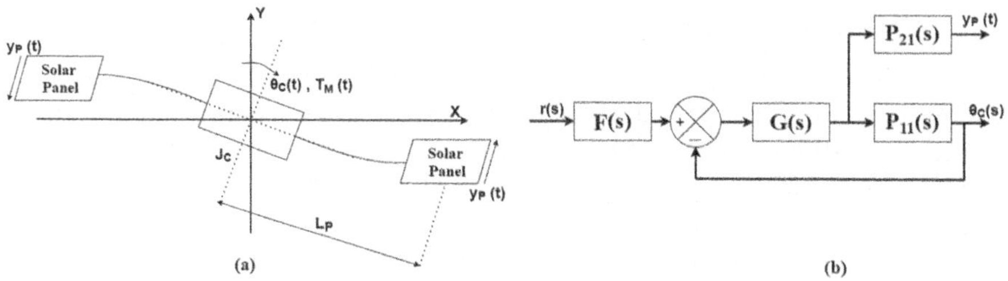

Fig. 1 **a** Model of satellite with solar panels. **b** 2-DOF feedback single input multiple output control structure for satellite with solar panels

- L_P, length (m) of each solar panel from satellite central body.
- K_P, solar panel stiffness coefficient (N/m).
- $J_T = J_C + 2ML_P^2$, total moment of inertia (kg m^2) of the system.
- B_P, solar panel damping coefficient (Ns/m).
- t, time in second (s).

The controllable parameters $(\theta_C(t))$ and solar panel deformation $(y_P(t))$ can be considered as generalized coordinates of total system body. As per the model define in [19], we have derived the state space representation of the system.

$$\dot{X} = AX + BU$$
$$Y = CX$$

where,

$$A = \begin{bmatrix} 0 & \frac{2L_P B_P}{J_T} & 0 & \frac{2L_P K_P}{J_T} \\ 0 & \frac{-B_P(J_C+2ML_P^2)}{MJ_C} & 0 & \frac{-K_P(J_C+2ML_P^2)}{MJ_C} \\ 1 & 0 & 0 & 0 \\ 0 & 1 & 0 & 0 \end{bmatrix}; B = \begin{bmatrix} \frac{1}{J_0} & 0 \\ \frac{-L_P}{J_0} & 0 \\ 0 & 0 \\ 0 & 0 \end{bmatrix};$$
$$C = \begin{bmatrix} 0\,0\,1\,0 \\ 0\,0\,0\,1 \end{bmatrix} \tag{1}$$
$$X = \begin{bmatrix} \dot{\theta}_C \\ \dot{y}_P \\ \theta_C \\ y_P \end{bmatrix}; U = \begin{bmatrix} T_M \\ 0 \end{bmatrix} \& Y = \begin{bmatrix} \theta_C \\ y_P \end{bmatrix}$$

Then, we have established the relation between Y and U as,

$$Y(s) = P(s)U(s)$$

Table 1 Parameter change range of transfer function $p_{11}(s)$ and $p_{21}(s)$

Parameters	Varying range	Grid point
J_C	432–720	72
B_P	0.24–0.48	3

where,

$$P(s) = C(sI - A)^{-1}B = \begin{bmatrix} p_{11}(s) & p_{12}(s) \\ p_{21}(s) & p_{22}(s) \end{bmatrix} \quad (2)$$

The transfer function $p_{11}(s)$ represents the relation between rotational torque and angular velocity of the satellite body.

$$\frac{\theta_C(s)}{T_M(s)} = p_{11}(s) = \left(\frac{1}{J_C}\right) \frac{(s^2+(B_P/M)s+(K_P/M))}{s^2(s^2+(B_P J_T/(MJ_C))s+(K_P J_T/(MJ_C)))} \quad (3)$$

Similarly $p_{21}(s)$, the relation between deviation of solar panel y_P and rotational torque of the system can be defined as;

$$\frac{y_P(s)}{T_M(s)} = p_{21}(s) = \left(\frac{1}{J_C}\right) \frac{(-L_P)}{(s^2+(B_P J_T/(MJ_C))s+(K_P J_T/(MJ_C)))} \quad (4)$$

and remaining $p_{12}(s) = p_{22}(s) = 0$.

Basically among these total six parameters of Eqs. 3 and 4, J_C and B_P play the most influential role regarding aerodynamic complexity due to any type of disturbances. So we consider those parameters within interval type of plant uncertainty defined in Table 1. Rest of the parameters have been defined as per [19]. As a result, the modified form of those transfer functions becomes look like;

$$\frac{\theta_C(s)}{T_M(s)} = p_{11}(s) = \left(\frac{1}{J_C}\right) \frac{(s^2+(B_P/20)s+16)}{s^2(s^2+(B_P(J_C+160)/(20J_C))s+(16(J_C+160)/J_C))} \quad (5)$$

$$\frac{y_P(s)}{T_M(s)} = p_{21}(s) = \left(\frac{1}{J_C}\right) \frac{-2}{(s^2+(B_P(J_C+160)/(20J_C))s+(16(J_C+160)/J_C))} \quad (6)$$

2.2 *Performance Criteria*

Performance descriptions are defined for the SIMO controlled plant regarding stability, input, output and external disturbance, as well as tracking considering $G(s)$ as controller and $F(s)$ as pre-filter, respectively.

1. Stability Specification:

$$\left|\frac{p_{11}(s)G(s)}{1 + p_{11}(s)G(s)}\right| \le 2, \ \forall\, \omega \in [0.03\ 5]\ \text{(rad/s)} \quad (7)$$

2. Input Disturbance Rejection Specification:

$$\left|\frac{p_{11}(s)}{1+p_{11}(s)G(s)}\right| \leq \frac{s}{s+0.5}, \ \forall\, \omega \in [0.03\ 0.1]\ \text{(rad/s)} \tag{8}$$

3. Output Disturbance Rejection Specification:

$$\left|\frac{1}{1+p_{11}(s)G(s)}\right| \leq \frac{s}{s+0.5}, \ \forall\, \omega \in [1\ 15]\ \text{(rad/s)} \tag{9}$$

4. External Disturbance (over $y_P(s)$ through $p_{21}(s)G(s)$) Rejection Specification:

$$\left|\frac{p_{21}(s)G(s)}{1+p_{11}(s)G(s)}\right| \leq 0.20, \ \forall\, \omega \in [1\ 5]\ \text{(rad/s)} \tag{10}$$

5. Reference Tracking Specification:

$$\delta_l(s) < \left|\frac{F(s)p_{11}(s)G(s)}{1+p_{11}(s)G(s)}\right| \leq \delta_u(s), \ \forall\, \omega \in [0.01\ 0.1]\ \text{(rad/s)}$$

where [20]

$$\delta_l(s) = \frac{0.0169}{s^2+0.26\,s+0.0169}, \ \delta_u(s) = \frac{10\,s+1}{39\,s^2+10\,s+1} \tag{11}$$

The QFT based SIMO controller with 2-DOF architecture is designed in the next two subsections for the uncertain plant described in (5), (6) to meet up the objectives defined in (7)–(11).

2.3 Controller Design

The selection of nominal plant as well as calculation of plant templates on the basis of design requirements, transient specification should be within closest proximity, in our desired frequency points. We can observe from 5 that system oscillates forever with initial zero damping due to double integrator in the transfer function. Also, it is notable from Fig. 2 that there is a tremendous resonance zone around $\omega = 4.5$ to $\omega = 5$ rad/s for both $p_{11}(s)$ and $p_{21}(s)$ respectively. It can be varied with the parametric uncertainty of transfer function and turned the system into verge of instability. Basically this resonance zone restrict the bandwidth of control structure. With all of these considerations, 108th plant has been chosen as nominal plant define in (12), (13) among total 216 plants that generated as per parametric uncertainty of Table 1.

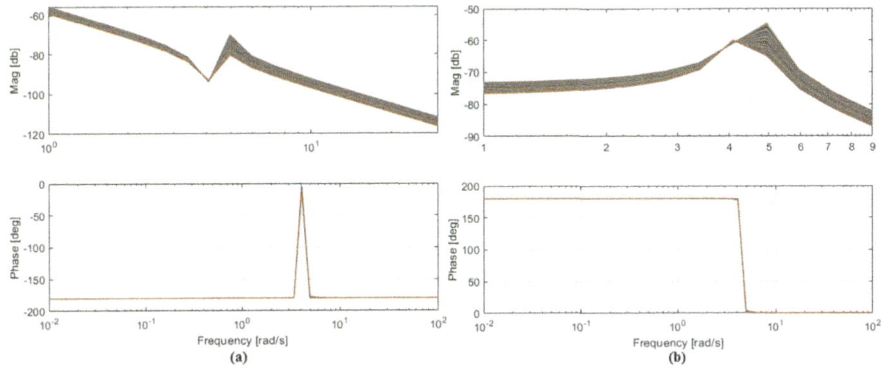

Fig. 2 **a** Bode diagram of $p_{11}(s)$ for all possible scenario, **b** bode diagram of $p_{21}(s)$ for all possible scenario

$$p_{11,n}(s) = \frac{s^2 + 0.024\,s + 16}{555.7\,s^4 + 17.18\,s^3 + 11450\,s^2} \tag{12}$$

$$p_{21,n}(s) = \frac{-2}{555.7\,s^2 + 17.18\,s + 11450} \tag{13}$$

The plant templates are formed by plotting the open loop magnitude-phase of all the possible plants as per uncertainty in Nichol's chart for required frequency range. It is followed by plotting of different controllable closed loop gain specifications for nominal plant by moving corresponding templates along the periphery of those specifications. These plots are popularly known as different type of bounds like stability bounds, disturbance bound, robust tracking bounds in QFT. All the bounds are then grouped together to find out common accessible zone for respective frequency. After that all the Nichol's plot of compensated plant are reshaped to lie in the region defined by those convergence bounds to meet all the closed loop specifications for respective frequency. It is called loop-shaping process of the controller. The modified controller transfer function is given below (14):

$$G(s) = \frac{250(s + 0.1)\left(s^2 + 0.6552\,s + 21.9\right)}{(s + 1)(s + 4.68)^2} \tag{14}$$

2.4 *Design of Pre-filter*

In 2-DOF control architecture, controller holds the responsibility of stability, sensitivity, output or external disturbances rejection explicitly. Pre-filter plays the role to improve the plant performance by keeping the magnitude of closed loop frequency

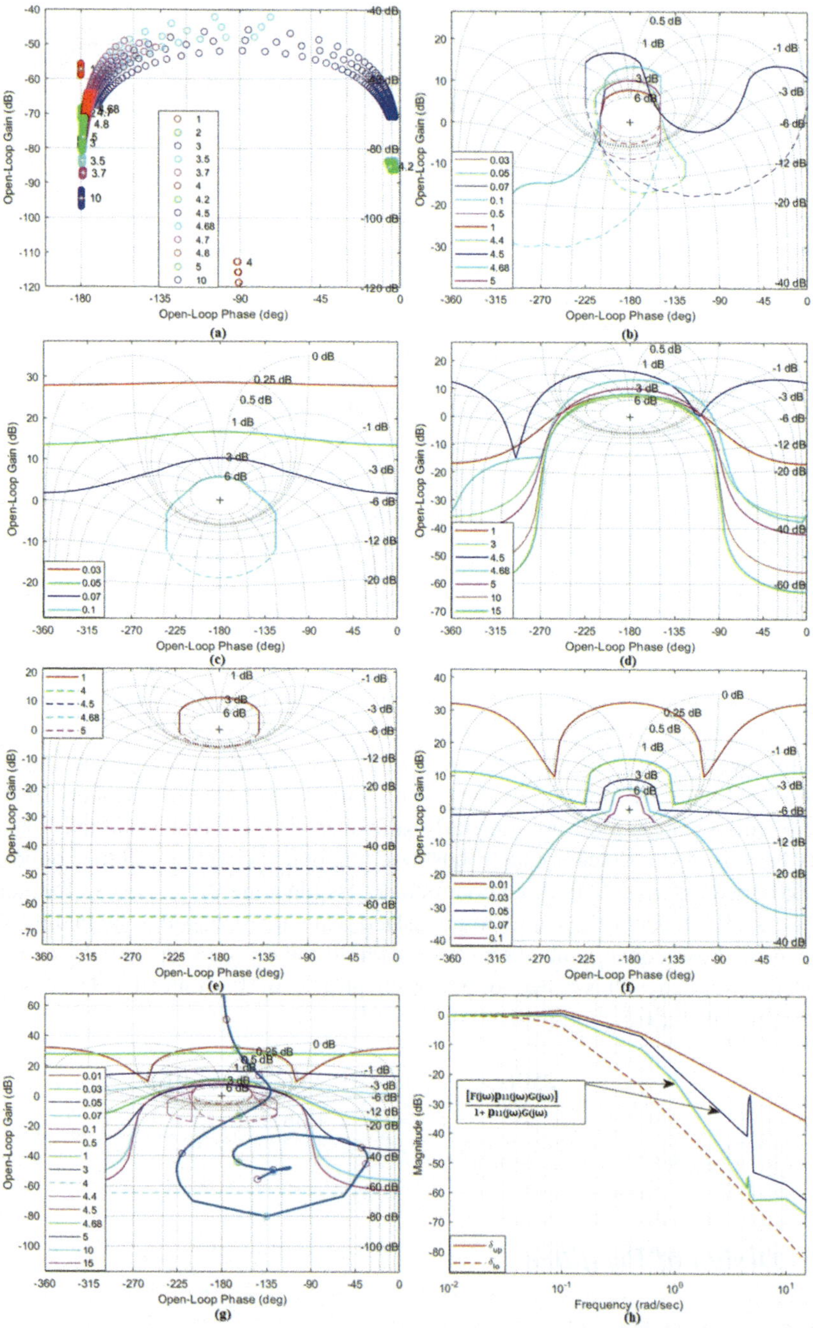

Fig. 3 **a** Plant templates of $p_{11}(s)$. **b** Stability bounds, **c** input disturbance rejection bounds, **d** output disturbance rejection bounds, **e** external disturbance rejection bounds, **f** tracking bounds, **g** controller loop shaping, **h** frequency response of pre-filter compensated system

response within the specified limit defined by robust tracking specification. The pre-filter transfer function (15) is synthesized as per [21] to meet up the requirements defined in Fig. 3f. The frequency response (magnitude) of closed loop control system with compensated pre-filter is shown in Fig. 3h.

$$F(s) = \frac{0.445(s + 0.2975)}{(s + 0.1325)} \tag{15}$$

3 Result Analysis and Discussion

The performance of single input multiple output controlled plant has been verified through numerical computation in MATLAB Simulink environment. Plant $p_{11}(s)$ is subjected to a variable step input at $t = 1$ s and $t = 150$ s along with a step type disturbance at $t = 85$ s and $t = 245$ s, respectively. The nominal plant response along with other selected plants within uncertainty of $p_{11}(s)$ are shown in Fig. 4a. The plants are settled before 30 s and 170 s, respectively, with negligible overshoot. The effect of disturbance is nullified within next 30 s for both of the cases. It indicates that proposed 2-DOF controller along with pre-filter can be given a satisfactory performance in terms of robustness, set point tracking, disturbance rejection with acceptable transients for plant $p_{11}(s)$ irrespective of plant uncertainty.

In addition, the same controller also plays a magnificent role to reduce the oscillation of solar panel plant model ($p_{21}(s)$) under different circumstances. Here, we provide a unitary step disturbance input in $p_{11}(s)G(s)$ closed loop block just after the $p_{11}(s)$ at $t = 85$ s. The disturbance that make the entry at the output of the $p_{11}(s)$ plant, travels through out the control loop before going to the open loop plant of $p_{21}(s)$. This is referred as external disturbance for $p_{21}(s)$ as per (10). Fig. 4b clearly indicates that the oscillations of solar panel significantly reduced to near about zero irrespective of external disturbance due to that controller *G*(*s*).

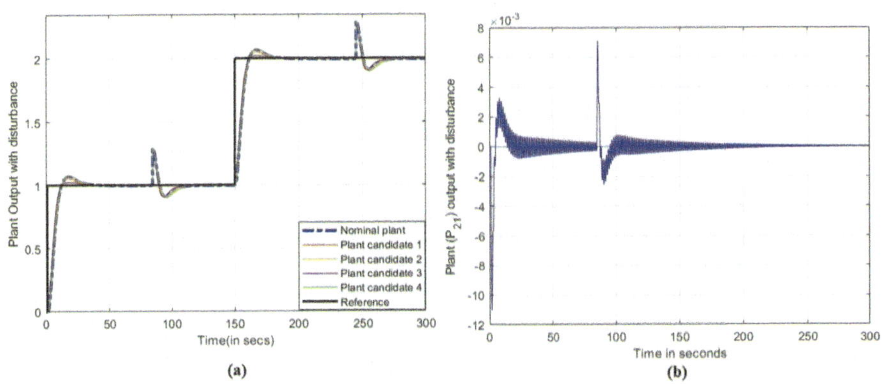

Fig. 4 **a** Plant $p_{11}(s)$ output response with disturbance rejection, **b** solar panel oscillations $y_P(t)$

4 Conclusion and Future Scope

Here, a Quantitative Feedback Theory (QFT) based 2-DOF control architecture is used for controlling the attitude and vibration of a satellite mounted with solar panel. The aerodynamic complexity of the system are dealt quite satisfactorily regarding multiple performance criterion consequently irrespective of model uncertainty and structure complexity of the system. The effectuate of 2-DOF proposed control architecture is validated through numerical simulation.

However, the quantification of robustness in case of unstructured uncertainty as well as incorporation of model following adaptive pre-filter may be considered as future scope of this work.

References

1. Horowitz IM (1963) Synthesis of feedback systems. Academic Press, New York
2. Horowitz IM, Sidi M (1972) Synthesis of feedback systems with large plant ignorance for prescribed time-domain tolerances. Int J Control 16(2):287–309
3. Horowitz IM (1992) Quantitative feedback theory (QFT). QFT Publications, Colorado, 4470 Grinnell Ave., Boulder
4. Sanz MG (2014) Encyclopedia of Systems and Control. Springer, Cleveland, USA
5. Zeng Y, Araujo AD, Singh SN (1999) Output feedback variable structure adaptive control of a flexible spacecraft. Acta Astronaut. 44(1):11–22
6. Singh SN, Zhang R (2004) Adaptive output feedback control of spacecraft with flexible appendages by modeling error compensation. Acta Astronaut. 54(4):229–243
7. Hu Q, Shi P, Gao H (2007) Adaptive variable structure and commanding shaped vibration control of flexible spacecraft. J. Guid. Contr. Dyn. 30(3):804–815
8. Bang H, Ha CK, Kim JH (2005) Flexible spacecraft attitude maneuver by application of sliding mode control. Acta Astronaut. 57(11):841–850
9. Hu Q (2009) A composite control scheme for attitude maneuvering and elastic mode stabilization of flexible spacecraft with measurable output feedback. Aerosp Sci Technol 13(2):81–91
10. Azadi M, Eghtesad M, Fazelzadeh SA, Azadi E (2015) Dynamics and control of a smart flexible satellite moving in an orbit. Multibody Syst. Dyn. 35:1–23
11. Lee KW, Singh SN (2012) L1 adaptive control of flexible spacecraft despite disturbances. Acta Astronaut. 80:24–35
12. Baghi B, Kabganian M, Nadafi R, Arabi E (2016) Three-axis attitude stabilization of a flexible satellite using non-linear PD controller. Trans Inst Meas Contr:1–15
13. da Fonseca IM, Rade DA, Goes LCS, Sales TP (2017) Attitude and vibration control of a satellite containing flexible solar arrays by using reaction wheels, and piezoelectric transducers as sensors and actuators. Acta Astronaut 139:357–366
14. Huo J, Meng T, Song R, Jin Z (2018) Adaptive prediction back-stepping attitude control for liquid-filled micro-satellite with flexible appendages. Acta Astronaut 152:557–566
15. Guo F, Zhang T, Zhang F, Gao L, Wang Z, Zhang S (2020) Event–triggered coordinated attitude control for satellite formation under switching topology. Adv Control Appl
16. Golzari A, Pishkenari HN, Salarieh H, Abdollahi T (2020) Quaternion based linear time-varying model predictive attitude control for satellites with two reaction wheels. Aerosp Sci Technol 98
17. Chakrabarti D, Narayanasamy. PD and PD^{β} based sliding mode control algorithms with modified reaching law for satellite attitude maneuver. Adv Space Res. https://doi.org/10.1016/j.asr.2019.11.005

18. Arefkhani H, Sadati SH, Shahravi M (2020) Satellite attitude control using a novel constrained magnetic linear quadratic regulator. Control Eng Practice 101
19. Sidi M (1997) Spacecraft dynamics and control; a practical engineering approach. Cambridge University Press, UK
20. García-Sanz M. Robust control engineering: practical QFT solutions. CRC Press, New York
21. Borghesani C, Yossi Chait OY (2003) The QFT frequency domain control design toolbox for use with MATLAB users guide. Terasoft, San Diego

Mathematical Model of the Disease Psoriasis: An Optimal Control-Based Approach

Ramashis Banerjee and Raj Kumar Biswas

Abstract In this article, mathematical model of Psoriasis is used and its control has been shown. The diseases, which are called autoimmune is generally occurred when there is an existence of non-uniform immune response of the human body in contradiction with the substances as well as the tissues generally presents in the human body. The disease Psoriasis is classified as an autoimmune chronic skin disease which is modified by the T-Cells arbitrated hyper-proliferation of the epidermal Keratinocytes. The important role is played by Dendritic as well as CD8+ T-Cells for the occurrence of Psoriasis. The mathematical model of Psoriasis, which is used here consisting of T-Cells, Dendritic Cells, CD8+ T-Cells and Keratinocyte cell populations and fractional differential equation (FDE) has been introduced in the mathematical model to analyse and observe the effects of memory in the biological system. After using fractional differential equation (FDE) in the mathematical model, it is tried to explore the memory effects which are associated with the biological system and the non-local behaviour of fractional derivative helps to find the position of Keratinocyte cell. So, it can be said that the dynamics of the disease Psoriasis may be predicted in a better way by using fractional differential equation (FDE) rather than its ordinary differential equation (ODE) counterpart. At last, drug has been introduced in the system by formulating fractional optimal control problem (FOCP) and it is seen that the disease Psoriasis has been controlled as drug effect restricts the interaction between T-Cells and Keratinocytes, which has been seen clearly in the numerical simulation.

Keywords Psoriasis · pMHC · Fractional differential equation (FDE) · Memory · Non-local · Fractional optimal control problem (FOCP)

1 Introduction

It is seen that in human body the immune system plays an important as well as imperative responsibility for the spread and progression of Psoriasis. For spreading

---

R. Banerjee (✉) · R. K. Biswas
National Institute of Technology Silchar, Silchar, Assam 788010, India

S. Suhag et al. (eds.), *Control and Measurement Applications for Smart Grid*,
Lecture Notes in Electrical Engineering 822,
https://doi.org/10.1007/978-981-16-7664-2_14

the disease T-Cells as well as Keratinocytes accomplish a crucial and pathogenic role. Dendritic Cells (DCs) are also responsible for performing a crucial function in the matter of Psoriatic pathogenesis. It supports when it is able to attend antigens by complex II molecules of principal major histo compatibility (pMHC). The concentration of T-Cells and Dendritic Cells are enlarged and they are observed in the dermal phase as well as epidermal phase of Psoriatic lesions [1–5]. CD8+ T-Cell and its population is considered as an inducers for the progression and the spread of Psoriasis. TNF-α as inflammatory Cytokines has a major pathogenic task in Psoriasis [6, 7]. Immune T-Cells, which are the cause of an uneven proliferation for the matter of few false signalling. Due to the method of the stimulation as well as the mutual accumulation of T-Cells, TNF-α and IFN-γ are free to go. At the same time the method of proliferation of Keratinocyte cell is inspired in the region of epidermal layer. The crucial and important thickening of the epidermal level is observed due to Psoriatic plaques.

The concept of fractional calculus came into picture three hundred years ago when the question was raised by Leibniz in 1695 [8]. Fractional differential equation is treated as an alternative to nonlinear differential equations [9]. On the other hand FDEs are also help to implement the effect of memory, which is a hereditary characteristic and presents in most of the biological systems. The mostly used fractional derivatives like Caputo and Riemann–Liouville derivatives signifies the memory effect lies between a derivative of integer order and a power of time [10, 11]. The main concept lies in biological systems as fractals that they are directly related to fractional differential equations (FDEs) [12, 13]. It is seen that any cytobiological system has the ability to show the memory effects so they are best expressed by means of FDEs.

Here in this article the disease Psoriasis has been studied deeply to understand the cyto biological aspects. The mathematical model of the disease Psoriasis was developed first by Roy and Bhadra [13]. After their formulation other researchers came up with the mathematical model of the disease Psoriasis and analysed it [14–18]. But it is observed in literature that in large number of cases, the system under study has been modelled through ordinary differential equation (ODE). Roy et al., showed the inclusion of FDEs in the mathematical model of the disease Psoriasis for the first time [19, 20]. Here in this article fractional optimal control problem has been formulated for the treatment of the disease Psoriasis. Here a set of FDEs are extensively used to explore and investigate the dynamics of the disease Psoriasis in presence of CD4+ T-Cells as well as CD8+ T-Cells. In [18] a mathematical model has been formulated after considering the effect of memory which generally presents in a biological system like the disease Psoriasis, it is seen that if Keratinocytes is produced in an excess amount then it helps to restrict the casual effects of the disease psoriasis. This whole process of restricting the disease effects can be well equipped with the help of optimal control strategy through the inclusion of optimal drug in the mathematical modelling of the disease psoriasis, modelled by using FDEs.

The remaining of the article is arranged as mentioned. In Sect. 2, different definitions of fractional derivatives have been discussed. In Sect. 3, mathematical model of the disease psoriasis has been described. In Sect. 4, fractional optimal control problem (FOCP) has been formulated to show the drug effect on the disease. In fifth Sect. 5, simulation results have been shown and the article ends with the conclusion which is described in Sect. 6.

2 Different Definitions of Fractional Derivative

Here different definitions of fractional derivative is mentioned which will be used to model the dynamics of the disease psoriasis. Caputo and Riemann–Liouville derivatives are mentioned here [21]. Here left and right-sided derivatives are also mentioned.

Left-sided Caputo's derivative is written as

$$ {}_{a}^{C}D_{t}^{\alpha} f(t) = \frac{1}{\Gamma(n-\alpha)} \int_{a}^{t} \frac{f^{(n)}(s)}{(t-s)^{\alpha-n+1}} \mathrm{d}s \tag{1} $$

Right-sided Caputo's derivative is written as

$$ {}_{t}^{C}D_{b}^{\alpha} f(t) = \frac{(-1)^{n}}{\Gamma(n-\alpha)} \int_{t}^{b} \frac{f^{(n)}(s)}{(t-s)^{\alpha-n+1}} \mathrm{d}s \tag{2} $$

where the order of the derivative is α and the condition $n-1 < \alpha < n$ follows, here n is an integer and Γ is called the gamma function.

Left-sided Riemann–Liouville derivative is written as

$$ {}_{a}^{RL}D_{t}^{\alpha} f(t) = \frac{1}{\Gamma(n-\alpha)} \frac{\mathrm{d}^{n}}{\mathrm{d}t^{n}} \int_{a}^{t} \frac{f(s)}{(t-s)^{\alpha-n+1}} \mathrm{d}s \tag{3} $$

Right-sided Riemann–Liouville is written as

$$ {}_{t}^{RL}D_{b}^{\alpha} f(t) = \frac{(-1)^{n}}{\Gamma(n-\alpha)} \frac{\mathrm{d}^{n}}{\mathrm{d}t^{n}} \int_{t}^{b} \frac{f(s)}{(t-s)^{\alpha-n+1}} \mathrm{d}s \tag{4} $$

where the order of the derivative is α and the condition $n-1 < \alpha < n$ follows, here n is an integer, Γ is called the gamma function and $a > 0, b > 0$ are the constants. Throughout the article ${}_{0}^{C}D_{t}^{\alpha}$ is used to signify left-sided Caputo's derivative operator.

3 Mathematical Model

In this present work, a fractional order mathematical model of the disease psoriasis is used which was formulated by Roy and Bhadra [13]. The model consists of four populations namely CD4+ T-cells, Dendritic Cell (DC), CD8+ T-cells and Keratinocytes. It is seen from the research that the T-Cells and the DCs are originated from bone marrow or thymus of human body. The rate of accumulation of both the cells in the area of closeness of the proper domain is a and b. The whole process of Psoriatic pathogenesis comes to an end through the phase of sensitization, silent phase and effector phase. During the phase of sensitization, DCs which are not developed are able to take antigen and transfer it to the area of secondary lymphatic organs in presence of cytokines. Next DCs helps to trigger the growth of T-cells which involve in skin infiltrating T-Cells and relocate into the tissues [22]. It is seen that that the small fraction of T-Cells and DCs which have already involved on the process of triggering will not be available in the future process of triggering [17]. It is seen that T-Cells as well as DCs are not similar types of immune cells and their features are not similar in the immune system of human. So, their triggering rates rate also different. The rate of triggering of Keratinocytes by the T-cells because of mediation of Cytokines by T-cells is denoted by γ_1 and rate of growth of Keratinocytes is denoted by γ_2. It is seen that when DCs reach matured stage then it will become lymph node. It is seen that the matured DCs are able to connect with CD8+ T-cells at a rate of qn in the location of lymph node where q is the T-Cells Receptor (TCR) based on mean peptide and n is denoted to specify the average number of pMHC complexes present per DC. Due to which DC population is reduced on the other hand CD8+ population is enhanced. The rate at which T-cells and DCs are removed is assumed as μ and μ'. The method of proliferation of CD8+ T-cell is triggered by similar antigen which helps to represent DCs at a rate of r. The interaction between both the cells are done at a rate θ. As a result the population of CD8+ T-cell is decreased and population of Keratinocyte cell is increased for contributing in this interaction as it has been observed that the population of Keratinocyte cell is formed after the interaction of above mentioned two cells at a rate of θ_1. The CD8+ T-cells and Keratinocyte cells are decayed at a rate of ξ and λ, respectively.

In last few years researchers started to focus to model a system using FDEs [11–13]. As FDEs are different from its integer order counterpart for their non-local behaviour and the power of describing memory which are the part of biological systems. Based on the assumptions the mathematical model of the disease psoriasis has been developed [13].

$$
\begin{aligned}
{}_0^C D_t^\alpha P &= a - \delta P Q - \gamma_1 P S - \mu P, \\
{}_0^C D_t^\alpha Q &= b - \beta P Q - qnQR - \mu' Q, \\
{}_0^C D_t^\alpha R &= rqnQR - \theta QR - \xi R, \\
{}_0^C D_t^\alpha S &= (\delta + \beta) P Q + \gamma_2 P S + \theta_1 QR - \lambda S, \qquad (5)
\end{aligned}
$$

where the initial conditions are denoted by.
$P(0) = P_0, Q(0) = Q_0,\ R(0) = R_0, S(0) = S_0.$

and ${}_0^C D_t^\alpha$ signifies left-sided Caputo's derivative operator. All the parameters, which are used in the mathematical model of the disease psoriasis are positive.

Parameters, which are used in the mathematical modelling of the disease psoriasis with their values are given below.

Parameters	Values (per day)
a	10 (mm^{-3})
b	15 (mm^{-3})
δ	0.02 (mm^{-3})
β	0.068 (mm^{-3})
γ_1	0.0004 (mm^{-3})
γ_2	0.0002 (mm^{-3})
r	0.7
q	0.002
n	9
θ	0.0008 (mm^{-3})
θ_1	0.0006 (mm^{-3})
μ	0.08
μ'	0.003
ξ	0.09
λ	0.6
Parameters	Values (per day)
A	0.00011
B	0.002

4 Formulation of Fractional Optimal Control Problem

Here in this section fractional optimal control problem has been formulated for the treatment of the disease psoriasis. Formulation of fractional optimal control means the inclusion of $u(t)$ in the system of equations of the disease psoriasis. Control $u(t)$ signifies the effect of drug for the treatment of the disease psoriasis. The application of drug on the disease psoriasis helps to reduce the population of Keratinocyte by reducing the interaction between CD4+ T-cells and Keratinocytes. After including optimal control in fractional order modelling of the disease psoriasis the system of equations become as mentioned below.

$${}_0^C D_t^\alpha P = a - \delta P Q - (1 - u(t))\gamma_1 P S - \mu P,$$

$$
\begin{aligned}
{}_0^C D_t^\alpha Q &= b - \beta PQ - qnQR - \mu' Q, \\
{}_0^C D_t^\alpha R &= rqnQR - \theta QR - \xi R, \\
{}_0^C D_t^\alpha S &= (\delta + \beta)PQ + (1 - u(t))\gamma_2 PS + \theta_1 QR - \lambda S,
\end{aligned} \tag{6}
$$

where the initial conditions are denoted by
$P(0) = P_0, Q(0) = Q_0,\ R(0) = R_0, S(0) = S_0.$

and ${}_0^C D_t^\alpha$ signifies left-sided Caputo's derivative operator. The system of equations can be written in matrix form as

$$ {}_0^C D_t^\alpha X = f(X(t), u(t)), \tag{7} $$

where X is a vector and it is represented by $X = \begin{bmatrix} P & Q & R & S \end{bmatrix}^T$ for minimizing the population of Keratinocyte the cost function with the weights A and B is defined as

$$ J(u) = \int_{t_0}^{t_f} \left(Au^2 + BS^2\right)dt \tag{8} $$

subject to the dynamics of the fractional order system given by (6).

The actual aim is to find the optimal control $u^*(t)$ that will be able to minimize the cost function $J(u)$. Here $u(t)$ signifies the effect of drug on the fractional order system described by (6) and the control is bounded between 0 and 1 [23].

For the formulation of fractional optimal control problem (FOCP) of the fractional order system described by (6), the Euler–Lagrange optimal conditions need to be understood. In [24], Agrawal showed a formulation of FOCP and the derivation of the optimality conditions. Formulation of the problem is mentioned below in steps.

The fractional order system including the control is given by

$$ {}_0^C D_t^\alpha x = f(x, u, t), x(0) = x_0 \tag{9} $$

In this equation, state vector is denoted by x and control is denoted by u and time is denoted by t. So, the cost function is defined as

$$ J(u) = \int_0^t g(x, u, t)\mathrm{d}t \tag{10} $$

The aim is to find the optimal control $u^*(t)$ by minimizing the above cost function, subject to the dynamics of the system given by (9).

The state vector of the system is written as

$$ {}_{0}^{C}D_{t}^{\alpha}x = f(x,u,t), x(0) = x_0 \tag{11} $$

where u is the control vector, the co-state vector p of the system can be written as

$$ {}_{t}^{C}D_{t_f}^{\alpha}p = \frac{\partial g}{\partial x} + p\frac{\partial f}{\partial x}, \; p(t_f) = 0. \tag{12} $$

The function which signifies optimal control $u^*(t)$ is able to satisfy the equation given below

$$ \frac{\partial g}{\partial u^*} + p\frac{\partial f}{\partial u^*} = 0. \tag{13} $$

The conditions of optimality (11)–(13) given by Euler–Lagrange are required for formulating FOCP. Another approach of formulating the FOCP using Hamiltonian function and the Hamiltonian function is defined as

$$ H(x,u,p,t) = g(x,u,t) + p^T f(x,u,t). \tag{14} $$

The state vector of the system can be obtained from the Hamiltonian function as

$$ {}_{0}^{C}D_{t}^{\alpha}x(t) = \frac{\partial H(t,x(t),u(t),p(t))}{\partial p} \tag{15} $$

The co-state vector can be obtained as

$$ {}_{t}^{C}D_{t_f}^{\alpha}p(t) = -\frac{\partial H(t,x(t),u(t),p(t))}{\partial x} \tag{16} $$

The optimal control can be obtained from the equation given below

$$ \frac{\partial H(t,x(t),u(t),p(t))}{\partial u} = 0 \tag{17} $$

And the transversality condition is

$$ p(t_f) = 0 \tag{18} $$

After using these equations, the co-state equations are written below as

$$ \begin{aligned} {}_{t}^{C}D_{t_f}^{\alpha}p_1 &= -p_1[\delta Q + (1-u(t))\gamma_1 S + \mu] - p_2\beta Q \\ &\quad + p_4(\delta+\beta)Q + p_4(1-u(t))\gamma_2 S, \\ {}_{t}^{C}D_{t_f}^{\alpha}p_2 &= -p_1\delta P - p_2(\beta P + qnR + \mu') \\ &\quad - p_3(\theta R + rqnR) + p_4[(\beta+\delta)P + \theta_1 R], \end{aligned} $$

$$
{}_{t}^{C}D_{t_f}^{\alpha} p_3 = -p_2 qnQ - p_3(\theta Q + \xi - rqnQ) + p_4 \theta_1 Q,
$$
$$
{}_{t}^{C}D_{t_f}^{\alpha} p_4 = -p_1(1-u(t))\gamma_1 P - p_4(\lambda - (1-u(t))\gamma_2 P) + 2BS \tag{19}
$$

The transversality condition is given as

$$
p_i(t_f) = 0, i = 1, 2, 3, 4. \tag{20}
$$

The expression for optimal control function is obtained from the Eqs. (12) and (16) as

$$
u^*(t) = \frac{PS(p_4\gamma_2 - p_2\gamma_1)}{2A} \tag{21}
$$

After considering the boundedness of optimal control, the optimal control function is written as

$$
u^*(t) = \min\left\{\max\left\{\frac{PS(p_4\gamma_2 - p_2\gamma_1)}{2A}, 0\right\}, 1\right\} \tag{22}
$$

When $u^*(t)$ is used instead of $u(t)$ in Eq. (6), then Eqs. (6) and (19) are able to represent a FOCP. The system of optimality helps to constitute a two-point boundary value problem (TPBVP) which includes a set of FDEs.

5 Simulation and Result

Here in this section the numerical simulation of the FOCP has been shown after the FOCP is formulated from Euler–Lagrange optimality conditions. The parameter values used in the simulation has been assumed after studying the literature. There are several numerical as well as analytical methods present to approximate the solution of FDEs. In this article, FOCP has been solved such a way that the state equations have been solved in forward iteration and the co-state equations have been solved in backward iteration. After approximation of all the FDEs the simulation has been carried out through MATLAB platform.

For approximating the nonlinear FDEs. Grunwald–Letnikov method is used which is found in [25].

The numerical approximation formula for αth derivative at the point kh in the sense of Grunwald–Letnikov is written as

$$
{}_{(k-\frac{L_m}{h})GL}D_{t_k}^{\alpha} f(t) \approx \frac{1}{h^\alpha} \sum_{j=0}^{k} (-1)^j \binom{\alpha}{j} f(t_{k-j}) \tag{23}
$$

In this equation, L_m signifies the length of the memory and $t_k = kh,(k = 1, 2, ...)$ h signifies the step of time–space of iteration and the binomial coefficients are signified by $(-1)^j\binom{\alpha}{j}$. Generally, the binomial coefficients are denoted by $c_j^\alpha(j = 0, 1, 2, ...)$ and found by the calculation given below as

$$\begin{aligned} c_0^\alpha &= 1, \\ c_j^\alpha &= \left(1 - \tfrac{1+\alpha}{j}\right)c_{j-1}^\alpha \end{aligned} \tag{24}$$

Consider a nonlinear FDE with initial condition has taken in terms of Grunwald–Letnikov sense and written as

$$\left({}_{GL}D_{a+}^\alpha f\right)(t) = g(f(t), t) \tag{25}$$

This equation can be numerically approximated as

$$f(t_k) = g(f(t_k), t_k)h^\alpha - \sum_{j=1}^{k} c_j^\alpha f(t_{k-j}) \tag{26}$$

The simulation has been carried out for the values of α lies between 0.7 and 1.

It is observed in Fig. 1, the behavioural pattern of the population of T-Cells, Dendritic Cells, CD8+ T-Cells and Keratinocytes for different values of α, where ${}_0^C D_t^\alpha$ signifies the left-sided Caputo fractional derivative, which is included in the mathematical model of the disease Psoriasis. The values of α has been reduced gradually such as 1, 0.9, 0.8 and 0.7 to observe the behaviour of the system. It has been observed in the case of integer order system (i.e. $\alpha = 1$) that the population of T-Cell enhances gradually for the sake of its production, which is constant but it

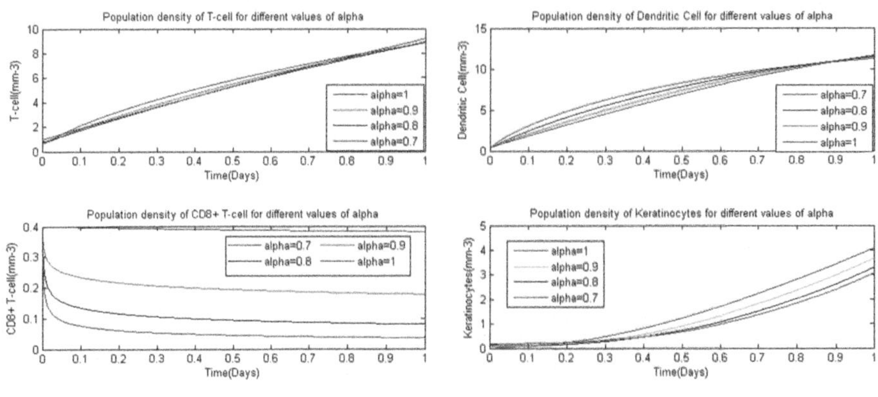

Fig. 1 Population density of T-cell, Dendritic Cell, CD8+ T-cell and Keratinocytes for different values of alpha

has also been seen that when the value of α starts to reduce to 0.8, the population of T-Cell has been reduced accordingly and at the same time it is able to reach more quickly to the stable situation. But when α starts to reduce to 0.7, then the population takes more time to reach its position which is stable. It has also been observed that the population of Dendritic Cells is inversely proportional to α. They are able to enhance correspondingly as the values of α are reduced. Population of Dendritic Cells moves so fast to stable position when the value of $\alpha = 0.8$ rather than 1 and 0.7. The population of CD8+ T-Cell as well as Keratinocyte also move to the stable position when the value of $\alpha = 0.8$.

It is observed in Fig. 2, (i.e. $u(t)$) is showed which signifies the effect of drug as optimal control. It has been observed that for the values of $\alpha = 0.8$, less drug is required as compared to the case when the value of $\alpha = 1$. It signifies when FOCP is used instead of OCP, amount of drug required is less for controlling the disease Psoriasis that has been studied under the perceptive of mathematical modelling to control its spread.

It has been observed that a model made of ODE which is able to describe an instantaneous process. The notion which is used to represent the instant is dependent on the time scale which has been considered, on the other hand it has been seen that FDE has the ability to represent the fading of memory which is largely dependent on the range of. It has also been observed that such memories have the ability to describe the present events with the collective information from the earlier events. In the present study, FDEs have been used to show the impact of memory which presents in most of the cell biological system. It has been seen that the model which has been represented here using FDEs are the modified form of ODE model but the effect of memory is characterized by the parameter of the model. It has been seen that the system which has been represented by FDE has the ability to gain the position of stability faster than its ODE model counterpart. Here the presence of memory (α) plays a very crucial role to understand the effect of memory presents in a cell biological system. The system is very sensitive to its parameters. It has been observed that when the value of is reduced, then the system becomes more stable. This article is all about the cell biological system which is having memory effect

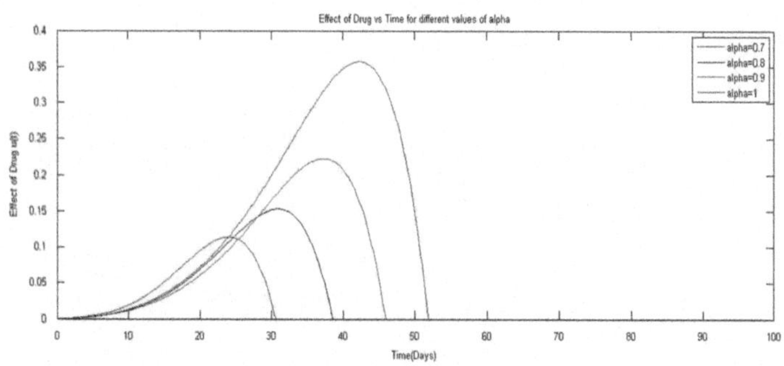

Fig. 2 Effect of drug with time for different values of alpha

and the disease Psoriasis is considered and it has been shown that the population of Keratinocyte will be decreased after considering the value of 0.8 through the effect of optimal drug dosing.

6 Conclusions

In this present work, the outcome is able to reveal that the memory effect of the cell is capable of maintaining the overall stability of the whole cell biological system for a long period of time. Thus it can be said that the activation of an intermediate memory is more beneficial for regularizing the excess production of Keratinocyte quickly. So, after regulating the memory of the cell, the cell biological system can be explored more accurately. At the same time if the control therapeutic approach is introduced with the help of optimal drug into the disease Psoriasis modelled as fractional order system, the population of Keratinocyte can be reduced more effectively. Hence, it can be said that FODE model helps in our understanding of the nature of the disease Psoriasis more effectively rather than its ODE counterparts, which is also unable to signify the effect of memory.

References

1. Prignano F, Ricceri F, Bianchi B, Lotti T (2009) Quantity, distribution and immunophenotypical modification of dendritic cells upon biological treatments in psoriasis. Int J Immunopathol Pharmacol 22(2):379–387
2. Traub M, Marshall K (2007) Psoriasis—pathophysiology, conventional, and alternative approaches to treatment. Alternative Med Rev 12(4)
3. Deng S, May BH, Zhang AL, Lu C, Xue CC (2013) Topical herbal medicine combined with pharmacotherapy for psoriasis: a systematic review and meta-analysis. Arch Dermatol Res 305(3):179–189
4. Carter PH, Zhao Q (2010) Clinically validated approaches to the treatment of autoimmune diseases. Expert Opin Investig Drugs 19(2):195–213
5. Li K, Zhou R, Jia WW, Li Z, Li J, Zhang P, Xiao T (2016) Zanthoxylum bungeanum essential oil induces apoptosis of HaCaT human keratinocytes. J Ethnopharmacol 186:351–361
6. Krueger, J. G., & Bowcock, A., Psoriasis pathophysiology: current concepts of pathogenesis. Annals of the rheumatic diseases, 64(suppl 2), ii30-ii36, (2005).
7. Li ZJ, Choi DK, Sohn KC, Lim SK, Im M, Lee Y, Seo YJ, Kim CD, Lee JH (2016) Induction of interleukin-22 (IL-22) production in CD4+ T cells by IL-17A secreted from CpG-stimulated keratinocytes. Ann Dermatol 28(5):579
8. Nomikos M, Pappas A, Kopaka ME, Tzoulakis S, Volonakis I, Stavrakakis G, Avgenakis G, Anezinis P (2011) Urothelial carcinoma of the urinary bladder in young adults: presentation, clinical behavior and outcome. Adv Urol
9. Abbas S, Benchohra M, N'Guérékata GM, Slimani BA (2012) Darboux problem for fractional-order discontinuous hyperbolic partial differential equations in Banach algebras. Complex Variable Elliptic Equations 57(2–4):337–350
10. Clayton NS, Griffiths DP, Emery NJ, Dickinson A (2001) Elements of episodic–like memory in animals. Philosoph Trans R Soc London Ser B Biol Sci 356(1413):1483–1491

11. Rana S, Bhattacharya S, Pal J, N'Guérékata GM, Chattopadhyay J (2013) Paradox of enrichment: a fractional differential approach with memory. Phys A 392(17):3610–3621
12. Ding Y, Ye H (2009) A fractional-order differential equation model of HIV infection of CD4+ T-cells. Math Comput Model 50(3–4):386–392
13. Ahmed E, Elgazzar AS (2007) On fractional order differential equations model for nonlocal epidemics. Phys A 379(2):607–614
14. Savill NJ, Weller R, Sherratt JA (2002) Mathematical modelling of nitric oxide regulation of rete peg formation in psoriasis. J Theor Biol 214(1):1–16
15. Sherratt JA, Weller R, Savill NJ (2002) Modelling blood flow regulation by nitric oxide in psoriatic plaques. Bull Math Biol 64(4):623–641
16. Roy PK, Datta A, Chatterjee AN (2011) Saturation effects on immunopathogenic mechanism of psoriasis: a theoretical approach. 13(3):310–318
17. Roy PK, Datta A (2012) Negative feedback control may regulate cytokines effect during growth of keratinocytes in the chronic plaque of psoriasis: a mathematical study. Int J Appl Math 25(2):233–254
18. Roy PK, Datta A (2013) Impact of perfect drug adherence on immunopathogenic mechanism for dynamical system of psoriasis. Biomath 2(1):1212101
19. Datta A, Roy PK (2014) Effect of half-saturation in psoriatic pathogenesis using fractional derivative: a mathematical study. Inflammation 2(3):4–5
20. Roy PK, Datta A, Rana S (2013) The fractional-order differential equation model of psoriatic pathogenesis: a mathematical study. African Diaspora J Math New Ser 15(2):35–46
21. Li C, Zeng,F (2015) Numerical methods for fractional calculus, vol 24. CRC Press
22. Sabat R, Philipp S, Höflich C, Kreutzer S, Wallace E, Asadullah K, Volk HD, Sterry W, Wolk K (2007) Immunopathogenesis of psoriasis. Exper Dermatol 16(10):779–798
23. Fleming WH, Rishel RW, Marchuk GI, Balakrishnan AV, Borovkov AA, Makarov VL, Rubinov AM, Liptser RS, Shiryayev AN, Krassovsky NN, Subbotin AN (1975) Applications of mathematics. Deterministic Stochastic Optimal Control
24. Agrawal OP (2008) A formulation and numerical scheme for fractional optimal control problems. J Vib Control 14(9–10):1291–1299
25. Noupoue YYY, Tandoğdu Y, Awadalla M (2019) On numerical techniques for solving the fractional logistic differential equation. Adv Diff Equations (1):108

Fault Detection in Floating PV System Using DC Leakage Current

G. Srinivasa Murthy and Suryanarayana Gangolu

Abstract The usage of non-conventional energy sources are rapidly increasing due to the shortage of conventional sources in which, the solar based energy has huge demand among the non-conventional energy sources since it is cost-effective and availability is high. Further, in recent trends of solar technology, the floating PV systems which are placed on the water bodies by means of a floating mechanism in which, the power is transferred from the panels to the inverter which is located at the banks of the power house by using DC cables. The problem here is that, the DC leakage current in the cables flows through its insulating material through ground, and this is often known as DC leakage. The faults in the DC side as well as AC side of the system may affect the DC leakage values. If these Leakage currents are unnoticed then they may flow to the equipment like metallic structures, inverter and may damage them in long term. The proposed algorithm utilizes these leakages to find the faults in the system to prevent the equipment damage.

Keywords Floating PV panels · DC leakages · Trojan horse

1 Introduction

The solar energy is the most demanding energy resource due to it's availability and other factors like cost, maintenance, etc. To push it's boundaries to further levels, there have been developments like Roof top PV, Building Integrated PV, Floating PhotoVoltaics (FPV), etc. Here, it is focused on the floating PV which are mounted on the water bodies using floating mechanisms to save the land and utilise the solar irradiance falling on the water bodies like Ponds, Lakes, other artificial water bodies (man made) like Dams, etc. Since, there are a lot of advantages in using this FPV system in power generation. Due to the presence of water below the PV panels, the panel temperature will stay low and this effect can lead to increase in PV efficiency by

G. Srinivasa Murthy (✉) · S. Gangolu
National Institute of Technology, Uttarakhand, Srinagar, India

S. Suhag et al. (eds.), *Control and Measurement Applications for Smart Grid*, Lecture Notes in Electrical Engineering 822,
https://doi.org/10.1007/978-981-16-7664-2_15

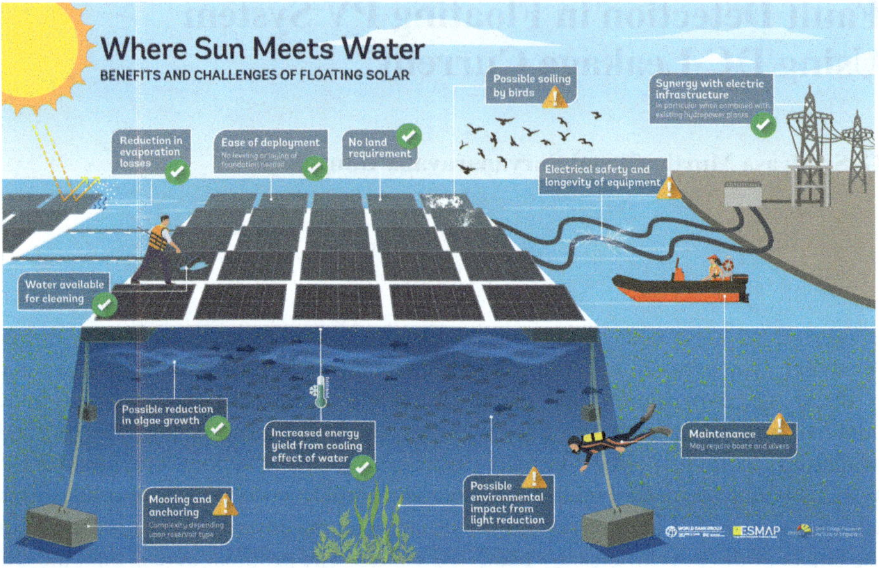

Fig. 1 FPV system with various advantages and disadvantages

15% annually [1, 2]. But due to the presence of the leakage currents in PV systems, i.e., the DC cable may exhibit some leakage currents during the power transfer from PV modules to the inverter.

In most of the cases, these leakage currents are very less and can be in some cases found negligible. But in ground mounted PV, the capacitive leakage currents have major effect on the system and in Floating PV, the length of the DC cables are more than the normal ground mounted or roof top mounted PV systems since the inverter and PV modules are kept far apart from each other (Inverter is kept at shore of the water bodies), therefore, there may be some effect of leakage currents on the Floating PV systems also. The leakage currents due to capacitors can be neglected in the Floating PV system because of galvanic isolation. It is also seen that the FPV system is connected to the grid, the power from grid may have effect on the FPV system during faults occurred in the grid system, since we preferred the grid connected system due to it's certain advantages compared to the Stand-alone FPV system [3]. The cables used must be mechanically fit, robust and able to withstand temperature changes of the water and also should be able to maintain the maximum power of the system [4]. As mentioned in [5], the author used a technique where the unattented DC leakages are treated as *Trojan horse* and used this in different modelling techniques to find the excess leakage currents, if these are not solved or not found during the fault conditions, there may be further damage to the inverter [6] and other structure of the PV system due to the imbalance caused by the faults (such as leakage currents) in the DC side of the system [7]. In general, the PV systems are in-built with the protection devices which detect and clears the grounding faults, such devices are called as Ground Fault Detection and Interrupt (GFDI) device [8]. In [9], it is mentioned that the faults in PV system may effect the PV power generation by nearly 18.9%. During faults in AC side of the system, the transition of currents may cause imbalance in the input (neutral wire) and output current (three phase wires) of the inverter. Therefore, the formed imbalance will indicate the presence of faults within the inverter as well as any ground faults (i.e., leakage currents) occurring in the DC side of the system [4].

The general diagram depicting fault, advantages and disadvantages for this FPV system [10] is shown in Fig. 1.

The PV modules in the simulink model is designed with Perturb and Observe MPPT technique used with Boost converter to extract maximum power from the FPV modules [11].

The inverter is designed as a three phase full bridge grid connected inverter. The gate pulses for the inverter are generated using PWM generator, for which voltage control mode is used. In this mode, different transformations are performed to the voltages, i.e., in voltage and current transformation block, the three phase voltages and currents are converted to dq0 form by using abc to alpha-beta and alpha-beta to dq0 transformations. A PI controller is used in current controller block. To get frequency from alpha-beta of voltage, a PLL block is used. From frequency and dq0 values from current controller block, three phase voltage reference is found in inverse transformation block. The gate signals for the inverter block are generated using the three phase voltage references in PWM generator block.

The main an objective of this paper is to find and clear the fault in the system using the DC leakage in the cables. However, according to [12], we cannot detect every type of fault in the PV system due to change in the fault impedance.

The remaining of the paper contains following sections

* DC Leakages in PV system
* Methodology of proposed scheme
* Flowchart of proposed scheme
* Setting of threshold limit
* Simulation results and Discussion
* Conclusion
* References.

2 DC Leakage Current in PV System

The leakage current in floating PV is defined as the flow of currents in non-ideal materials of PV modules, cables, other supporting structures. If the distance between FPV modules and inverter increases, the leakage current will be more and it may effect on the operation of the system. In Figs. 2 and 3 [7], the flow of leakage current is shown with two different cases. In Fig. 2, the leakage current is passing through the inverter via the ground. In Fig. 3, the leakage current is flows from the cable

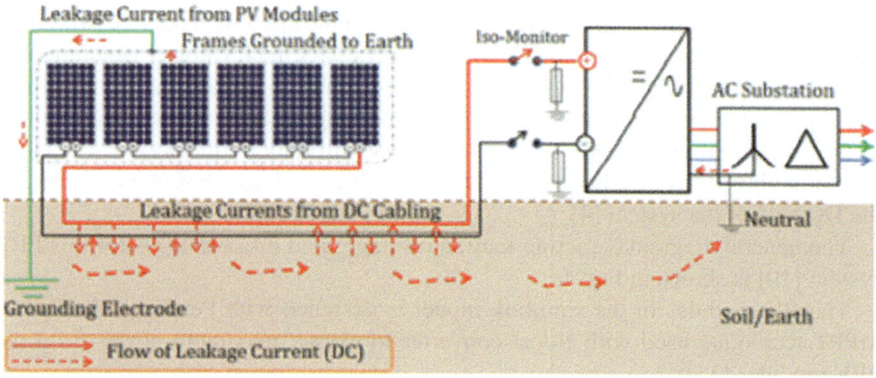

Fig. 2 DC leakage current through inverter

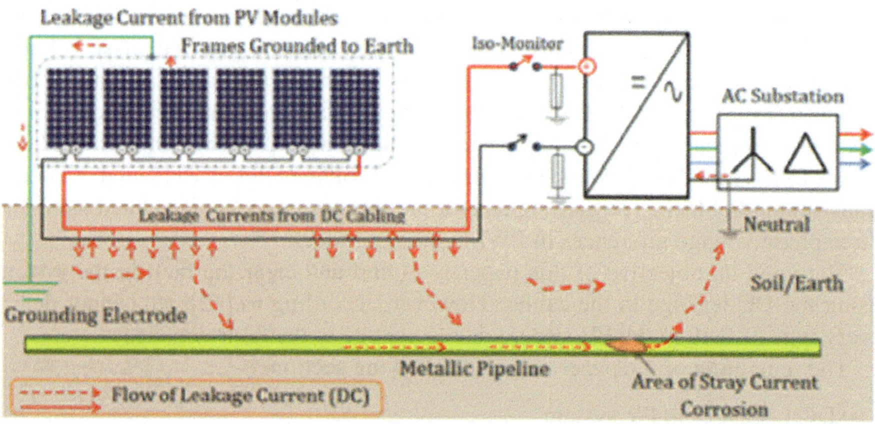

Fig. 3 DC leakage current through metallic pipes

to the inverter via metallic pipelines which results corrosion of the metallic pipes. Hence, in this paper, the new methodology is proposed based on the calculation of leakage current in the cable.

3 Methodology of Proposed Scheme

3.1 Normal Operating Condition

Table 1 shows all the necessary components in Figs. 4 and 5.

Figure 4 shows the FPV system in normal operating mode connected to the grid via dc-dc boost converter, DC cables and inverter. Here, by measuring the input and

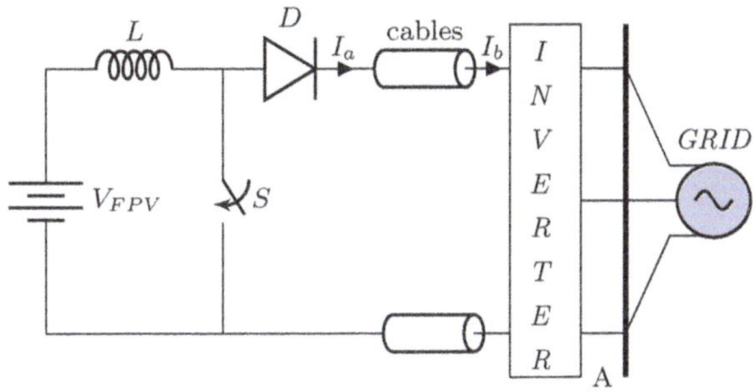

Fig. 4 Grid connected floating PV source

Table 1 Nomenclature and their description

Parameters	Description
V_{FPV}	Voltage from FPV modules
L D S	Boost converter components
Cables	DC cables from FPV to inverter
I_a, I_b	Input and output currents of cables
Inverter	3-phase full bridge inverter
A	Bus bar
I_f	Fault at GRID
GRID	AC grid connected system

output currents of the cable, the absolute value of difference of the both end currents is calculated under the healthy and faulty conditions. By using the mean/average value of the difference, a threshold value is set and compared with the value of every cycle.

$$D = I_a - I_b \tag{1}$$

$$D_{abs} = |D| \tag{2}$$

$$\text{Mean} = \frac{\sum_{n=1}^{N} D_{abs_n}}{N} = M \tag{3}$$

D is the difference between the cable input current I_a and output current I_b. D_{abs} is the absolute value of the difference found in Eq. 1. N is the number of samples per cycle. From Eq. 3, we check M (Mean) value with threshold value for each cycle to find the abnormality in the system.

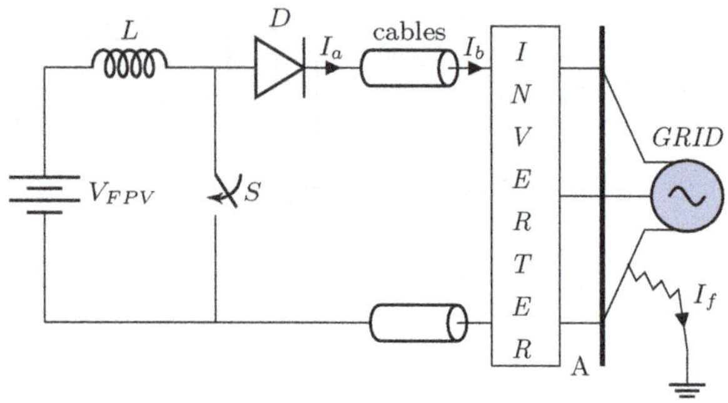

Fig. 5 Grid connected floating PV source during fault at grid/bus bar

3.2 *During Fault Condition*

Figure 5 depicts the system with fault (LG, LLG, LLLG) in the AC side/Grid system, the current difference calculated at the DC cables will intend to change. Depending on the type of the fault the difference may vary since the circulating currents from faults may affect the system [5], for every fault and fault impedances the differences may not be equal to each other. Further, due to the faults in the grid/bus bar the entire system may get affected since the PV is connected in series with the grid [13].

4 Flowchart

1. The start of the algorithm.
2. Extract the input and output currents of DC cables.
3. Compute the Difference between the input and output currents.
4. Compute the absolute value of the Difference.
5. Calculate the mean of the Difference.
6. Further, the mean is compared with threshold value to find the fault as shown in Fig. 6.

5 Setting of Threshold Limit

In normal operating condition of the system, the mean/average value of the difference between input and output current of the DC cable is computed and found 16.789 mA, in addition this threshold value is compared with the mean of difference between input and output currents from DC cable for every cycle. Therefore, with this in mind, the above value is considered as the threshold value (16.789 mA) for achieving better

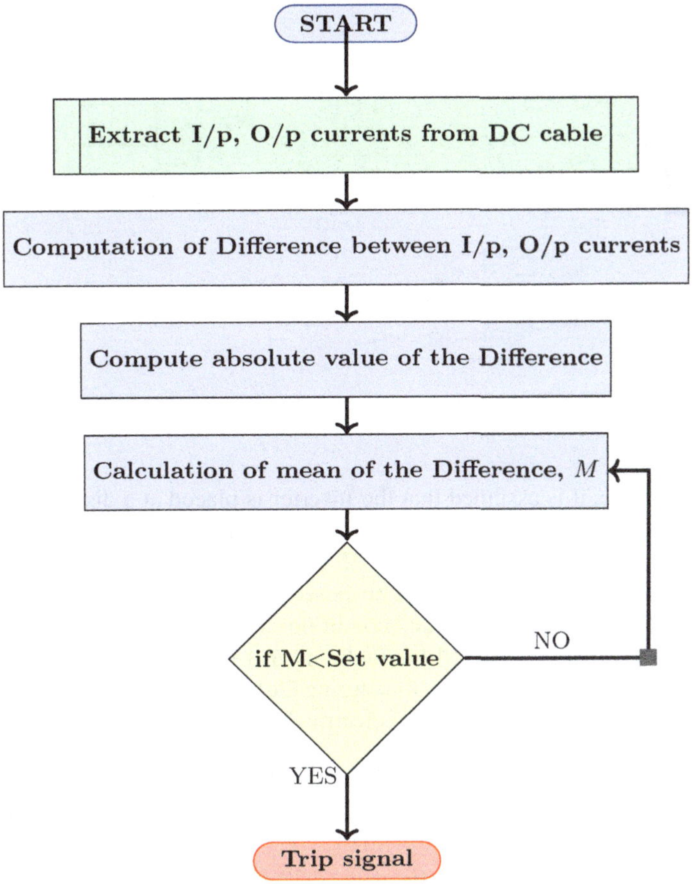

Fig. 6 Flowchart depicting proposed scheme

results from the logic. Further, different type of faults are introduced into the system at different cycle durations, for every condition the said threshold value is works perfectly (Here each cycle comprises 200 samples).

6 Simulation Results

The proposed scheme is simulated on a grid connected Floating PV system as in Fig. 5 and is modeled MATLAB/SIMULINK. In this 1Soltech 1STH-215-P module composed of 40 parallel strings with 10 series connected modules per string, the modules were introduced with an illumination of 1000 W/m^2 PV capacity, the system has open-circuit voltage of 36.3 V and short-circuit current of 7.84 A at 25 °C panel

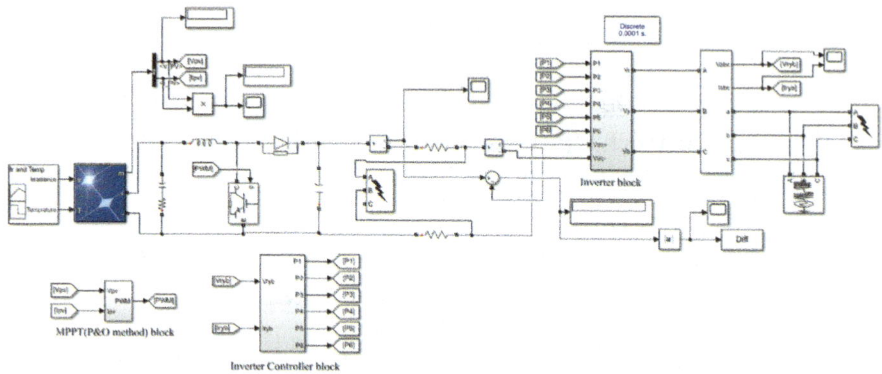

Fig. 7 Simulink model for validation of proposed system

temperature. Here, it is assumed that the Inverter is placed at a distance of 1 km (at the banks of the water body) and the DC cable resistance is considered 340 MΩ per km. The AC grid is modelled with 25 kV, 50 Hz, system. Figure 7 shows the SIMULINK model for verification of proposed scheme.

It is observed that for different fault conditions, the DC leakage at cables is varying differently. The change observed for faults in grid is that the leakage currents are reducing in the cables or we can say that during faults the leakage currents are flowing from cables through the ground to the electrical devices(like inverter, other metallic structures) as shown in Figs. 2 and 3 [7].

Figures 8, 9 and 10 shows the DC leakage wave forms of LG, LLG and LLLG faults occurred in grid with fault at 0.1 s (6th cycle) and fault clearing at 0.15 s. The related data are shown in below.

When LG fault (with a fault resistance of 70 Ω) occurs in the system as shown in Fig. 8, there will be increase in leakage current flow to the ground, i.e., the value at which they tend to cross the threshold value is observed to be 16.784 mA at 6th cycle, which is close to the threshold value. After fault clearance they quickly reach to the normal condition.

During LLG fault (with a fault resistance of 25 Ω) as shown Fig. 9, at 6th cycle the value of the leakage current is found to be 15.923 mA. Even after the clearance of the fault, the leakage current value reaches normal condition slowly, while staying in the relay operating region.

When LLLG fault (with a fault resistance of 18 Ω) take place as shown in Fig. 10, the value for which the relay sends signal is measured to be 14.883 mA at 6th cycle. As seen during LLG fault, the same phenomenon is observed in this case, i.e., the value of leakage current stays for a while even after the fault removal from the system.

The reason for continuous flow of the leakage current even after the fault is cleared is due to the DC link voltage, i.e., whenever there is a fault in the grid, the dc-link voltage increases in the following order: LL fault, LLG fault, and three phase fault. After the fault clearance it decreases in reverse order. However, there is an exception

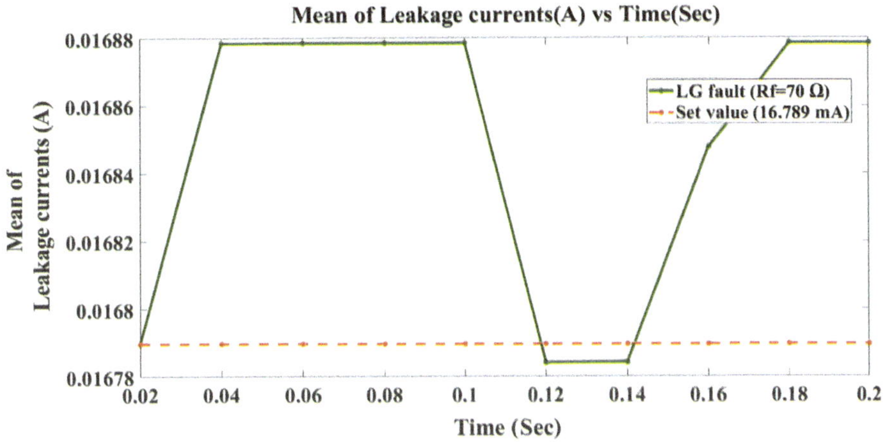

Fig. 8 Leakage current waveform for LG fault

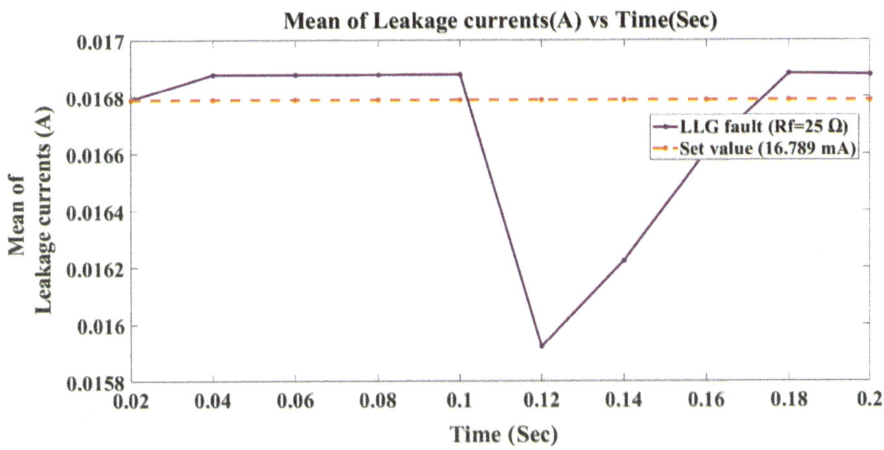

Fig. 9 Leakage current waveform for LLG fault

Table 2 Values of fault resistance R_f, mean M for different type of faults occurred in grid

	Type of fault	R_f Ω	Mean M (mA)
Unsymmetric	LG	70	16.784
	LLG	25	15.923
Symmetric	LLLG	18	14.883

for LG fault since the increase in dc link voltage is less (5–8 V) and after fault removal it goes back to its original value in less duration (0.05 s) [14].

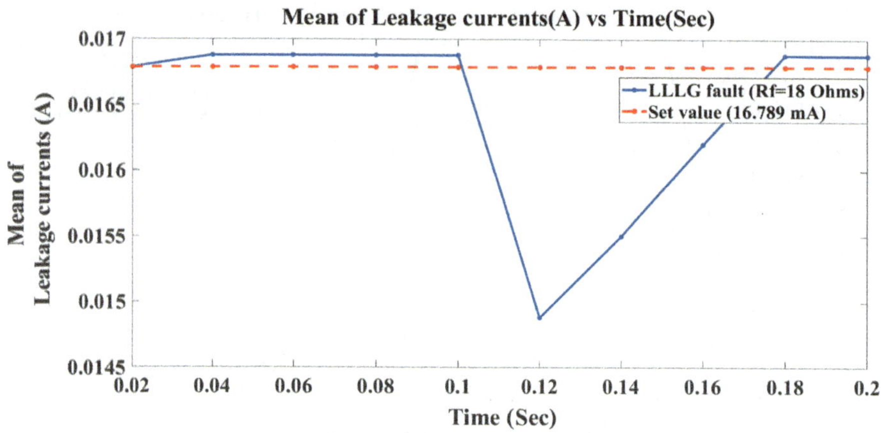

Fig. 10 Leakage current waveform for LLLG fault

Table 3 System parameters and their values

System parameters	Description
DC cable resistance	340 MΩ/km
Grid voltage	25 kV
Input voltage	250–350 V
Output voltage	600 V
Rated power	100 kW
Switching frequency	5 kHz

Table 2 shows different faults and values at which trip signal is issued to the protection devices. Where M is the mean Leakage current value at which the trip signal is given to the protection devices.

7 Conclusions

From the above results, we can observe that when fault is at AC side, as the fault resistance is increasing in different fault conditions, the mean of the difference (input current—output current of the cables for every 200 samples) is also increases. From the graphs, compared to LLG and LLLG faults, during LG fault, the trip signal is close to the threshold value, in other words LLG and LLLG faults effect the DC leakages in cables more than the single line to ground fault. Based on the fault occurrence in AC side, the proposed logic works perfectly. Whereas the drawback is that when the fault is in DC side this logic will not work perfectly since the measuring devices at the both ends of the cable will measure the high fault currents flowing through them instead of leakage currents in the cables.

The necessary parameter values (both system and boost converter) are mentioned in Table 3. The cable length is taken as 1 km.

References

1. Enaganti PK (2020) Performance analysis of submerged polycrystalline photovoltaic cell in varying water conditions. IEEE J Photovolt 10(2):531–538
2. Lanzafame R (2011) Field experience with performances evaluation of a single-crystalline photovoltaic panel in an underwater environment. IEEE Trans Ind Electron 57(7):2492–2498
3. Ghiassi-Farrokhfal Y, Kazhamiaka F, Rosenberg C, Keshav S (2015) Optimal design of solar PV farms with storage. IEEE Trans Sustain Energy 6(4):1586–1593
4. Jain S, Agarwal V (2007) A single-stage grid connected inverter topology for solar PV systems with maximum power point tracking. IEEE Trans Power Electron 22(5):1928–1940. https://doi.org/10.1109/TPEL.2007.904202
5. Dimitriou A, Lazari AL, Charalambous CA (2017) Understanding of DC leakage and faults in floating photovoltaic systems: trojan horse to undermine metallic infrastructure and safety. In: IEEE Manchester PowerTech, Manchester, pp 1–6
6. Safety of power converters for use in photovoltaic power systems—part 2: particular requirements for inverters. In: IEC standard 62109-2
7. Dimitriou A, Buxton D, Charalambous CA (2018) Stray current DC corrosion blind spots inherent to large PV systems fault detection mechanisms: elaboration of a novel concept. IEEE Trans Power Deliv 33(1):3–11
8. Hernández JC, Vidal PG (2009) Guidelines for protection against electric shock in PV generators. IEEE Trans Energy Convers 24(1):274–282. https://doi.org/10.1109/TEC.2008.2008865
9. Firth SK, Lomas KJ, Rees SJ (2010) A simple model of PV system performance and its use in fault detection. Sol Energy 84(4):624–635. https://doi.org/10.1016/j.solener.2009.08.004
10. Where sun meets water: floating solar market report (2019) In: World Bank group; energy sector management assistance program. Solar Energy Research Institute of Singapore
11. Soedibyo SA, Hafidz I, Zulkarnain GR, Ashari M (2017) MPPT design on solar farm using perturb and observe technique considering tilt angle and partial shading in Giligenting Island. In: International seminar on intelligent technology and its applications (ISITIA), pp 222–226
12. Abdulmawjood (2018) Detection and prediction of faults in photovoltaic arrays: a review. In: IEEE 12th international conference on compatibility, power electronics and power engineering (CPE-POWERENG), Doha, pp 1–8
13. Badi LWA, Zakaria Z, Nordin AHM, Mustapa RF (2014) Unbalanced faults analysis in grid—connected PV system. In: IEEE international conference on power and energy (PECon), Kuching, pp 360–365
14. Banu IV, Istrate M (2014) Study on three-phase photovoltaic systems under grid faults. In: IEEE international conference and exposition on electrical and power engineering (EPE), Iasi, pp 1132–1137

Effect of Temperature and Substrate Biasing of Titanium Thin Film on Normal and Patterned Silicon Wafer

B. Rajagobalan and G. Meenakshi

Abstract Magnetron sputtering can be the most powerful and quick way to extend multiple films. Changes can be made to operating parameters such as the substrate's pressure and temperature, physical characteristics, and film morphology. The ripple-patterned base changed the evolution of the models as well, and this can be seen through the numerous films that appeared in that case. Changing the orientation of the substrate radically increases the frequency distribution causes the re-refining of the films. Also, concerning this, we have done a research with the deposition of a very thin titanium film on usual wafers in the annealing temperature and on substrate distortion.

Keywords DC magnetron sputtering · X-ray diffraction · Scanning electron microscopy

1 Introduction

Titanium, also known as the "Space Metal Era," is a chemical element with the atomic number 22. In addition to [as previously mentioned], it has a silver color. This metal, named for the Greek mythological Titans, was found to be very strong and possessing magical powers. Ti is the ninth-following mentioned above, of alum, and after that, iron and magnesium, it is the fourth most common element on the Earth. In almost every form of life and the world's water, the oxygen is present. Until the late 1930s, pure titanium was not manufactured in a commercial process due to its great affinity for oxygen and nitrogen. In the early 1980s, Sponge titanium production already exceeded 100,000 tons per year and was still steadily growing.

The weight power ratio of titanium is the higher. The strength is as heavy as steel, but 45% lighter. The costs in stainless steel have been 1.3 times the price. The module

B. Rajagobalan (✉)
Department of Physics (S&H), IFET College of Engineering, Villupuram, Tamil Nadu, India

G. Meenakshi
Department of Physics, Kanchi Mamunivar Centre for P.G. Studies, Puducherry, India

S. Suhag et al. (eds.), *Control and Measurement Applications for Smart Grid*,
Lecture Notes in Electrical Engineering 822,
https://doi.org/10.1007/978-981-16-7664-2_16

Fig. 1 BCC

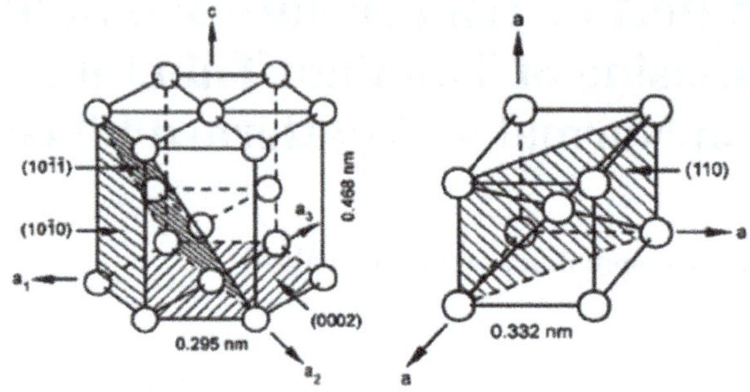

Fig. 2 HCP

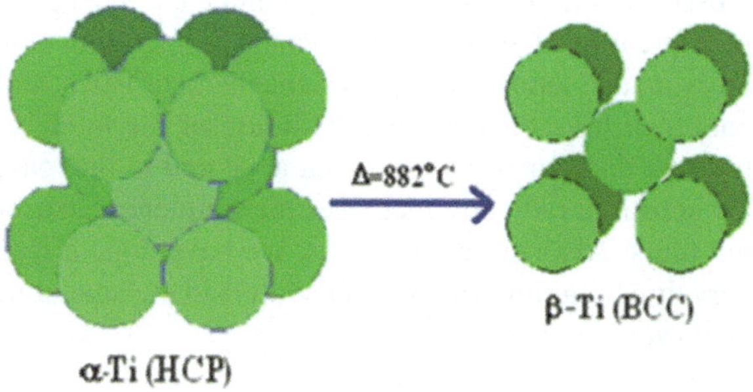

of steel is 55%. In most settings, including the human body, titanium is extremely corrosion-resistant, even higher than that of stainless steel. Titanium has two simple crystal structures [1] (Figs. 1 and 2).

1.1 Thin Film

Thin films vary on their thickness from solid materials. For the sake of simplicity, "a thin film" is an atomic layer of material, or a single-layer of material with an approximate thickness of one to five micrometers can be used. It is divided into two types, namely thin and thick films.

A thin film represents a type of small dimension material produced through the random nucleation and growth processes of atomically/ionically/molecular species on a substrate, which are individually condensed/reacted.

Narayandass et al. [2] found that titanium (Ti) thin films with different cathode strength have been deposited with DC magnetron sputtering under conventional conditions. Ahmed et al. [3] showed that the titanium nitride (TiN) film deposition on the substrate of Si (100) using DC magnetron sputter technique has been documented. Chawla et al. [4] displayed that titanium (Ti) thin films deposited in glass substrate were characterized by XRD, SEM, and AFM in magnetron under various deposition parameters. On silicon surfaces Xe+ bombardment, there are very few studies published, but the magnetron sputtering of Ar+ was less. Ponon et al. [5] studied the material and electrical properties of reactively sputtered TiNx following post-deposition annealing in an oxygen-free environment. As the nitrogen flow rate increased during deposition, the as-deposited TiNx resistivity decreased. Wang [6] suggested that Al and F co-doped ZnO thin films were deposited on glass substrates at room temperature, 100, and 200 °C via radio frequency magnetron sputtering using a 1.5 weight percent ZnF_2 and 1 weight percent Al_2O_3 co-doped ZnO target. Iqbal and Mohd-Yasin [7] determined that the film's specific crystal orientations are determined by the ions' landing kinetic energies and the adatoms' surface mobility. Arunachalam et al. [8] investigated that TiO_2 Thin Zn-doped films have been deposited by spray pyrolysis onto glass substrates. The uncontaminated TiO_2 and doped films were polycrystal-like in nature with a tetragonal structure, which was oriented preferentially along the (101) plane.

Ohtsuka et al. [9] studied the impact of surface morphology, crystalline quality, and residual stress sputtering pressure on aluminum nitride (AlN) films deposited with pulsed reacting DC at 823 K has been investigated on high quality *c*-plane nitride sapphire substrates AlN thin layers. Wang et al. [10] demonstrated that using DC-sputtering with the Al target, *c*-axis oriented aluminum nitride (AlN) films were successfully deposited on glass substrates without the use of buffer layers. We investigated the crystalline quality, deposition rate, surface morphology, and residual stress of AlN films grown with various sputtering parameters.

According to Jiao et al. [11], their research employs two distinct approaches for studying aluminum nitride (AlN) thin films via radiofrequency magnetron sputtering in a water-cooled environment. After depositing Si onto the AlN films, this (or another suitable substrate), the first goal is to explore how different types of Si substrate interact with the film properties. The second part of the experiment will be to determine the influence of RF power and N_2 flow on AlN thin-sput sputtered-on-SiO_2. It varies with flow rate between N_2 and Ar for expanded AlN, but the crystal peak is formed at a maximum flow rate. How the deposition rate and structural properties of AlN thin films are influenced by the substrate temperature, such as preferred *c*-axis and surface structure, were evaluated by Jin et al.

Moreira et al. [12] investigated the effects of AlN thin films formed via DC reactive magnetron sputtering with a range of sputtering forces and N_2 concentrations. To explore the influence of these parameters on structural and morphological properties, it was necessary to study the effects. and the most promising DC magnetron sputtering method is based on gasechanic aluminum deposition which is based on the use of N_2 and is suggested by them as being the most practical to use as reactive for films is a DCM, this being said (50%). Following air oxidation, AlN films were mainly formed by the AlN after a surface was exposed to the air, with a layer of a less-coherent AlON thinning out on top. A more complete study has now been conducted by Chiu et al. [13] on the case of the AlN thin film deposition on Pt electrodes using RF sputtering, in which many in-depth measurements have been performed. While all of these parameters significantly influenced the crystal orientation and texture of the AlN films, they also had an effect on the strength, flow rate, N_2 concentration, and target distance. Venkataraj et al. [14] reported on the use of reactive DC magnetron sputtering to deposit aluminum nitride films on glass, Si (100), and graphite substrates at varying nitrogen flow rates but maintaining the same overall strain (0.8 Pa). Unlike oxide sputtering, the Al target does not exhibit significant hysteresis, as determined by quartz crystal microbalance studies. Duquenne et al. [15] demonstrated the use of reactive magnetron sputtering to fabricate aluminum nitride films. Magnetic structure has a significant impact on film production and origination selection. Two distinct processes can be identified based on the magnetic structure. Second, when the AlN films are scanned with a balanced magnetron, they exhibit a low tensile residual tension, a small grain size, and an oxygen content of approximately 5%. Second, without the need for substrate heating, AlN films with larger grain sizes in the 80–100 nm range, a higher density, and a low oxygen content of less than 0.5% are produced using an unbalanced magnetron.

Andreas et al. [16] determined that the scattering and diffraction of ion beams in crystalline silicon indicates the presence of multiple ion species. When compared with Ar+ ion, the dimensions of samples of different parameters like wave periods, unstructured layer thickness, and altitude of the amplitude was improved. In this respect, we agreed to research the effect of annealing temperature and substrate biasing of the titanium thin film deposited on silicon wafers in normal and ripple-patterned substrate.

2 Experimental Procedure

Nowadays, for thin-film repositories on metals, ceramics, and polymers, magnetrons are commonly used. Sputtering the magnetron is one of the easiest options to generate the requisite material for a fast thin film. Different types of magnetrons are available, depending on the requirement. The present study is aimed at studying the planar magnetron sputtering system. Profiling, pressure, and other plasma conditions of the magnetron depend on the magnetron output. The majority of large coaters currently use sputter deposition planar magnetrons. The planar magnetron current–voltage properties are found to obey the relation

$$I = kvn \tag{1}$$

where k denotes constant and n denotes exponential range between 3 and 15 [17, 18]. It is clear that the magnetic field (B) profile is an integral part of the planar magnetrons' output.

2.1 Normal Substrate

The samples are synthesized externally in a commercial way. Preparation of the standard substrate clean and unscraper normal silicon (substrate), quantity-2, was

Fig. 3 Target Ti (titanium)

held on the sample holder of the DC planar sputter in a shutter closed on top of the sample stick. By locking the face lid using screws to prevent any kind of leakage, the system was locked. With the aid of vacuum pumps, vacuum was created inside the chamber, and water was also used to provide cooling for pumps. Our samples can be oiled if cooling is not given, and the entire device is impaired. With the aid of gauges attached to the device, the pressure was tested. Argon gas was injected inside the chamber after 2–3 h of arrival with appropriate pressure, i.e., exceeding a pressure of 5–6 $\times$ 10^{-5} mbar, and the flow was tested with the intake valve before the pressure fluctuations came to an end, and it is getting persistent at 3 $\times$ 10^{-3} mbar. Following pressure maintenance, the DC power has been turned on by steadily raising the voltage from 500 to 550 V, in which purple argon gas plasma can be visualized from the glass chamber window.

The sputtering process of the negative target, i.e., titanium, commences deposition as soon as plasma is produced. Shutter over the holder of the sample was exposed to the silicon substrate, and time was noticed by stop watching for exposure to plasma sputtering. The goal atoms thrown on a substrate (silicon) by the argon ions sputter and form an atomic layer of titanium. After 14 min, the DC supply was stopped, so that when thick titanium films were created, the sputtering process stopped, if more were deposited. The system facial lid was opened after 1 h of a vacuum explosion, and thin films were deposited on the sample of silicone using the tweezers from the holder without scratching the surface and packaged into a suitable plastic container and marked off.

The characterization of silicon samples can be done. The normal silicon substrate is thus made. Fundamental parameters used for the magnetron sputtering process were as follows; base pressure is 5.7 $\times$ 10^{-5} mbar, operated pressure is 3 $\times$ 10^{-3} mbar, voltage is 550 V, current is 0.90 A, and time of deposition is 14 min (Figs. 3 and 4).

2.2 *Ripple Patterned Substrate*

The preparation of a ripple-patterned substrate is described below. These ions produce regular-periodic ripples with a source energy of 67° below the surface conductive

Fig. 4 Si (silicon) substrate

anode are formed at a 500 eV (100 eV) source voltage on a Kaufman-type surface area substrate. The original root mean square is less than 0.1 nm beneath the surface. During the exposure to the sample is maintained at a constant temperature of 15 °C and under working pressure of 2 × 10 to 4 bar. Then after the film was created, the nanoparticles were deposited, they were used for post-exposure maintenance. AFM non-contact for ripple patterned surface topography modules has been mainly characterized for performance. Its Plasmon resonance has been observed in reflecting spectroscopy and SEM-characterized nanoparticles.

Si (001) was grown using an in-depth explanation of the technique native-oxide-covered rippled nanostructures in the Helmholtz-Zentrum Dresden-Rossendorf [1]. Concisely, the Si (001) nanoscale substrate was irradiated at room temperature at a beam of 500 eV Arians aligned with an azimuth angle of 67°, at a nano-scale of nano-scale period between ~20 and ~50 nm, and an amplitude of ~0.8 to ~1.5 nm, determined by AFM; The range of fluences from ion was ~1016 cm^{-2} to ~1019 cm^{-2}. The error in the frequency of the AFM images was measured at $\mp15\%$ of the energy spectral density. The as prepared patterned substrate is shown below with the same basic parameters as shown in the above method. Two paths, one parallel and one perpendicular, are selected for the ripple patterned substrate (Figs. 5 and 6).

3 Characterization

The following are explained and recorded in the following steps. A titanium film prepared on silicon wafer in the standard and ripple pattern substrate.

3.1 XRD Analysis

Powder diffraction is a science of crystallographic, crystallite (grain size) and choice polycrystalline or solid powder samples (XRD). The schematic diagram of powder diffraction is shown in Fig. 7. Powder diffraction is used in a widespread way by contrasting diffraction data with an ICDD database to classify unknown substances.

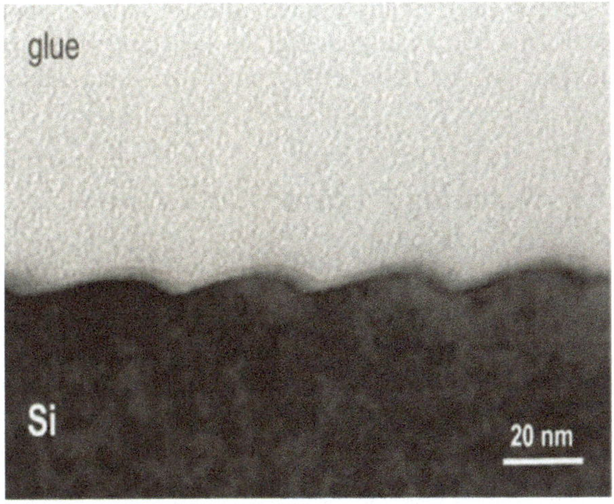

Fig. 5 Cross-sectional TEM image of rippled Si (001)

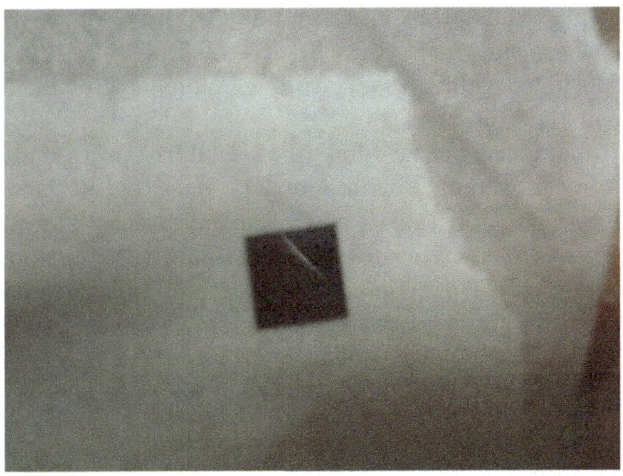

Fig. 6 Prepared ripple patterned image

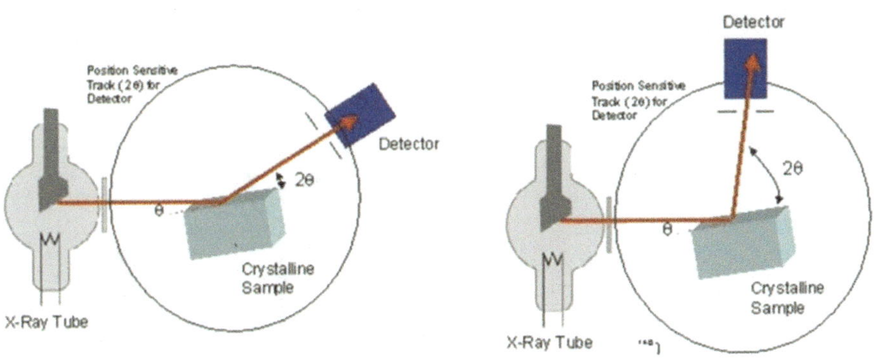

Fig. 7 Schematic diagram of powder diffraction

It can also be employed to classify heterogeneous solid mixtures in order to assess the relative abundance and structural information on unknown materials in combination with grid refining methods such as Rietveld refining [19].

3.2 Scanning Electron Microscopy

Scanning electron microscopy (SEM) involves scanning the surface areas with a pulsed electron beam in *r*-direction and performing high-energy sample staining with a continuous-stripping, interrupted electrons in a raster pattern. Electrons come into contact with the sample generators, containing information about the topography, composition, and other properties, such as electrical conductivity (Fig. 8).

In contrast with the perpendicular direction and normal substrate, the crystallite size increases significantly in the parallel direction.

From these two contrasts, we can conclude that the ripple-patterned substrate displays anisotropy as a result of the normal ripple templates in the substrate. SEM analysis shows also that the deposited titanium thin film is condensed in a typical substrate, while the very simple columnar growth structure is observed in ripple-patterned substrate (Fig. 9).

Inorganic plastic films (that is, films not formed on polymers such as plastics or metals) contrast with polymeric ones (plastic- and metal-based films). From the information on Table 1 and visual appearance information, it can be concluded that strain is released in annealing. Crystallite reduction pressure is decreased when stress is released. This helps increase the mobility of the film's adatoms on the substrate by increasing the substrate temperature.

Fig. 8 Schematic diagram of scanning electron microscope

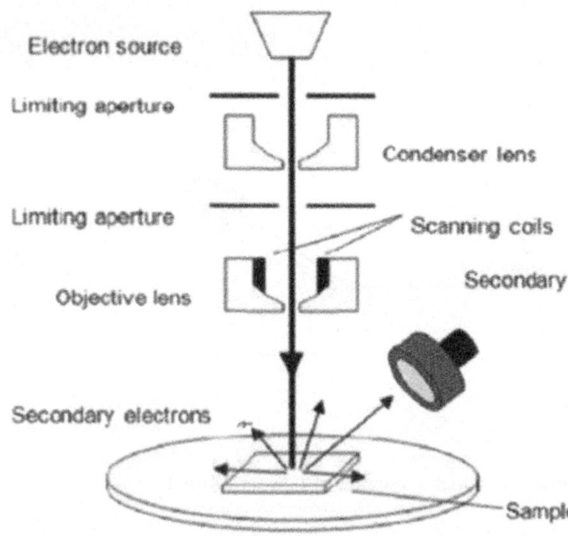

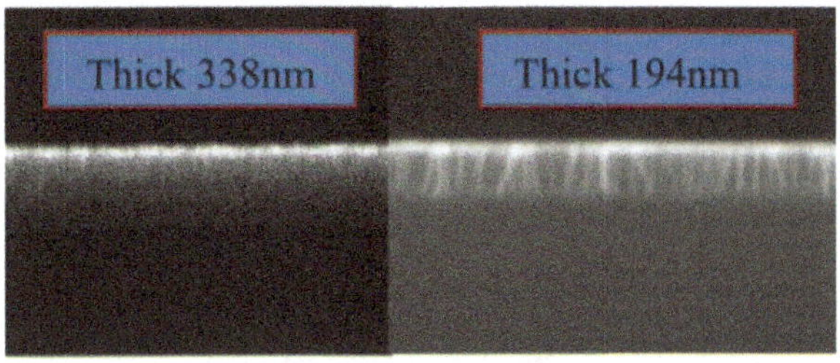

Fig. 9 Analysis of SEM in normal and ripple pattern substrate

Table 1 Parameters versus slope

Parameters	Slope (η)
Annealing—500 °C	1.8437
Annealing—300 °C	1.5787
−200 V biasing	−7.1464
−50 V biasing	0.7060

Finally, the film made on a typical substrate is compared to a film with different voltages. From Table 1, the strain is compressed at −200 V, resulting in crystallite size reduction, and the bias at −50 V has lower impact on the strain. The strain is released as the substrate is heated. We may assume, therefore, that the bias of the substrate helps improve the film quality.

4 Results and Discussion

4.1 X-Ray Diffraction Analysis

The pattern of X-ray diffraction of titanium thin films grown on different substrates is discussed below.

4.1.1 Effect of Annealing Temperature on XRD Spectra of Ti Film

XRD spectra of the titanium film grown at room temperature and at different substrate temperatures like 300 and 500 °C is shown in Fig. 10a, b. The observed peaks with its corresponding <hkl> values are compared with the normal substrate shown in Fig. 10b.

The peaks are indicated by the JCPDS (ICDs#44-1294) standard software, which confirms the titanium presence. As indicated in the following data, titanium films (100), (002), (101), and (103) were observed to show preferred diffuse scattering properties, which makes (110) rare titanium more readily apparent at 45° and intense peaks unlikely candidates for interpretation. Along with titanium peaks, planes of

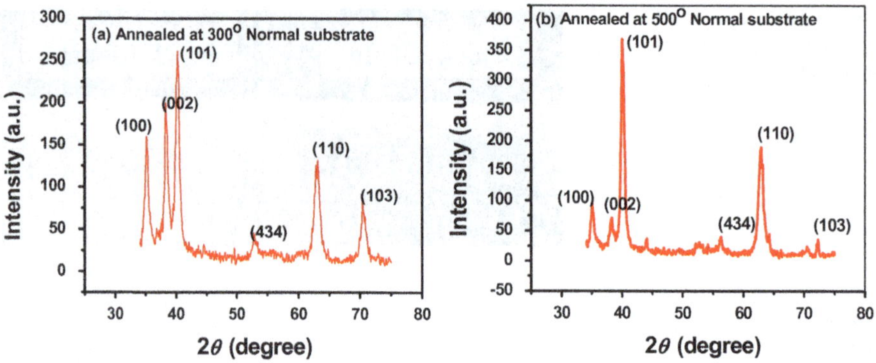

Fig. 10 **a, b** Intensity versus 2θ graphs of Ti films annealed at 300 and 500 °C temperatures

Silicon oxide peak (434) is also observed. One can clearly see the normal substrate without annealed gives high intensity peak around (100) plane. However, the original (100) subsidies and (101) appear as the desired orientation when the substrates are renewed with some temperature. The peak shift due to the rise in the temperature of the substrate increases orientation.

Even if we compare the graph, the peaks (101) and (110) observed at 500 °C give better intensity than at 300 °C. This is because the higher substrate temperature could facilitate the enhanced mobility of adatoms in the film surface and favored the formation of orientation of grains. The slope shown in Table 1 also clearly tells us that the normal substrate without annealed gives negative slope, whereas the substrate with annealed gives positive slope, meaning that the strain is released as the substrate is heated. Thus substrate temperature plays a key role in strain development. Thus, we analyzed all our XRD data files using the origin graph and smoothed it. The range is shown after smoothing in Fig. 10a, b. So, in the rest of the discussion, we just showed smooth spectrums.

4.1.2 Effect of Substrate Biasing on XRD Spectra of Ti Film

The observed XRD spectra of the films prepared at a biased substrate of −50 and −200 V with their corresponding <hkl> values are shown in Fig. 11a, b. XRD spectra of normal substrate without biasing are shown in Fig. 11b. The ions which are deposited by an elevated by a negative substrate bias and/energy provide a great momentum to the atoms enough to improve crystallinity and support the growth of thin-film growth both carry momentum and position anions. Strictly so, eliminating all prejudices results in a better video image. Electric potential between the partial layer and substrate, both produced by accelerating charges, results in expansions. Bombarding the target creates a compact film and pushes molecules away from the film that are not well connected. This can also be seen from Table 1 that the slope

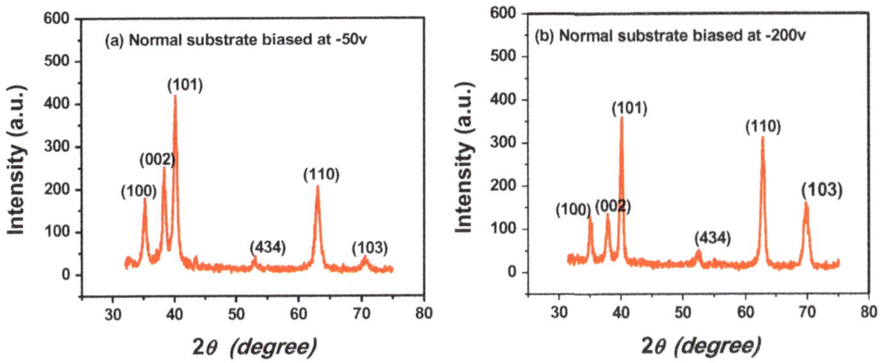

Fig. 11 **a, b** Intensity versus 2θ graphs of Ti film prepared at a substrate biasing of −50 and −200 V

Table 2 Parameters versus crystallite size

Parameters	Crystallite size (nm)
Annealing—500 °C	24
Annealing—300 °C	30
−200 V biasing	8
−50 V biasing	25

has higher value with less substrate bias and vice versa. This clearly tells us lower strain will have less crystallite size and higher strain will have more crystallite size.

In Table 2, the crystallite size of the deposited Ti thin film is shown in the range 8–25 nm, and thus the synthesized Titanium (Ti) thin film is nanostructured.

Looking at the graph, we conclude that the ripple substrate patterns are isotropic because they have constant template ripple spacing. Normal growth also reveals the abundance of the titanium thin film and the simple columnar structure, while the columnar growth is decreasing.

A comparison was made between the film synthesis growth method on a normal substrate and an annealed substrate. When the substrate is under strain, the normal strain and annealed, they determine that they are entirely separate from each other. The adatoms are effectively raised in temperature, and this allows for a broader separation on the film surface.

. Table 2 shows that the crystallite size is negatively affected at −200 V, and the amorphous is also at −50 V, so we can assume that the substrate is neutral at this voltage. Based on the assumptions on the substrate bias, we can understand that film quality is enhanced by the formation of orientation of grains.

4.2 XRD Spectra of Titanium Nitride

The XRD data values were taken from thesis "Nanostructured Titanium Nitride Thin Film Deposition by Cylindrical Magnetron Sputtering" by Prerna Sharma at FCIPT, Institute for Plasma Research, Gandhinagar. With the help of these XRD values, I have calculated strain and crystallite size.

4.2.1 XRD Spectra of TiN for Various Pressures

The same set of experiments was performed to deposit titanium nitride on normal sample under various pressures. As mentioned earlier to find the crystallite size and slope, we have to analyze the XRD spectrum. After analyzing the spectra the final form of the graph is Fig. 12a, b.

Tables 3 and 4 show the deposited TiN thin film for different pressure in normal substrate, and so, TiN thin, synthesized film, with a nanostructure, shows the XRD analysis. The crystallite size ranges from 38 to 120 nm.

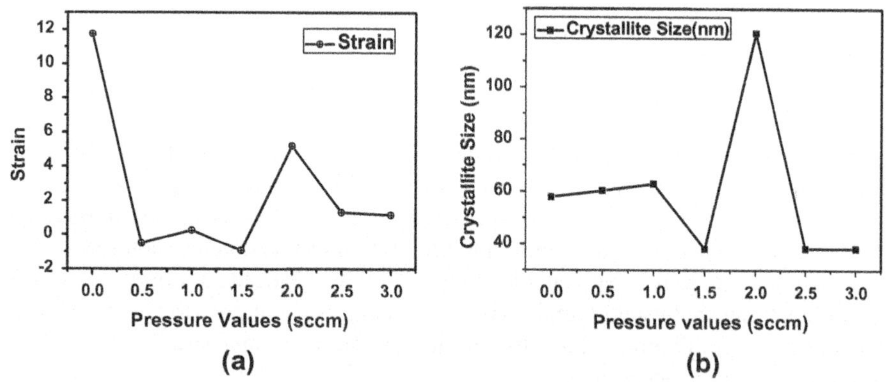

Fig. 12 **a** Strain versus pressure values. **b** Crystallite size versus pressure values

Table 3 Pressure versus crystallite size

Pressure values (sccm)	Crystallite size (nm)
0	57
0.5	60
1	62
1.5	38
2	120
2.5	38
3	38

Table 4 Pressure versus slope

Pressure values (sccm)	Slope (η)
0	11.7425
0.5	−0.5086
1	0.2699
1.5	−0.9159
2	5.1959
2.5	1.3264
3	1.1786

Table 3 and the graph indicate that the strain is decreased with increase in crystallite size. This is because the rate of formation of TiN depends on the amount of Ti atom being sputtered from the target and the rate at which they get nitride by reacting with nitrogen. Thickness of the film is decreasing with increase in nitrogen flow.

4.2.2 XRD Spectra of TiN for Different Biasing Voltages

The same set of experiments was performed to deposit titanium nitride on normal sample under different voltages. The final form of the analyzed XRD graph is shown in Fig. 13a, b.

The deposited TiN thin film on normal substrate for various biasing voltages is shown in Tables 5 and 6, respectively. Therefore, the synthesized TiN thin film displays the nanostructure in the range of 27–44 nm, represented by XRD analysis.

Deposition of pure titanium on normal silicon on biasing the substrate has effect from lower voltage to higher voltage. That is, the substrate biased at lower voltage has less strain and vice versa.

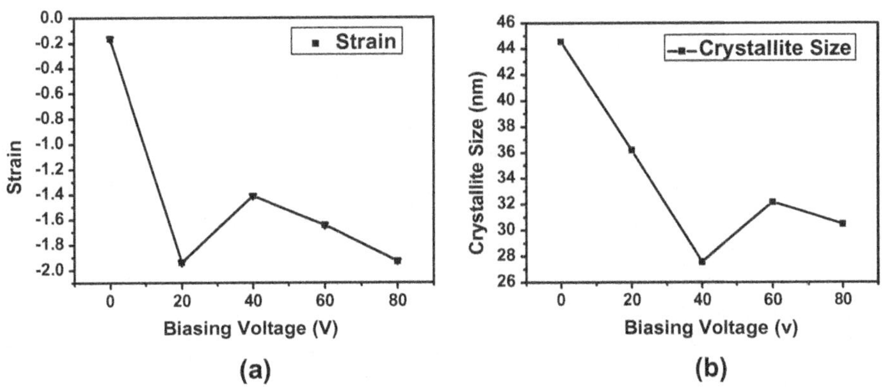

Fig. 13 **a** Strain versus biasing voltage. **b** Crystallite size versus biasing voltage

Table 5 Table showing biasing voltage versus crystallite size

Biasing voltage (V)	Crystallite size (nm)
0	44
20	36
40	27
60	32
80	30

Table 6 Biasing voltage versus slope

Biasing voltage (V)	Slope (η)
0	−0.1641
20	−1.9385
40	−1.4135
60	−1.6436
80	−1.9288

In case of deposition of titanium nitride on the same sample have no effect, it means that the strain is compressed and the crystallite size become less as we increase the biasing voltage from 0 to 80 V.

The reason may be the nitrogen content in the film. Higher substrate bias values contribute to coarse grains and cell dislocation limits and thus less coating hardness. For better coating properties, an optimal value of substrate distortion is therefore needed. Thickness of the thin film is decreasing at higher substrate bias values which can be attributed to re-sputtering mechanism of the growing film. The crystallite dimensions and strain are shown for different parameters in Tables 5 and 6, respectively.

4.3 Scanning Electron Microscopy Analysis

Figure 14a, b shows titanium (Ti) surface morphology studies on the SEM silicon (Si) substrate (LEO—430VP) for various parameters.

Surface morphology analysis of the silicon substrate with SEM shows that the film has inhomogenic growth, as shown in Fig. 14a. The growth is approximately 350 nm in size. It looks as though the surface of the specimen is thick and very silver. The picture shown nearby shows the growth of the film clearly.

In contrast with ripple patterned substrate, the growth of a regular substrate is achieved. We can clearly see that in normal substrates growth is greater than any other substrate.

Figure 14b shows the SEM growth of titanium (Ti) on silicon (Si) in normal substrate at 300 °C. The size of the range lays around 300 nm. When compared to

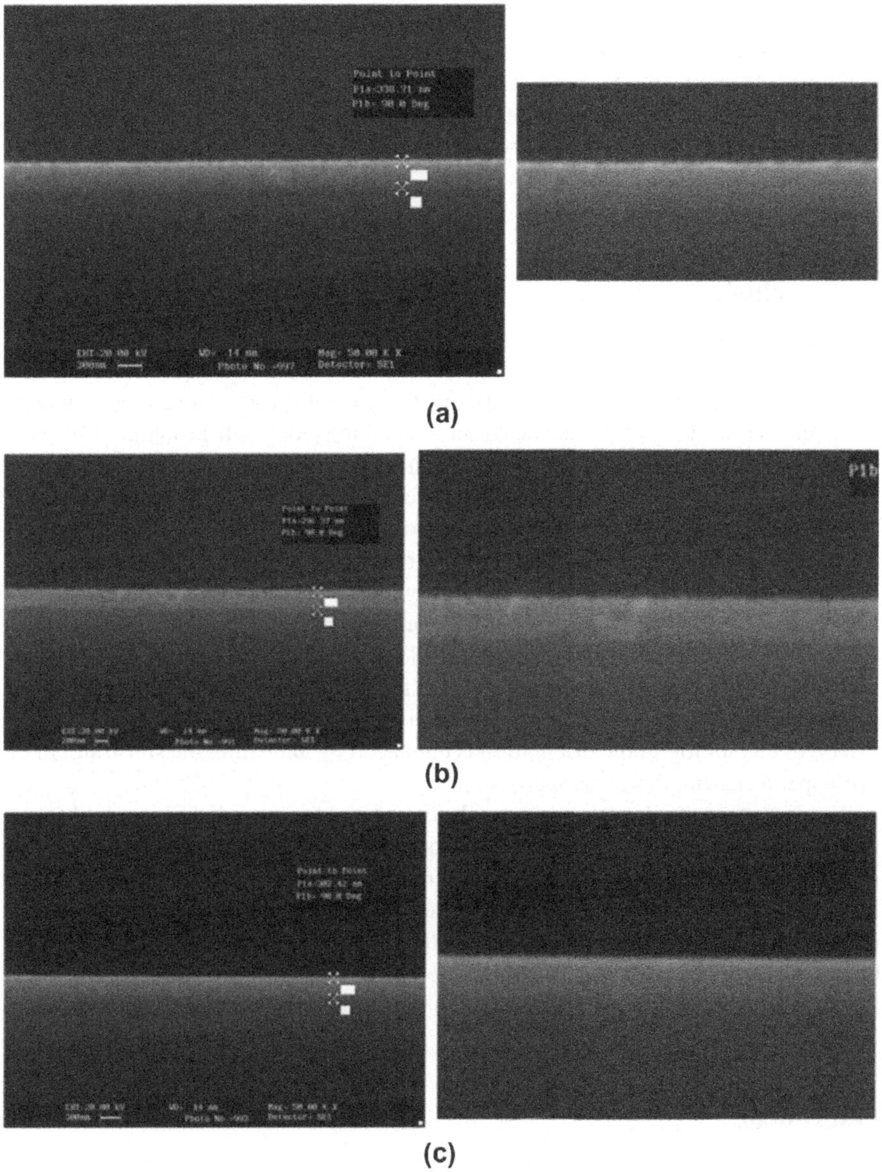

Fig. 14 **a** SEM image growth—normal substrate. **b** SEM image growth at 300 °C. **c** SEM image growth at −200 V

normal substrate, the thickness of the film is decreased, but the quality of the film is increased as we heat the substrate.

Figure 14c shows the SEM growth of Ti on Si biased at −200 V. The size of the range lies in the range of around 300 nm. If we look into the strain measurement, the strain is compressed with less crystallite size when compared to normal substrate without biasing.

5 Conclusion

To study the physical aspects of plasma coated material is an excellent research work which has been carried out by scientist for the past 50 years. The earlier scientists who were working in this field faced an exceedingly difficult problem. The paths followed were torturous indeed the amount of effort expended by generation of brilliant scientist was considerable. Because of these scientist' efforts, one could able to continue the experimental study in the field. Hence, the author got interest in this field, did research work and got fruitful results with necessary support.

The most effective and rapid way to grow different films is through magnetron sputtering. There is a possibility to adjust operating parameters such as pressure and temperature of the sheet, and physical characteristics of the film. Patterns in substrate often change the growth of films as seen. The substrate's major bias contributes to the film being re-sputtered. Thinner film in a certain parameter was therefore observed. Crystallographic orientation is enhanced by annealing the substrate. In the XRD-spectrum, this film was seen.

Ripple patterned substrate strain can be calculated for annealed film growth There are ways to bias the film measurement on the ripple-patterned substrate. It is possible to obtain a cross-section from the film grown on ripple-patterned-developed plastic of the high-atomica microtome. X-Ray diffraction spectrum measurements in various azimuthal angles in order to learn about the film-patterned-substrate anisotropy strain. The presence of ripple patterned substrate has to be checked after the removal of conduction of atomic force microscopy on the ripple patterned substrate.

References

1. Meenakshi G et al (2015) Investigation of anisotropic strain induced in Ti thin film grown on patterned substrate. Mater Sci 04(01). ISSN (online) 2394-1537
2. Narayandass SK et al (2010) The effect of thickness of titanium nitride coatings on bacterial adhesion 2. Trends Biomater Artif Organs 24:90–93. http://www.sbaoi.org
3. Arshi N, Lu J, Koo BH, Lee CG, Ahmed F (2012) Effect of nitrogen flow rate on the properties of TiN film deposited by e beam evaporation technique. Appl Surf Sci 258:8498–8505
4. Chawla AK et al (2010) Effect of ambient gas on structural and optical properties of titanium oxynitride films. Appl Surf Sci 256(13):4129–4135

5. Ponon NK, Appleby DJR, Arac E, King PJ, Ganti S, Kwa KSK, O'Neill A (2015) Effect of deposition conditions and post deposition anneal on reactively sputtered titanium nitride thin films. Thin Solid Films 578:31–37
6. Wang F-H (2016) Effect of substrate temperature on transparent conducting Al and F co-doped ZnO thin films prepared by rf magnetron sputtering. Appl Surf Sci 370:83–91
7. Iqbal A, Mohd-Yasin F (2018) Reactive sputtering of aluminum nitride (002) thin films for piezoelectric applications: a review. Sensors 18
8. Arunachalam A, Dhanapandian S, Manoharan C, Sivakumar G (2015) Physical properties of Zn doped TiO_2 thin films with spray pyrolysis technique and its effects in antibacterial activity. Spectrochim Acta Part A Mol Biomol Spectrosc 138:105–112
9. Ohtsuka M, Takeuchi H, Fukuyama H (2016) Effect of sputtering pressure on crystalline quality and residual stress of AlN films deposited at 823 K on nitrided sapphire substrates by pulsed DC reactive sputtering. Jpn J Appl Phys 55:05FD08
10. Wang J, Zhang Q, Yang G, Yao C, Li Y, Sun R, Zhao J, Gao S (2016) Effect of substrate temperature and bias voltage on the properties in DC magnetron sputtered AlN films on glass substrates. J Mater Sci Mater Electron 27:3026–3032
11. Jiao X, Shi Y, Zhong H, Zhang R, Yang J (2015) ALN thin films deposited on different Si-based substrates through RF magnetron sputtering. J Mater Sci Mater Electron 26:801–808
12. Moreira M et al (2011) Electrical characterization and morphological properties of AlN films prepared by dc reactive magnetron sputtering. Microelectron Eng 88:802–806
13. Chiu K-H et al (2007) Deposition and characterization of reactive magnetron sputtered aluminum nitride thin films for film bulk acoustic wave resonator. Thin Solid Films 515:4819–4825
14. Venkataraj S et al (2006) Structural, optical and mechanical properties of aluminium nitride films prepared by reactive dc magnetron sputtering. Thin Solid Films 502:235–239
15. Duquenne C et al (2008) Impact of magnetron configuration on plasma and film properties of sputtered aluminum nitride thin films. J Appl Phys 104:063301
16. Andreas et al (2014) Mater Sci Eng R Rep 86:1–44
17. Ranjan M, Oates TWH, Facsko S, Möller W (2010) Opt Lett 35:2576
18. Oates TWH, Ranjan M, Facsko S, Arwin H (2011) Opt Express 19:2014
19. Thornton JA (1989) Stress-related effects in thin films. Thin Solid Films 171:5–31

IoT-Based Electric Monitoring System for Smart Electric Meter Application

Muhammad Faiz Bin Mohd Kamal, Ranjit Singh Sarban Singh, Siti Aisyah Anas, and Mahaveerakannan Renganathan

Abstract This paper describes about the proposed Internet of things-based electric monitoring system for smart electric meter application which is a supporting device that provides information that can be checked with the actual monthly electricity bill that is provided by the respective energy provider. This system is proposed as currently smart electric meters are being progressively installed at individual residential in Malaysia to reaffirm the electricity consumption measurement is accurate and does not lead to any kind of inaccuracy in the measured electricity consumption. With that, looking at the smart electric meter advancement, consumers should be happy that their residential is equipped with such a great technology, but some have shown unhappiness due to increase in their actual monthly measured electricity consumption. Many consumers complained that there was an increment in their electricity billing after the conventional electric meter replacements. Hence, the consumers lodged reports to the authority which encouraged the energy provider to investigate on the raised issue. An investigation report was presented, and the report explains that the installed smart electric meters have no faulty and inaccuracy in performing the electricity consumption reading, but it sparks the unhappiness among the consumers. Looking into this aspect and to allow the consumers to self-monitor their electricity consumption, an Internet of things-based electric monitoring system for smart electric meter application is proposed. This system is developed to real-time measure and monitor the electricity consumption as well as to provide a secondary consumed electricity information check with the actual monthly power usage bill from the energy provider.

M. F. B. M. Kamal · R. S. S. Singh (✉) · S. A. Anas
Fakulti Kejuruteraan Elektronik dan Kejuruteraan Komputer (FKEKK), Centre for Telecommunication Research and Innovation (CeTRI), Universiti Teknikal Malaysia Melaka (UTeM), Hang Tuah Jaya, 76100 Durian Tunggal, Melaka, Malaysia
e-mail: ranjit.singh@utem.edu.my

M. Renganathan
Department of Computer Science and Engineering, Saveetha School of Engineering, Thandalam, Chennai 602105, India

S. Suhag et al. (eds.), *Control and Measurement Applications for Smart Grid*, Lecture Notes in Electrical Engineering 822,
https://doi.org/10.1007/978-981-16-7664-2_17

Keywords Smart electric meter · Internet of Things meter application · Internet of Things electricity consumption

1 Introduction

This paper describes about the proposed Internet of things-based electric monitoring system for smart electric meter application which is a supporting device that provides information that can be checked with the actual monthly electricity bill that is provided by the respective energy provider. This system is proposed as currently smart electric meters are being progressively installed at individual residential in Malaysia to reaffirm the electricity consumption measurement is accurate and does not lead to any kind of inaccuracy in the measured electricity consumption. With that, looking at the smart electric meter advancement, consumers should be happy that their residential is equipped with such a great technology, but some have shown unhappiness due to increase in their actual monthly measured electricity consumption. Many consumers complained that there was an increment in their electricity billing after the conventional electric meter replacements. Hence, the consumers lodged reports to the authority which encouraged the energy provider to investigate on the raised issue. An investigation report was presented, and the report explains that the installed smart electric meters have no faulty and inaccuracy in performing the electricity consumption reading, but it sparks the unhappiness among the consumers. Looking into this aspect and to allow the consumers to self-monitor their electricity consumption, an Internet of things-based electric monitoring system for smart electric meter application is proposed. This system is developed to real-time measure and monitor the electricity consumption as well as to provide a secondary consumed electricity information check with the actual monthly power usage bill from the energy provider.

2 Literature Review

Implementation and installation of Internet of things smart electric meter has become the focus and received attention over the years [1]. The installation of the Internet of things smart electric meter into houses has drastically increased due to the urbanization as well as to improve the sustainability, man power effectiveness, smart governance [2], and quality of data reading as well as measurement [3]. Research on Internet of things developed and implementation of smart electric meters such as [4–8] have been developed and implemented widely which especially tackles the user concern, reduces the human error, and provides effective energy consumption monitoring. Looking at the advantages of developing and implementation of Internet of things smart electric meters, the presented research does not mention the implementation of these smart electric meters from the user's perspective especially providing the quality of read consumed electricity data. Hence, the proposed Internet of things-based electric monitoring system for smart electric meter application has been developed and implemented into local houses to look into issues such as [9, 10]. The issues that were raised by the consumers when the smart electric meters were installed were the electricity billing cost started to increase, and this gave the negative impression about the benefits of installing the smart electric meters into houses. Therefore, in the following section, the development and implementation of the proposed Internet of things-based electric monitoring system for smart electric meter application is developed to be used as secondary device to measure and record the consumed electricity which later can be used to validate the actual received electricity bill from the electricity provider.

3 Methodology

This paper describes the method to monitor the electric consumption registered by AMR enabled energy meter in real-time using SCT-013-000 YHDC 100 A non-invasive alternating current (AC) sensor split core type current transformer (CT). This type of sensing current method using the principle of converting the primary current measured to secondary current according to turn ratio. As electric meters are owned by the utility/energy provider company which prohibited from any modification or tampered. This type of CT sensor is used to build-up with primary winding, magnetic core and secondary winding which allow to sense various range of current pass through the hole of current sensor to compute the electrical consumption. Moreover, the computed consumed electricity data is uploaded via internet on MySQL database as the cloud storage to store information according to date and time, current, power and house id.

3.1 Overview of Proposed Project

Figure 1 shows the proposed IoT based electric monitoring system for smart electric meter application for smart electric meter that is developed using power electronic devices. This can be categorized into three (3) parts which is measurement unit known as the smart electric meter, processing unit and management unit.

The measurement part is designed to obtain the analog data from the power supply line of smart electric meter. The front-end power measurement system processes the data by read input and calculate the pulse receive from the smart electric meter. A management data at the back-end known as storage system is used to record and stored the data output of consumed electricity into the cloud storage from front-end power measurement system. The recorded data is transferred wirelessly from the integrated microcontroller. The website application which is known as dashboard is

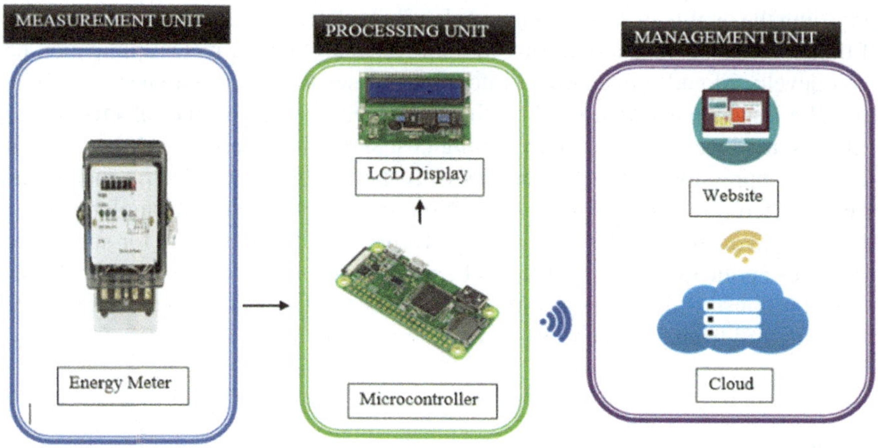

Fig. 1 Overview of proposed Internet of Things based monitoring system for smart meter application for smart electric meters

developed in order to allow the consumers or users to access the information on the consumed electricity and billing information.

3.2 *Energy Measurement*

In this project, the current sensor is used to measure and sense the current flow from installed smart electric meters. This type of sensor known as split core current transformer which responding to the magnetic field around a current carrying conductor. A signal sensor circuit is build-up to design the circuit element such as burden resistor and capacitor to carry the output signal of current transformer (CT). Since microcontroller un-build with General Purpose Input–Output (GPIO) analog input pin, the output of signal circuit is fed to an Analog to Digital Conversion (ADC) module. Then, analog current signal from the power supply is convert to digital form by integration of ADC as shown in Fig. 2.

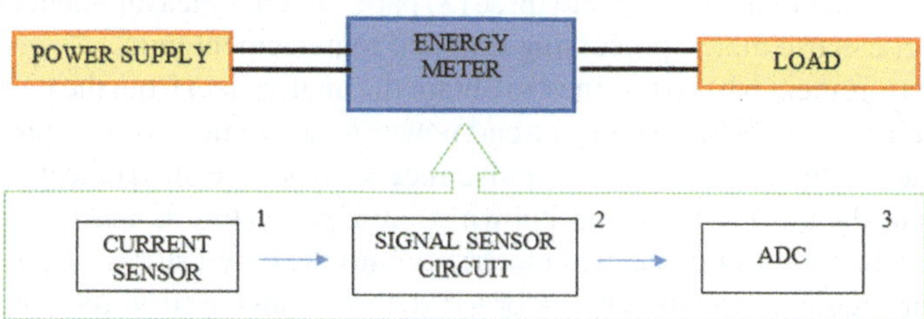

Fig. 2 Energy measurement system

3.3 Power Measurement and Recording System

An embedded code in Python Programming which is written to conduct measurement at the front-end is upload into Raspberry Pi Zero Wireless (RPW0). The process of measurement is shown in the flow chart shown in Fig. 3. Data signal is processed after is received from the ADC module which convert signal from analog to digital input. Firstly, RPW0 collects the input signal from ADC module once it is powered up. Data signal is converted into higher resolution of using the 16-bit conversion. The formula to gain power consumption in kWh is then implemented. Data input is display on LCD Display that include current and power consumption for every usage. After that data input is recorded into google sheet as a backup data recorded if the connection to database is failed or unstable. The data is updated every 1 s according to interval setting on RPW0. The explained work and operation occur completely in the RPW0, respectively.

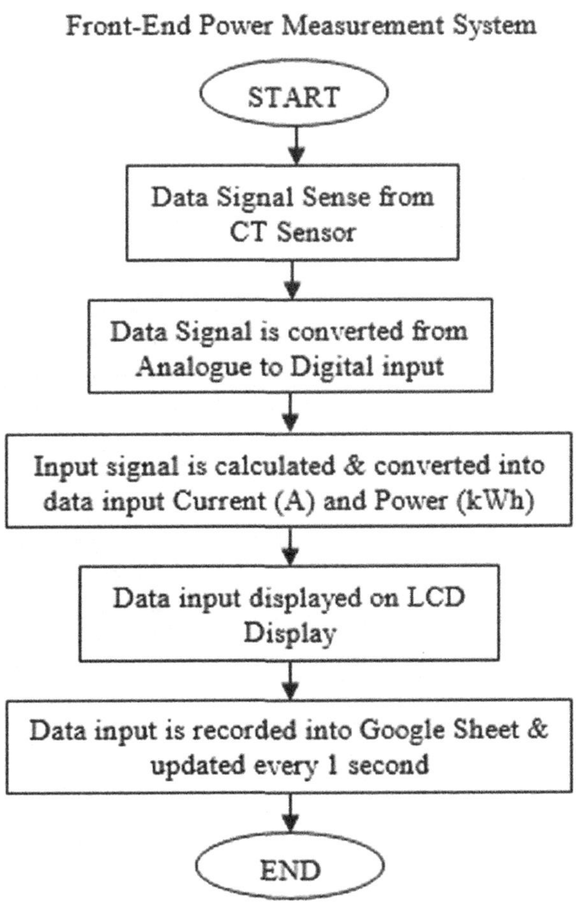

Fig. 3 Flow chart of front-end power measurement

The flowchart of back-end stored system is shown in Fig. 4 which consist of several parts which is front-end system, database and website application as dashboard. Every part mainly records the data input and send through the connection via the database. Firstly, the data input from the front-end measurement system is send and stored into database simultaneously. The provided input data is insert to database at every 1 s which the RPW0 sends the data to a simple PHP script. This script connects to the MySQL database and stores the received data in the database.

Meanwhile, the database will insert the input data into the website in real-time and update the data at every 1 s. The website will show all the data input from the database in every update. Lastly, the current and power data are recorded into history website for consumers or users view.

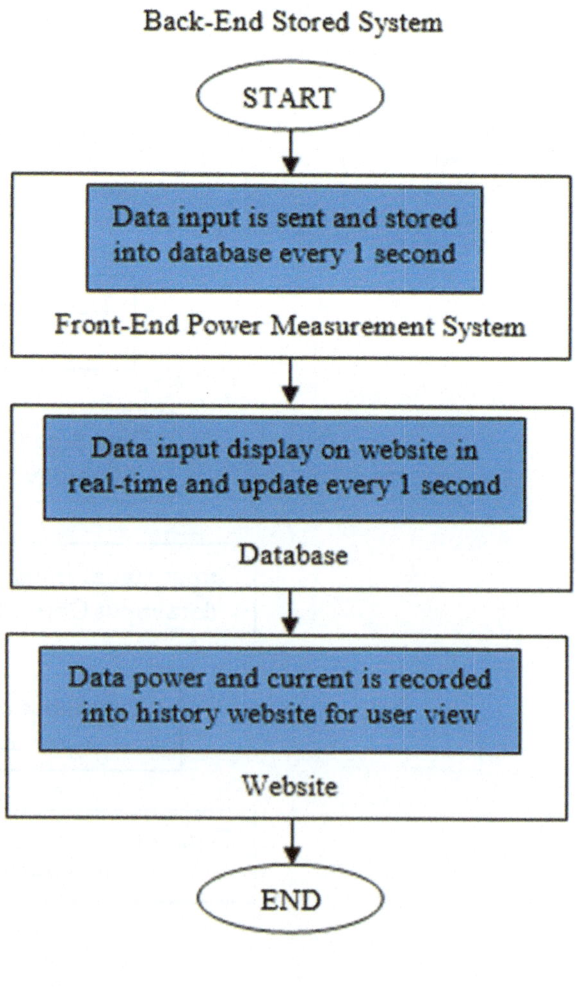

Fig. 4 Flow chart of back-end stored system

Fig. 5 Constructed IoT based monitoring system for smart meter application

4 Result and Discussion

4.1 Hardware Implementation

Figure 5 shows the completely constructed IoT based electric monitoring system for smart electric meter application. The proposed system shows the accuracy of the collected data at the LCD display. As the data collect, the current and power display show in two decimal places to give an accurate reading measurement of the system proposed. Reading data from the system to keep updating at every 1 s as this to allow the system to provide a real-time data of the measurement. The obtained result from the front-end system is uploaded into the back-end system in the database to store the read data and present the data at the website according to real-time data.

4.2 Website

Though the readings transfer from the RPW0 including current (A) and power (kWh) can be displayed in tabular form in the browser, for having a better view of the pattern of power consumption. Initially, the measurement of reading data from the front-end system is stored in the MySQL database server via internet. This communication establishes through sending the data to a simple PHP script. This script connects to the MySQL database and stores the data in the database. RPW0 upload the power consumption in unit of kWh into MySQL database every seconds. The latest data in the database is called whenever the user open the website. As the result, the latest data is shown in website in real-time according to collecting data of MySQL database. The real-time data from raspberry pi is shown through website referring to Fig. 6. The current and power in kW/h will keep updating if there is an increasing of power consumption in every usage.

The reading data stored in MySQL database can be shown in website which known as data logging. Figure 7 shows the data logging from MySQL database. This page available for the user to set the start date and end date to shows the information

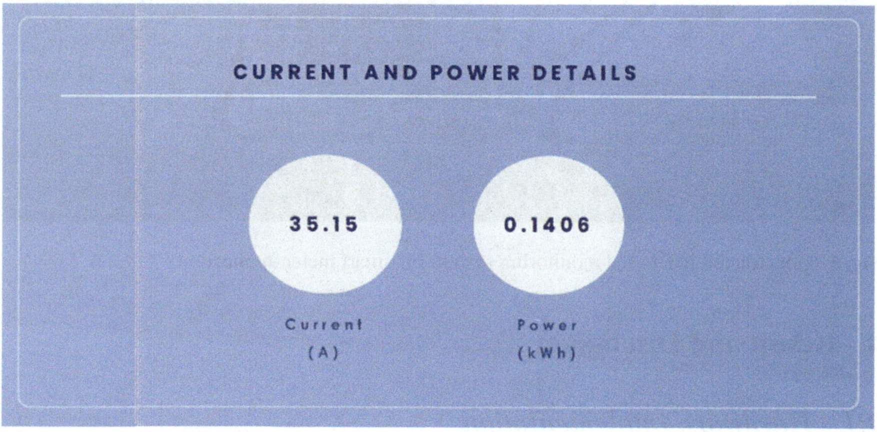

Fig. 6 Real-time current and power details

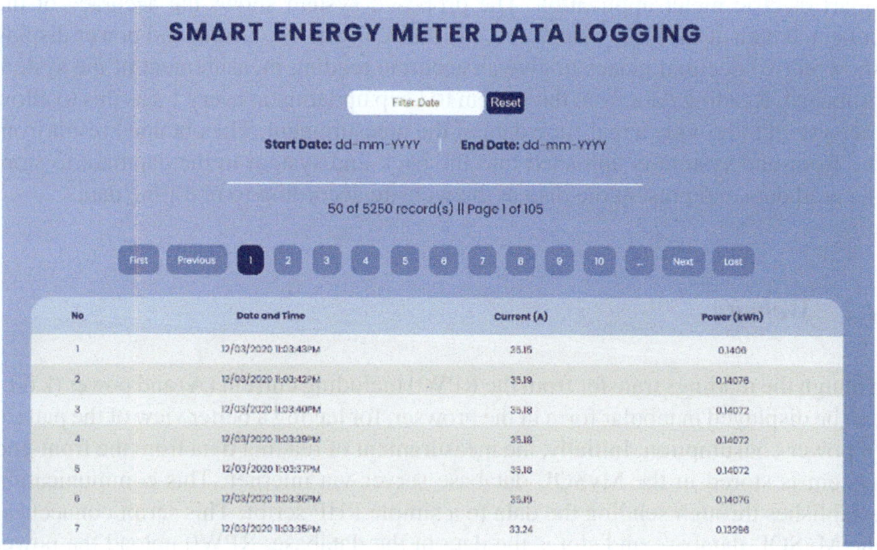

Fig. 7 Real-time smart energy meter data logging

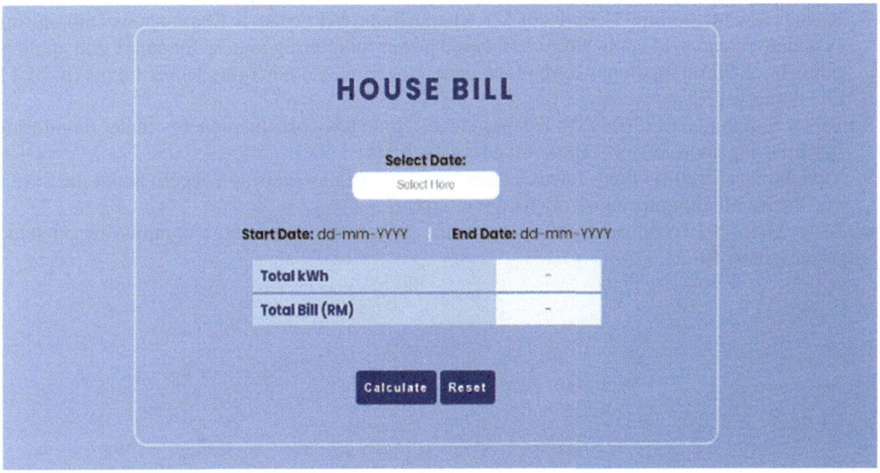

Fig. 8 Real-time billing details

of measuring current and power in kWh from MySQL database. Hence, data of current and power will show accurately according to period time selected by user from database.

Figure 8 shows the billing details which can be set according to user demand using date in calendar. It will show the total power in kWh and billing payment must be paid. For example, the set date is start on January 01, 2020 and end on January 31, 2020. The total of power consumption is 514 kWh and total bills is RM 187.42 according to the period set by user.

References

1. Adhya S, Saha D, Das A, Jana J, Saha H (2016) An IoT based smart solar photovoltaic remote monitoring and control unit. In: 2016 2nd international conference control. Instrumentation, energy communication. CIEC 2016, no October, pp 432–436
2. VS, KSK (2017) Smart energy metering and power theft control using arduino & GSM. In: 2017 2nd international conference for convergence in technology (I2CT), pp 858–861
3. Aboelmaged M, Abdelghani Y, Abd MA, Ghany E (2017) Wireless IoT based metering system for energy efficient smart cites. In: 2017 29th international conferene on microelectronics (ICM), no Icm, pp 3–6
4. Barman BK, Yadav SN, Kumar S, Gope S (2018) IoT based smart energy meter for efficient energy utilization in smart grid. In: 2018 2nd international conference on power, energy and environment: towards smart technology (ICEPE), pp 1–5
5. Chooruang K, Meekul K (2018) Design of an IoT energy monitoring system. In: Design of an IoT energy monitoring system, pp 3–6
6. Devadhanishini AY, RK M, Nandinipriya N, V S, Gowri PGP (2019) Smart power monitoring system using IoT. In: 2019 5th international conference on advanced computing & communication systems (ICACCS 2019), no Icaccs, pp 2019–2022

7. Khan F, Abu M, Siddiqui B, Rehman AU, Khan J, Sadiq MT (2020) IoT based power monitoring system for smart grid applications IoT based power monitoring system for smart grid applications. In: 2020 international conference on engineering and emerging technologies (ICEET), no February, pp 1–5
8. Sarker S, Rahman M (2019) An IoT based real - time low cost smart energy meter monitoring system using android application. Int J Eng Tech 5(3):1–7
9. Thambirajah S (2021) Press release: smart meter consumers raising concern about the device and the installation procedure of changing. 2021:22–24
10. MalayMail (2019) Minister: public can lodge complaints with energy commission on smart meter issues. 19–22

Numerical Approximation of Performance of Pneumatic Artificial Muscles with Change in Altitude

Bitopan Das, Polash Pratim Dutta, and Partha Pratim Dutta

Abstract Pneumatic artificial muscles have gained popularity because of their lightweight nature, high force to weight ratio and several other advantages. As a result, they have seen applications in robotics, prosthetics and morphing aircrafts. However, the performance of PAMs are dependent on the operating pressure and muscle contraction. Prosthetic and robotic muscles using PAMs which are intended for use in different environment conditions like high altitudes might see a variation in their performance. This paper aims to understand the effects of the altitude of operation on the pulling force exerted by pneumatic artificial muscles. A numerical approximation of the static characteristics of PAMs with respect to altitude is done using software tools and the variations of force with changes in altitude are plotted and analysed.

Keywords Pneumatic artificial muscles · McKibben muscles · Altitude

1 Introduction

1.1 Pneumatic Artificial Muscles

Pneumatic artificial muscles (PAMs) are a relatively new type of actuators. PAMs have gained interest in the recent times primarily because of the fact that they are light weight, have a high force to weight ratio and are flexible. McKibben muscles which are the most widely used PAMs were invented in the 1950s by a physicist, McKibben [1]. But at that time, they did not appeal as serious contenders in the search for new

B. Das · P. P. Dutta (✉) · P. P. Dutta
Department of Mechanical Engineering, Tezpur University, Napaam, Assam 784028, India
e-mail: polashd@tezu.ernet.in

B. Das
e-mail: bitopan_meb18@agnee.tezu.ernet.in

P. P. Dutta
e-mail: ppdutta@tezu.ernet.in

S. Suhag et al. (eds.), *Control and Measurement Applications for Smart Grid*,
Lecture Notes in Electrical Engineering 822,
https://doi.org/10.1007/978-981-16-7664-2_18

actuation methods as material technology put limitations on their performance. In the 1980s, the Bridgestone Company made the Rubbertuator, a McKibben muscle which was intended for use in painting applications. Since then, researchers have started considering PAMs as a capable new actuation method. These lightweight actuators have found applications in robotics, prosthetics and industrial applications [2, 3]. Recently, they have also garnered interest for potential use in morphing aircrafts [4–7].

A lot of PAM actuated robots and prosthetics are being proposed for use in different environments owing to several advantages other than its lightweight nature like ease of control and operation, moderate actuation pressure ranges, high tolerance against misalignment and impulse loading [2]. Also, PAMs have low impacts on the environment and are virtually hazard-free [5, 8, 9]. However, PAMs are very much pressure sensitive and the environment, especially the altitude may impact the performance of PAMs which in turn could affect the output otherwise desired. This would particularly be an issue if rescue and surveillance robots using PAM actuators [10] and people using PAM actuated prosthetics have to operate or commute through a range of altitudes. Works focusing on this aspect of altitude effects on PAM performance have not been done in literature yet. This paper aims to understand the variation in the performance of PAMs with changes in altitude by using a static model. A numerical approximation is performed on PAMs manufactured by FESTO which is the most recommended PAM manufacturer in the market [11].

1.2 Static Modelling of McKibben Muscles

McKibben muscles consist of an elastic bladder and a non-extensible braided shell which is sealed at both ends with an air fitting on one end to allow for pressurization of the muscle through inflation. When the muscle is inflated, the elastic bladder, which is typically latex, expands, pushing against the walls of the outer shell. In an unloaded state, as the bladder gets inflated, the volume of the muscle starts increasing and then stops at a certain point because of the non-extensible nature of the shell: this is the maximum volume of the muscle. The increase in the volume results in its length to decrease, effectively providing a contraction similar to biological muscles while the tensile force of the braided shell leads to the pulling force on the other end [12]. Much like biological muscles, PAMs exhibit non-linear behaviour [13]. The non-linear behaviour in PAMs is due to the compressibility of air and the properties of materials used [14, 15].

Chou and Hannaford [16] showed that the relation between tension, length and pressure can be produced by following a virtual work method. The work input upon inflation of the muscle can be expressed as the pressure applied which changes the volume of the PAM, i.e.

$$\mathrm{d}W_{\mathrm{in}} = (P - P_0)\mathrm{d}V$$

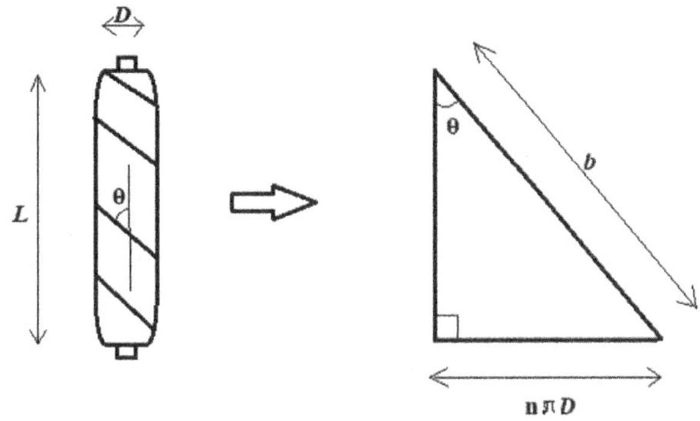

Fig. 1 Geometry of McKibben muscles. θ is the braid angle with n turns of the thread; the total thread length is b. The triangle shows the correlation between braid angle θ, diameter D, number of turns of the thread n and total thread length b

$$= P'\mathrm{d}V \tag{1}$$

Here, P_0 is the ambient pressure, P is the absolute internal pressure, dV is the change in volume and $P^{'}$ is $P - P_0$, i.e. the pressure differential. This work is equivalent to the work done by the muscle in pulling a load with force F through a distance dl [8]. Therefore,

$$P'\mathrm{d}V = -F \cdot \mathrm{d}l \tag{2a}$$

$$F = -P'\mathrm{d}V/\mathrm{d}l \tag{2b}$$

Assuming that the muscle is cylindrical initially, it can be shown that

$$F = \left(2 \cdot \pi \cdot r \cdot l \cdot P'\right) \cdot \mathrm{d}r/\mathrm{d}L - \left(\pi \cdot r^2 \cdot P'\right) \tag{3}$$

where $(2{\cdot}\pi{\cdot}r{\cdot}l{\cdot}P^{'}){\cdot}\mathrm{d}r/\mathrm{d}l$ is the radial component of the force and $(\pi{\cdot}r^2{\cdot}P^{'})$ is the axial component of the force (Fig. 1).

The initial radius, initial braid angle and the initial (resting) length of the PAM are r_0, α_0 and l_0, respectively. When inflated, there occurs reorientation of the braid angle. From Fig. 1, it can be seen that $l/l_0 = \cos\alpha/\cos\alpha_0$ and $r/r_0 = \sin\alpha/\sin\alpha_0$. A correlation can be established between r and l as

$$r = r_0 \cdot \frac{\sqrt{1 - \cos^2\alpha}}{\sin\alpha_0} = r_0 \cdot \frac{\sqrt{1 - \left(\cos\alpha \cdot \frac{l}{l_0}\right)^2}}{\sin\alpha_0} \tag{4}$$

Using the force equation (Eq. 3) and the expression from $\frac{\mathrm{d}r}{\mathrm{d}l}$, F can be expressed as

$$F = \pi \cdot r_0^2 \cdot P' \cdot \left(\frac{3}{\tan^2\alpha_0} \cdot \frac{l^2}{l_0^2} - \frac{1}{\sin^2\alpha_0}\right) \tag{5a}$$

If we take $a = \frac{3}{\tan^2 \alpha_0}$, $b = \frac{1}{\sin^2 \alpha_0}$ and $k = \frac{l_0 - l}{l_0}$, F can be expressed as a function of P' and k as [17]

$$F(P', k) = (\pi \cdot r_0^2) \cdot P' \cdot (a \cdot (1-k)^2 - b) \tag{5b}$$

Equation 5b assumes that the maximum contraction is same for different values of pressure, but this assumption does not yield results in accordance with experimentally obtained results [18]. A correction function $\varepsilon(P')$ is therefore added to adjust to the correlation between P' and k. Although this modified equation works well for larger values of pressures, there are disparities seen between theoretical and model forces at lower pressures. For this reason, a function $\mu(k)$ is added which is correlated with k [19]. The final modified equation for force then becomes

$$F(P', k) = \mu(k) \cdot (\pi \cdot r_0^2) \cdot P' \cdot \left(a \cdot \left(1 - \varepsilon(P')k\right)^2 - b\right) \tag{5c}$$

where

$$\mu(k) = a_k \cdot e^{-k \cdot c_k} - b_k \tag{6}$$

$$\varepsilon(p) = a_\varepsilon \cdot e^{-p} - b_\varepsilon \tag{7}$$

This model has been worked on and perfected by numerous previous works.

2 Numerical Approximation of Force Characteristics with Respect to Change in Altitude

Using Eq. 5c, the variation of force with the change in pressure can be calculated. For this, a_k, b_k, c_k, a_ε and b_ε need to be found. The static characteristics of models of PAMs manufactured by the company FESTO, which is the most trusted PAM manufacturer in the market, are used. By using the least square curve fitting method in MATLAB, a_k, b_k, c_k, a_ε and b_ε for each of the muscles are calculated [20, 21]. A flowchart of the procedure is given below:

A look at the variation of force with contraction for the static characteristics of the FESTO muscles shows that for all the DMSP models starting from the DMSP-10 series, the maximum operating contraction is about 25% [11]. The maximum operating pressure for the DMSP-10-100N is 8 bar, whereas for the DMSP-20-200N and DMSP-40-400N, it is 6 bar. Within the permissible operating range, the maximum force which occurs at 0% contraction for each muscle is about 630 N for the DMSP-10-100N, 1500 N for the DMSP-20-200N and 6000 N for the DMSP-40-400N. As suggested in the operating instructions, the maximum pressure values for

Fig. 2 Flowchart of the procedure

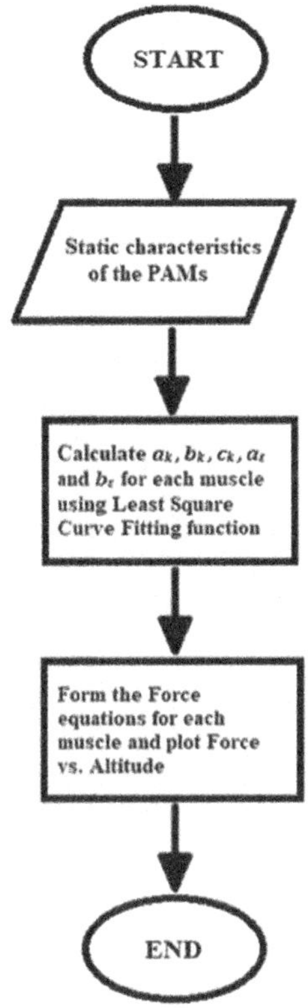

each muscle will be considered for the static modelling as deviations from ambient conditions can then be made up for.

Now, the changes in the force of the muscles with the change in altitude have to be calculated. With an increase in altitude the ambient pressure decreases. The maximum contraction for the muscles is 25% in each case for safe and efficient operation of the muscles. The value of the initial diameter (r_0) is 10 mm for DMSP-10-100N and DMSP-10-250N; 20 mm for DMSP-20-200N and DMSP-20-400N; 40 mm for DMSP-40-200N and DMSP-40-400N. The initial braid angle (α_0) is 36.67° for each muscle.

To find the coefficients a_k, b_k, c_k, a_ε and b_ε in each case, the least square curve fitting method in MATLAB is used [19]. The least squares method is a procedure for approximating a curve with the best fit for a given set of data points which is done by minimizing the square of the residuals.

The values of a and b for each muscle are 5.4 and 2.8, respectively. Using the plot function in MATLAB, the changes in force with altitude for each of the PAMs are plotted.

3 Results

For the initial static modelling, data at the maximum permissible operating pressure are used to reduce deviations. The range of altitude taken for each muscle is from 0 to 5000 m.

In the DMSP-10-100N, the force is the highest at the surface at 630 N which drops in a non-linear way with increasing altitude. At about 1750 m, the value of the force drops to half of that on the surface (Fig. 3).

In the DMSP-10-250N, the maximum value of the force is 480 N which is at the surface and then drops until it becomes zero beyond 4000 m (Fig. 4).

In the DMSP-20-200N, the force is maximum at zero altitude which is about 1500 N and then it drops in a non-linear way; reaching half its maximum value at 2220 m (Fig. 5).

The DMSP-20-400N, like all the previous PAMs has a maximum value of the force at the surface which undergoes a non-linear drop in value (Fig. 6).

In the DMSP-40-200N, the peak force is at the surface with the value being 5200 N after which it drops (Fig. 7).

In the DMSP-40-400N too, the peak force is at the surface which is about 6000 N; which at 5000 m drops to about 1000 N (Fig. 8).

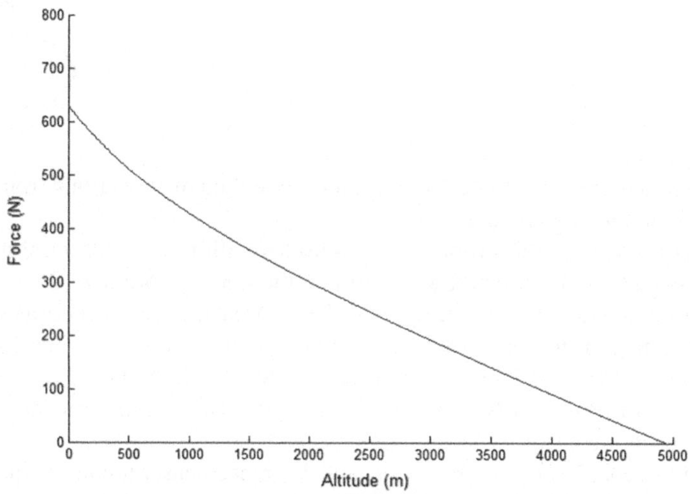

Fig. 3 Plot of force versus altitude for DMSP-10-100N

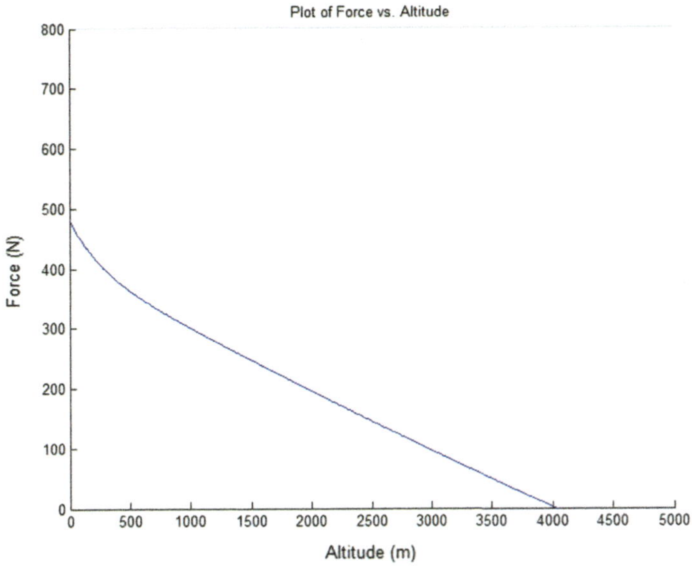

Fig. 4 Plot of force versus altitude for DMSP-10-250N

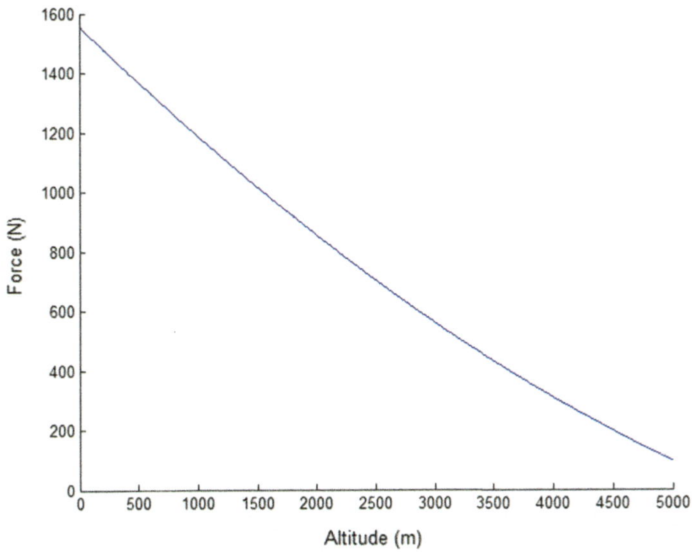

Fig. 5 Plot of force versus altitude for DMSP-20-200N

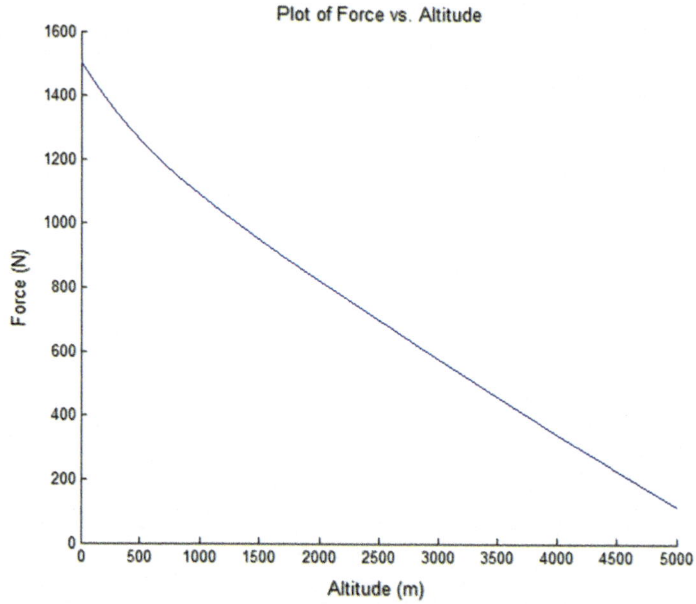

Fig. 6 Plot of force versus altitude for DMSP-20-400N

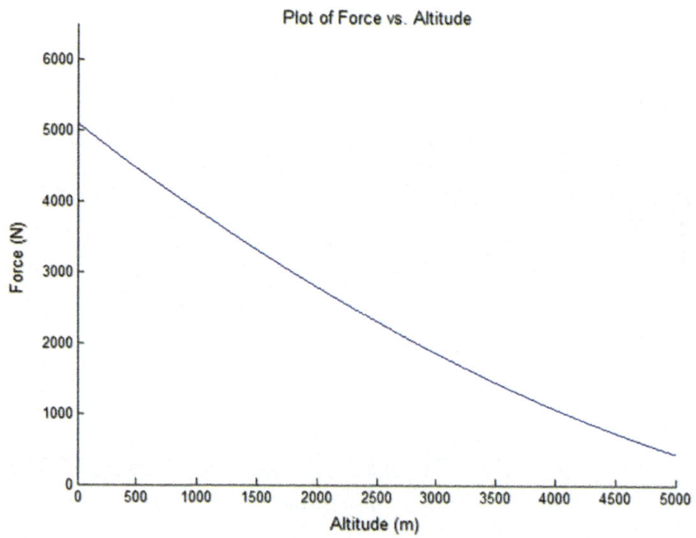

Fig. 7 Plot of force versus altitude for DMSP-40-200N

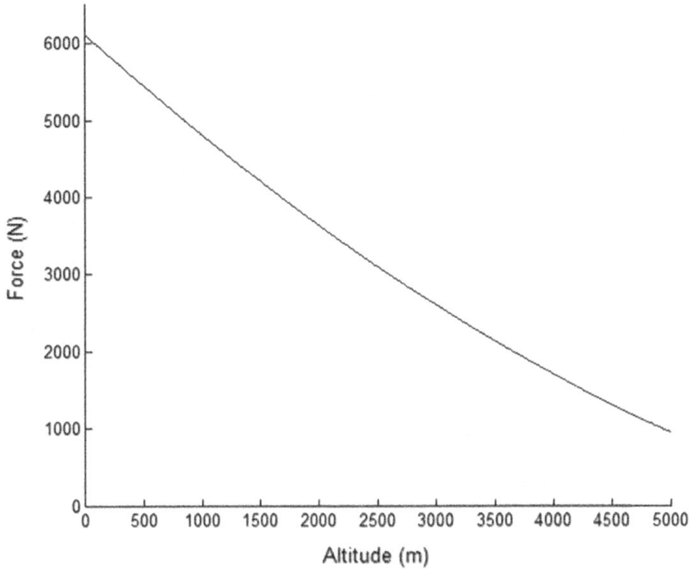

Fig. 8 Plot of force versus altitude for DMSP-40-400N

4 Conclusion

From the plots of force versus altitude for each of the muscles: DMSP-10-100N, DMSP-10-250N, DMSP-20-200N, DMSP-20-400N, DMSP-40-200N and DMSP-40-400N, it can be seen that the pulling force exerted by these muscles are at their maximum at the surface, i.e. at zero altitude. As the altitude increases, the force decreases. This is the effect of the overall drop in the atmospheric pressure with increase in altitude.

The numerical approximation of the force dependence of the FESTO McKibben muscles on altitude shows that the atmospheric pressure drops with an increase in altitude and as can be seen from Eq. 1, P_0 decreases when P' decreases. This study is a static analysis and the results show that with increasing altitude, the force exerted by a PAM will drop in a non-linear manner. This implies that PAMs used as actuators in a device will exert less force than intended at higher altitudes. This could lead to a loss of desirable control over such devices. For the force to be kept constant as desired, P' needs to be maintained at a constant value. Since P_0 cannot be manipulated, the absolute internal pressure has to be adjusted to keep P' constant, i.e. the pneumatic power source has to provide the extra pressure required to keep P' at a constant value.

References

1. Schulte H (1961) The application of external power in prosthetics and orthotics. In: The characteristics of the McKibben artificial muscle. National Research Council, p 784
2. Andrikopoulos G, Nikolakopoulos G, Manesis S (2011) A survey on applications of pneumatic artificial muscles. In: 19th Mediterranean conference on control and automation. IEEE, pp 1439–1446
3. Barbarino S, Bilgen O, Ajaj RM, Friswell MI, Inman DJ (2011) A review of morphing aircraft. J Intell Mater Syst Struct 823–877
4. Abdulrahim M, Lind R (2006) Using avian morphology to enhance aircraft maneuverability. In: AIAA atmospheric flight mechanics conference and exhibit, p 6643
5. Woods BKS, Gentry MF, Kothera CS, Wereley NM (2012) Fatigue life testing of swaged pneumatic artificial muscles as actuators for aerospace applications. J Intell Mater Syst Struct 327–343
6. Sfakiotakis M, Laue DM, Davies BC (2001) An experimental undulating-fin device using the parallel bellows actuator. In: Proceedings 2001 ICRA. IEEE international conference on robotics and automation. IEEE, pp 2356–2362
7. Vocke III RD, Kothera CS, Wereley NM (2018) Chordwise implementation of pneumatic artificial muscles to actuate a trailing edge flap. Smart Mater Struct 075040
8. Tondu B, Lopez P (2000) Modeling and control of McKibben artificial muscle robot actuators. IEEE Control Syst Mag 15–38
9. Daerden F, Lefeber D, Verrelst B, Ham RV (2001) Pleated pneumatic artificial muscles: actuators for automation and robotics. In: International conference on advanced intelligent mechatronics proceedings. IEEE/ASME, pp 738–743
10. Hildebrandt A, Sawodny O, Neumann R, Hartmann A (2002) A flatness based design for tracking control of pneumatic muscle actuators. In: 7th international conference on control, automation, robotics and vision. ICARCV 2002. IEEE, pp 1156–1161
11. Fluidic muscle DMSP. https://www.festo.com/cat/enus_us/data/doc_enus/PDF/. Accessed 21 Dec 2020
12. Wickramatunge KC, Leephakpreeda T (2009) Study on mechanical behaviors of pneumatic artificial muscles. Int J Eng Sci 188–198
13. Hošovský A, Havran M (2012) Dynamic modeling of one degree of freedom pneumatic muscle-based actuator for industrial applications. Tehn vjesn 673–681
14. Daerden F, Lefeber D (2014) The concept and design of pleated pneumatic artificial muscles. Int J Fluid Power 41–50
15. Ramasamy R, Juhari MR, Sugisaka M (2005) Conceptual view of pneumatic artificial muscles. In: Proceedings of the 1st international workshop on artificial life and robotics, pp 14–15
16. Chou CP, Hannaford B (1996) Measurement and modeling of pneumatic artificial muscles. IEEE Trans Robot Autom 90–102
17. Tóthová M, Pitel J, Hošovský A, Sárosi J (2015) Numerical approximation of static characteristics of McKibben pneumatic artificial muscle. Int J Math Comput Simul 228–233
18. Sárosi J (2012) New approximation algorithm for the force of fluidic muscles. In: 7th IEEE international symposium on applied computational intelligence and informatics. IEEE, pp 229–233
19. Kerscher T, Albiez J, Zollner JM, Dillmann R (2006) Evaluation of dynamic model of fluid muscles using quick-release. In: The first IEEE/RAS-EMBS international conference on biomedical robotics and biomechatronics. Biorob 2006. IEEE, pp 637–642
20. Itto T, Kogiso K (2011) Hybrid modeling of McKibben pneumatic artificial muscle systems. In: 2011 IEEE international conference on industrial technology. IEEE, pp 65–70
21. Hiramitsu T, Nabae H, Suzumori K, Endo G (2018) Prototyping of cylindrical structures made of helical artificial muscles. Trans JSME

Variable Selection Technique for Solar Global Irradiance Prediction

Isha Arora, Tarlochan Kaur, and Jaimala Gambhir

Abstract Sun's energy is variable and intermittent by nature and has straight influence on output yielded by photo voltaic (PV) systems. Solar global irradiance prediction is crucial for conducting various research projects in emerging field of renewable energy sources (RESs). Selecting the most significant input variables that influence the solar irradiance prediction is important as it can reduce computational time and burden, enhance the prediction accuracy, increase the convergence speed and simplify the structure of the model. This work has been aimed at finding input variables that affect the prediction of solar global irradiance the most based on backward elimination technique integrated with Pearson correlation coefficient approach. 14 input variables—3 geographical variables, 9 meteorological and 2 calendar variables have been considered for the analysis. Feed forward neural network (FFD) technique has been trained on 25 different climatic regions of India and has been used to predict irradiance for Chandigarh, India. The proposed approach has yielded that surface pressure (SP), wind speed (WS), year (Y) are the most insignificant variables, whereas Clearness Index (CI), maximum temperature (MaT) are most influential ones.

Keywords Error · Meteorological parameters · Prediction · Solar irradiance

1 Introduction

The conventional fossil fuels have been utilized over many decades to meet worldwide energy demand and are swiftly depleting due to ever growing need of electric energy for domestic, industrial and commercial needs [1]. For commercialization, urbanization, economic development and improving living standard of any nation, energy is quite crucial component. The shift towards RESs has been witnessed in recent years as fossils are not only on the edge of exhaustion but are also a major contribution towards environmental pollution, global warming and dangerous greenhouse gases emissions [2]. Government policies, subsidies, advancement in RES

I. Arora (✉) · T. Kaur · J. Gambhir
Punjab Engineering College, Chandigarh, India

S. Suhag et al. (eds.), *Control and Measurement Applications for Smart Grid*,
Lecture Notes in Electrical Engineering 822,
https://doi.org/10.1007/978-981-16-7664-2_19

technology and efficiency and need of sustainable future have also promoted to the shift towards RESs and development of various research activities in the field of renewable technologies [3].

Renewable energy is the energy extracted from naturally existing sources that can be replenished such as solar energy, wind energy, tidal energy, biomass energy and geothermal energy. Prime advantage of RES are no carbon emissions, no associated fuel cost, hence, resulting in substantial fuel costs.

Solar insolation received at the Earth's upper atmosphere is around 174,000 TW. The amount of Sun's energy received by Earth's surface each hour is ample to satisfy electricity requirements of the entire population over complete year [4].

India comes off to be a tropical nation with more than 300 sunny days per year, has massive solar energy prospect, as it lies north to the Equator, between 8° 4′ and 37° 6′ N latitude and 68° 7′ and 98° 5′ E longitude. With its population growing swiftly, there exists huge need of energy to meet the load demand. As of November 2020, 38% of India's installed electrical power generation capacity 136 GW out of 373 GW, is met by RESs [5]. About 5000 trillion kWh per year energy is incident over India's surface with major areas receiving 4–7 kWh/m^2-day, and can reach up to 7.5 kWh/m^2-day, in northern parts in the month of May [6, 7]. Solar-installed capacity of India has reached 35,739 MW as of June 2020 as quoted in "Solar power in India" [8].

PV system is one of the rapidly emerging clean and green energy producing substitutes. PV system is devoid of any moving parts; therefore, they need less of maintenance and provide long period service. They are simpler in construction, can work well as standalone or islanded systems. PV systems can generate power ranging from the levels of micro to mega [9].

Solar radiation intensity is dependent on climatic attributes (local meteorological conditions, clouds, humidity and precipitation), and location specific conditions, has inherent variability and volatility and seasonal changes in energy generation. The performance of panels depends significantly on climatic conditions, optimal and geometric properties, time of use, geographical location, orientation, tilt angle etc. So, there are numerous issues that affect the power output available from panels.

There exist various difficulties in integration of RES with grid. Due to the intermittent and uncertain character of the resource, it becomes difficult to strike the balance between generated PV power and the load demand. Solar power prediction is described as more economically efficient and more reliable way out for managing renewable and distributed energy resources [10, 11].

Poor prediction of solar power output can be worrisome and a setback for large scale PV plant owners when taking part in power trading markets can also lead to fiscal penalties and hence, more grid integration costs and poor economic decisions due to forecasting errors [12]. So, it is vital to have forecasting models with reasonably high accuracy over differing time horizons to decrease uncertainty in renewable energy and to improve economical profits.

Weather parameters such as temperature, solar irradiance greatly affect the power produced by PV panels. The amount of solar radiation striking the Earth's surface can be measured by installing various devices such as pyrheliometer, pyranometer, photoelectric sunshine recorder and quantum sensors. with the data logging system. It is not possible to employ these measuring devices at every site, at residential or small-scaled commercial consumers because of various economic restraints, and/or technical factors (maintenance and calibration of device being the crucial ones), that gives rise to inconvenience in solar radiation data accessibility for major section of the world. Forecasting of solar radiation intensity in remote, distant and alienated areas has been regarded as tedious task due to no meteorological stations installed in these regions. Solar radiation prediction has garnered interest lately due to its utilization in various RESs integrated researches and applications [13, 14].

The first step of forecasting PV power is to find out amount of solar irradiance on the surface of Earth. Nonlinear relation can be developed between PV power output and solar irradiance.

Recent methodologies in field of prediction includes application of artificial learning algorithms which can be employed on vastly varying time horizons that comprises of artificial neural network (ANN), fuzzy model, support vector regression (SVR), decision trees, etc. [15]. A convolutional NN model has been used to forecast 5–20 min ahead irradiance from total-sky images and RMSE results have been compared with the persistence model, and former has been observed as better performing for short-term prediction [16]. Fuzzy–NN integrated model has been used to predict hourly and day-to-day solar irradiance. Sky and temperature parameters have been allocated as divergent fuzzy sets based on fuzzy rules [17].

SVR approach is developed to predict daily and monthly solar power across six regions of Bangladesh. Sunshine duration, Earth temperature, wind speed, humidity, precipitation, atmospheric pressure, tilt angle and elevation have been taken as input attributes. Radial basis function (RBF) kernel used for model validation results in lower levels of MSE [18]. Various parameters such as geographical information that play a significant part in solar power production can also be included.

Yadav et al. have applied Waikato Environment for Knowledge Analysis (WEKA) software to determine most influential factors for solar radiation prediction. Temperature, maximum temperature, minimum temperature, altitude and sunshine duration are considered as most influential factors in prediction [19].

The variability in weather forecasting in analyzed for solar PV generation foretelling. Sensitivity test is conducted to determine most effective weather parameter. Results depict that relative humidity plays most impactful role in foretelling [20].

Prediction and estimation of solar global irradiance based on various machine learning (ML) techniques require various parameters—geographical parameters such as longitude and latitude meteorological parameters likewise temperature, sunshine hours, clearness index, precipitation etc., and calendar variable-month of the year etc. Data acquisition is huge investment incurring typical and tedious task especially with regard to meteorological parameters [21]. So, determining most influential parameters for prediction becomes quite significant.

2 Artificial Neural Network (ANN) Prediction

ANN—a parallel information processing technique based on biological neural networks, neuron/node being the basic building block, where training of numerous input patterns is carried out to approximate the output. They can predict output with significant levels of accuracy, handle nonlinear relations between input and output very well, can solve multivariate problems, controls enormous amount of data patterns parallelly and efficiently and are effectively good at generalization [22]. They are constituted of layers: input, hidden and output, and are connected via weighted links that are utilized to transmit the information signals.

2.1 Multilayer Feed Forward Network (FFD)

FFD networks are oftenly used ANN structure in realistic applications. Numerous layers of neurons are attached in forward direction and information signals transmit from input layer towards output layer. FFD structure is free of any feedback connections or loops. It undertakes supervised learning and is employed to predict output based on historical input database [23].

Activation Function

Transfer function or activation function is used to compute output of a specific layer from its net input. In this work, bipolar sigmoidal transfer function (TANSIG) has been employed.

TANSIG—continuous differential function that maps the value lying in −1 to + 1 is expressed as

$$f(\text{net}) = \frac{2}{1 + e^{-2*\text{net}}} - 1 \quad (1)$$

Here, net represents additive value attained after multiplication of input patterns with their corresponding weights, and $f(\text{net})$ represents transfer function.

Levenberg Marquardt (LM) Learning Algorithm

LM or damped least squares method has been employed in this work to conduct NN learning process. It is a robust technique with quick and stable convergence that works towards reducing loss function often expressed as sum of squared errors.

Weights of the network are revised after each epoch. The output is then compared with target value, and error is then calculated based on a defined error function. Error signal is then propagated backwards to amend the weights for upcoming iteration, aimed at reducing the error function. The process goes on until terminal condition or least error condition is attained [24].

2.2 Decimal Normalization

Apt processing of database is crucial for better prediction accuracy. Data normalization is the process of transforming database from one domain to other; it makes parametric values dimensionless, making it simpler to handle, consuming lesser training time. ANNs are sensitized to data normalization [25]. Here, decimal normalization technique has been utilized to normalize database as it has been contemplated that it results in better accuracy in comparison to Min–max and Z-score normalization techniques [26]. It normalizes the data in the range of [0, 1], shifting decimal point of a parametric value depending on maximal absolute value of that particular parameter, expressed as

$$Z_t = \frac{Z_{i,t}}{10^n} \tag{2}$$

Here n is the smallest integer such that (Max $|Z_t'|$) < 1.

3 Backward Elimination Method and Pearson Correlation Coefficient Method

Variable selection is the process of removing irrelevant or less significant features from data which do not contribute towards output variable prediction, resulting in higher accuracy model. Less significant or irrelevant input features can affect the performance of model in a negative manner. Variable selection reduces model overfitting, decreases computational time and training time, and enhances accuracy of the model.

It is usually a two-step process—applying apt method for computing significance, estimating the cost function that is loss function (prediction error) in this case. Following technique has been used to perform variable selection.

Pearson Correlation Coefficient

Pearson Correlation Coefficient is powerful means to depict the relation between 2 variables. Its value (r) ranges from -1 to $+1$, and it measures linear correlation between the variables. It calculates the degree of change in one variable as a result of other, assuming Gaussian distribution of data [27].

r value approaching 1 depicts a strong positive relation between 2 variables, r value closer to -1 means negative relation between 2 variables, that is, if value of 1 variable rises, that of other will fall. $r > 0$, infers 2 variables have a direct relation, on the other hand, $r < 0$ denotes an inverse relation between 2 variables. Value of r closer to 0 means random or no relation between 2 variables. Here, it has been used to compute the degree of relation between different input variables and solar irradiance.

$$r_{x,y} = \frac{\sum_{i=1}^{n}(x_i - \overline{x})(y_i - \overline{y})}{\sqrt{\sum_{i=1}^{n}(x_i - \overline{x})^2}\sqrt{\sum_{i=1}^{n}(y_i - y)^2}} \tag{3}$$

where $\overline{x}$ is the mean of X, and $\overline{y}$ is the mean of Y.

Backward Elimination Method

Technique used above gives the significance of various input features with solar irradiance. But, how many features of all are requisite for predicting the irradiance still remains a question. So, to answer this, a method termed as backward elimination has been used here.

In this technique, the prediction procedure begins with using all input features and computing error statistics. The most insignificant variable is eliminated at every step from input dataset, prediction is executed and error is computed. If a variable is crucial to target variable prediction, error value will significantly rise. Process is repeated to remove the input variables that are least important to carry out prediction, until termination condition is reached. The prime objective of developing a prediction model is to attain higher levels of accuracy. So, beyond that step, all input features remaining in the model are considered as requisite for predicting the target variable [28].

Figure 1 shows the flowchart of backward elimination method employed integrated with R-value statistical variable selection technique.

4 Simulation and Result

4.1 Region Under Consideration

Solar irradiance striking horizontal surface for Chandigarh city with geographical coordinates as 30.74° N, 76.79° E at 321 m above sea level situated in India has been predicted. It is identified as humid subtropical climatic region with monthly averaged temperature ranging from 10.84 to 33.79 °C. Chandigarh has extensive solar potential which can be procured via different solar technologies efficiently. Hence, high accuracy solar irradiance prediction is quite worthwhile as well as important for numerous research projects and realistic installations.

Two-year database has been extracted from different public websites to conduct training of prediction model for 25 different regions all over India, with vastly varying climatic conditions. The climatic database for these areas have been extracted from National Aeronautics and Space Administration (NASA) [29, 30]. The input database has been normalized using decimal normalization technique. Following variables have been considered as input neurons -

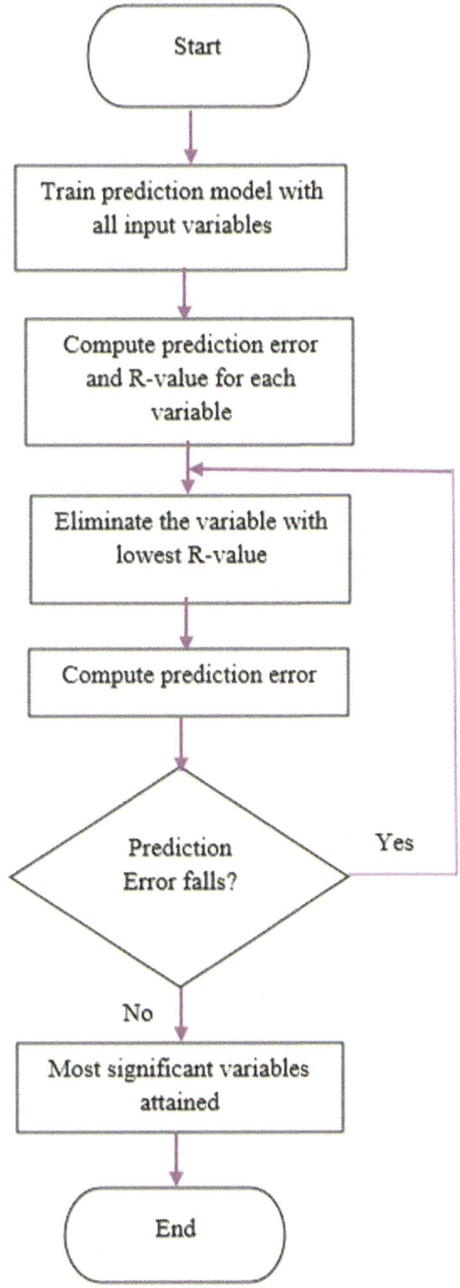

Fig. 1 Process flowchart for backward elimination variable selection

Geographical parameters—latitude of the region (°), longitude of the region (°), altitude of the region (m), monthly averaged meteorological parameters—temperature (°C), minimum temperature (°C), maximum temperature (°C), relative humidity (%), wind speed (m/s), sunshine hours (minutes), clearness index, surface pressure (kPa), precipitation (mm/day) of the region and calendar variables—month of the year and year for prediction analysis of irradiance (kWh/m^2/day). Models have been employed to foretell solar irradiance based on previous two-year satellite database and performance has been evaluated for Chandigarh city. All model simulations have been conducted in Jupyter Notebook environment.

For ANN prediction model, input layer has been composed of 14 neural units, and there is one neuron in the output layer that is irradiance. Predicted values have been compared with actual values to validate the accuracy of the prediction model.

4.2 Performance Evaluative Measures

Performance of solar irradiance prediction model have been computed in terms of mean squared error (MSE), mean absolute percentage error (MAPE), mean absolute error (MAE) and root mean square error (RMSE).

Mean Absolute Error (MAE)—It computes the mean of absolute prediction errors [31].

$$\text{MAE} = \frac{1}{n}\sum_{i=1}^{n}\left|y_{i,a} - y_{i,p}\right| \tag{4}$$

Mean Squared Error (MSE)—It computes the squared difference between predicted output and target output.

$$\text{MSE} = \frac{1}{n}\sum_{i=1}^{n}(y_{i,a} - y_{i,p})^2 \tag{5}$$

Mean Absolute Percentage Error (MAPE)—It computes the absolute difference between predicted and target output I percentage format.

$$\text{MAPE} = \frac{1}{n}\sum_{i=1}^{n}\left|\frac{y_{i,a} - y_{i,p}}{y_{i,a}}\right| * 100 \tag{6}$$

Root Mean Square Error (RMSE)—It calculates the standard deviation of prediction errors.

$$\mathrm{RMSE} = \sqrt{\frac{1}{n}\sum_{i=1}^{n}(y_{i,a} - y_{i,p})^2} \tag{7}$$

where $y_{i,a}$, $y_{i,p}$ represent actual target and predicted output values for ith pattern, $i = 1, \ldots, N$ stands for pattern count, N depicts total number of input patterns. Error values given in the results are pertaining to the original database, after normalized data have been scaled back using inverse of normalization technique.

4.3 Results

Input variables features considered for testing their significance with GHI using Pearson Correlation Coefficient method are three geographical parameters—latitude (Lat), longitude (Long), altitude (Alt), 9 meteorological variables—precipitation (P), surface pressure (SP), average temperature (T), maximum temperature (MaT), minimum temperature (MiT), relative humidity (RH), wind speed (WS), clearness index (CI) and sunshine hours (SH); two calendar variables—month (M) and year (Y).

On conducting the study using Eq. (3), correlation coefficient of clearness index, maximum temperature has the largest value that infers that large value of these variables will yield larger irradiance values. Variable such as latitude (Lat), longitude (Long), altitude (Alt), precipitation (P), surface pressure (SP), month (M), relative humidity (RH), year (Y) have negative r-values with irradiance, that is, increase in value of any of these or all variables would result in falling of irradiance value. Order of correlation is—CI, MaT, T, SH, MiT, M, RH, P, Long, Lat, Alt, WS, Y, SP. Figure 2 shows the R-values as attained for different input variables.

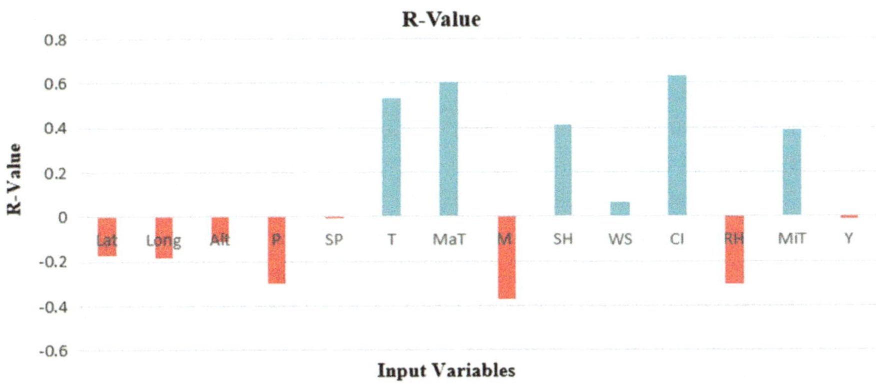

Fig. 2 R-value as attained for input variables

Structure learning has been performed on ANN–FFD network to find out number of hidden layers and number of hidden neurons that yield least error metrics. On performing training process, minimal error has been attained with FFD structure as 14–5–5–1, that is, two hidden layers and five neurons in each hidden layer. So, error results pertaining to only this FFD topology has been reported in this paper.

Further, backward elimination process has been used and followed as per the process flowchart depicted in Fig. 1 to find out what number of variables are requisite for irradiance prediction.

It has been observed that beyond case tenth as shown in Table 1, that is, after input variables: surface pressure, year, wind speed, altitude, latitude, longitude, precipitation and relative humidity are eliminated step-by-step, prediction error (MSE, MAE, RMSE and MAPE) begins to rise. So, it has been found out that month, sunshine hours, Clearness Index, minimum temperature, average temperature and maximum temperature are most significant input variables requisite for solar irradiance prediction.

The effect of backward elimination technique integrated with Pearson Correlation Coefficient approach can be very well observed from Figs. 3, 4, 5 and 6 that on eliminating the insignificant variables, overall prediction error reduces. Accuracy enhancement has been attained as 15.34, 8.77, 7.97, 8.1% improvement in terms of MSE, MAPE, RMSE and MAE, when only significant variables are used to when all variables have been used as input to prediction model. So, eliminating variables not only reduces computational effort and time, it enhances model accuracy as well.

Table 1 MSE attained on backward elimination application

S. No.	Case under consideration	No. of input variables	MSE
1	All input variables considered	14 variables	0.0782
2	Surface pressure variable eliminated (SP)	13 variables	0.0711
3	Year variable eliminated (Y)	12 variables	0.0708
4	Wind speed variable eliminated (WS)	11 variables	0.0706
5	Altitude variable eliminated (Alt)	10 variables	0.068
6	Latitude variable eliminated (Lat)	9 variables	0.0673
7	Longitude variable eliminated (Long)	8 variables	0.0672
8	Precipitation variable eliminated (P)	7 variables	0.0666
9	Relative humidity variable eliminated (RH)	6 variables	0.0662
10	Month variable eliminated (M)	5 variables	0.08325
11	Minimum temperature variable eliminated (MiT)	4 variables	0.08623
12	Sunshine hours variable eliminated (SH)	3 variables	0.2586
13	Temperature variable eliminated (T)	2 variables	0.4039
14	Maximum temperature variable eliminated (MaT)	1 variable	0.6319

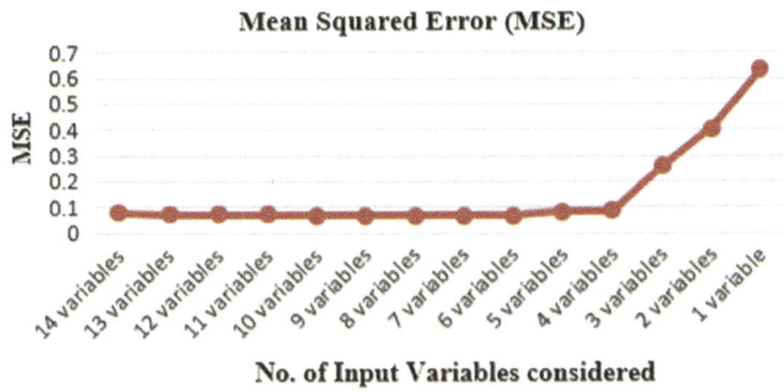

Fig. 3 Change in MSE with elimination of input variables

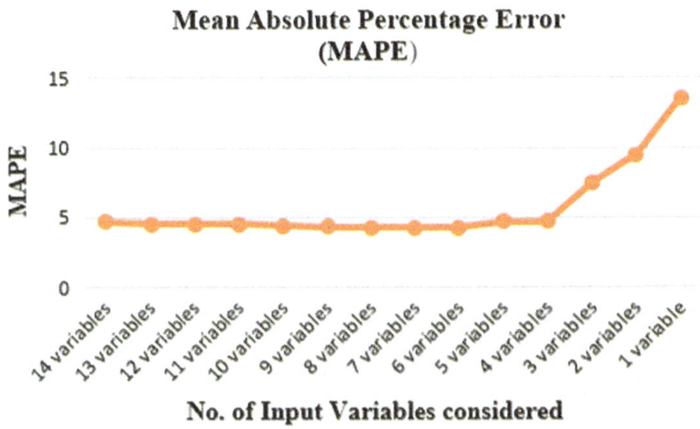

Fig. 4 Pattern of MAPE (%) with elimination of input variables

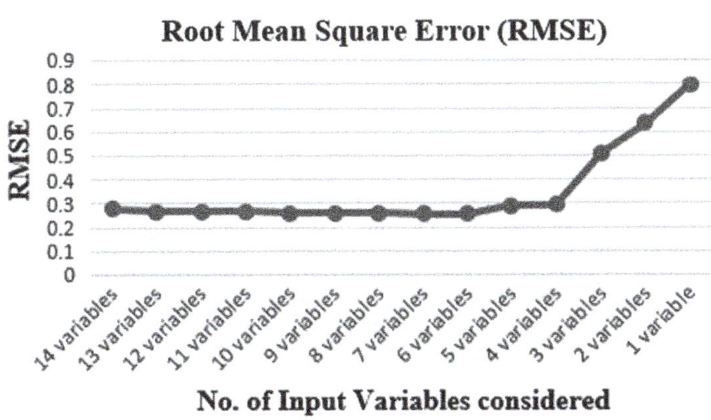

Fig. 5 Effect of variable elimination on RMSE

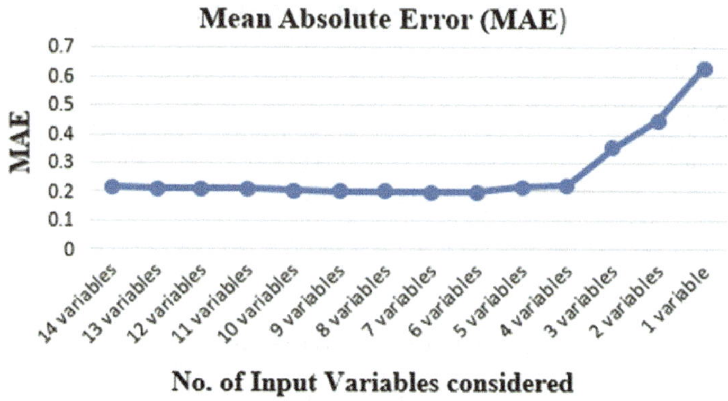

Fig. 6 Effect of variable elimination on MAE

5 Conclusion

In this work, most significant variables for solar global irradiance prediction has been found out. Irradiance has been predicted for Chandigarh using database of 25 different regions of India with divergent climatic conditions. Backward elimination with Pearson Correlation Coefficient technique has been used for variable selection and artificial neural network (multilayer feed forward) model for conducting prediction. At each step, input variable with least significance and least contribution to solar irradiance prediction is eliminated. Accuracy enhancement has been observed as there has been reduction in MSE, MAPE, RMSE and MAE by 15.34%, 8.77, 7.97 and 8.1%.

So, this work shows that backward elimination along with Pearson Correlation Coefficient can be used to find the most significant variables and eliminate the least significant ones with reduction in computational burden and time, and enhancement in prediction accuracy.

References

1. Zhen Z, Liu J, Zhang Z et al (2020) Deep learning based surface irradiance mapping model for solar PV power forecasting using sky image. IEEE Trans Ind Appl 56(4):3385–3396
2. Sumithira TR, Kumar AN (2012) Prediction of monthly global solar radiation using adaptive neuro fuzzy inference system (ANFIS) technique over the state of Tamil Nadu (India): a comparative study. Appl Sol Energy 48:140–145
3. Rizwan M, Jamil M, Kirmani S, Kothari DP (2014) Fuzzy logic based modeling and estimation of global solar energy using meteorological parameters. Energy
4. Moosa A, Shabir H, Ali H, Darwade R, Gite B (2018) Predicting solar radiation using machine learning. In: Proceedings of the second international conference on intelligent computing and control systems, pp 1693–1699
5. Renewable energy in India. Wikipedia. https://en.wikipedia.org/wiki/Renewable_energy_in_India. Accessed 30 Dec 2020

6. Jamil B, Siddiqui AT, Akhtar N (2016) Estimation of solar radiation and optimum tilt angles for south-facing surfaces in humid subtropical climatic region of India. Eng Sci Technol Int J 19(4):1826–1835
7. Arora I, Gambhir J, Kaur T (2020) Decimal normalisation based artificial neural network model for daily averaged all sky solar irradiance prediction. In: ICIIIE conference proceedings
8. Solar power in India. Wikipedia. https://en.wikipedia.org/wiki/Solar_power_in_India. Accessed 30 Dec 2020
9. Hosenuzzaman M, Rahim NA, Selvaraj J, Hasanuzzaman M (2014) Factors affecting the PV based power generation. In: CEAT IEEE conference proceedings
10. Sanders S, Barrick C, Maier F, Rasheed K (2017) Solar radiation prediction improvement using weather forecasts. In: 16th IEEE international conference on machine learning and applications, pp 499–504
11. Shihabudheen KV, Pillai GN (2018) Wind speed and solar irradiance prediction using advanced neuro-fuzzy inference system. In: 2018 international joint conference on neural networks
12. Aggarwal SK, Saini LM (2014) Solar energy prediction using linear and non-linear regularization models: a study on AMS (American Meteorological Society) 2013–14 Solar Energy Prediction Contest. Energy 78:247–256
13. Islam SN, Baig Z, Zeadally S (2019) Physical layer security for the smart grid: vulnerabilities, threats, and countermeasures. IEEE Trans Ind Inf 15(12):6522–6530
14. Arora I, Gambhir J, Kaur T (2020) Solar irradiance forecasting using decision tree and ensemble models. In: 2020 second international conference on inventive research in computing applications (ICIRCA), India, pp 675–681
15. Jufri FH, Oh S, Jung J (2018) Day-ahead system marginal price forecasting using artificial neural network and similar-days information. J Electr Eng Technol 14(2):561–568
16. Kaur T, Kumar S, Kaur R, Gera A (2018) ANN based global solar radiation prediction: a case study. Kaav Int J Sci Eng Technol 5(4):54–63
17. Muhammad A, Lee JM, Hong SW et al (2019) Deep learning application in power system with a case study on solar irradiance forecasting. In: IEEE 2019 international conference on artificial intelligence in information and communication (ICAIIC), pp 275–279
18. Howlader M, Howlader M, Khan SA, Nur U, Amin A (2017) GIS-based solar irradiation forecasting using support vector regression and investigations of technical constraints for PV deployment in Bangladesh. In: ICAEE IEEE, pp 675–680
19. Yadav AK, Malik H, Chandel SS (2014) Selection of most relevant input parameters using WEKA for artificial neural network based solar radiation prediction models. Renew Sustain Energy Rev 509–519
20. Sangrody H, Sarailoo M, Zhou N, Tran N, Motalleb M, Foruzan E (2017) Weather forecasting error in solar energy forecasting. IET Renew Power Gener 11(10):1274–1280
21. Kumar S, Kaur T (2020) Efficient solar radiation estimation using cohesive artificial neural network technique with optimal synaptic weights. Proc Inst Mech Eng Part A J Power Energy 234(6):862–873
22. Arora I, Gambhir J, Kaur T (2019) Monthly averaged all sky solar irradiance prediction using artificial neural networks for Chandigarh region. In: National systems conference on innovative and emerging trends in engineering systems. Springer Nature
23. Laopaiboon T, Ongsakul W, Panyainkaew P, Sasidharan N (2018) Hour-ahead solar forecasting program using back propagation artificial neural network. In: International conference and utility exhibition on green energy for sustainable development (ICUE)
24. Yu H, Wilamowski BM (2011) Levenberg-Marquardt training. In: Industrial electronics handbook, volume 5—intelligent systems. CRC Press, pp 12-1–12-15
25. Roy S, Sharma P, Nath K et al (2018) Pre-processing: a data preparation step. In: Encyclopedia of bioinformatics and computational biology, vol 1. Elsevier, Oxford, pp 463–471
26. Arora I, Gambhir J, Kaur T (2020) Data normalisation-based solar irradiance forecasting using artificial neural networks. Arab J Sci Eng
27. Shi X, Huang Q, Li J, Lei X (2018) Study on short-term predictions about photovoltaic output power from plants lacking in solar radiation data. In: 11th IEEE international conference on intelligent computation technology and automation, pp 75–78

28. Wang X (2018) A new variable selection method for soft sensor based on deep learning. In: 2018 5th IEEE international conference on cloud computing and intelligence systems (CCIS), China, pp 674–678
29. http://www.timeanddate.com/
30. https://power.larc.nasa.gov/data-access-viewer/
31. Suresh V, Janik P, Rezmer J, Leonowicz Z (2020) Forecasting solar PV output using convolutional neural networks with a sliding window algorithm. Energies 13:723

ZigBee-Based Health Monitoring System

G. Rajesh Gowd and M. P. R. Prasad

Abstract In the present COVID-19 situation, remote monitoring of a patient is preferred over contact monitoring, here Using ZigBee we can achieve it. Covid patient's oxygen levels are deteriorating suddenly so need to setup an alarming system for low-blood oxygen levels (hypoxemia). By monitoring the heartbeat of patients with certain pre-medical conditions, we can save them from sudden heart attacks. This paper describes the working of a wireless-based health monitoring system that monitors parameters like SpO2 (oxygen saturation in the blood), heart rate of a human with the help of Max30100 integrated pulse oximetry and heart-rate monitor sensor. Most of the systems that are used nowadays work in offline mode (storing data and sending it later), but this system is designed for real-time monitoring of a patient. A lot of conventional systems use wired communication technology but here the data is transmitted wirelessly from one system to another through two Xbees (Co-ordinator Xbee and Router Xbee). Abnormalities in the patient can be monitored and indications can be sent to medical officials/doctors. The implementation can be achieved by using an Arduino Uno Microcontroller and ZigBee.

Keywords Arduino UNO · Max30100 · Zigbee · Health monitoring

1 Introduction

Health plays an important role in everyone's life, so we need to take care of one's health condition by employing regular monitoring systems. Always monitoring patients by humans is impossible and with regular health monitoring systems, the patient is restricted to a bed with many wires and equipment. So the solution lies in the utilization of better monitoring systems with less complexity and more accuracy.

G. Rajesh Gowd (✉) · M. P. R. Prasad
NIT Kurukshetra, Kurukshetra, India

M. P. R. Prasad
e-mail: mprprasad@nitkkr.ac.in

S. Suhag et al. (eds.), *Control and Measurement Applications for Smart Grid*,
Lecture Notes in Electrical Engineering 822,
https://doi.org/10.1007/978-981-16-7664-2_20

Nowadays, wireless sensor networks (WSN) play an important role in the research, technological community leading to the development of different sensing systems. Most of the research is mainly focused on improving the quality of human life by designing and fabricating well-equipped integrated sensor modules, which are very accurate in measuring health parameters that are useful in the identification of symptoms of a particular disease.

In this paper, a prototype of a wireless ZigBee-based health monitoring system is implemented that can measure health parameters like SpO2, BPM. We can also measure blood glucose and temperature using different sensors like continuous glucose monitor (CGM) and LM35 (range: −55 °C to 150 °C), respectively. These parameters are remotely monitored from a different location and send to doctors in some other location. If these parameters are abnormal doctors will give suggestions or treatment based on the situation. This type of contactless monitoring is useful in Monitoring Covid-19 patients.

Heart rate is defined as how many times the heart contracts and relaxes in a unit time (generally 1 min). Heart rate varies from person to person and for different age groups. Normal heart rate for an adult—60–100 beats per minute (bpm). Children: 80–120 bpm. Like heart rate, the temperature of a body also differs from one person to other. Average body temperature for Adults ranges from 97 to 99 °F (36.1–37.2 °C). Babies and children range from 97.9 to 99 °F (36.6–37.2 °C). For old age people (age above 65), it is lower than 98.6 °F (36.2 °C). A normal oxygen level in the lungs for a healthy person range between 80 and 100 mm of mercury (mm Hg). Saturation oxygen level in the blood (SpO2) for a normal healthy person is ranges 95–100% [1]. A Max30100 is a sensor used to measure health parameters like SpO2 and heart rate measurement.

Examples of Zigbee-based applications in renewable energy systems are smart grid electricity systems, the smart grid allows two-way communication to sense demand of electricity correspondingly generation happened, when we are using renewable energy sometimes generation may not meet demand in those cases works with the normal electricity grid.

2 Zigbee Specification

Generally, many wireless technologies such as Wi-Fi, Bluetooth, LoRaWAN are used for medical data transmission. Based on characteristics such as data rate, range, and power consumption particular application has to be decided. In designing a system, above-mentioned characteristics are considered. ZigBee is more suitable in comparison with others due to the following features [2]:

- ZigBee consumes less power.
- The range is very large (300 ft to 40 miles) which is an essential feature so that patients can be monitored within hospitals or clinics or somewhere outside the buildings like ambulance services.

- ZigBee has a low data rate (250 kbps) but is quite sufficient for measuring and sensing important signs.
- ZigBee costs less.

3 Review of Earlier Technique

Wired patient monitoring systems are huge in size, and are not easily portable. Emerging wireless technologies like Wi-Fi, Bluetooth, LoRaWAN, and ZigBee have advantages like low power consumption and mobility, wearable sensors, so these wireless technologies are used in sophisticated and future health monitoring systems. Nowadays, there are many wireless health gadgets available in markets based on Bluetooth like the apple watch.

Johan presented [3, 4] that by using a non-contact capacitive-based electrode, respiration rate (RR) and electrocardiogram (ECG) are measured and are transmitted through Bluetooth v4.0 or BLE wireless communication method. Gazell is the protocol built on BLE which is used communication between the two devices. ECG and RR are measured and are transmitted in a burst to save power in continuous transmission. After processing data through filters, ECG data observed clearly. By using burst transfer needed large bandwidth and Bluetooth consumes more power compared to ZigBee.

With the advantages of ZigBee over Bluetooth and Wifi as listed in Table 1. And ZigBee specifications mentioned above, the proposed system is based on ZigBee technology. This system mainly focuses on solving the drawbacks of the above existing systems in a way that it overcomes the power consumption, range, real-time monitoring issues.

Table 1 Comparison between current wireless standards [5]

Parameter	Standard		
	Wi-Fi	Bluetooth	ZigBee
Bandwidth	Up to 54 Mbps	1 Mbps	250 kbps
IEEE spec	IEEE 802.11b	IEEE 802.15.1	IEEE 802.15.4
Range	100 m	10 m	300 ft to 40 miles
Power consumption	400+ mA TX	40 mA TX, standby 0.2 mA	30 mA TX, standby 3 μA
Best for	High data rate	Interoperability, cable replacement	Long battery life, low cost
Applications	High-data rate networking, file transfers	Wireless USB, headsets, handsets	Remote control battery-operated products

4 System Architecture

The functional block diagram of a transmitter and receiving section is as shown in Figs. 1 and 2. The transmitter is the starting point of the device where information regarding the health parameters are taken with the help of the MAX30100 sensor module. Sensed values are fed to Arduino UNO as input and are transmitted through the ZigBee module.

Before usage of these modules in transmission and receiving sections, transmitting ZigBee and receiving ZigBee has to be configured as Router and co-ordinator by using XCTU software. The information in the transmitting section is reached to the receiving section with the help of these configurations of Xbees. After receiving this vital information, the Arduino has to display it in LCD or Serial monitor of Arduino and an indication of seriousness in a patient is indicated by a buzzer [2].

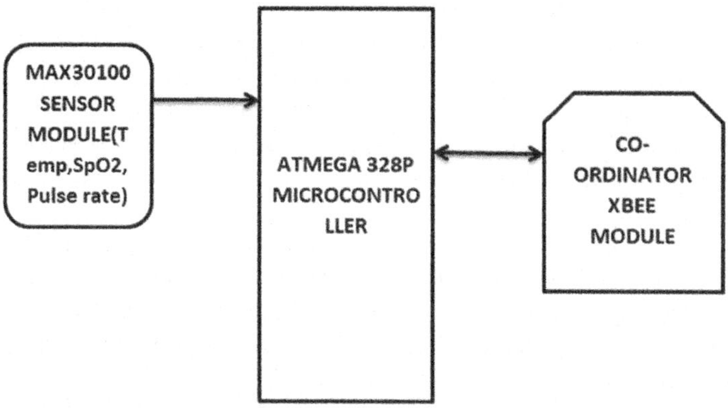

Fig. 1 Block diagram of transmitter section

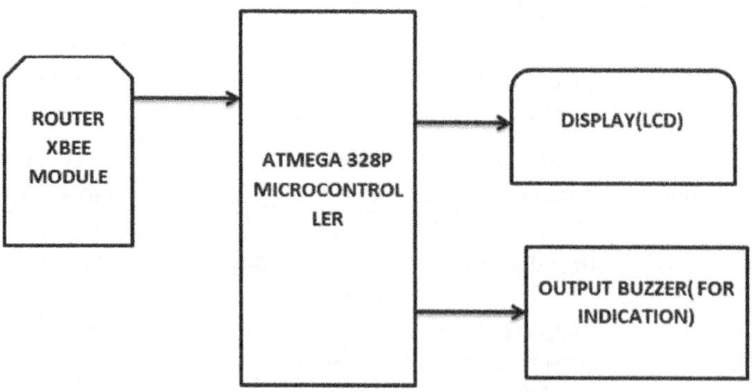

Fig. 2 Block diagram of receiver section

Fig. 3 MAX30100

4.1 Hardware Details

4.1.1 MAX30100

The MAX30100 (Fig. 3) is a heart rate monitoring sensor with integrated pulse oximetry. It consists of a photodetector, optimized optics, two LEDs, high-performance analog signal processing to detect heart-rate signals, level of oxygen in the blood (SpO2). The MAX30100 operates between power supplies 1.8–3.3 V.

4.1.2 ZigBee

XBee XB24CZ7WIT-004 module (Fig. 4) is from Digi. Series 2 which improves the data protocol and power output. These modules allow simple reliable communication between computers, microcontrollers, systems having a serial port. Series 2 modules support the creation of complex mesh networks based on the XBee ZB ZigBee mesh firmware and also supports point to point and multi-point networks are also pin

Fig. 4 Xbee

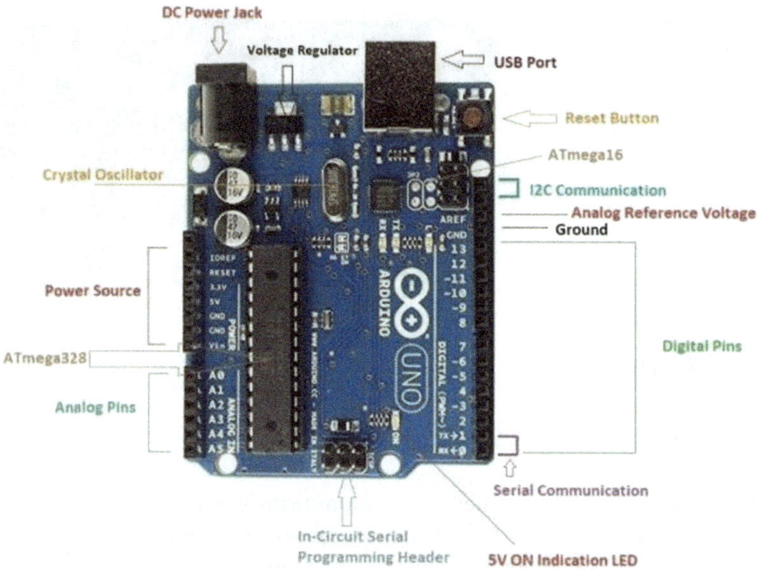

Fig. 5 Arduino UNO

configuration of Series 1 and 2 XBee modules are the same but, Series 1 modules cannot communicate with Series 2 modules [6].

4.1.3 Arduino UNO

The Arduino Uno (Fig. 5) is an open-source microcontroller. The board has 14 digital I/O pins (6 of which are PWM outputs), 6 analog I/O pins, and is programmable with the Arduino Integrated Development Environment (IDE), via a USB A to B cable. By a 9-V external battery or by USB cable it can be powered. While the Uno communicates using the original STK500 protocol [6].

4.1.4 LCD Display

Since the display has 16 Columns and 2 Rows it is named as 16×2 LCD (Fig. 6) There are many combinations like 16×1, 8×1, 8×2, 10×2 etc. are available. So, it has a total of 32 characters ($16 \times 2 = 32$), and each character is made of 5×8 Pixel Dots [5].

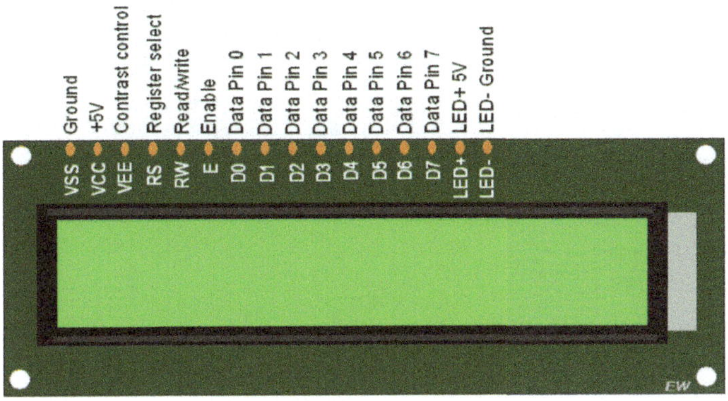

Fig. 6 LCD display

4.2 *Software Details*

4.2.1 XCTU Software

XCTU is a Test Utility and Configuration Software. It is a free multi-platform application designed to enable developers and researchers to interact and configure with Xbee RF modules through a simple GUI. Through this, it easy to configure, set-up, and test Zigbee/XBee RF modules [7].

Steps to configure ZigBee/Xbee:

- Add a radio module manually, Open the XCTU software and then click on the search on the top.
- Select the port where the radio module is connected, Leave the port parameters to default (Data bit-8, Baud rate-9600, stop bit-1, Parity-None).
- Upgrade the latest firmware.
- Configure Xbee modules as COORDINATOR and ROUTER.

Figure 7 shows after configuration of the transmitter as co-ordinator and receiver as a router using XTCU software, using configuration setting as shown in Table 2. Communicating packets from the transmitter (Blue) to receiver (red).

4.2.2 Arduino IDE

The Arduino Integrated Development Environment (IDE) is developed with the help of functions of C and C++. It is a cross-platform application (for Linux, Windows, Mac OS). By using this IDE, we can write and upload the programs/code to Arduino compatible boards to work as per our requirement.

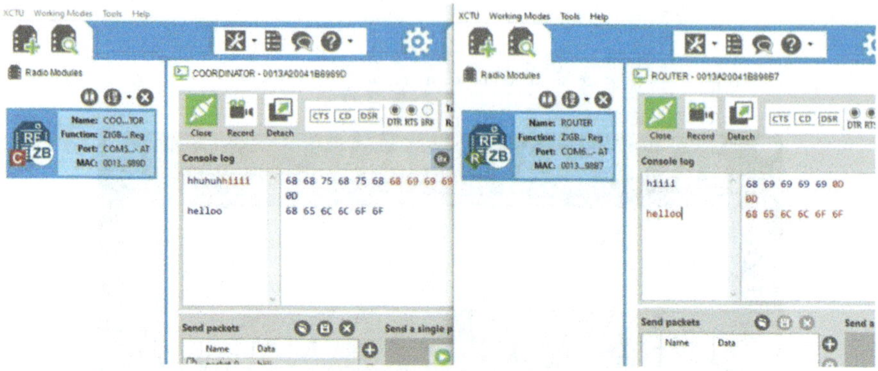

Fig. 7 Co-ordinator and router configuration

Table 2 Configuration settings of co-ordinator and router

	Coordinator	Router
Pan ID	1234 (any hex value from 0 to FFFF)	1234 (same as coordinator)
DL (destination address)	FFFF (to operate the radio in the broadcast mode)	0000
Node identifier (NI)	Coordinator	Router
CE coordinator	Enable	Disable
JV channel verification	Disable	Enable
AP API enable	Transparent mode	Transparent mode

4.3 System Installation/Modules Interfacing

The proposed system works based on the Zigbee communication protocol used for the transmission of medical data from one place to another place. Here transmitter consists of sensors, the Arduino UNO microcontroller, and the Zigbee RF module and the receiver end consists of Zigbee, Arduino UNO, and LCD. Health parameters are sensed with a Max30100 sensor and are transmitted through Co-ordinator (Transmitter) Zigbee and are received by a router (receiver) Zigbee and values are displayed on 16 × 2 display. Figures 8 and 9 show the circuit diagram of the transmitter and receiver section.

Connections of MAX30100 sensor with Arduino UNO as follows:

Vin pin of Max30100 with output 3.3 V of Arduino UNO

Gnd to Gnd

SDA of MAX30100 to SCL of Arduino UNO

SCL of MAX30100 to SDA of Arduino UNO.

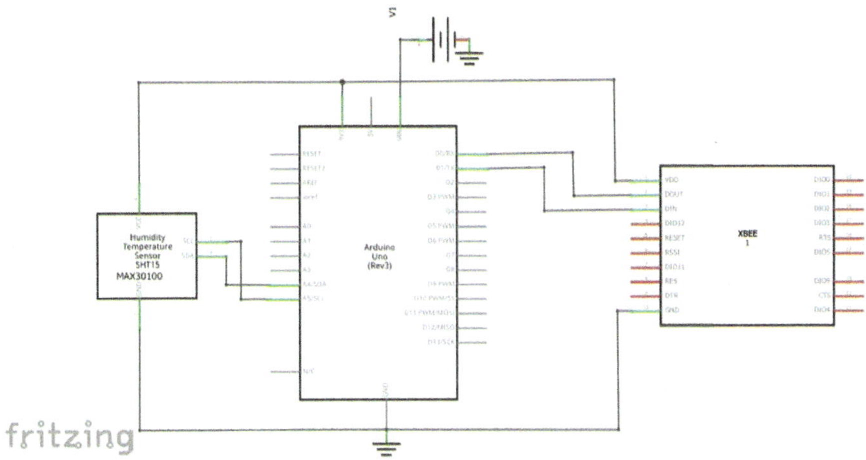

Fig. 8 Transmitter circuit diagram

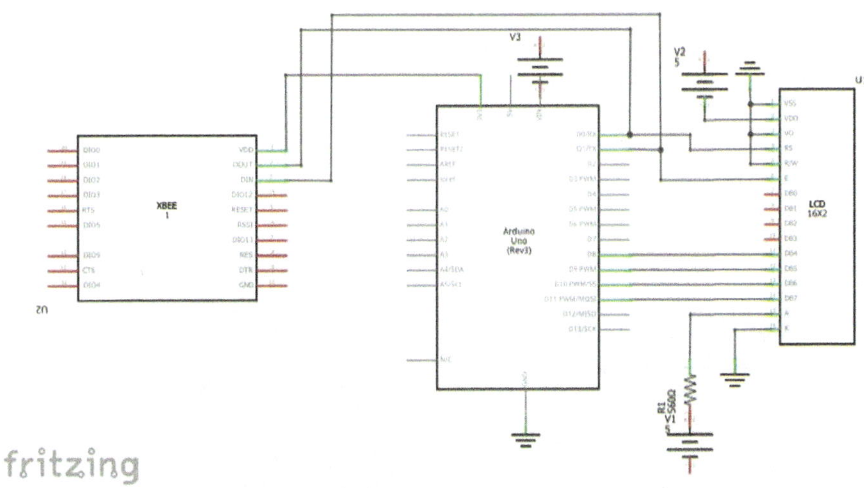

Fig. 9 Receiver circuit diagram

Connections of LCD with of ARDUINO UNO as follows:

VSS of LCD to ground of ARDUINO UNO

VDD or VCC to +5 V power of ARDUINO UNO

VEE to ground of ARDUINO UNO (gives maximum contrast)

RS (Register Selection) to PIN0 of ARDUINO UNO

RW (Read/Write) to ground LCD in read mode)

E (Enable) to PIN1 of ARDUINO UNO

D4 to PIN8 of ARDUINO UNO

D5 to PIN9 of ARDUINO UNO
D6 to PIN10 of ARDUINO UNO
D7 to PIN11 of ARDUINO UNO.

Connections of ZigBee with Arduino UNO as follows:

VCC of ZigBee is connected to 3.3 V of Arduino
GND of ZigBee is connected to the GND of Arduino
Transmitter and receiver pin of ZigBee are connected to the receiver and transmitter pin of Arduino, respectively.

5 Setup and Results

Information about health parameters like heart rate, SpO2 from a patient sensed by using MAX30100, and by using Arduino IDE we have to write and dump a code on to microcontroller to collect sensed data and transmit it to the serial port of ZigBee Module. At the receiver, the microcontroller collects the receiving data from the serial port of ZigBee and displays those values on LCD.

Figure 10 shows transmitter and Fig. 11 shows receiver and sampled medical parameters of a person is measured from one location and transferred to another location through ZigBee and displayed on the LCD at receiver end as shown in Fig. 11.

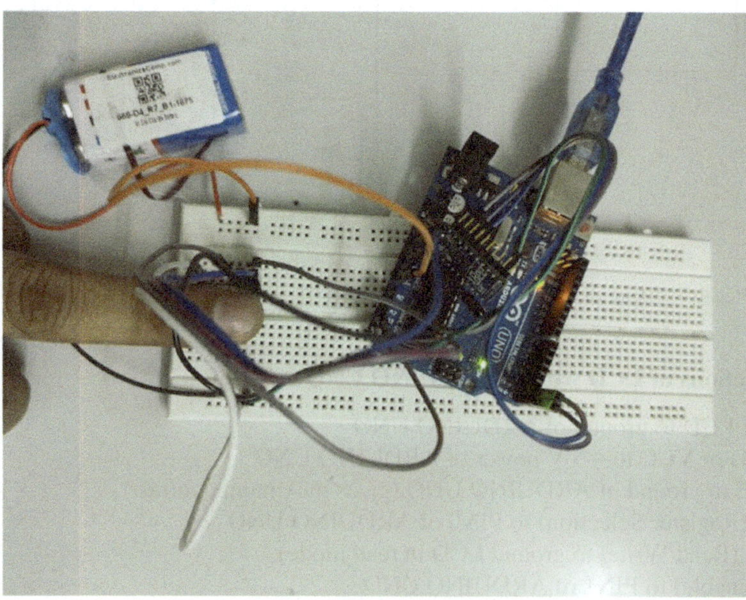

Fig. 10 Transmitter

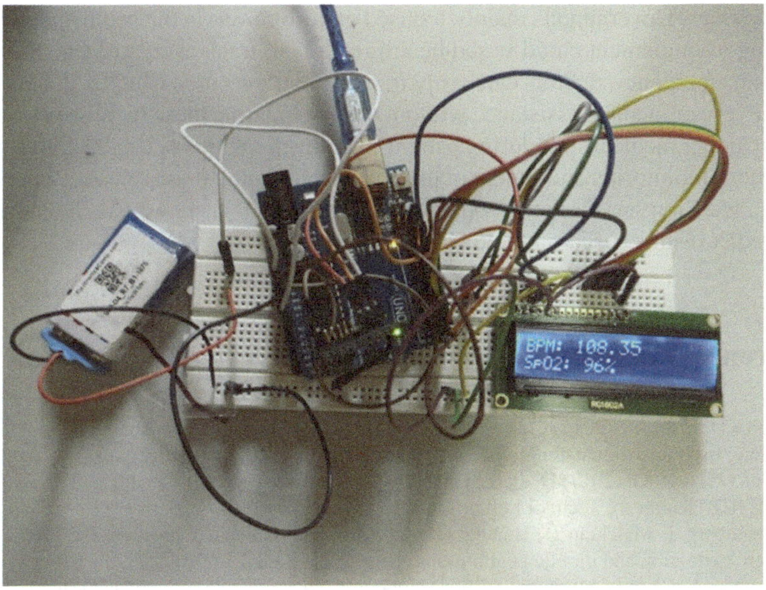

Fig. 11 Receiver

6 Conclusion

In this paper, a power-efficient, low-cost, and long-range wireless ZigBee-based health monitoring system is implemented that measured the health parameters like SpO2, BPM from a patient in one place and transmitted to another place via Zigbee modules and displayed in LCD. Wireless transmission of health parameters from patients to doctors is made simple, easier, and faster with ZigBee communication and an Arduino Uno Microcontroller with very less circuitry. It is useful for a wide range of people with heart-related problems and warns them and informs personal doctor, and also family in advance about the dangerous health situation ahead to give them immediate medical assistance.

7 Applications and Future Scope

ZigBee communication can be extended with the use of IoTs, used in different applications like home automation, in detecting hazardous gases in mines, low-power digital radios, telecommunication applications, radiation monitoring at nuclear facilities, wireless remote sensing applications during natural calamities like floods to monitor the flow of water in dams, in agriculture to measure soil details remotely, etc.

Nowadays, Governments mainly focused on resolutions in the health care system, planning to implement cloud-based health records to track every citizen's health so that in an emergency, doctors can easily track patient's records which will help in the fast delivery of medical assistance without delay. Updating these records can be done through either medical staff during patient's health check-ups or can be done through continuous monitoring equipment of patients and updating these sensed medical data through web/app services and by using Artificial intelligence, we can diagnose treat and predict results based on trends in data of a patient.

References

1. Singh KD, Joshi AK (2017) Cost effective open source wireless body sensor networking through ZigBee. In: International conference on communication and signal processing, 6–8 Apr 2017
2. Alwan OS, Rao KP (2017) Dedicated real-time monitoring system for health care using ZigBee. IEEE IET Healthcare Technol Lett 4(4):142–144
3. Wannenburg J, Malekian R, Hancke GP (2018) Wireless capacitive-based ECG sensing for feature extraction and mobile health monitoring. IEEE Sens J 18(14)
4. Majumder S, Chen L, Marinov O (2018) Noncontact wearable wireless ECG systems for long-term monitoring. IEEE J Biomed Eng 11
5. Singh J, Tiwari M, Shelar M (2012) ZigBee based patient monitoring system. Int J Comput Appl 51(22). 0975-8887
6. Hoshang A, Tang L (2011) Wireless network for health monitoring: heart rate and temperature sensor. In: Fifth international conference on sensing technology. IEEE
7. XCTU user guide available at https://www.digi.com/resources/documentation/digidocs/PDFs/90001458-13.pdf

Time Series Forecasting for Electricity Consumption Using ML and DL Algorithms

Neeraj Kumar, Sumit Mishra, Tanmay Baweja, Ashutosh Dubey, and Abhishek Dhiman

Abstract As the world exhausts its non-renewable energy reservoirs, building predictive models for household energy consumption becomes important a fortiori. In this study, we examine how XGBOOST, LSTM, and CNN perform on big data trends in a time series forecasting. The data used incorporates 2,075,259 measurements that were gathered from a residence situated in Sceaux, which is 7 km from Paris, France. The performance has been calculated on the parameters of mean absolute error (MAE) and root mean square error (RMSE). The period of data taken for analysis is from December 2006 to November 2010, which sums up for 47 months. The dataset provides a measurement of power consumption in a single household with one minute of the sampling rate. Exploratory data analysis and statistical tests were performed to produce stationarity of the time series data. Predictions are made by identifying previously captured data and further processing null values, which could be used to forecast future circumstances. This can further be used to calculate the active power consumption of a household.

Keywords Time series forecasting · XGBOOST · LSTM · CNN · Standard scalar · Matplotlib

1 Introduction

Perturbs in available resources and the continuous catastrophic collapse of nature have raised alarms and gained worldwide attention. Almost every state and national energy regulating body in their cumulative data has reflected an increase in consumption of energy, in both commercial and residential sectors [1]. This has led to an imbalance between resources and consumption and excessive load demands on random

N. Kumar (✉) · S. Mishra · T. Baweja · A. Dubey · A. Dhiman
Department of Electrical and Electronics Engineering, Bharati Vidyapeeth's College of Engineering (Approved by AICTE, New Delhi-Affiliated to GGSIP University, New Delhi), New Delhi, India
e-mail: neeraj.kumar@bharatvidyapeeth.edu

S. Suhag et al. (eds.), *Control and Measurement Applications for Smart Grid*, Lecture Notes in Electrical Engineering 822,
https://doi.org/10.1007/978-981-16-7664-2_21

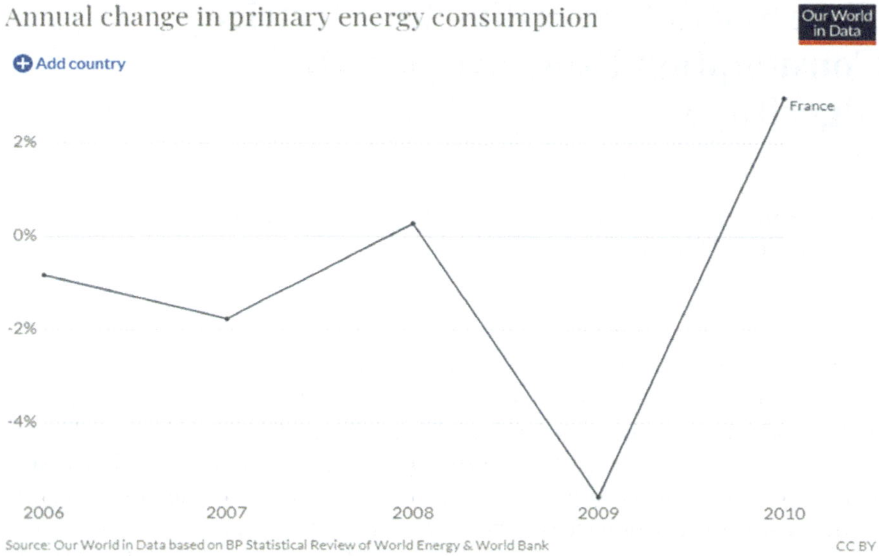

Fig. 1 Annual change in energy consumption in France (2006–10) [2]

occasions. Also, in France, the annual change in energy consumption is not persistent as shown in Fig. 1. This trend makes it evident that there is an incidental rise in demand.

Therefore, it becomes a topic of national importance to predict energy consumptions, with, of course, different sectors segregated and measured separately. This is achievable with a concrete structure of advanced control strategies. Availability of existing information surely helps to build edifice management to take required actions on limiting consumption and reducing peak loads [3]. One of the most preferred techniques to improve energy efficiency is to apply machine learning. With the help of artificial intelligence and training models, predicting future analysis of any parameter becomes easy. The same has been proved effective with energy consumption measurements based on past incidents and data available associated with historic events.

The major challenge this sector faces is unorganized variations. A power system is only good for as long as its inputs are organized [4]. Therefore, it becomes important to carefully understand and predict short patterns over a significantly large period. With higher accuracy in predicting these patterns, the overall efficiency of the energy sector of a nation or state can be increased.

The only solution to overcome these issues and produce a highly efficient structure is to accurately forecast the energy consumption of individual households. Although DNN, ANN, and various other methods have already been implemented, the efficiency was not very fascinating. Therefore, our research focus, out of multiple machine learning models, is on XGBOOST, LSTM, and CNN.

2 Project Overview

2.1 Electricity Dataset Description

Dataset is from the UCI Machine Learning Repository and is contributed by George Hebrail for study purposes. The name of the dataset is Individual Household Electric Power Consumption data. It has been recorded for over 47 months, from December 2006 to November 2010. This data reflects electricity consumption readings that have been recorded every minute in every house situated at a target location, Sceaux 7 km away from Paris, France. The dataset has 2,075,259 measurements of various electrical quantities, including 3 sub-metering values as well. This archive also has 1.2518% null values. A null value is indicated by a missing value between the two adjacent semicolon separators.

The data used in the time series analysis should be in the same units as changing the units will result in the wrong trends capturing and overall doom of the model. For example, if the model is used to forecast the reactive power in kilowatt and in the past data the reactive power is in watt. This condition will lead to the overall failure of the model. The trends can be secular positive trends and negative secular trends. This variation in the data may vary from data to data as well. So, the analysis should be done carefully.

2.2 Time Series Analysis

The time series analysis approach is an approach to forecast future data according to the past and the present data. The time series have different patterns and trends which are cyclic, seasonal, and random. The time series can be inefficient when there is a random trend in the dataset as the machine learning model will not be able to catch the trend. The seasonal trend is the trend that occurs after a particular interval [5]. The cyclic trend is also like the seasonal trend but the period can be whatever insists of the season.

Time series analysis can be used to forecast the sales which can be very useful for the business perspective as different approaches and decisions should be made to accomplish the best result. The time series is different from machine learning in a direct sense as the time series analysis approach is used to forecast but not to predict. The terms forecast and prediction might be the same in English, but they are no same from the analytics perspective.

Time series data possesses an ordering of a natural temporal. This capability makes it different from cross-sectional studies. This makes time series analysis distinct from cross-sectional studies with absolute nil natural ordering of the parameters (e.g., wage layout of people with concerning the educational qualification, where these data can be put in any hierarchy) [6]. Time series analysis, at the same time, is different from spatial data analysis. In that, observations have a link with the geographical locations (e.g., defining the price of a house based on location and its intrinsic characteristics). In a time, series, a stochastic model would normally reflect an observation more closely in time than observations further apart. Also, the natural one-way ordering of time is used in time series models so that values for a particular interval period will be reflected as deriving when compared to the past values instead of the future values.

Time series analysis has been studied by a wide number of people and a host of research has already been done. It can be divided into two types, univariate models and a multivariate model [7]. An assumption is usually made while applying time series techniques is that the data remains stationary. This ensures that parameters like mean, autocorrelation, and variance stay stationary throughout the analysis. Various techniques are available for time series modelling.

Time series is steadily used to deal with real-valued discrete numeric data, continuous, or even symbolic observations. Time series provides the agility to be plotted using run charts. Time series has a host of uses ranging from finances, statistics, forecasting, weather predictions, control engineering, communications, recognition of patterns, early warning of earthquakes, cosmology, signal processing, etc. Time series has a significant value in machine learning and data science involving temporal observations.

2.3 Feature Engineering

The process of extracting features out of primary data using data mining procedures is termed feature engineering. To improve the overall performance of features of data, featuring engineering is used. Feature engineering comes under the ambit of machine learning [8]. All dependent units in the data on which the modelling is to be conducted share property or an attribute. Any property can be a feature unless it is irrelevant to the model.

The aim of this feature could easily be understood with the help of a real-life problem. This feature may come in handy while solving a problem. To achieve influencing results for the predictive models, features play a vital role. It is a widely accepted fact about feature engineering that it has a vital role to play as part of Kaggle competitions and specifically in deciding machine learning projects' fate.

Certain features are engineered based on the date and time features; the features engineered are:

(1) Day of the week—Which is the day of the week ranging from 1 to 7.
(2) The month of the year—Which is the month of the year ranging from 1 to 12.
(3) Leap year—This feature tells if the year is a leap year or not, 0 if the year is not a leap year, and 1 if the year is a leap year.
(4) A quarter of the year—It talks about which quarter of the year is considered, ranges from 1 to 4.
(5) Week of the Year—It is the week number of the year ranging from 1 to 52.

2.4 Preprocessing

The dataset contains 2,075,259 measurements and eight different features named Global Active Power, Global Reactive Power, Voltage, Global Intensity, Energy Sub-Metering No. 1, Energy Sub-Metering No. 2, Energy Sub-Metering No. 3, Date, and Time. All these features are explained in the Data Description Section [9].

The dataset consists of some null values but these null values are in the form of strings like "?" and empty strings. So, these values are firstly replaced and changed to NumPy "NaN" for further preprocessing. The dataset contains around 25,979 Null values in each of the columns. Around 1.251% of the data was null. To fill the null values, Pandas Interpolation is used with the method of "slinear."

The dataset has date and time but in the string format but for the further feature engineering part, date and time need to be in the Datetime64 format, so the date and time were changed to this format. Our target variable was the active energy consumption which was not given in the dataset and has to be calculated; hence, a target variable is made named "Active energy" which has to be predicted further shape, and analysis of the features are given in the exploratory data analysis part. The formula used for the calculation of the active energy consumption is:

$$\text{Active Energy} = (\text{global\_active\_power} * 1000/60 - \text{sub\_metering\_1} - \text{sub\_metering\_2} - \text{sub\_metering\_3})$$

Normalization is a rescaling of the data from the original range so that all values are within the new range of 0 and 1 [1]. The normalization of the initial features which are given in the dataset are done so now the features lie between 0 and 1. Normalization is done using the Min-Max Scaler of the Pandas Library. Every single piece of data is then sorted according to the date and time they were recorded.

The dataset is then divided into training and testing. The data from 2006 to 2009 are used for training purpose, and the data from 2010 is used for the testing purposes. The 1,600,236 data points are present in the training set which is about 77.11% of the data, and there are about 475,023 data points in the testing set which is about 22.89% of the data.

2.5 Exploratory Data Analysis

The bar graph depicts the distribution of the dataset among the years. It can be inferred that the data is evenly distributed every year except 2006, where it has a minimum number of data points (Fig. 2).

This line plot shows the active energy utilized at an interval of 1 min in watt-hour. The confidence interval here is 97%. We can see that the data is seasonal. The seasonality is regular. There is a seasonal trough around July and August, and

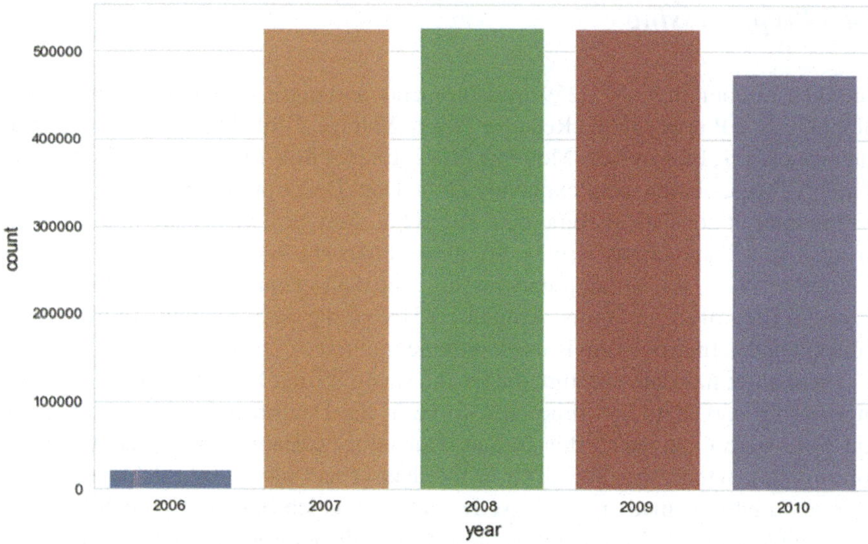

Fig. 2 Distribution of data over the years

the seasonal peaks are at the starting of the year around January and February. The maximum active energy consumption given is 124.83 Wh [3] (Fig. 3).

The Global Active Power and Global Reactive power line plot is given; here, we can see the seasonality in the global active power with not much trend over the year. The global reactive power does not show any seasonality, and the trend over the year is constant. Both global active power and global reactive power are used to calculate the active power consumption [4]. The confidence interval taken for the plotting is 95% for both the quantity (Fig. 4).

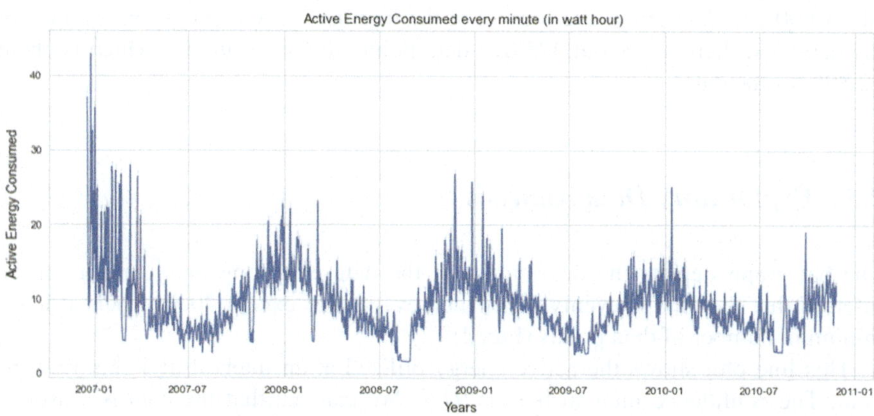

Fig. 3 Active energy consumption

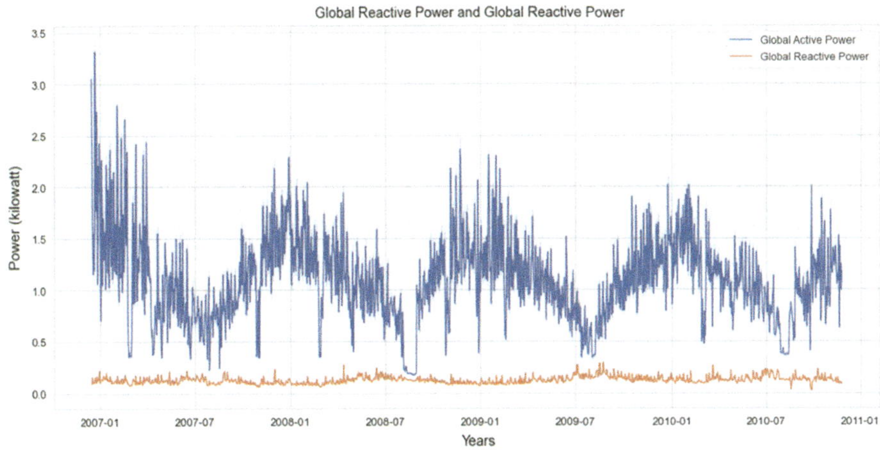

Fig. 4 Global active and reactive power

These line plots show the Energy Sub-Metering numbers 1, 2, and 3 [5]. We can see that the maximum power for Energy Sub-Metering No. 1 and 2 is 88 and 80 Wh, but for the Energy Sub-Metering No. 3, the maximum power is 33 Wh. We can see that the Energy Sub-Metering has some seasonality where we can see the trough and peaks as we saw in the active power consumption [6]. The confidence interval used here is 95% (Fig. 5).

This line plot shows the moving average, i.e., rolling mean and rolling standard deviation. In statistics, a running or rolling average is a term used for the calculative analysis of data points by building a streak of mean on complete data that is available. This can also be referred to as the moving mean [7]. This means it is an example of a finite impulse response filter. Multiple variations incorporate either simple, cumulative, or weighted forms. The prominent aim of using these averages is to omit fluctuations that occur for a short period and smoothens them [8]. It further outlines a better picture of long-term trends. The parameters of the mean are set according to the threshold between the short and long term, depending on the operation [9] (Fig. 6).

3 Methodology

The major challenge the energy sector faces is the unorganized variations in consumption and supply. A power system is only good for as long as its inputs are organized. Therefore, it becomes important to carefully understand and predict short patterns over a significantly large period. With higher accuracy in predicting these patterns, the overall efficiency of the energy sector of a nation or state can be increased [10].

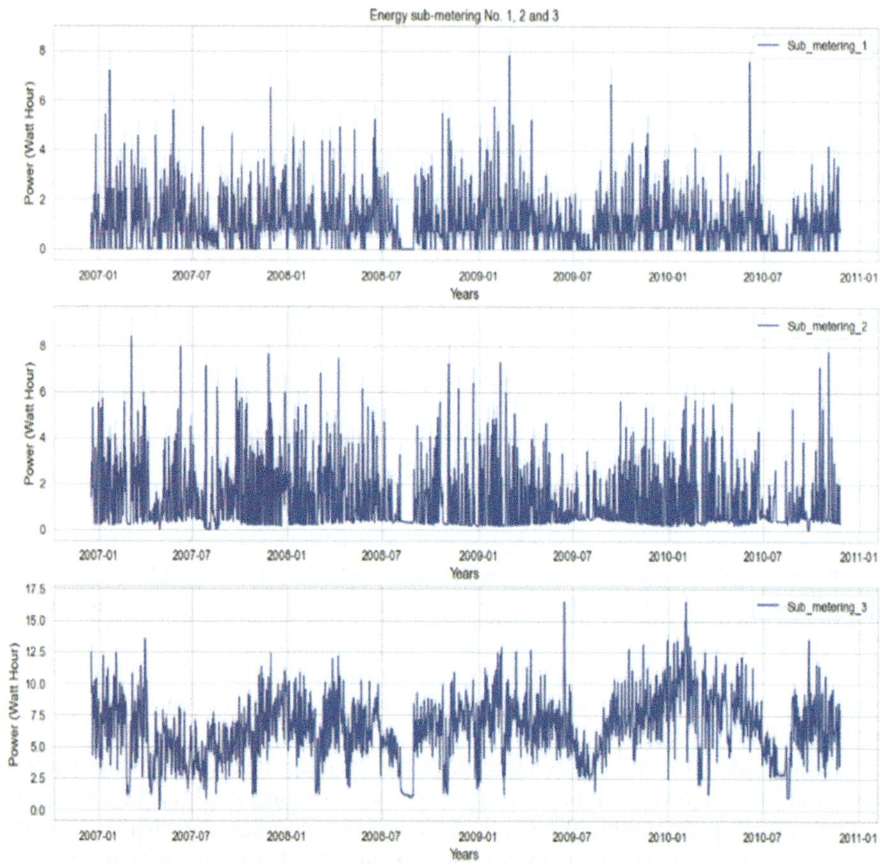

Fig. 5 Energy sub-metering 1, 2, 3

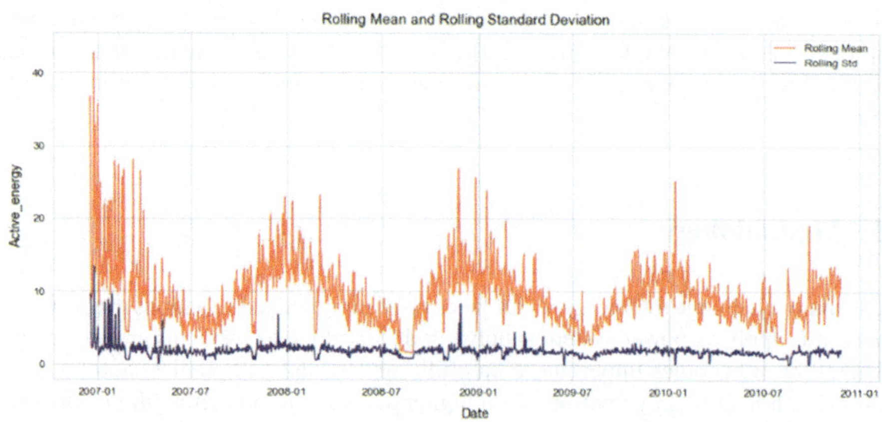

Fig. 6 Rolling mean and rolling standard deviation

The only solution to overcome these issues and produce a highly efficient structure is to accurately forecast the energy consumption of individual households.

One such way is to implement machine learning, train models artificially based on the data available, and predict future trends [11]. The domain of machine learning provides the capability of teaching machines to achieve operations where tasks seem far-fetched. Based on viable answers already available, the future course of the study can be decided. The computer is trained in such a way that it improves its algorithms to act upon future courses in search of potential answers [12].

The machine learning domain provides a significantly good number of optimizations as well. Multiple problems can be curated as a reduction of loss function on examples of a training dataset. The differences between the model prediction while training and the actual case can be reflected with the help of loss functions [13].

Creation of a dedicated model based on training data available and further optimizations and additional operations to predict or forecast future outcomes are involved in machine learning modelling. There are different types of models based on the uses and efficacy yield in the machine learning domain [14]. Out of multiple models and methods, we chose to stick to the one which has the best efficiency out of the many available. Therefore, we conducted a comparison by training models using three different modes, namely convolutional neural network (CNN), extreme gradient boost (XGBOOST), and long short-term memory (LSTM). Our major focus would be to pick the most efficient model and train our final model on that [15] (Fig. 7).

4 Modelling

4.1 LSTM

Long short-term memory networks also famously known as the LSTMs are an advanced version of the recurrent neural networks as they are proficient enough to learn long-term dependencies. Hochreiter and Schmidhuber invented this architecture in 1997, and it was further polished by many people in the following work. This architecture performed extremely well on a variety of problems, and that is why they are now widely used. LSTMs are specially designed to eliminate the long-term dependency issue. Their common trait is to remember information for long periods, in which they are quite efficient. A recurrent neural network involves the repetition of neural network modules. For a typical RNN, this repeating module comprises a very simple structure, such as a single layer.

The flow of information in an RNN is regulated with the help of gates. These gates help the network in remembering the essential information and thus leaving the unimportant information aside. So, this allows the network to keep the crucial information through long chains of sequences which aid in making predictions (Fig. 8).

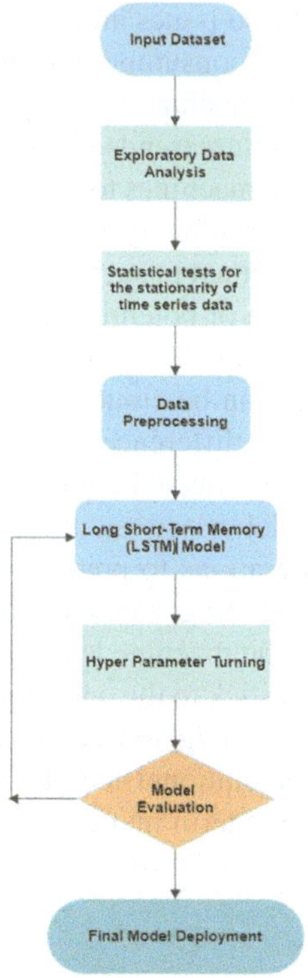

Fig. 7 Methodology

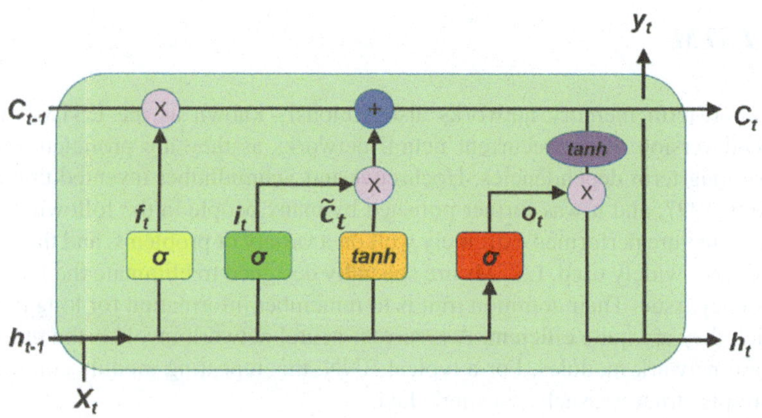

Fig. 8 LSTM architecture [16]

$$\widetilde{c_t} = \tanh\left(w_c\left[h_{t-1}, x_t\right] + b_c\right)$$

$$c_t = f_t * c_{t-1} + i_t * \widetilde{c_t}$$

$$h_t = o_t * \tanh\left(c^t\right)$$

4.2 CNN

A convolutional neural network also known as CNN is an advanced algorithm in the field of deep learning. It is specifically used for imagery analysis. It extracts the critical features from an input image by appointing importance by weights and biases relative to the distinctive objects found in an image. This makes the preprocessing of an image way smoother than the earlier manual feature engineering process as CNN learns them automatically.

Multilayer perceptron consists of fully connected dense layers that means that each neuron of the presentation layer is connected to every other neuron of its next layer. This tends to make the network over fit and so they require regularization. Now, CNN can be called this regularized version of an artificial neural network. They incorporate the use of hierarchical patterns that the input data offers and create more complicated patterns using these simple and smaller patterns. This makes a CNN less complex than an ANN and requires lesser computation power than an artificial neural network. In particular, unlike a regular neural network, the layers of a ConvNet have neurons arranged in three dimensions: width, height, depth.

The input shape of the tensor that is given to a CNN is (number of images) × (image height) × (image width) × (image depth). But after passing it through a convolutional layer, the image becomes abstracted to a feature map, with a new shape as (number of images) × (feature map height) × (feature map width) × (feature map channels).

4.3 XGBoost

XGBoost is an open-sourced gradient boosting framework available for C++, Java, Python, R, Julia, Perl, and Scala. All operating systems such as Linux, Windows, and macOS can be used for working on XGBoost. From the project description, it intends to produce a "Scalable, Portable and Distributed Gradient Boosting (GBM, GBRT, and GBDT) Library." The distributed processing frameworks such as Apache Hadoop, Apache Spark, and Apache Flink as well as a single system can run it easily.

Recently, XGBoost has earned a lot of popularity and became the choice of algorithm for many winning teams of machine learning competitions. It is an optimized gradient boosting machine learning library. The API is available in many other languages whereas the algorithm itself was originally implemented in C++. The algorithm offers parallelization within a single tree. Some of the advantages of using XGBoost are:

1. High speed and excellent performance are showcased by it.
2. Complete utilization of the processing power of modern multicore computers is done.
3. Large datasets can be easily trained with the help of XGBoost.
4. Its performance is consistently better among other single algorithm methods.

5 Results

We utilized the DATASET for complete work shown in this paper. This is a publicly available dataset at the UCI machine learning repository. Results of various machine learning as well as deep learning algorithms were recorded. The training was mainly conducted on three algorithms, namely CNN, LSTM, and XGBoost. The graphs are plotted to compare the performance, i.e., the prediction with the actual data. Figures 9, 10, and 11 showcases the results attained by all three algorithms. A total of three metrics were considered for the assessment of the performance of these models. These metrics are RMSE, MSE, and MAE. These results are given in Table 1. The best performing model is the XGBoost among the three with a very good overall result. The feature importance plot for XG boost is also plotted to see the importance of the feature as XGBoost was the best performing model.

We made use of root mean square error (RMSE) as a parameter to evaluate the methods. From our results, we can conclude that XGBoost received the lowest RMSE value and thus performed the best among all three algorithms.

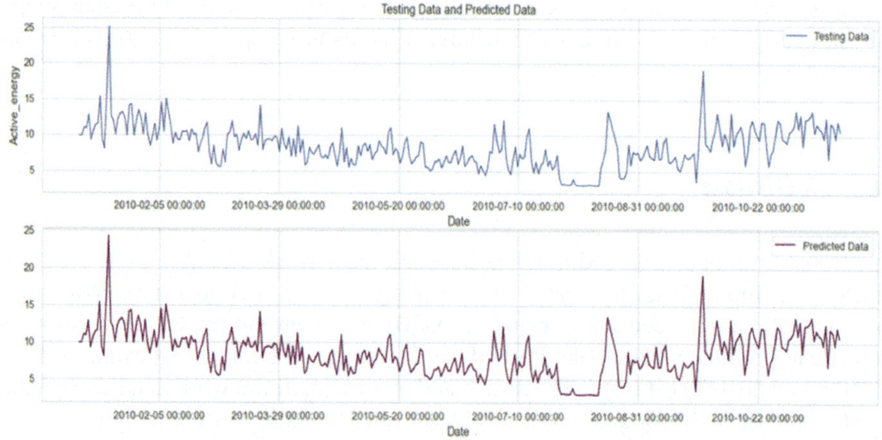

Fig. 9 XGB result (testing and predictions)

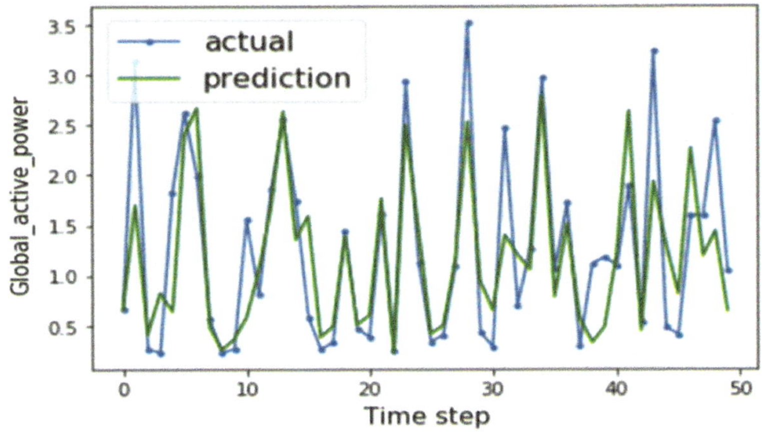

Fig. 10 CNN result (testing and predictions)

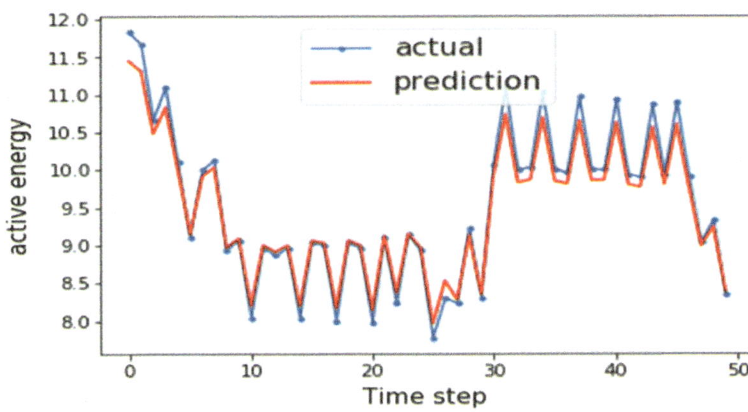

Fig. 11 LSTM result (testing and predictions)

Table 1 Comparison of results

MODEL	RMSE	MSE	MAE
XGBOOST	0.227	0.051	0.072
LSTM	0.698	0.350	0.405
CNN	0.436	0.333	0.396

6 Conclusion

In the paper, we did the exploratory data analysis to gain insights about the data, and preprocessing is done as explained. Feature engineering is also enhancing the performance of the model as the features are equally important. From our observations, we found that XGBOOST outperforms both LSTM and CNN models for time series forecasting with an RMSE of 0.227. XGBoost is a machine learning algorithm, but in this forecasting of electricity consumption, it overtook the performance of the neural networks.

Although, it is valuable to note that this was before any meticulous hyperparameter tuning, and there lies always room for improvement in building deep learning techniques viable.

7 Future Scope

Implementation of AI is everywhere nowadays in every possible domain. AI in electrical engineering is majorly used for forecasting. Forecasting electricity consumption in an area is a very important application of AI that gives the transportation and distribution of electricity a certain edge. In forecasting, majorly machine learning algorithms topple the deep learning algorithms in many instances like this one, where the machine learning algorithm is performing much better than the deep learning algorithms. For future scope, new machine learning and deep learning algorithms can be implemented with different hyperparameters. Hyperparameter tuning can also be done. The implementation of big data can play a very big role in this as the data collected over the years is very large and cannot be saved only as a comma-separated file. Different big data algorithms can be implemented, and forecasting can be done in a proper format with streaming data as well.

References

1. Parate A, Bholte S (2019) Individual household electric power consumption forecasting using machine learning algorithms. Int J Comput Appl Technol Res
2. https://ourworldindata.org/energy/country/france?country=~FRA
3. Chujai P, Kerdeprasop N, Kittisak K (2013) Time series analysis of household electric consumption with ARIMA and ARMA models. In: Proceedings of the international multiconference of engineers and computer scientists, vol 1
4. Navani JP, Sharma NK, Sapra S (2016) Technical and non-technical losses in power system and its economic consequence in Indian economy. Int J Electron Comput Sci Eng 1(2):757–761
5. Azadeh A, Faiz ZS (2011) A meta-heuristic framework for forecasting household electricity consumption. Appl Soft Comput 11(1):614–620
6. Dzobo O et al (2014) Multi-dimensional customer segmentation model for power system reliability-worth analysis. Int J Electr Power Energy Syst 62:532–539
7. Min B, Golden M (2014) Electoral cycles in electricity losses in India. Energy Policy 65:619–625
8. Rathod RR, Garg RD (2016) Regional electricity consumption analysis for consumers using data mining techniques and consumer meter reading data. Int J Electr Power Energy Syst 78:368–374

9. Rasanen T, Kolehmainen M (2009) Feature-based clustering for electricity use time series data. Springer, pp 401–412
10. Benitez I, Diez JL, Quijano A, Delgado I (2016) Dynamic clustering of residential electricity consumption time series data based on Hausdorff distance. Electr Power Syst 140
11. Zhang GP (2011) Time series forecasting using a hybrid ARIMA and neural network model. Neurocomputing 50:159–175
12. Tsai CL, Chen WT, Chang CS (2016) Polynomial-fourier series model for analyzing and predicting electricity consumption in buildings. Energy Build 127:301–312
13. Gozalez PA, Zammarreno JM (2005) Prediction of hourly energy consumption in buildings based on a feedback artificial neural network. Energy Build 37:595–601
14. Box, Jenkins (1986) Forecasting and time series analysis using the statistical system. UTS module
15. Sandels C, Widen J, Nordstrom L, Andersson E (2015) Day-ahead prediction of electricity consumption in a Swedish office building from weather, occupancy and temporal data. Energy Build
16. http://colah.github.io/posts/2015-08-Understanding-LSTMs/

A Comparison of Different Methodologies for Short Term Load Forecasting

Neeraj Kumar, Apoorva Jain, Shalini Sati, Kushagra Kapoor, and Pratham Garg

Abstract Load request determining has been a genuine test for power framework planning for various degrees of energy areas. Different computational insight strategies and techniques have been utilized in the power market for momentary burden anticipating, thinking about the sort of information and other expected elements. This work presents a logical, specialized reasonings behind transient burden expecting techniques in the energy field. Significant advantages and downsides of these strategies are examined to speak to the proficiency of each approach in different conditions. At last, utilizing long short-term memory network (LSTM), recurrent neural network (RNN) and feed forward neural network (FFNN), short-term load forecasting have been performed.

Keywords Short-term load forecasting · Deep learning · Time-series

1 Introduction

Short-term load forecasting (STLF) is an essential piece of the resource arranging area. Planning of period's load market requires request booking with different energy areas, to be specific of the load consumed and needed, age, transmission, and circulation. STLF helps energy framework administrators with a different dynamic approach in load management, including flexibly arranging, framework security, dispatching machines, request side administration etc. While STLF is exceptionally fundamental for the time-ahead load multidivisional activity, mistaken interest estimating will cost the utility a huge monetary weight.

As of late, progressions in various regions of environmentally friendly power, Data and Correspondence Innovation (ICT) empowered more proficient load consuming, compelling following of force use and age at each part of the system. The coordination between forecasting framework and sustainable power lattice by utilizing ICT could accomplish compelling utilization of principle matrix, produce lesser bills both for

N. Kumar (✉) · A. Jain · S. Sati · K. Kapoor · P. Garg
Bharati Vidyapeeth's College of Engineering, New Delhi, India

S. Suhag et al. (eds.), *Control and Measurement Applications for Smart Grid*, Lecture Notes in Electrical Engineering 822,
https://doi.org/10.1007/978-981-16-7664-2_22

specialist corporate sector as well as public and private sector units. Likewise, the worldwide carbon yield can be decreased fundamentally.

Practically all areas require electrical ability to meet their work prerequisites in either structure. Contrasting the information increment of force request every year, one can see that the chart is remarkable. Considering the way that petroleum derivatives (which are a significant wellspring of force creation) will run out in the not-so-distant future, research on elective fuel sources has started.

Even though the sun based, wind and atomic energies appear to be encouraging, they are right now incapable to meet the present and anticipated demand of load. To satisfy the necessary electrical need across the heap, a great deal of examination is being done, referencing one of it. Load prediction is a viable technique for understanding the demand before the day with fulfilling precision. There are three unique sorts of dividing them based upon their use: Momentary figure which is worried about forecasting the interest from not many hours to days. Medium figure which is tied in with foreseeing the interest from not many weeks to months. Long values of figures which is tied in with foreseeing the load from not many months to years.

Here, the emphasis is on a particular type of load forecasting: short-term forecasting, which is about predicting the demand from a few hours to days.

Different methodologies can be used to perform load forecasting. The determination of various strategies can be utilized to perform load determining. The assurance of a determining strategy depends on different components including the pertinence and accessibility of past information and figures, the estimate esteems, the degree of exactness for climate information conditions, required forecast precision, and considerably more. As needs be, choosing the best technique at first relies upon the time and estimation of the skyline of the expectation.

The organization of this paper is as follows. In Sect. 2, various researchers' work on various time-series problems have been discussed, with more emphasis on STLF. Section 3 describes the methodology used here. In Sect. 4, results have been discussed, while also discussing the possible future works. In Sect. 5, this paper has been concluded.

2 Related Work Done

Load forecasting holds an extraordinary saving capability for electric utility enterprises since it decides its principal type of revenue, especially on researchers. Exact load estimating encourages the electric utility to settle on unit responsibility choices, decrease turning hold limit and timely unit upkeep plan appropriately. It is in this manner fundamental that the power producing associations sought to have earlier information on future interest with incredible exactness of values of load. Some information including calculations assume the more relatable load value to foresee the heap forecasting [1]. STLF is especially basic for the time-ahead activity, off base interest estimating will cost the utility a huge monetary weight [2]. Since the ANN was starting in that time, to eliminate the reluctance among the contemporary specialists Henrique Steinhart Hipper has given an audit and assessment on neural networks in STLF [3]. Regression is a technique for displaying an objective worth dependent on autonomous indicators. It is generally utilized for anticipating and discovering circumstances and logical results connection between various factors. Relapse methods generally vary dependent on the quantity of autonomous factors and the sort of connection between the free and ward factors. Straightforward LR and multiple LR actualizes a relapse examination between the free factors ($x1$, $x2$ … xn) and subordinate variable (y). Moghram et al. have investigated about MLR to figure 24 h load utilization dependent on temperature and wind speed for winter and summer [4]. Strategies and methods are significant with regards to exact assessment. In any case, restricted writing is accessible for STLF systems. Most overviews in the writing are given to the examination of various STLF strategies [4–7]. A study introduced by Soares et al. uncovers the conceivable outcomes of the viable group ML calculations in taking care of relapse issues [8].

3 Methodology

In this research, three different methodologies are compared.

3.1 Recurrent Neural Networks

Recurrent neural networks, (Fig. 1) otherwise called RNNs, are a class of neural organizations that permit past yields to be utilized as information sources while having hidden states. They are as follows:

For each time-step, the activation and the output are expressed as follows:

$$a^{\langle t\rangle} = g1(Waa\ \ a^{\langle t-1\rangle} + Wax\ x^{\langle t\rangle} + ba \tag{1}$$

Fig. 1 Recurrent neural network architecture

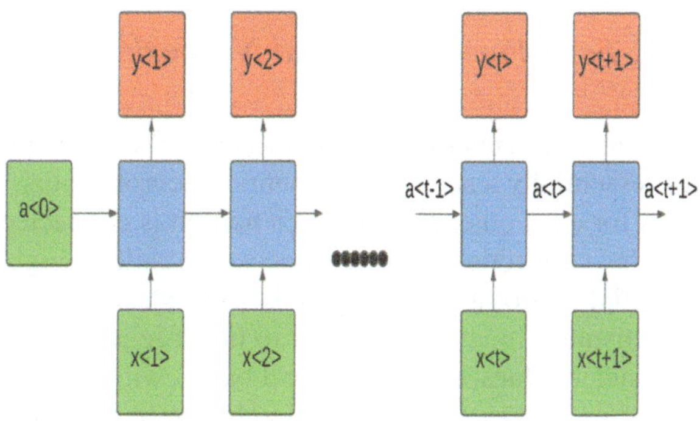

$$y^{\langle t\rangle} = g2\left(Wya\ a^{\langle t\rangle} + by\right) \tag{2}$$

where *Wax, Waa, Wya, ba, by* are coefficients that are shared temporally and $g1$, $g2$ activation functions (Fig. 3).

3.2 Long Short-Term Memory Network

Long short-term memory networks—typically called "LSTMs"—are an extraordinary sort of RNN, fit for learning long haul conditions. They function admirably on a vast assortment of issues and are currently broadly utilized.

LSTMs are expressly intended to keep away from the drawn-out reliance issue. Recalling data for extensive periods is essentially their default conduct, not something they struggle to learn.

The neural network architecture of the proposed long short-term memory model has been discussed here. The model comprises of three types of layers named as the inner layer which takes the input, then LSTM layer and the final outer dense layer that gives us the output. Input layer is used to create sequential data for LSTM layer. Each sequence is the number of time-steps. A vector is denoted by and meteorological parameters, namely Temp, RH, SR, and WS. These sequences are then fed to the LSTM layer. Each LSTM layer consists of several storage memory blocks (Fig. 3). These blocks carry memory cells which are interconnected along with a multiplicative unit which is known as gates. A memory cell keeps the temporary part of the network and gates are considered to balance the information A LSTM cell has three gates—which are one which gives the input, other provides the output and last is forget gate. The flow of entering information in memory is done by the input gate, while output gate takes care of the output flow of cell activation in rest all part of a network. The forget gate introduces a disadvantage [8] of LSTM models by avoiding them from taking continuous input which are not divided into subsequence (Fig. 2).

3.3 Feed Forward Neural Network

A feed forward neural network (FFNN) is the most basic neural network architecture. It is essentially a stack of rudimentary neuron layers, with different activation functions for each other neuron. Each neuron is connected to a neuron in the next layer, and that connection has a numeric value, called weight. The activation function is an essential part of a neural network, as it introduces non-linearity. An FFNN without an activation function is similar to linear regression. This neural network learns the parameter values, i.e. weights and biases with each forward and backward propagation. Here, hyperparameter tuning has been performed as well, where an optimum combination of learning rate and momentum has been used to get the best results (Fig. 3).

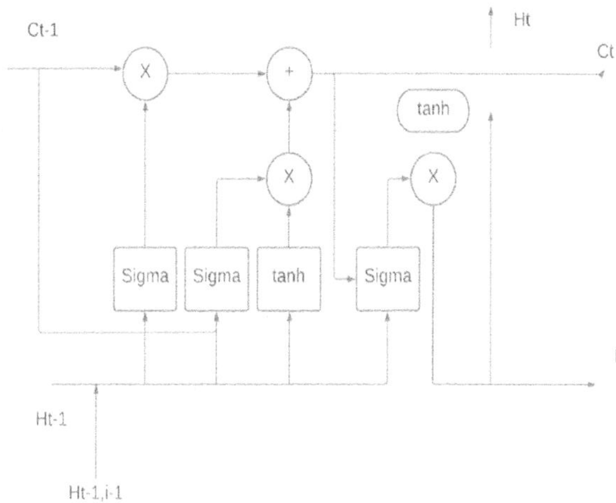

Fig. 2 Long short-term memory network architecture

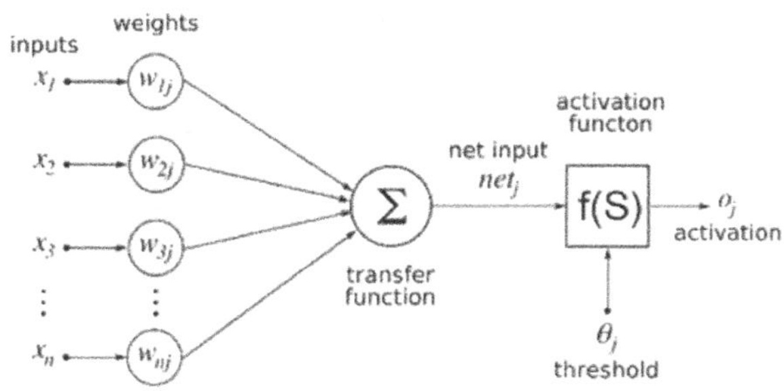

Fig. 3 Feed forward neural network single layer network

$$Yi = \sum_{i=1,j=1}^{i=n,j=h} (Xi)(Wij) \tag{3}$$

$$Yo = f(Yi) \tag{4}$$

Here, *Xi* is the input feature, *Wij* are the weights, and *f* is the activation function.

4 Results and Discussion

FFNN model had a mean squared loss of 0.0208. LSTM model had a mean squared loss of 0.0072 and performed the best out of all the models. The mean squared error of the RNN model was 0.0175.

Figures 5, 6, and 7 illustrate the performance of the models by comparing the predicted and the actual values by each of these models.

5 Conclusion

In this paper, we have presented a system which can propose a specific solution for load forecasting. To prove the reliability of the system we have tested the system with real world data sets. We have conducted several experiments to evaluate the performance of three algorithms, i.e. recurrent neural network (RNN), long short-term memory network (LSTM), feed forward neural network (FFNN) and concluded the experiment with a comparison of actual and predicted data.

After training the above three models on the dataset, it can be observed that LSTM performs the best out of them, followed by RNN and FFNN. To compare the three models and analyzing their performance, see Figs. 4, 5, 6, 7 and Table 1.

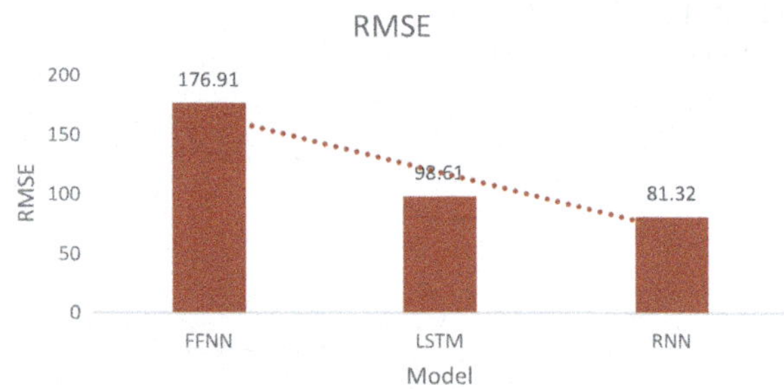

Fig. 4 Comparison of RMSE of the models used

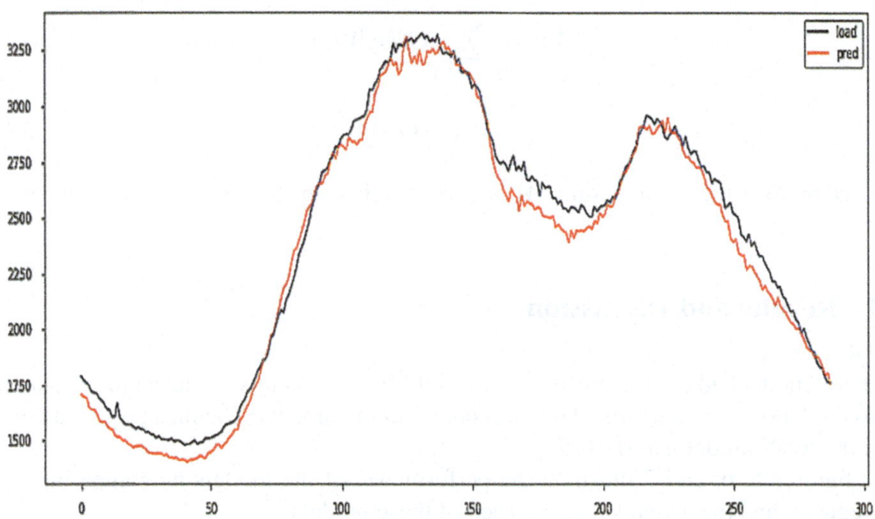

Fig. 5 Comparison of actual and predicted by RNN

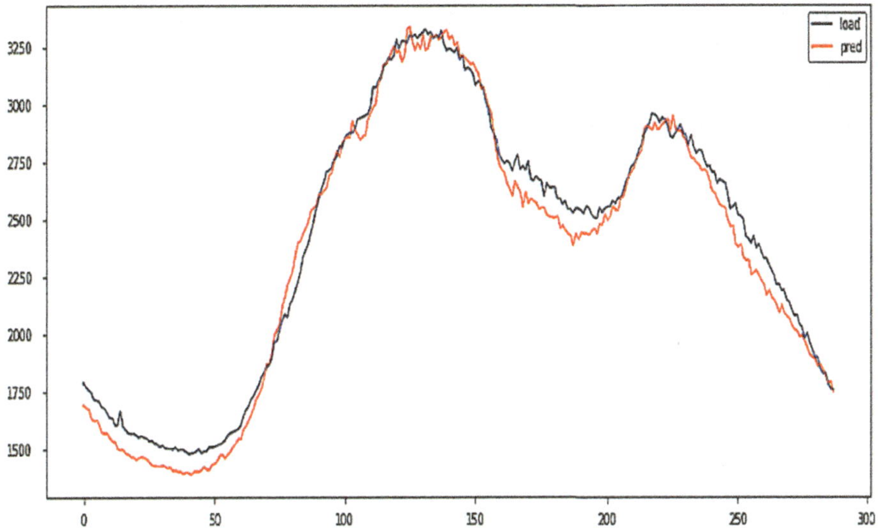

Fig. 6 Comparison of actual and predicted by LSTM

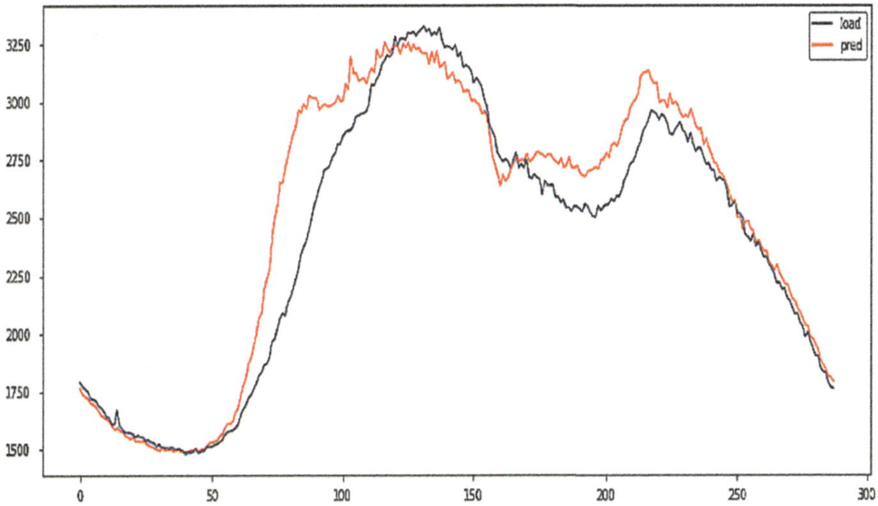

Fig. 7 Comparison of actual and predicted by FFNN

Table 1 Evaluating each model by RMSE and MAPE metrics

Model	RMSE	MAPE
LSTM	81.32	2.63
RNN	98.61	2.82
FFNN	176.91	5.96

Acknowledgements We express our deep gratitude to Neeraj Kumar, Assistant Professor, Department of Electrical and Electronics Engineering for his valuable guidance and suggestion throughout our project work. We are thankful to Mrs. Shashi Gandhar, Project Coordinator, for her valuable guidance. We would like to extend our sincere thanks to Head of the Department, Mrs. Kusum Tharani for her timely suggestions.

References

1. Sarangi PK, Singh N, Chauhan RK, Singh R (2009) Short term load forecasting using artificial neural network: a comparison with genetic algorithm implementation. ARPN J Eng Appl Sci 4(9) (2009)
2. Shahidehpour M, Yamin H, Li Z (2003) Market operations in electric power systems: forecasting, scheduling, and risk management. Wiley, Hoboken, NJ, USA
3. Hippert HS, Pedreira CE, Souza RC (2001) Neural networks for short term load forecasting: a review and evaluation. IEEE Trans Power Syst 16(1)
4. Moghram I, Rahman S (1989) Analysis and evaluation of five short-term load forecasting techniques. IEEE Trans Power Syst 4(4):1484–1491
5. Raza MQ, Khosravi A (2015) A review on artificial intelligence based load demand forecasting techniques for smart grid and buildings. Renew Sustain Energy Rev 50:1352–1372
6. Fallah S, Deo R, Shojafar M, Conti M, Shamshirband S (2018) Computational intelligence approaches for energy load forecasting in smart energy management grids: state of the art, future challenges, and research directions. Energies 11:596
7. Hippert HS, Pedreira CE, Souza RC (2001) Neural networks for short-term load forecasting: a review and evaluation. IEEE Trans Power Syst 16:44–55
8. Mendes-Moreira J, Soares C, Jorge AM, de Sousa JF (2012) Ensemble approaches for regression: a survey. ACM Comput Surv 45(1):1–10

Design and Development of Phase Lock Loop (PLL) Premised Islanding Investigative Techniques

Mohan P. Thakre and Nishant P. Matale

Abstract The day-to-day use of distributed generation (DG) in the electrical grid is now significantly increased owing to its advantages. As infiltration of distributed sources increases within the electricity grid, there are many multiple difficulties with the grid interface which need to be acknowledged. The established network is comprised of central power generators; while the next power system will involve that several distributed energy resources (DERs). Utility operators regulated a centrally controlled power generation system, but several DERs would not be regulated by utility operators. Intelligent control strategies should be used by DG for highly reliable and ideal grid functionality. This article addresses an insight into the issues of an anti-islanding grid interface, together with the different approaches to island identification. The ability to detect within the non-detection zone (NDZ) along with financial impact are major determinants for the interpretation of each technique. Two new cases with the interpretation which occur throughout the electrical power system (EPS) are also the impact of various means in parallel and NDZ as well as the likelihood of a false trip precipitated by a load step has been presented in this article.

Keywords Islanding detection · Grid synchronization · Grid-connected inverter · PLL

1 Introduction

Nowadays, the use of DERs in EPS is rising due to the higher cost of conventional energy sources and the environmentally friendly nature of renewable energy sources (RES). There are several RESs, a few of which have been mentioned in this: solar, wind, biomass, geothermal [1, 2]. The DERs are capable of supplying power to the EPS. Power electronics is indeed the functionality between the DER and EPS [3, 4]. This same inverter uses to still be switched off when the voltage or frequency

M. P. Thakre · N. P. Matale (✉)
K. K. Wagh Institute of Engineering Education and Research, Nashik, India

S. Suhag et al. (eds.), *Control and Measurement Applications for Smart Grid*,
Lecture Notes in Electrical Engineering 822,
https://doi.org/10.1007/978-981-16-7664-2_23

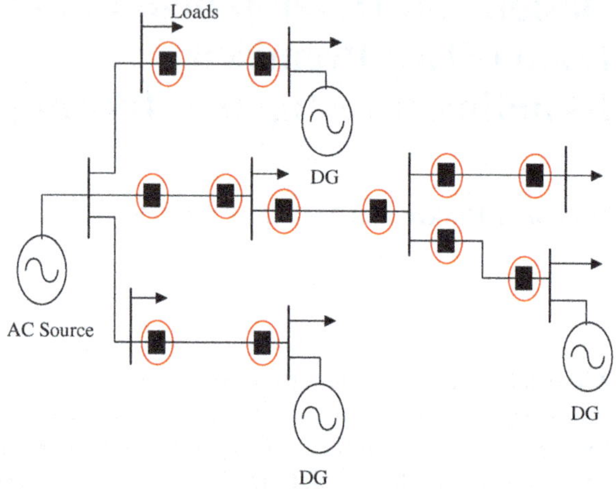

Fig. 1 EPS with circuit breakers and DGs

exceeds the specified values and therefore is capable of detecting when the centralized generator is not attached [5, 6]. Disassociation for service purposes or sometimes attributable to human error and any circuit breaker on EPS trips as shown in Fig. 1 are all the reasons for islanding. The IEEE std 1547 claims that DERs should be disconnected within 2 s whenever the island occurs [7]. There are many reasons for the disconnection of DERs from the EPS, such as to avoid the electrical grid from reconnecting with the DERs out of phase creating a huge voltage increase which damages the loads and is also harmful to the line operator [8–10].

Throughout the coming years, DER operations can be switched to properly controlled island mode. Due to the higher use of DG, massive issues arise, such as how to successfully move from centrally controlled power generation toward several different DERs. The requirement is presented in this article where even the EPS has been separated and the island has been established. Different methods of island identification have been used in literature including variations in voltage, impedance, and frequency along with injecting harmonics, reactive power, jump in frequency, and SCADA [4, 11–13]. Utilizing output change, the position of the smart grid is evaluated by passive methods [12]. Active models were developed to predict the status of the grid connection. Both methods have been implemented in a hybrid approach. Various strengths and limitations have been linked to each method. This same assessment of each method has been conducted based on three factors: NDZ, price, and quality of power. Two further factors are suggested in this paper, such as that the anti-islanding technique should not cause false trips due to switching on or off local loads and should be capable of detecting different methods throughout parallel within the NDZ. This article shows that such circumstances must be taken into consideration when designing the detection of islanding. Simulation has been performed employing three techniques and testing has been executed on a prototype DC-DC converter linked to the inverter as seen inside Fig. 2. A grid is formed by using a 120 V_{rms} channel.

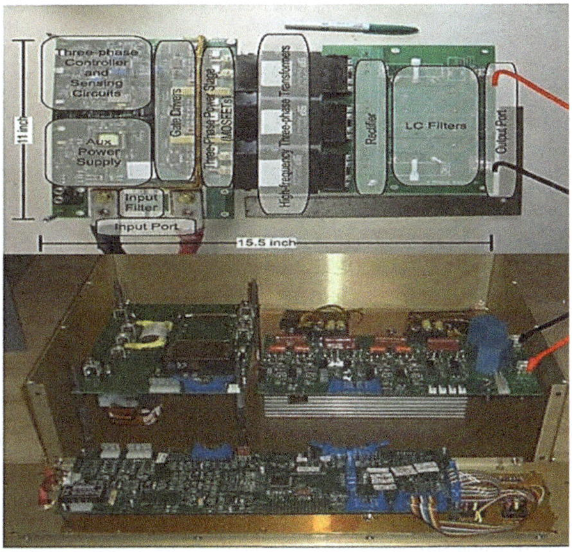

Fig. 2 Prototype circuits used for the experiments

A commercial unit has been used to perform a particular measure. The relevant information on the patent shall be used for the simulation and testing of this method [14]. The NDZ has been the first case to pose the technique to be undetectable. In this situation, the load present in the island matches the generated power, implying that the current from the EPS has been zero. The parallel inverter case might be the next condition wherein multiple methods have been used on the same island [15]. The market has different DERs because they will have patented their anti-islanding technique. The last situation is that the anti-islanding technique must not cause a false trip during the loading step from local loads. The voltage changes on point of common coupling (PCC) due to local load switching on and off. The phase shift in the current is caused by the inductors and capacitors of a load switching on or off, and the DER control loops would then display a step response. Owing to these considerations, the DERs should also not unacceptably turn off.

2 Passive Methods

The frequency and voltage limits specified by the IEEE standard 1547 shown in Fig. 3 are used by the passive method for detecting the island. For inverter operating voltage range is between 88 and 110% of the nominal voltage at PCC. The frequency range for the operating inverter is 59.3–60.5 Hz. The conditions which do not follow IEEE 1547 must disconnect within the clearing times shown in Fig. 3.

The NDZ is testing conditions in this method. The condition in which the current from the electric grid to the local loads goes is small and the generation from the

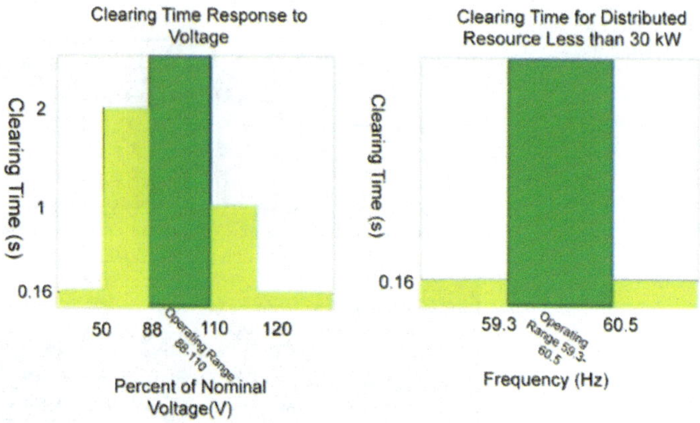

Fig. 3 Voltage and frequency limits specified in IEEE Std. 1547

Fig. 4 Islanding test experimental setup

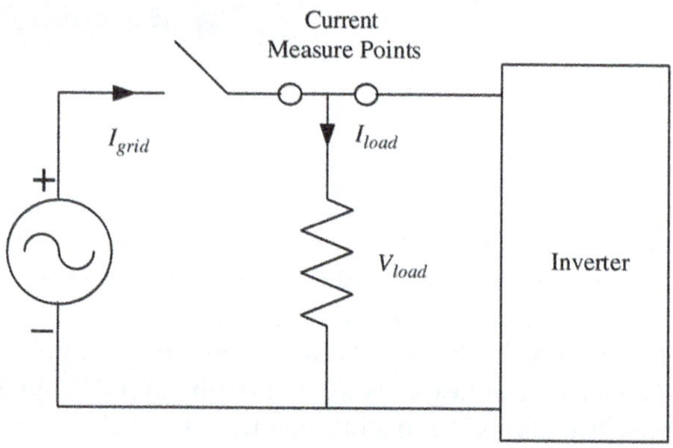

DGs match the local loads is known as NDZ. Figure 4 shows the test setup. The zero value of the current I_{grid} indicates that the active power of the resistive load matches with the inverter active power. The testing of the same method is done within NDZ shown in Fig. 5. There is no change in voltage and frequency during the island. As the limits specified by the IEEE std. 1547 are not fulfilled by this method it does not suitable for the application.

3 Active Methods

The PLL is used by active methods to detect an island. A disturbance is injected by active methods to determine the status of the grid connection.

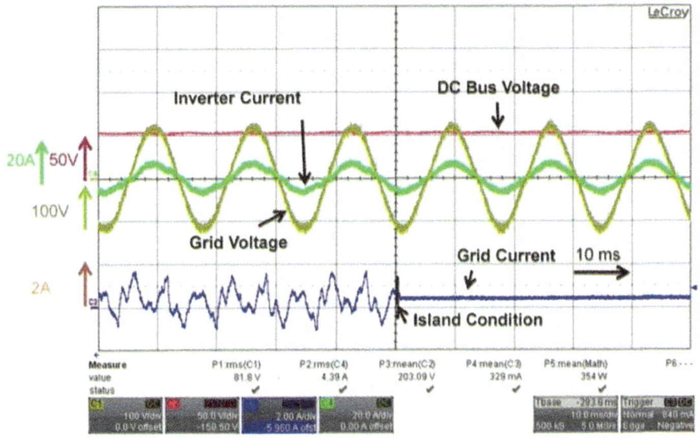

Fig. 5 Passive method in NDZ

3.1 PLL

The PLL is used as the foundation for detecting the island condition in this paper. The synchronization of an oscillating waveform with a measured waveform is the principle of PLL. The PLL generates an oscillating waveform. The components of PLL are a voltage-controlled oscillator (VCO), an integrator, and a phase detector (PD). Figure 6 shows the simplified structure of PLL. Equation (1) gives the magnitude of grid voltage. The signal at V_e is equal to the measured waveform (V_{grid}) multiplied by the PLL output (V_f) shown in Eq. (1). Where the angle of V_{grid} and V_f is denoted by θ_1, θ_2.

$$V_e = V_{grid} V_f = -\cos(\theta_1)\sin(\theta_2) = \frac{1}{2}[\sin(\theta_1 + \theta_2) + \sin(\theta_1 - \theta_2)] \tag{1}$$

If $\theta_2 = \theta_1 + \phi$ then Eq. (1) can be rewritten as Eq. (2),

$$V_e = \frac{1}{2}[\sin(2\theta_1 + \phi) + \sin(\phi)] \tag{2}$$

Fig. 6 Phase lock loop

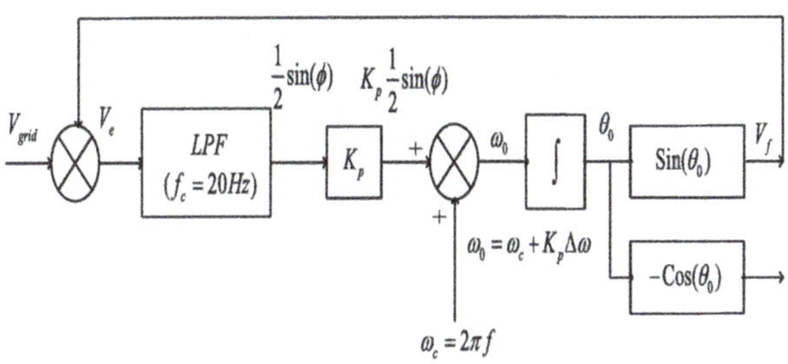

The design of a low pass filter (LPF) is done in such a manner so that it will significantly attenuate the second harmonic signal in the above equation so only sin (ϕ) is left which at small values of ϕ is equivalent to ϕ. The ϕ denotes the difference between two θ terms and having K_p as gain which is included to expected angular frequency (ω_0), is given by Eq. (3).

$$\int \omega_0 \mathrm{d}t = \omega_0 t = \theta_0 \tag{3}$$

3.2 Method Based on Jump in Frequency

The extra frequency is added to the PLL in the frequency jump method shown in Fig. 7. In this paper, 10 Hz is used as an additional frequency. The PSIM has been used for simulation purposes to verify the detection capability of this technique.

The output voltage, as well as grid current, has been seen in Figs. 8 and 9, before and after an island condition, respectively. The simulation is within the NDZ as the

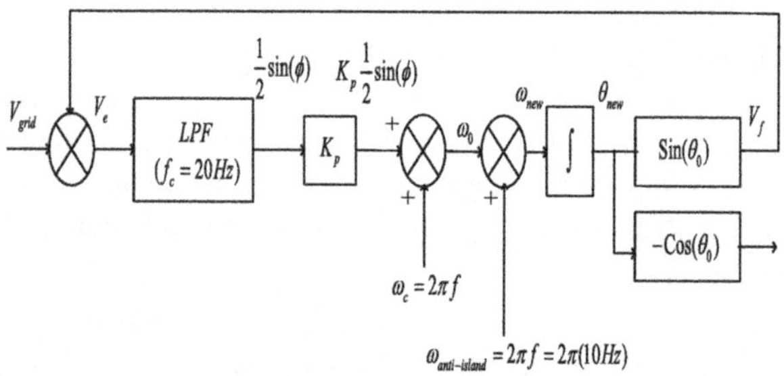

Fig. 7 Frequency jump method

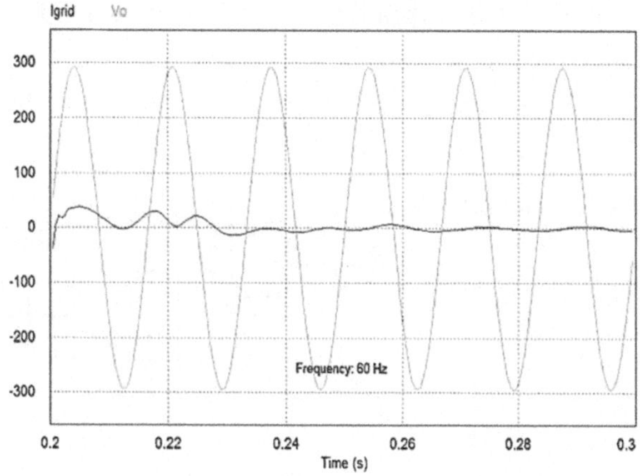

Fig. 8 V_o with I_{grid} before island

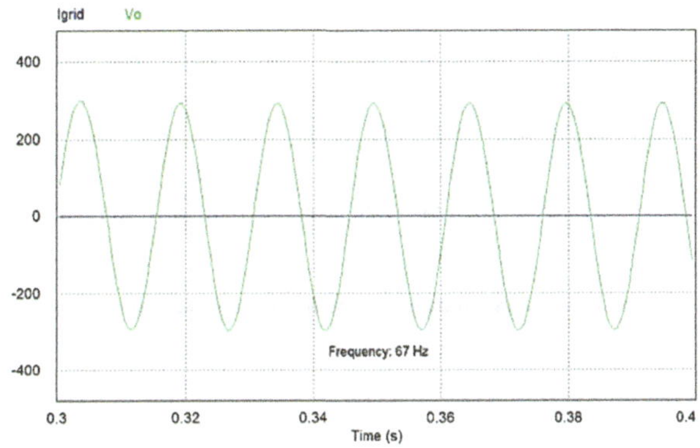

Fig. 9 V_o with I_{grid} later to the island

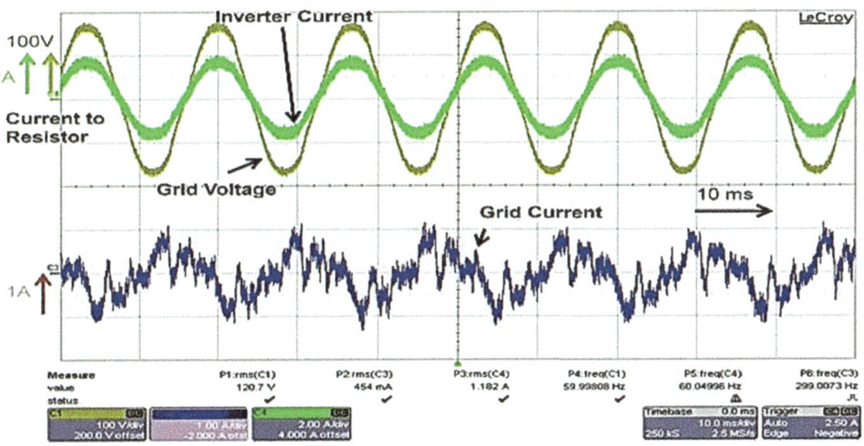

Fig. 10 V_o with I_{grid} before islanding

grid current reaches zero as shown in Fig. 11. The jump in frequency is 67 Hz after the island is formed and then the frequency reaches 70 Hz. So, this method can be observed in NDZ. The experimental results are shown in Figs. 10 and 11.

3.3 *Method Based on Harmonic Injection*

About the existing theta value, the harmonic injection method disrupts the sine wave PLL. Equation (4) gives a new PLL output (V_f). Due to the small value of K, cos and sin term reach unity and K, respectively, converting this method equivalent to an injection of 2nd harmonics.

$$V_f = \sin(\theta_0 + K \sin \theta_0) = \sin \theta_0 \cos(K \sin \theta_0) + \cos \theta_0 \sin(K \sin \theta_0) \quad (4)$$

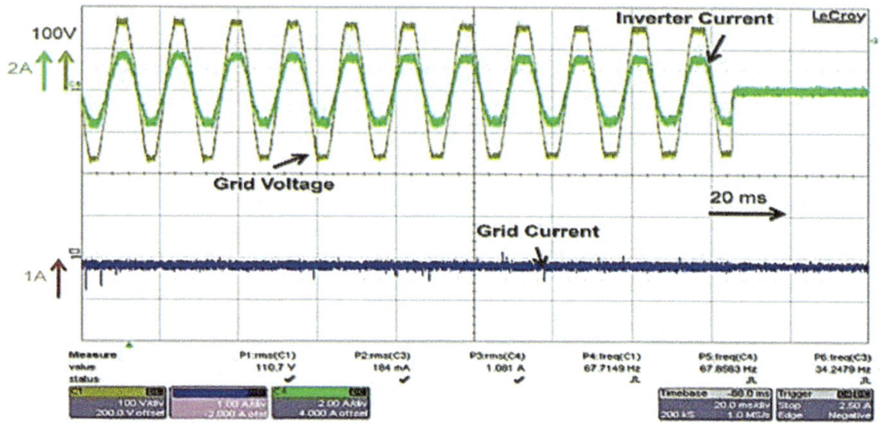

Fig. 11 V_o with I_{grid} later to islanding

Fig. 12 Harmonic injection PLL

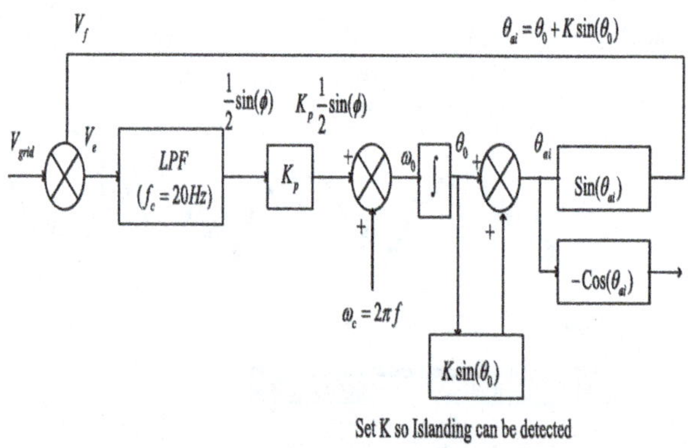

Fig. 13 Simulation results of the method based on harmonic injection

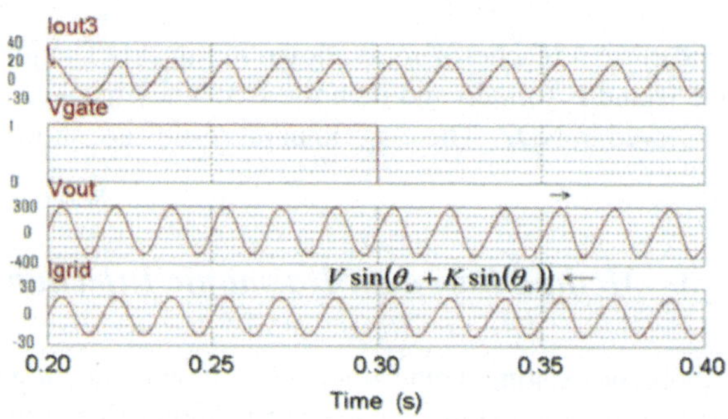

Figure 12 shows this new PLL method. Figure 13 shows simulation results. Due to continuous injection of harmonics by inverter current which results in degradation

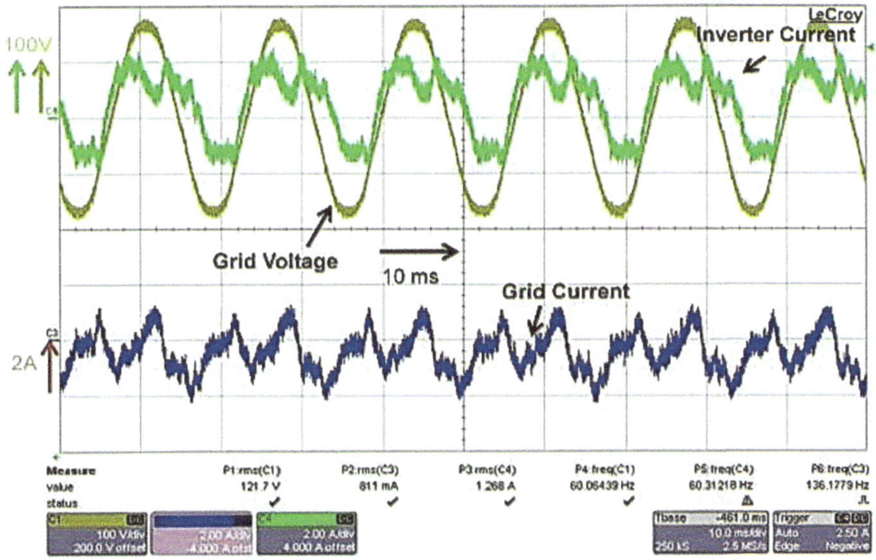

Fig. 14 Harmonic injection before islanding

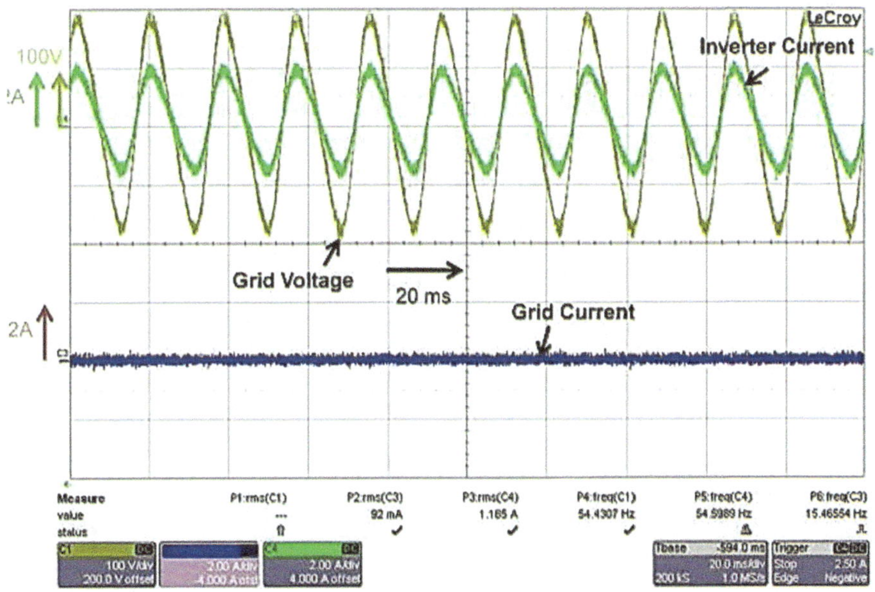

Fig. 15 Harmonic injection later to islanding

in power quality. Figures 14 and 15 shows the experimental results verifying that for single inverter case harmonics is detectable.

3.4 A Method Based on Enphase

The patent is used as the investigation of the anti-islanding technique in commercial inverters. The phase shift islanding detection technique has been used in the Enphase micro-inverter. Figure 16 demonstrates the Enphase technique. This method does not use change in voltage for identification of islanding condition rather it uses the change in PLL error from the grid connection under the islanding condition. When the value of PLL output and signal which is measured is equal then the error value becomes zero.

Figure 17 shows the simulation of the grid-connected case in the Enphase technique. The Enphase error will increase equally whenever the phase shift has been started and falls later to the phase shift. Figure 18 demonstrates the significance after islanding; the error signal decreases but unable to rise to overvalue in steady-state.

Later to islanding, the Enphase method at zero-crossing will be switched off. The verification is performed by doing the same experiment several times for the case of a single inverter as shown in Fig. 19. The phase shift injection is a sign for the detection of the insular condition suggested in the patent. Figure 19 shows that a phase shift occurs later to the island, which causes 57.8 Hz frequency at peak. These results indicate that the Enphase methods also recognize a phase shift which results in inverter turn off next to zero crossings. The capacitor or inductor of high value is used for further investigation. The output capacitor of the Enphase technique tends to cause a phase shift.

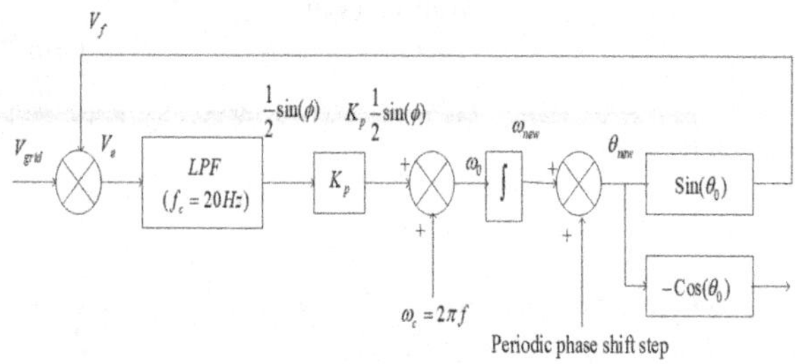

Fig. 16 Enphase method PLL

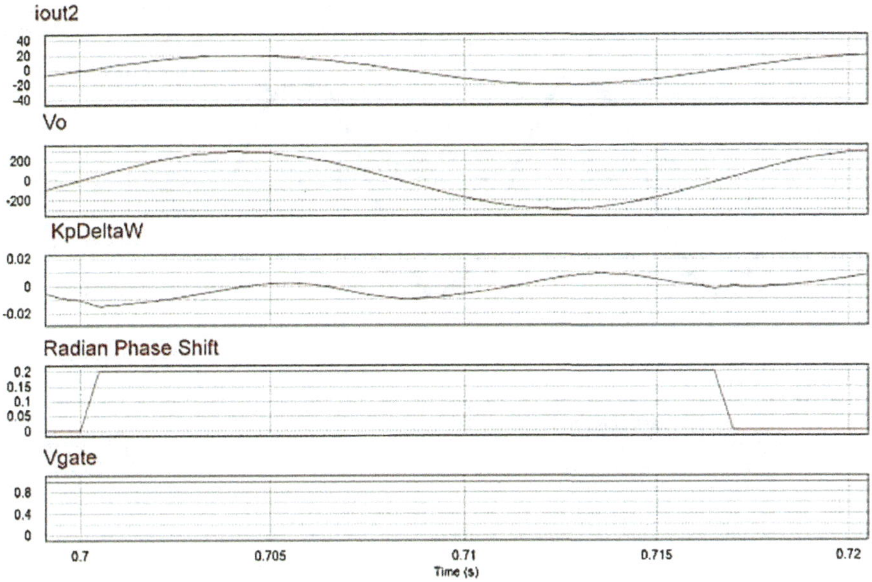

Fig. 17 Simulation results of Enphase method before the island

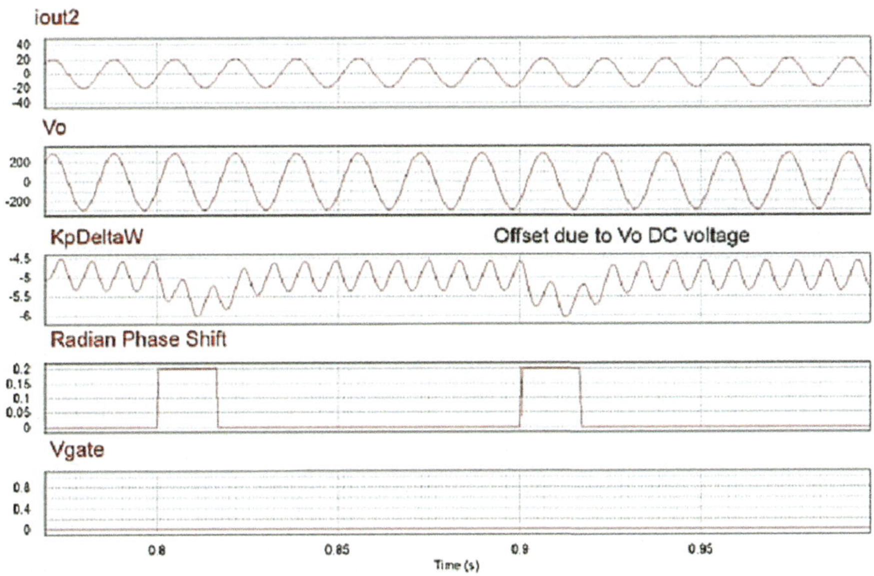

Fig. 18 Simulation results of Enphase method later to the island

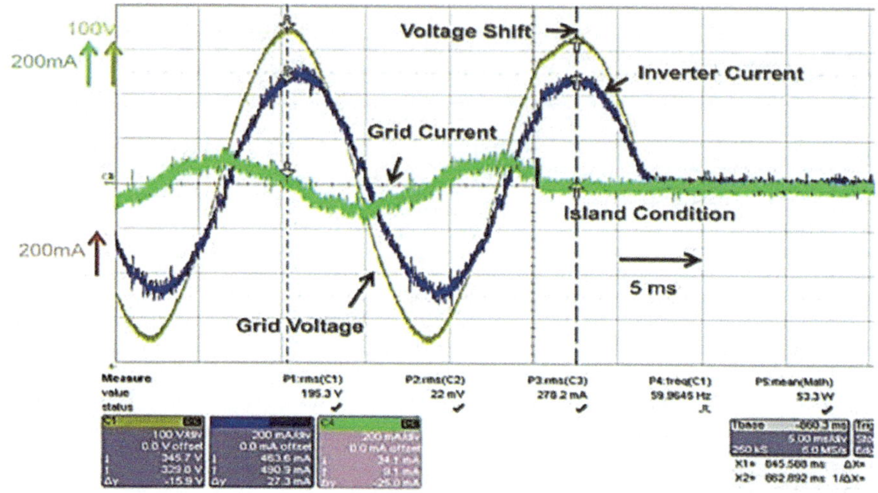

Fig. 19 Experimental results of Enphase method

4 Parallel Condition in Multiple Methods

The frequency jump techniques, as well as the harmonic injection method, have been evaluated in parallel at the same power level and different power levels shown in Fig. 20.

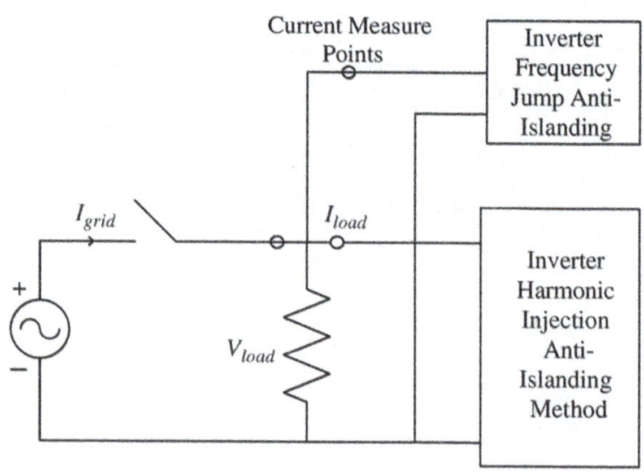

Fig. 20 Test setup for parallel inverters with various identification techniques of islanding

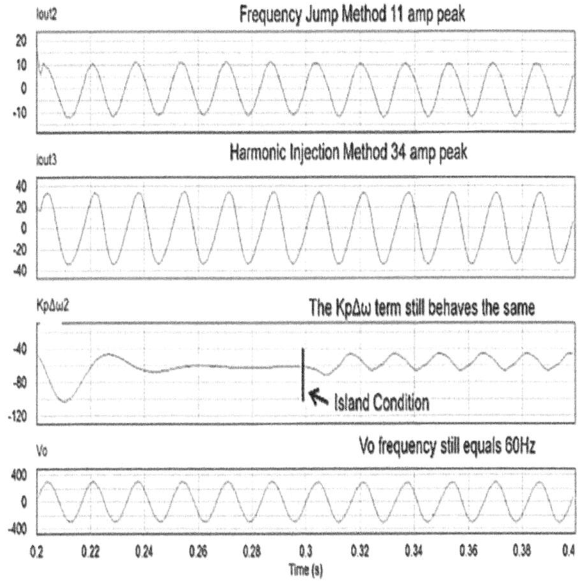

Fig. 21 A method based on frequency jump and harmonic injection at various power levels in parallel

4.1 Method Based on Frequency Jump and Harmonic Injection at Various Power Levels in Parallel

In this case, the power level is lower in the frequency jump method compared to the harmonic injection method. Figure 21 demonstrates that the jump in frequency technique having a peak of 11 A whereas the harmonic injection method does have a peak of 34 A. To identify an island, the output voltage also is not sufficiently significantly impacted by the frequency jump technique. In this situation, the harmonic injection method will be disconnected before the frequency jump technique could be detected.

4.2 Method Based on Frequency Jump and Harmonic Injection at Same Power Levels in Parallel

In parallel conditions of two approaches at the same power level, the output voltage will have considerably less visible variation at the time of the island scenario that is shown in Fig. 22. That further implies that this phase implies a low enough threshold for disconnection throughout this condition. False island identification could cause other EPS happenings when this method does have a low threshold. Current disturbance has an effect on voltage, which would be measured by islanding detection systems. If the current is lower, the impact on voltage would be in the NDZ of multiple units throughout parallel.

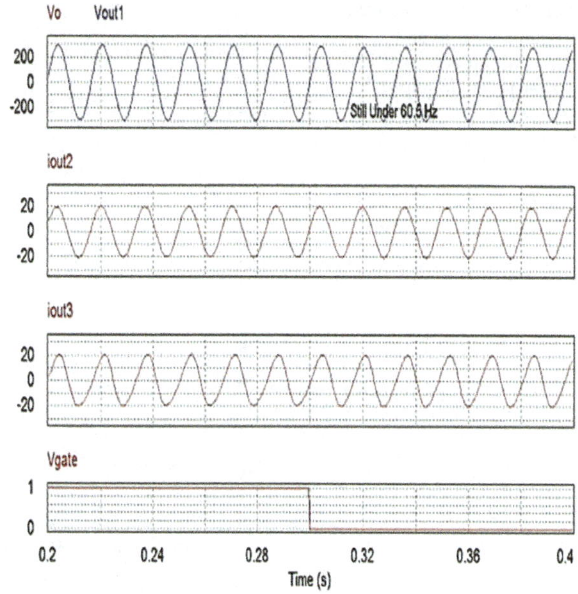

Fig. 22 Parallel condition of two methods at the same power

4.3 Parallel Operation of Enphase Technique and Passive Technique

Simulation shall be performed for parallel operation of the Enphase technique and the passive technique shown in Fig. 23. Realize that the error signal fluctuated till 0.49. The whole technique needs an accurateness of around 0.02.

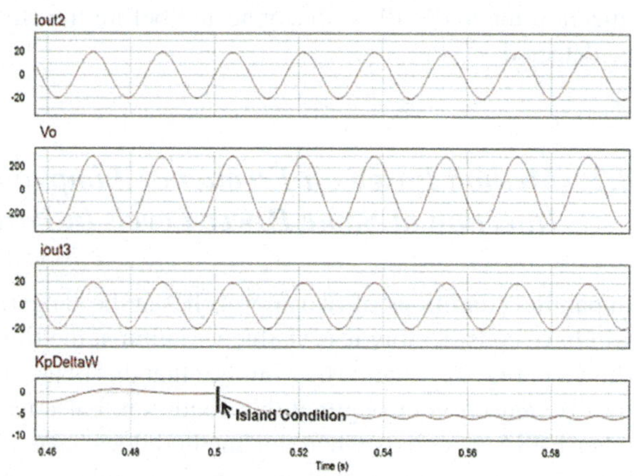

Fig. 23 Parallel operation of the Enphase technique and the passive technique

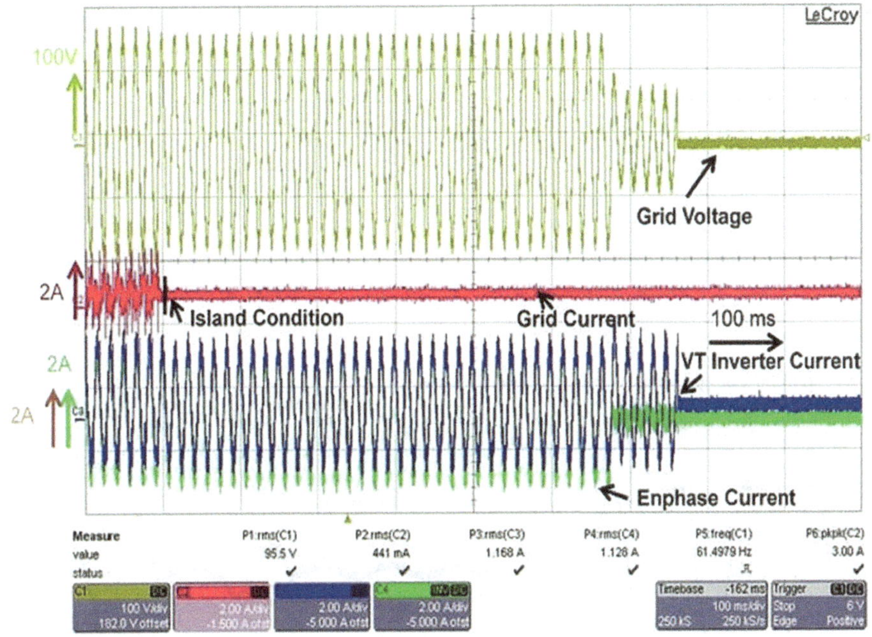

Fig. 24 Enphase and passive method in parallel

Using error signal for identification of an island that results in better power quality without influencing the voltage. Precise measurements are required in this technique. Figure 24 shows that the Enphase method will remain allied in a parallel case.

5 Load Step

A load step has been set up where the output of the Enphase is parallel to the load of 300 Ω (160 W) and the EPS. The EPS link comes from a 208 V_{rms} outlet that is connected to a 208 V_{rms} to a 240 V_{rms} transformer. The switch is connected to the 6 Ω (9.6 kW) resistor. Figure 25 shows the schematic of the load step test.

Due to the load step, more current will flow from the EPS resulting in a high voltage drop over the line impedance. As a result an inverter voltage drop to 91–93% of the nominal voltage shown in Fig. 26. The inverter remains on according to IEEE 1547 Std. as this voltage is above the lower voltage. As the voltage drops further, there may be a disconnection of circuit breakers on the electrical grid resulting in a blackout. The frequency monitoring in PLL has been significantly impacted by the load step. The 14 Ω load which is parallel to the 7 Ω load can be seen in Fig. 27. Owing to this change in load, high variation is taking place in PLL. Frequency monitoring

Fig. 25 Test setup of load step

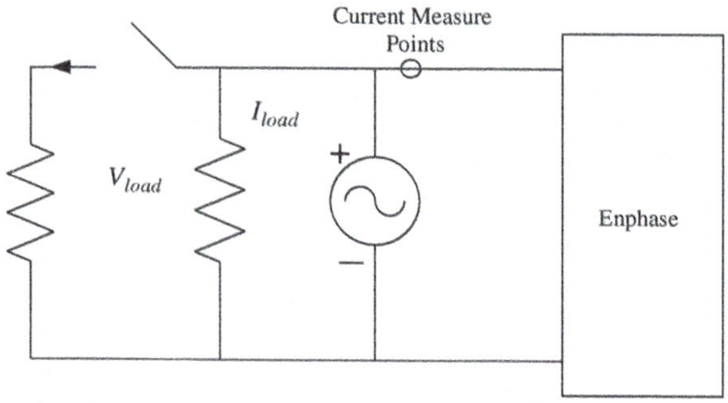

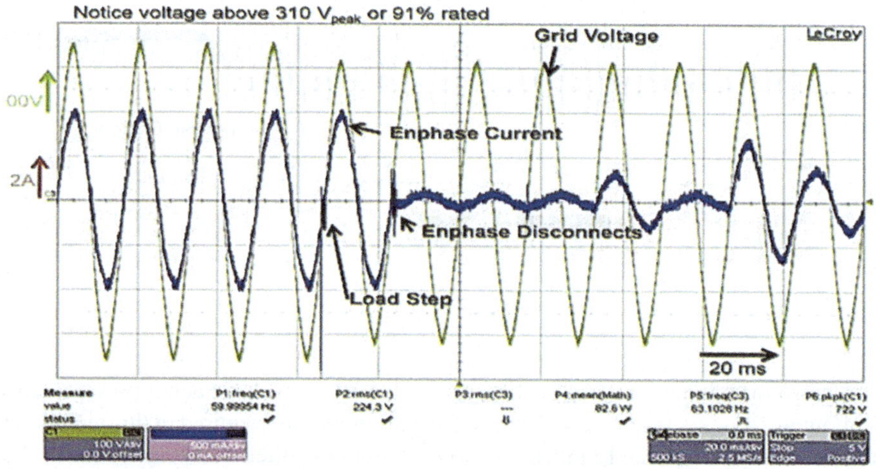

Fig. 26 Load step results

might be taken out of range although there is no island where PLL frequency has been used. The current phase change with a 500 μF capacitor load seen in Fig. 28.

6 Method Based on Communication in Power Lines

Detection of islanding has been done by the use of power line communications. The biggest issue associated with this system has been the power lines behaving similar to the inductor and the capacitor that filters the communication signal. Temperature, cable size, the length between transmitting and receiving lines, as well as other factors influenced the value of inductance and capacitance of the line.

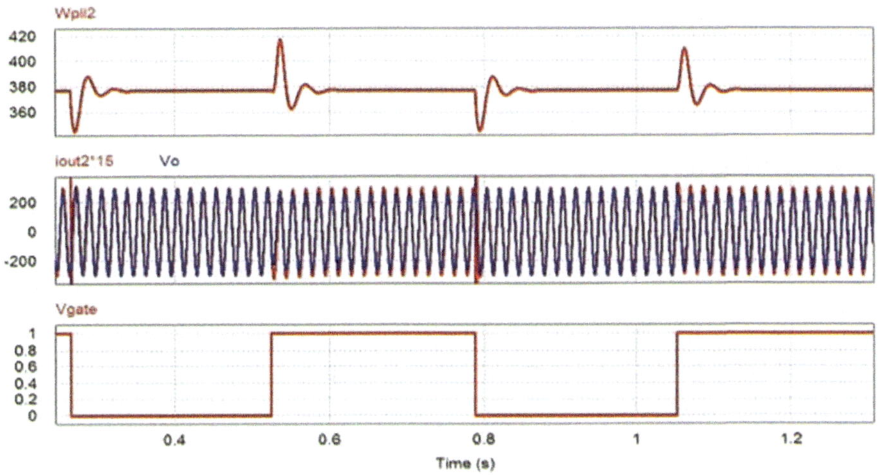

Fig. 27 PLL results for load step

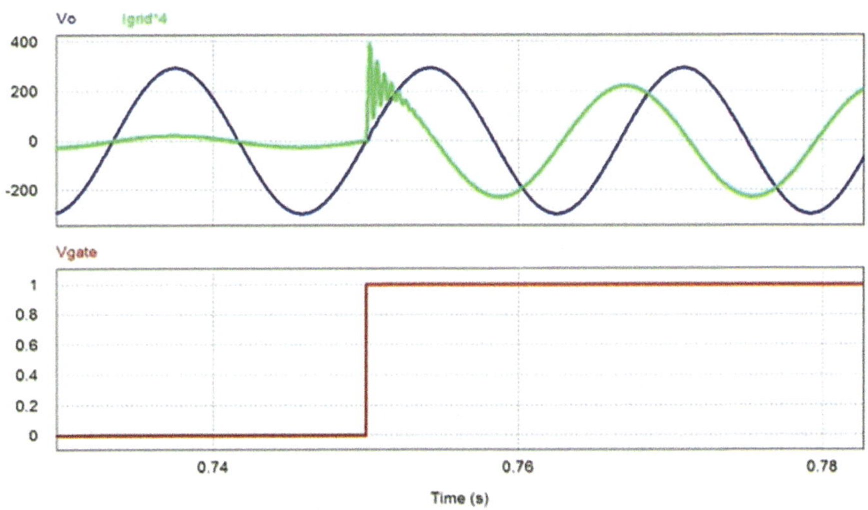

Fig. 28 Change in current phase at load step

7 Supervisory Control and Data Acquisition (SCADA)

More monitoring and information exchange facilities have been needed to enhance reliability and to monitor DERs in the future. High investment is required for the security and installation of communication facilities. There is a chance of loss of communication. The islanding detection techniques discussed throughout this article would still be crucial in event of failure of interaction between both the DERs and the utility operator.

8 Conclusion

Alternative techniques of identifying islanding and conditions are discussed in this paper. In which passive techniques are not applicable for NDZ. The active techniques are appropriate for NDZ. Power quality has been degraded throughout the case of a harmonic injection method. In the parallel case of multiple methodologies, when the output current is far less in one method than the other, it will not be detected. The whole situation requires a high inverter sensitivity. The island has been identified by the Enphase technique and will be switched off in the next zero crossings. Owing to its ability, it enhances the effectiveness of the power and unnecessary tripping would be an issue on the other side. The Enphase system is especially sensitive to transient outputs. These are validated by switching large loads connected to an electrical grid.

If islanding detection techniques are made more sensitive, there will be a problem of unnecessary tripping, however, if such methodologies are made less sensitive, the island will not be detected in parallel inverter cases. In these circumstances, therefore, an appropriate balance must be maintained. A very low transmission frequency has been needed for power line communications to avoid long-distance signal filtration by transmission lines. There is a noise issue in the communication signal at lower frequencies. Because as frequency is low, the transmission of data is also affected, attempting to make this technique uneconomic. In regards, the SCADA methodology requires inverter-based techniques in the case of communication loss.

References

1. Puttgen HB, MacGregor PR, Lambert FC (2003) Distributed generation: semantic hype or the dawn of a new era. IEEE Power Energy Mag 1:22–29
2. Sayali SS, Thakre MP (2020) Functionality of smart grid with RE enables the generation of solar energy: opportunity to come. In: International conference on power, energy, control and transmission systems (ICPECTS), pp 1–7
3. Borse PS (2020) Analytical evaluation of FOC and DTC induction motor drives in three levels and five levels diode clamped inverter. In: 2020 international conference on power, energy, control and transmission systems (ICPECTS), pp 1–6
4. Hadke VV (2019) Integrated multilevel converter topology for speed control of SRM drive in plug in-hybrid electric vehicle. In: 3rd IEEE international conference on trends in electronics and informatics (ICOEI 2019), pp 1013–1018
5. Xiang H, Yan Y, Jiang H (2009) A two-stage PV grid-connected inverter with optimized anti-islanding protection method. In: Sustainable power generation and supply
6. Maurya DS (2020) A detailed comparative analysis of different multipulse and multi-level topologies for STATCOM. In: International conference on electronics and sustainable communication systems (ICESC), pp 1112–1117

7. IEEE standard for interconnecting distributed resources with electric power system. IEEE Standard 1547 (2003)
8. De Mango F (2006) Overview of anti-islanding algorithms for PV systems. Part II: active methods. In: 12th international power electronics and motion control conference, pp 1884–1889
9. Walling R (2005) Distributed generation islanding implications on power system dynamic performance. In: Power engineering society summer meeting, Chicago, pp 92–96
10. Gaidhani TS (2020) Analysis of MMC coordination with hybrid dc breakers for HVDC grid protection. In: 4th international conference on electronics, communication and aerospace technology (ICECA), pp 333–338
11. Mahat P (2008) Review of islanding detection methods for distributed generation. In: International conference on electric utility deregulation and restructuring and power technology, pp 2743–2748
12. Mishra PP (2017) A passive islanding detection technique with reduced complexity for distributed generations. In: 7th international conference on power systems (ICPS), pp 830–835
13. Matale NP (2020) Alleviation of voltage sag-swell by DVR based on SVPWM technique. In: International conference on power, energy, control and transmission systems (ICPECTS), pp 1–6
14. Martin F, Bolfan T (2009) Method and apparatus for anti-islanding of distributed power generation systems. In: US Patent no. 0021877
15. Chen C (2009) Design of parallel inverters for smooth mode transfer microgrid applications. In: Annual IEEE applied power electronics conference and exposition, pp 1288–1294

Genetic Algorithm Based Sliding Mode Controller for Underactuated 2-DoF Gyroscope

Manish Patel and Bhanu Pratap

Abstract This paper proposes two sliding mode controllers (SMC) for trajectory tracking of the inner and outer gimbal of a highly coupled double-gimbal underactuated control moment gyroscope. First, the dynamics of the underactuated 2-DoF Gyroscope have been derived using the Euler-Lagrange equation then independent nonlinear control laws are derived for actuated inner gimbal and unactuated outer gimbal. Thereafter, conventional switching law combined with an adaptive disturbance estimation term is obtained in the sense of Lyapunov for bounded disturbance rejection. In this document, to deal with the highly coupled interaction between the unactauted outer gimbal and the rotating disc optimally, controller parameters are optimized using genetic algorithm for the operating condition. Lyapunov stability criteria with Barbashin-Krasvoskii theorem has been used to prove the asymptotic stability of the closed loop system. Numerical simulations are carried out to validate the proposed genetic algorithm based nonlinear control law for trajectory tracking of both the gimbals. The proposed controller nullified the effect of bounded input disturbance and parametric uncertainties considered and achieved the tracking task effectively.

Keywords Sliding mode control · Genetic algorithm · Underactuated gyroscope · Disturbance estimation · Uncertainty

---

M. Patel · B. Pratap (✉)
Department of Electrical Engineering, National Institute of Technology Kurukshetra, Kurukshetra, Haryana136119, India
e-mail: bhanu@nitkkr.ac.in

M. Patel
e-mail: manish_31904107@nitkkr.ac.in

S. Suhag et al. (eds.), *Control and Measurement Applications for Smart Grid*, Lecture Notes in Electrical Engineering 822,
https://doi.org/10.1007/978-981-16-7664-2_24

1 Introduction

The dynamics of the control moment gyroscope (CMG) have been considered as one of the most interesting problems of mechanics [1]. A CMG has a rotating disc mounted inside gimbals which are actuated and tilt its angular momentum. Gyroscope is widely used as an actuator to control orientation in sea, air and space vehicles, such as attitude control of miniaturized spacecraft [2], the stability of posture of a small inverted pendulum robot [3], as an indirect actuator for the rotational torque generation [4], scanning and tracking in combat vehicles [5], actuation of dual-arm space robots and several more [6] (Fig. 1).

In this document, an underactuated Gyroscope is considered to increase the tolerance if an actuator fails and to manage space and weight constraints in small systems. Trajectory tracking of unactuated outer gimbal with a constant angular speed of the disc and bounding actuated inner gimbal near the origin is proposed in [7, 8] using feedback linearization, and neural-network and regressor-based adaptive controller and in [9] using adaptive radial basis function neural network-based controller. Wide trajectory tracking task of two outermost unactuated gimbals utilizing torques of the disc and actuated inner gimbal is achieved using sliding mode controller in [10], using feedback linearization in [11] and using passivity-based controller in [12]. Decoupling control of inner and outer gimbal has been proposed in [13] for a double-gimbaled variable speed CMG but the system considered was fully actuated.

To the best of the authors' knowledge, trajectory tracking of unactuated outer gimbal and actuated inner gimbal has not received attention for the underactuated gyroscope. This paper deals with the design of independent nonlinear control laws for trajectory tracking of unactuated outer gimbal using actuator of the disc and actuated inner gimbal. The system considered is underactuated multi-input multi-output (MIMO) hence, two subsystems namely inner and outer gimbal is formed leaving internal dynamics in the form of disc subsystem to be dealt with separately. The design of a robust control law using the sliding mode approach is quite mature and hence, it has been used to derive control laws for each subsystem separately and

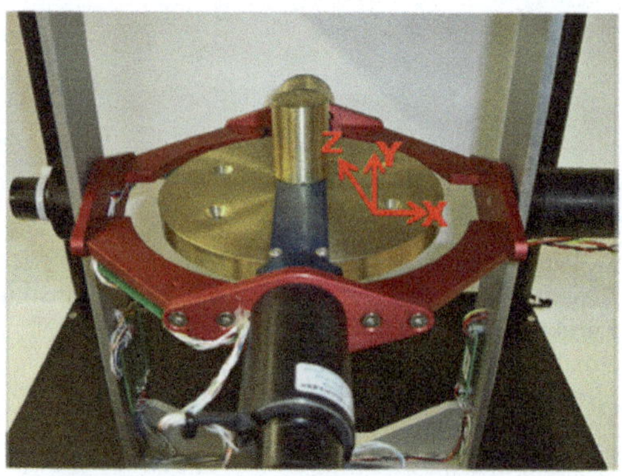

Fig. 1 Gyroscope

conventional switching law is combined with a disturbance estimation term which estimates bounded disturbances online.

Due to coupling present in the outer gimbal and the rotating disc, the control law designed for the outer gimbal to track the desired trajectory affects the angular velocity of the disc in the transient phase of control. To deal with this it is intuitive that optimizing controller parameters using some optimization algorithm will accomplish the task of tracking without affecting the internal dynamics much. Optimization of sliding mode controller using genetic algorithm is well covered in [14–17]. Hence, genetic algorithm has been used to optimize the controller parameters of the control law for the outer gimbal subsystem considering upper bound of input disturbance and worst case in uncertainty of system parameters. Global asymptotic stability of the two subsystems is proved using the Barbashin-Krasvoskii theorem [18], whereas the internal dynamics are proved to be bounded for bounded reference tracking.

The remaining portion of the paper is organized as follows. The problem statement is formulated in Sect. 2, the control laws are derived and optimization of its parameters is done in Sect. 3, and the system stability is proved in Sect. 4. Section 5 covers the numerical simulations and observations and the conclusion is given in Sect. 6 which is followed by the references.

2 Problem Formulation

2.1 Dynamic Model

The dynamics of the 2-DoF underactuated gyroscope are derived using Euler-Lagrange's equation [19] as below [1]

$$\ddot{\alpha} = -\frac{C}{J(\beta)}\left(\dot{\beta}\cos\beta\,(\dot{\gamma}+\dot{\alpha}\sin\beta)\right) - \frac{J'(\beta)}{J(\beta)}\dot{\alpha}\dot{\beta} - \frac{\sin\beta}{J(\beta)}\tau \tag{1}$$

$$\ddot{\beta} = \frac{C}{H}\left(\dot{\alpha}\dot{\gamma}\cos\beta + \dot{\alpha}^2\sin\beta\cos\beta\right) + \frac{J'(\beta)}{2H}\dot{\alpha}^2 + \frac{1}{H}\tau_1 \tag{2}$$

$$\begin{aligned}\ddot{\gamma} = {} & \frac{C}{J(\beta)}\dot{\beta}\sin\beta\cos\beta\,(\dot{\gamma}+\dot{\alpha}\sin\beta) \\ & + \frac{J'(\beta)}{J(\beta)}\dot{\alpha}\dot{\beta}\sin\beta - \dot{\alpha}\dot{\beta}\cos\beta + \left(\frac{\sin^2\beta}{J(\beta)} + \frac{1}{C}\right)\tau\end{aligned} \tag{3}$$

where,

$$J(\beta) = A_2 + (A_1 + A)\cos^2\beta + C_1\sin^2\beta \tag{4}$$

$$H = B_1 + B \tag{5}$$

A_2, A_1, A are the moment of inertia of outer gimbal, inner gimbal and disc about x-axis, respectively, B_1, B are the moment of inertia of inner gimbal and disc about y-axis, respectively, and C_1, C denotes the moment of inertia of inner gimbal and disc about z-axis, respectively [1]. α, β, γ denotes the angular position of the outer gimbal, inner gimbal and disc about x-axis, y-axis and z-axis, respectively. τ and τ_1 are actuating torques at the disc and the inner gimbal, respectively, to achieve underactuation torque at outer gimbal $\tau_2 = 0$.

2.2 *Control Problem*

It is desired to derive two control laws τ_1 and τ such that the angular position of the inner gimbal β tracks the desired trajectory β^d and the angular position of the outer gimbal α tracks α^d, respectively, whilst keeping angular velocity of the disc bounded around the operating point.

Tracking errors are defined as below

$$e_1 = \beta - \beta^d \tag{6}$$

$$e_2 = \alpha - \alpha^d \tag{7}$$

and control laws are to be derived so that $e_1(t)$, $\dot{e_1}(t)$, $e_2(t)$ and $\dot{e_2}(t)$ are globally asymptotically stable which can be written analytically as
$e_1(t) \to 0$, $\dot{e_1}(t) \to 0$, $e_2(t) \to 0$ and $\dot{e_2}(t) \to 0$ when $t \to \infty$.

3 Control Law Design

Design of robust nonlinear control law based on sliding mode control is proposed to accomplish the control objective stated in the document in the presence of uncertainty in input and model parameters. There are two modes in sliding mode control namely, reaching mode and sliding mode. The sliding surface should be designed carefully since phase trajectories follow the dynamics of the sliding surface in the second mode, i.e. sliding mode [20].

To design the sliding surface, order of the system should be known which is relative degree of the output with respect to the input of the system. It can be concluded from (1) and (2) that relative degree is two since inputs are appearing explicitly and maximum second derivative of the output is in the equations.

3.1 Generalized Sliding Mode Controller

First, a generalized control law will be derived for a second order system then it will be used to derive controller for the subsystems.

Let a second order system be :

$$\ddot{x} = (f(x, \dot{x}) + \Delta f) + (g(x, \dot{x}) + \Delta g)(u + ud) \tag{8}$$

where Δf and Δg denotes parametric uncertainties and ud is input disturbance. The system can be rewritten in terms of lumped disturbance 'd' as

$$\ddot{x} = f(x, \dot{x}) + g(x, \dot{x})u + d \tag{9}$$

The sliding surface is chosen as below

$$s = \dot{x} + cx \tag{10}$$

where c is a positive constant and can be chosen as per the required dynamics of the sliding surface. The constant plus proportional rate reaching law [20] given below will be utilized to derive the control law

$$\dot{s} = -\mathrm{Qsgn}(s) - Ks \tag{11}$$

$$\begin{aligned} \ddot{x} + c\dot{x} &= -\mathrm{Qsgn}(s) - Ks \\ f(x, \dot{x}) + g(x, \dot{x})u + c\dot{x} &= -\mathrm{Qsgn}(s) - Ks \end{aligned}$$

$$u = -\frac{1}{g(x, \dot{x})}(f(x, \dot{x}) + c\dot{x} + \mathrm{Qsgn}(s) + Ks) \tag{12}$$

control law (12) will accomplish the desired control objective but it should be taken care that the operating region is chosen in a way to avoid singularity problem due to $g(x, \dot{x})$. Also, to terminate the effect of uncertainties on the performance Q should be chosen in the sense of Lyapunov but that requires knowledge of upper bound of uncertainty. Upper bound of system uncertainty due to variations in model parameters changes with different operating conditions so to avoid efforts in estimating upper bound adaptive estimation of lumped disturbance is found effective [14].

For the lumped disturbance given in (9), $\hat{d}$ estimates and estimation error is defined as

$$\tilde{d} = d - \hat{d} \tag{13}$$

Updating (12) with estimation term gives updated control law as below

$$u = -\frac{1}{g(x,\dot{x})}(f(x,\dot{x}) + c\dot{x} + Q\text{sgn}(s) + Ks + \hat{d}) \tag{14}$$

Adaption law [21] mentioned below and control law (14) achieves our control objective by choosing Q, K and γ appropriately.

$$\dot{\hat{d}} = \gamma s \tag{15}$$

3.2 *Sliding Mode Controller for Outer and Inner Gimbal*

Rewriting (1) and (2) in the form of (9) as below

$$\ddot{\alpha} = f_1(\psi,\dot{\psi}) + g_1(\psi,\dot{\psi})\tau + d_1 \tag{16}$$

$$\ddot{\beta} = f_2(\psi,\dot{\psi}) + g_2(\psi,\dot{\psi})\tau_1 + d_2 \tag{17}$$

where $\psi = [\alpha\ \beta\ \gamma]$, d_1 and d_2 accounts for the input disturbance and parametric uncertainties for the inner and outer gimbal systems, respectively. Defining sliding surfaces for the inner gimbal and the outer gimbal in terms of tracking errors mentioned in (6) and (7)

$$s_1 = \dot{e_1} + c_1 e_1 \tag{18}$$

$$s_2 = \dot{e_2} + c_2 e_2 \tag{19}$$

Using (11), (14), (15), (16) and (17) control laws τ_1 and τ and adaption laws are as follows

$$\tau_1 = -\frac{1}{g_2(\psi,\dot{\psi})}(-\ddot{\beta}^d + f_2(\psi,\dot{\psi}) + c_1\dot{e_1} + Q_1\text{sgn}(s_1) + K_1 s_1 + \hat{d_1}) \tag{20}$$

$$\tau = -\frac{1}{g_1(\psi,\dot{\psi})}(-\ddot{\alpha}^d + f_1(\psi,\dot{\psi}) + c_2\dot{e_2} + Q_2\text{sgn}(s_2) + K_2 s_2 + \hat{d_2}) \tag{21}$$

$$\dot{\hat{d_1}} = \gamma_1 s_1 \tag{22}$$

$$\dot{\hat{d_2}} = \gamma_2 s_2 \tag{23}$$

Adaption laws (22), (23) and control laws (20), (21) are able to accomplish the control objective while nullifying the effect of disturbance on the performance by choosing suitable controller parameters.

Parameters Q_1, K_1 and γ_1 can be tuned independently without affecting the performance of other subsystems much but tuning Q_2, K_2 and γ_2 to make the transient response of the outer gimbal subsystem faster affects the angular velocity of the disc adversely because of underactuation and high coupling so, these parameters are required to be tuned optimally.

3.3 Optimization Using Genetic Algorithm

Genetic Algorithm is a population-based probabilistic search and optimization technique, which works based on the mechanisms of natural genetics and natural evaluation. The algorithm begins with a set of solutions, which is denoted as population, is initialized then genetic material of two or more solutions is combined to generate a diverse set of next solutions [22]. Based on fitness evaluation, next generation population is generated and the cycle repeats itself until the convergence criteria are achieved (Fig. 2).

In this document, *real value encoding* has been used to represent an individual that does not require any intermediate encoding and decoding steps [23] and the population is created randomly with a uniform distribution.

Fitness function or the function to be optimized is

$$J = 5|\dot{\gamma} - \dot{\gamma}^d| + \frac{1}{2}\int_0^t (e_2^2 + \tau^2)\mathrm{d}t \tag{24}$$

which is a shifted sphere function [24]. Selection scheme used to generate next generation is *tournament selection* which exhibits moderate population diversity and high selection pressure [23]. *Scattered* crossover operator is used which combines genetic material of two or more solutions and *gaussian* mutation technique is used to explore other solutions. Algorithm termination depends on either the maximum number of generations reached or the highest ranking solution fitness is reached.

Fig. 2 Basic algorithm cycle

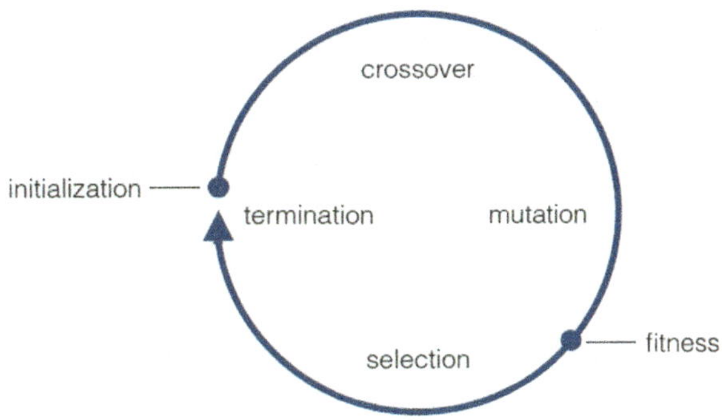

The maximum number of generations has been taken equal to 25 and population size equals to 20.

Since the function to be optimized is a unimodal shifted sphere function of dimension two [24], the operators and functions used would be able to find global optima effectively.

4 Stability Analysis

Assumption 1 It is assumed that the desired trajectory α^d, β^d their first derivative $\dot{\alpha}^d$, $\dot{\beta}^d$ second derivative $\ddot{\alpha}^d$, $\ddot{\beta}^d$ are available and bounded.

Assumption 2 $\bar{d}_1$ and $\bar{d}_2$ are the known upper bound for d_1 and d_2.

Theorem 1 *Let $x = 0$ be an equilibrium point for $\dot{x} = f(x)$. Let $V: R^n \rightarrow R$ be a continuously differentiable function such that*

$$V(0) = 0 \; and \; V(x) > 0, \; \forall \, x \neq 0 \tag{25}$$

$$||x|| \rightarrow \infty \Longrightarrow V(x) \rightarrow \infty \tag{26}$$

$$\dot{V}(x) < 0, \;\; \forall \, x \neq 0 \tag{27}$$

then $x = 0$ is globally asymptotically stable [18].

4.1 For Inner and Outer Gimbal

First, stability of the generalized system (9) will be analyzed then it will be used to get the conditions on the controller parameters for stability analysis of the inner and the outer gimbal subsystems.
Choosing Lyapunov candidate in terms of sliding surface (10) and estimation error (15)

$$V = \frac{1}{2}s^2 + \frac{1}{2\gamma}\tilde{d}^2 \tag{28}$$

If (28) satisfies (25), (26) and (27) then it can be inferred that $s = 0$ and $\tilde{d} = 0$ are globally asymptotically stable. The Lyapunov candidate chosen satisfies (25) and (26) next, conditions will be established such that it satisfies (27)

$$\dot{V} = s\dot{s} + \frac{1}{\gamma}\tilde{d}\dot{\tilde{d}}$$
$$\dot{V} = s(\ddot{x} + c\dot{x}) - \frac{1}{\gamma}\tilde{d}\dot{\hat{d}}$$
$$\dot{V} = s(f(x,\dot{x}) + g(x,\dot{x})u + d + c\dot{x}) - \frac{1}{\gamma}\tilde{d}\gamma s$$
$$\dot{V} = s(-Q\text{sgn}(s) - Ks + d - \hat{d}) - \tilde{d}s$$
$$\dot{V} = -Q||s||_1 - Ks^2$$

Similarly choosing Lyapunov candidates for the inner and outer gimbals.

$$V_1 = \frac{1}{2}s_1^2 + \frac{1}{2\gamma_1}\tilde{d_1}^2 \tag{29}$$
$$V_2 = \frac{1}{2}s_2^2 + \frac{1}{2\gamma_2}\tilde{d_2}^2 \tag{30}$$
$$\dot{V_1} = -Q_1||s_1||_1 - K_1 s_1^2 \tag{31}$$
$$\dot{V_2} = -Q_2||s_2||_1 - K_2 s_2^2 \tag{32}$$

For $Q_1 > 0$, $Q_2 > 0$, $K_1 > 0$ and $K_2 > 0$ (29) and (30) satisfies all the three conditions (25), (26) and (27) inferring both the sliding surfaces and estimation errors are globally asymptotically stable also on tracing sliding surfaces it can be concluded that $e_1(t)$, $\dot{e_1}(t)$, $e_2(t)$ and $\dot{e_2}(t)$ are globally asymptotically stable.

4.2 Disc Subsystem

Stability analysis of the disc rotating at the desired angular velocity will be studied considering the angular position of the outer gimbal is tracking the desired trajectory α^d and the inner gimbal is tracking the desired trajectory β^d.

Dynamics of the disc given in (3) can be rewritten as below using (21)

$$\ddot{\gamma} + \dot{\beta}^d \cot\beta^d \dot{\gamma} = -\frac{J'(\beta^d)}{C\sin\beta^d}\dot{\alpha}^d\dot{\beta}^d - 2\dot{\alpha}^d\dot{\beta}^d\cos\beta^d + (\sin\beta^d + \frac{J(\beta^d)}{C\sin\beta^d})(\hat{d_2} - \ddot{\alpha}^d) \tag{33}$$

Variable of interest, i.e. angular velocity of the disc $\dot{\gamma}$ can be computed by transforming (33) into a linear differential equation which can be solved but to avoid complex derivation, numerical simulation has been used to analyze the behaviour of the internal dynamics on changing the range of variables on which it is dependent.

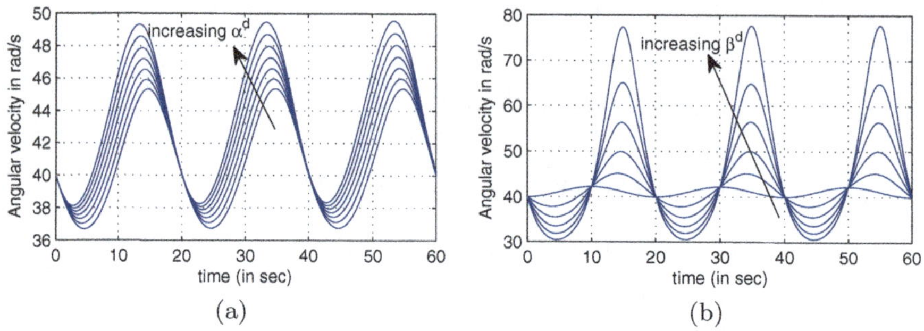

Fig. 3 Angular velocity of the disc on varying **a** α^d, **b** β^d

α^d is varied from $10^\circ \sin wt$ to $70^\circ \sin wt$, whilst β^d is varied from $40^\circ + 0^\circ \sin wt$ to $40^\circ + 20^\circ \sin wt$.

From Fig. 3 it can be concluded that the internal dynamics is bounded for bounded reference trajectory but inner gimbal can be tracked for smaller range and that too away from origin because of singularity present in (1).

5 Simulation Results

To validate the proposed control strategy for trajectory tracking of both the gimbals numerical simulations are done and presented in this section. Reference trajectory to be tracked is taken as:

$$\begin{aligned}\alpha^d &= 25^\circ \sin 2\pi f_1 t,\\ \beta^d &= 40^\circ + 5^\circ \sin 2\pi f_1 t, \text{ where } f_1 = 0.05\,\text{Hz}.\\ \dot{\gamma}^d &= 40\,\text{rad/s}\end{aligned}$$

Initial conditions are taken as $\beta(0) = 40^\circ$, $\dot{\gamma}(0) = 40$ rad/s and $\alpha(0) = -100^\circ$. Initial condition of α is so taken to address the coupling effect, whilst driving the β and $\dot{\gamma}$ to initial conditions.

Profile of input disturbances is given in Fig. 5b and parametric uncertainty is taken as $\pm 20\%$ for all model parameters which changes randomly online in the given range. Controller parameters for the outer gimbal subsystem are found out using genetic algorithm with the operators and parameters mentioned in Sect. 3.3, whilst controller parameters of the inner gimbal subsystem are decided by trial and error process.

To reduce the chattering effect *tanh* has been used instead of *sgn* [25]. From Fig. 4 it can be said that the proposed controller has been able to reject the input and

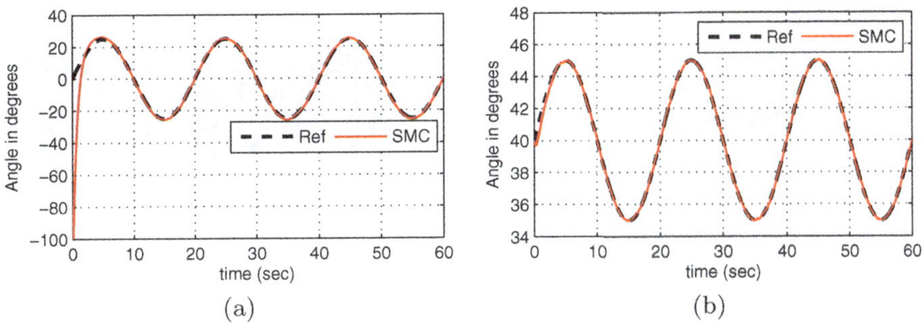

Fig. 4 Angular position of **a** outer gimbal, **b** inner gimbal

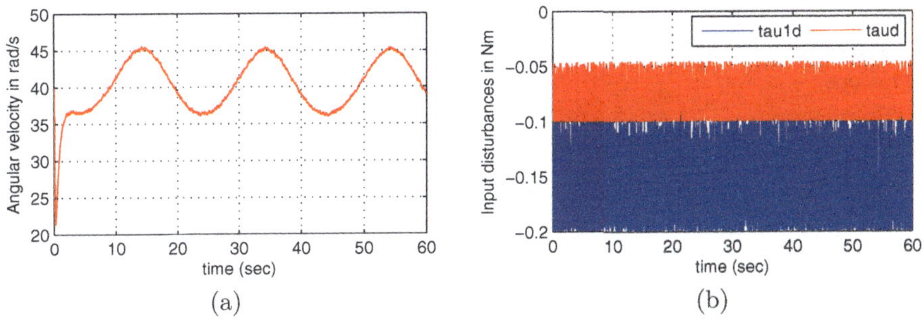

Fig. 5 **a** Angular velocity of disc, **b** input disturbances

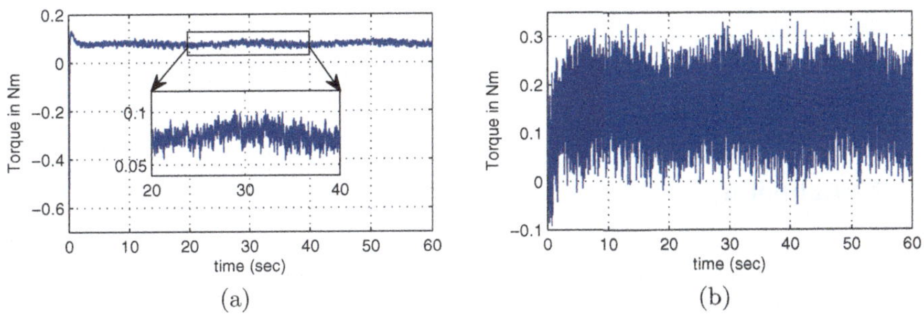

Fig. 6 Control input for **a** outer gimbal, **b** inner gimbal

parametric uncertainties and can follow the desired trajectory globally asymptotically and Fig. 5a shows that the angular velocity of the disc is bounded. Control inputs and sliding surfaces have been shown in Figs. 6 and 7, respectively (Tables 1 and 2).

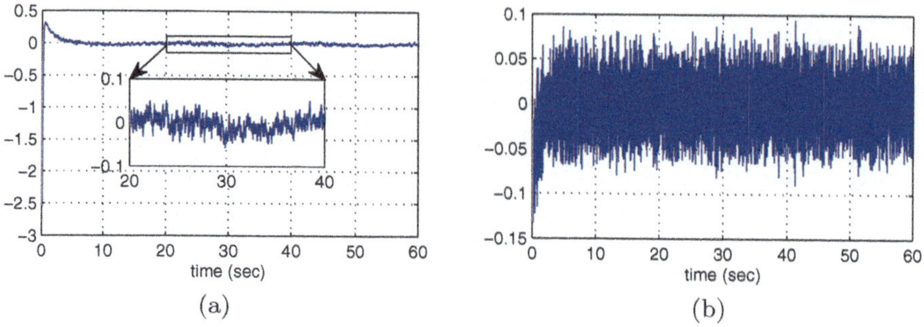

Fig. 7 Sliding surface for **a** outer gimbal, **b** inner gimbal

Table 1 Moment of inertia of quanser gyroscope

Subsystem of gyroscope	Moment of inertia (in kg m^2) about		
	x-axis	y-axis	z-axis
Red gimbal	0.00762058	0.02879104	0.02344554
Blue gimbal	0.00388552	0.00744682	0.00555680
Flywheel disc	0.00284584	0.00563843	0.00284582

Table 2 Controller parameters

Sliding surface constants	$c_1 = 4$, $c_2 = 1.38$
Other parameters	
For inner gimbal	$K_1 = 4$, $Q_1 = 150$, $\gamma_1 = 150$
For outer gimbal	$K_2 = 9.39$, $Q_2 = 4.83$, $\gamma_2 = 8.80$

6 Conclusion

In this work, robust nonlinear control laws have been designed for underactuated 2-DoF Gyroscope system to achieve MIMO tracking task, whilst keeping angular velocity of the disc bounded. The proposed control laws do not require the information of the upper bound of the input and parametric uncertainties. The controller parameters have been tuned via genetic algorithm to deal with the interaction between the unactuated state and the internal dynamics. The numerical simulations validate that the proposed control laws are effective in trajectory tracking, terminated the uncertainties, and optimization of controller parameters via genetic algorithm has been found effective in bounding the internal dynamics within the desired range. This work can be extended for tracking of the inner gimbal around origin, whilst avoiding the singularity problem.

References

1. Polo MFP, Molina MP (2004) Regular self-oscillating and chaotic behaviour of a pid controlled gimbal suspension gyro. Chaos Solitons Fractals 21(5):1057–1074
2. Biggs JD, Livornese G (2020) Control of a thrust-vectoring cubesat using a single variable-speed control moment gyroscope. J Guidance Control Dyn 43(10):1865–1880
3. Tanaka K, Nagasawa S (2020) Posture stability control of a small inverted pendulum robot in trajectory tracking using a control moment gyro. Adv Robot 34(9):610–620
4. Kim HW, Jung S (2020) Design and control of a sphere robot using a control moment gyroscope actuator for navigation. Int J Control Autom Syst 18(12):3112–3120
5. Krzysztofik I, Takosoglu J, Koruba Z (2017) Selected methods of control of the scanning and tracking gyroscope system mounted on a combat vehicle. Ann Rev Control 44:173–182
6. Jia Y, Misra AK (2017) Robust trajectory tracking control of a dual-arm space robot actuated by control moment gyroscopes. Acta Astronautica 137:287–301
7. Montoya-Cháirez J, Santibáñez V, Moreno-Valenzuela J (2019) Adaptive control schemes applied to a control moment gyroscope of 2 degrees of freedom. Mechatronics 57:73–85
8. Moreno-Valenzuela J, Montoya-Cháirez J, Santibáñez V (2020) Robust trajectory tracking control of an underactuated control moment gyroscope via neural network-based feedback linearization. Neurocomputing (2020)
9. Montoya-Cháirez J, Rossomando FG, Carelli R, Santibáñez V, Moreno-Valenzuela J (2020) Adaptive rbf neural network-based control of an underactuated control moment gyroscope. Neural Comput Appl:1–14
10. Toriumi FY, Angélico BA (2019) Sliding mode control applied to a multivariate underactuated control moment gyroscope. In: 2019 IEEE 58th conference on decision and control (CDC). IEEE, pp 4928–4933
11. Toriumi FY, Angélico BA (2019) Nonlinear controller design for tracking task of a control moment gyroscope actuator. IEEE/ASME Transactions on Mechatronics 25(1):438–448
12. Toriumi FY, Angelico BA (2020) Passivity-based nonlinear control approach for tracking task of an underactuated cmg. IEEE/ASME Trans Mechatron
13. Li H, Ning X, Han B (2016) Composite decoupling control of gimbal servo system in double-gimbaled variable speed cmg via disturbance observer. IEEE/ASME Trans Mechatron 22(1):312–320
14. Ye M, Wang H (2019) A robust adaptive chattering-free sliding mode control strategy for automotive electronic throttle system via genetic algorithm. IEEE Access 8:68–80
15. Farahmandrad M, Ganjefar S, Talebi HA, Bayati M (2020) Design of higher-order sliding mode controller based on genetic algorithm for a cooperative robotic system. Int J Dyn Control 8(1):269–277
16. Pazooki M, Mazinan A (2018) Hybrid fuzzy-based sliding-mode control approach, optimized by genetic algorithm for quadrotor unmanned aerial vehicles. Complex Intell Syst 4(2):79–93
17. Laware A, Talange D, Bandal V (2018) Evolutionary optimization of sliding mode controller for level control system. ISA Trans 83:199–213
18. Khalil HK (1996) Nonlinear systems. Macmilllan, London, UK
19. Goldstein H, Poole C, Safko J (2002) Classical mechanics
20. Hung JY, Gao W, Hung JC (1993) Variable structure control: a survey. IEEE Trans Indust Electron 40(1):2–22
21. Rudra S, Kumar Barai R, Maitra M, Mandal D, Ghosh S, Dam S, Dutta A, Bhattacharyya P (2013) Stabilization of furuta pendulum: a backstepping based hierarchical sliding mode approach with disturbance estimation. In: 2013 7th international conference on intelligent systems and control (ISCO). IEEE, pp 99–105
22. Kramer O (2017) Genetic algorithm essentials, vol 679. Springer, Berlin
23. Samanta D (2017) Nptel course: introduction to soft computing

24. Hussain K, Salleh MNM, Cheng S, Naseem R (2017) Common benchmark functions for metaheuristic evaluation: a review. JOIV Int J Inform Vis 1(4–2):218–223
25. Mishra A, Sinha NK (2019) Design and real time implementation of sliding mode and adaptive fuzzy control on quanser gyroscope. In: 2019 fifth Indian control conference (ICC). IEEE, pp 312–317

Higher Order Sliding Mode Controller Design for 2-DoF Torsion System with Genetic Algorithm Based Tuning

Priyam Rai and Bhanu Pratap

Abstract This paper presents the design of a robust controller for stabilization of the 2-DoF torsion system, which is an underactuated system, characterized by a higher number of degrees of freedom than the number of actuators. The output of the given servo dynamical system with two flexible torsional couplings should be positioned properly with minimum vibrations and response time. The control scheme presented here, to achieve the control requirement, is higher order sliding mode control with genetic algorithm tuning based on the hierarchical sliding mode. This control scheme provides robustness to matched disturbances, a finite time convergence like its conventional counterpart, along with the ability to overcome the problem of chattering. In addition, the choice of the hierarchical sliding mode has the added benefit of having a simpler subsystem-wise sliding mode design, which makes it more convenient to apply to the underactuated system. Genetic algorithm is applied to optimize the controller parameters. The study of the performance of the proposed scheme is based on bounded matched disturbance, randomly varying viscous damping coefficient, parameter uncertainties, and varying frictional force. MATLAB simulations have been carried out to verify the above proposition.

Keywords Higher order sliding mode control · Hierarchical sliding mode · Genetic algorithm · Torsion system · Matched uncertainties

---

P. Rai · B. Pratap (✉)
Department of Electrical Engineering, National Institute of Technology Kurukshetra, Kurukshetra, Haryana136119, India
e-mail: bhanu@nitkkr.ac.in

P. Rai
e-mail: priyam_31910113@nitkkr.ac.in

S. Suhag et al. (eds.), *Control and Measurement Applications for Smart Grid*, Lecture Notes in Electrical Engineering 822,
https://doi.org/10.1007/978-981-16-7664-2_25

1 Introduction

Physical-world control applications of torsional dynamics are well represented by the 2-DoF torsion system. It is an underactuated system that represents the effect of a flexible coupling among the actuator and load encountered in complex processes in industries. Numerous control methodologies that have been applied to the torsion system for the control of vibration are detailed in the literature. Fuchs et al. presented a novel high-accuracy numerical integration based state feedback method for control of multi-torsion chain in 2013 in [1]. A linear quadratic regulator based on the algebraic Riccati equation with the Lagrange multiplier method has been applied on the 2-DoF torsion system in [2]. A 2-DoF fractional order PID is employed to control the angular position of the 1-DoF torsion system in [3]. A dynamical sliding-mode control is used to drive the rotary velocities of drill string components to a constant positive value in a conventional vertical oil well drill string, which is represented by a discontinuous lumped parameter torsional model having four degrees of freedom in [4].

Variable structure control with sliding mode control(SMC) is a powerful nonlinear feedback control technique. The innate stability and robustness of this technique, because of the complete rejection of disturbances and insensitivity of parameter variations, during the sliding mode, is the result of the discontinuous nature of its fast switching control action. Although it might be sensitive to them during the reaching mode [5]. These properties of SMC along with finite-time convergence are offset by chattering, caused mainly by unmodelled cascade dynamics, which can cause extensive damage to the actuator [6]. Higher-order SMC was introduced in the Ph.D. dissertation of Arie Levant [7]. Since then various works have been published on higher-order SMC(HOSMC). Integral sliding based HOSMC has been presented for uncertain nonlinear systems in [8]. An HOSMC algorithm is proposed for hovercraft vessel control in [9] demonstrates the advantage of the strategy. A robust and adaptive controller based on the HOSMC approach has been investigated in [10] for the air feed nonlinear system with bounded uncertainty.

The torsion system which is being dealt with in this paper represents an underactuated system. The relative degree of this system with respect to the output function is greater than one. Therefore, the derivation of the equivalent control by using the conventional method becomes cumbersome and unmanageable. Hierarchical SMC can be applied to counter such a problem. In this paper, the hierarchical sliding mode-based HOSMC is developed for the 2-DOF torsion system, which consists of three subsystems. Therefore, three individual sliding surfaces are defined for each subsystem and then they are integrated for the entire system as shown in [11]. The stability analysis of the system is carried out using Lyapunov and Barbalat's theorem [12].

Further, for setting the numerical values of the aforementioned controller Genetic Algorithm based tuning is employed in this paper. GA-based optimization has been employed in various papers. Some of the recent papers are [13–16]. The introduction to GAs, starting with basic concepts like evolutionary operators and continuing with an overview of strategies for tuning and controlling parameters is discussed in books [17, 18].

According to the literature survey undertaken by the authors, this approach has not received any attention in regards to the 2-DoF torsion system discussed in this paper. Further matched disturbances and parametric uncertainties are considered for testing the validity of the proposed controller for the given system.

The rest of the paper is divided into six sections. The control problem is devised in Sect. 2, further, the aforementioned controller is designed in Sect. 3. Section 4 consists of GA-based tuning of controller parameters and Sect. 5 contains the stability analysis of the proposed controller. The simulation results are presented in Sect. 6, with the conclusion in Sect. 7. The references are mentioned at the end of the paper.

2 Problem Formulation and Prelimnaries

2.1 System Modelling

The torsion system comprises a chain of electrical and mechanical subsystems. Equations of motion derived below constitute the dynamical behaviour of the system. Figure 1 represents the 2-DOF torsion system.

As demonstrated below, the torsion assembly consists of a torsional spring elements with stiffness $k_s i$, and moment of inertia J_i where $i = 1, 2, 3$. A servo drive produces a torsional torque, τ, which acts as input for the torsion system. This torque is controlled by input voltage V given to the servo drive. This is described by the set of equations given below.

$$J_1\ddot{\theta}_1 = rk_m \frac{V - rk_m\dot{\theta}_1}{R_m} - \tau \tag{1}$$

where; motor resistance, $R_m = 2.6\,\Omega$, motor gyrator constant, $K_m = 0.00767$ Nm A^{-1} and transmission ratio, $r = 70$. When conservation of angular momentum is applied on the assembly of torsion elements, the following equation is obtained:

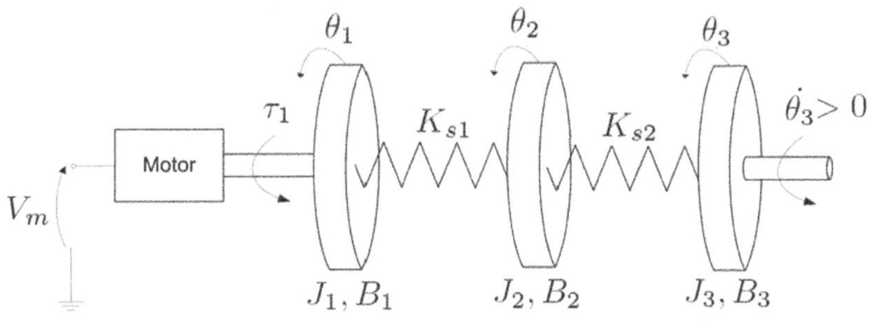

Fig. 1 Rotary 2-DOF torsion system

$$J_i\ddot{\theta}_i = k_{s(i-1)}[\theta_{(i-1)} - \theta_i] + B_{(i-1)}[\dot{\theta}_{(i-1)} - \dot{\theta}_i] + k_{si}[\theta_{(i+1)} - \theta_i] + B_i[\dot{\theta}_{(i+1)} - \dot{\theta}_i] \quad (2)$$

The following system equations are obtained using Eqs. (1) and (2):

$$\ddot{\theta}_1 = \frac{1}{J_1}\left[k_{s1}(\theta_2 - \theta_1) + B_1(\dot{\theta}_2 - \dot{\theta}_1) + V\frac{k_m r}{R_m} - \dot{\theta}_1\frac{k_m^2 r^2}{R_m}\right] \quad (3)$$

$$\ddot{\theta}_2 = \frac{k_{s1}}{J_2}[\theta_1 - \theta_2] + \frac{k_{s2}}{J_2}[\theta_3 - \theta_2] + \frac{B_1}{J_2}[\dot{\theta}_1 - \dot{\theta}_2] + \frac{B_2}{J_2}[\dot{\theta}_3 - \dot{\theta}_2] \quad (4)$$

$$\ddot{\theta}_3 = \frac{k_{s2}}{J_3}[\theta_2 - \theta_3] + \frac{B_2}{J_3}[\dot{\theta}_2 - \dot{\theta}_3] \quad (5)$$

2.2 State Space Model

The following state space model is obtained from the system Eqs. (3), (4) and (5) (Table 1):

$$\dot{x} = Ax + Bu \quad (6)$$

$$u = V \quad (7)$$

$$x = \begin{bmatrix}\theta_1 \; \theta_2 \; \theta_3 \; \dot{\theta}_1 \; \dot{\theta}_2 \; \dot{\theta}_3\end{bmatrix}^T \quad (8)$$

$$A = \begin{bmatrix} 0 & 0 & 0 & 1 & 0 & 0 \\ 0 & 0 & 0 & 0 & 1 & 0 \\ 0 & 0 & 0 & 0 & 0 & 1 \\ a & b & 0 & c & d & 0 \\ e & f & g & h & i & j \\ 0 & k & l & 0 & m & n \end{bmatrix} \quad (9)$$

$$B = \begin{bmatrix}0\;0\;0\;o\;0\;0\end{bmatrix}^T \quad (10)$$

The angular position of the third torsional disc θ_3 is taken as the output:

$$y = Cx \quad (11)$$

$$C = \begin{bmatrix}0\;0\;1\;0\;0\;0\end{bmatrix} \quad (12)$$

Table 1 Moment of inertia (MI) ($\times 10^{-4}$ kg m^2), Torsional stiffness (TS) (Nm/rad) and torsional damping (TD) ($\times 10^{-3}$ N ms^{-1})

MI	J_1	J_2	J_3	TS	k_{s1}	k_{s2}	TD	B_1	B_2
Value	19.261	3.2281	3.2281	Value	1.4927	1.3639	Value	8.2	7.8

where;

$$a = -\frac{k_{s1}}{J_1}; b = \frac{k_{s1}}{J_1}; c = -\frac{{k_m}^2{k_{s1}}^2 + B_1 R_m}{J_1 R_m}; d = \frac{B_1}{J_1}; e = \frac{k_{s1}}{J_2};$$
$$f = -\frac{k_{s1} + k_{s2}}{J_2}; g = \frac{k_{s2}}{J_2}; h = \frac{B_1}{J_2}; i = -\frac{B_1 + B_2}{J_2}; j = \frac{B_2}{J_2};$$
$$k = \frac{k_{s2}}{J_3}; l = -\frac{k_{s2}}{J_3}; m = \frac{B_2}{J_3}; n = -\frac{B_2}{J_3}; o = \frac{k_m r}{J_1 R_m}$$

2.3 *Desired Control Requirement*

The 2-DoF torsion system discussed here finds application in systems like powertrain of wind-mills, oil drills, robotics, etc. The system consists of three modules, the servo drive and two torsion modules. The above-mentioned applications require the positioning of the final output which is the position variable of the second torsion module, θ_3 at the desired position with minimum vibration and response time. Since the problem considered here is a regulation problem, the goal of the designed control is to reduce the error to zero asymptotically. Analytically, θ_3 and $\dot{\theta}_3$ should tend to zero.

3 Higher Order Sliding Mode Control Derivation

Let the SISO dynamical system given by Eqs. 3, 4 and 5 be represented by following set of equations which are affine in control variable:

$$\dot{x}(t) = A(x, t) + B(x, t)u(t); \; y(t) = s(x, t) \tag{13}$$

where, the state vector and the input control are described by $x \in \mathbb{R}^n$ and $u \in \mathbb{R}$, respectively. The sliding variable s: $\mathbb{R}^n \times \mathbb{R} \rightarrow \mathbb{R}$ is a sufficiently smooth output-feedback function.

It is assumed that the relative degree 'r' of the system represented by (13) with respect to output function 's' is well defined, uniform and time-invariant.

For HOSMC let the sliding surface be defined as:

$$s^{(i)} = \frac{\mathrm{d}^i y}{\mathrm{d}t^i} \tag{14}$$

The control purpose is to satisfy the constraint described by $s(x, t) = 0$ in minimum possible time by discontinuous feedback control. The rth order sliding mode for the system given in Eq. (13) with the given discontinuous feedback can be defined in following manner: on the condition that s, $\dot{s}, \ldots, s(r-1)$ are not discontinuous functions, and the set $S^r = \{x | s^{(0)}(x, t) = s^{(1)}(x, t) = \cdots = s^{(r-1)}(x, t) = 0\}$, called "$r$th-order sliding set", is not empty and an integral set in the Filippov sense, the motion on S^r is called "rth-order sliding mode" with reference to the sliding surface s.
The given system consists of various subsystems. Due to this physical structure, the hierarchical structure of SMC is defined. The sliding surface of each subsystem is defined separately and then layered together till all the subsystems are included. Defining the sliding surfaces of each subsystem using the equation (14):

$$s_1 = \theta_1(t) \tag{15}$$

$$s_2 = \theta_2(t) \tag{16}$$

$$s_3 = \theta_3(t) \tag{17}$$

The combined sliding surface of the complete system is obtained as below:

$$s(t) = \epsilon_1\theta_1(t) + \epsilon_2\theta_2(t) + \epsilon_3\theta_3(t) \tag{18}$$

Thereafter,

$$\dot{s}(t) = \epsilon_1\dot{\theta}_1(t) + \epsilon_2\dot{\theta}_2(t) + \epsilon_3\dot{\theta}_3(t) \tag{19}$$

$$\begin{aligned} \ddot{s}(t) = {} & (\epsilon_1 a + \epsilon_2 e)\theta_1(t) + (\epsilon_1 b + \epsilon_2 f + \epsilon_3 k)\theta_2(t) + (\epsilon_2 g + \epsilon_3 l)\theta_3(t) \\ & + (\epsilon_1 c + \epsilon_2 h)\dot{\theta}_1(t) + (\epsilon_1 d + \epsilon_2 i + \epsilon_3 m)\dot{\theta}_2(t) + (\epsilon_2 j + \epsilon_3 n)\dot{\theta}_3(t) \\ & + \epsilon_1 lou(t) \end{aligned} \tag{20}$$

From the Eq. (20), the relative degree of the 2-DoF torsion system concerning the output function given by (18) is obtained to be 2. Therefore, third-order SMC needs to be designed for chattering elimination.

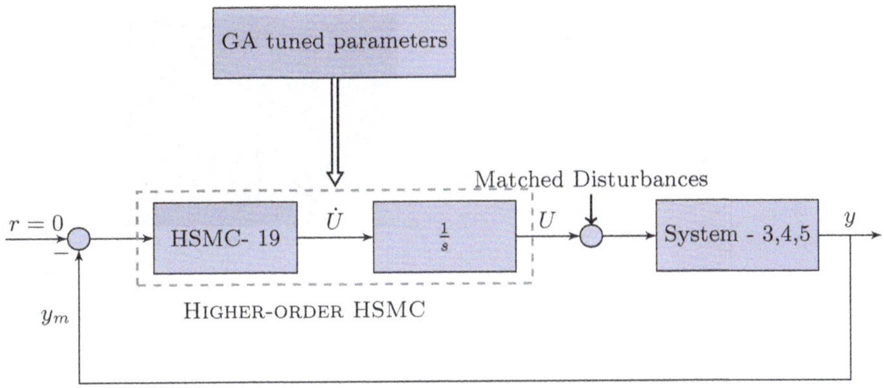

Fig. 2 Block diagram of the proposed controller

$$
\begin{aligned}
\dddot{s}(t) =& \{a(\epsilon_1 c + \epsilon_2 h) + e(\epsilon_1 d + \epsilon_2 i + \epsilon_3 m)\}\theta_1(t) \\
&+ \{b(\epsilon_1 c + \epsilon_2 h) + f(\epsilon_1 d + \epsilon_2 i + \epsilon_3 m) + k(\epsilon_2 j + \epsilon_3 n)\}\theta_2(t) \\
&+ \{g(\epsilon_1 d + \epsilon_2 i + \epsilon_3 m) + l(\epsilon_2 j + \epsilon_3 n)\}\theta_3(t) \\
&+ \{(\epsilon_1 a + \epsilon_2 e) + c(\epsilon_1 c + \epsilon_2 h) + h(\epsilon_1 d + \epsilon_2 i + \epsilon_3 m)\}\dot{\theta}_1(t) \\
&+ \{(\epsilon_1 b + \epsilon_2 f + \epsilon_3 k) + d(\epsilon_1 c + \epsilon_2 h) + i(\epsilon_1 d + \epsilon_2 i + \epsilon_3 m) \\
&+ m(\epsilon_2 j + \epsilon_3 n)\}\dot{\theta}_2(t) \\
&+ \{(\epsilon_2 g + \epsilon_3 l) + j(\epsilon_1 d + \epsilon_2 i + \epsilon_3 m) + n(\epsilon_2 j + \epsilon_3 n)\}\dot{\theta}_3(t) \\
&+ (\epsilon_1 c + \epsilon_2 h)ou(t) + \epsilon_1 o\dot{u}(t)
\end{aligned}
\tag{21}
$$

where $\dot{u}(t) = v(t)$ is the fictitious control designed in discontinuous mode, such that the stabilization to the equilibrium occurs with a finite time convergence (Fig. 2).

$$
\begin{aligned}
v(t) =& -\{1/(\epsilon_1 o)\} \times [\{a(\epsilon_1 c + \epsilon_2 h) + e(\epsilon_1 d + \epsilon_2 i + \epsilon_3 m)\}\theta_1(t) \\
&+ \{b(\epsilon_1 c + \epsilon_2 h) + f(\epsilon_1 d + \epsilon_2 i + \epsilon_3 m) + k(\epsilon_2 j + \epsilon_3 n)\}\theta_2(t) \\
&+ \{g(\epsilon_1 d + \epsilon_2 i + \epsilon_3 m) + l(\epsilon_2 j + \epsilon_3 n)\}\theta_3(t) \\
&+ \{(\epsilon_1 a + \epsilon_2 e) + c(\epsilon_1 c + \epsilon_2 h) + h(\epsilon_1 d + \epsilon_2 i + \epsilon_3 m)\}\dot{\theta}_1(t) \\
&+ \{(\epsilon_1 b + \epsilon_2 f + \epsilon_3 k) + d(\epsilon_1 c + \epsilon_2 h) + i(\epsilon_1 d + \epsilon_2 i + \epsilon_3 m) \\
&+ m(\epsilon_2 j + \epsilon_3 n)\}\dot{\theta}_2(t) \\
&+ \{(\epsilon_2 g + \epsilon_3 l) + j(\epsilon_1 d + \epsilon_2 i + \epsilon_3 m) + n(\epsilon_2 j + \epsilon_3 n)\}\dot{\theta}_3(t) \\
&+ (\epsilon_1 c + \epsilon_2 h)ou(t) + k1s(t) + k2\dot{s}(t) + k3\ddot{s}(t) + k4sign(\dddot{s}(t))]
\end{aligned}
\tag{22}
$$

$$
u(t) = \int v(t)\mathrm{d}t \tag{23}
$$

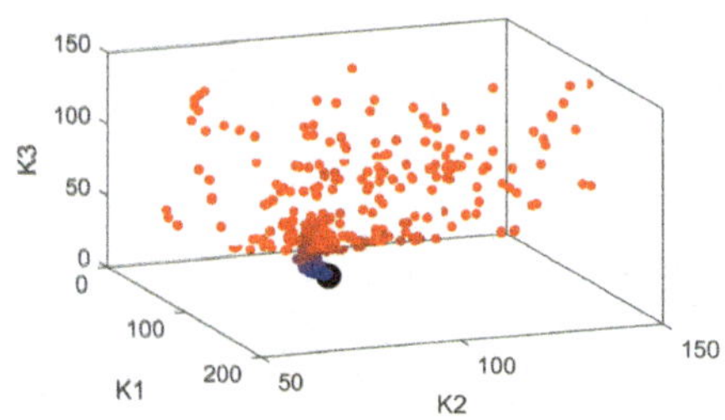

(a) Controller parameters distribution

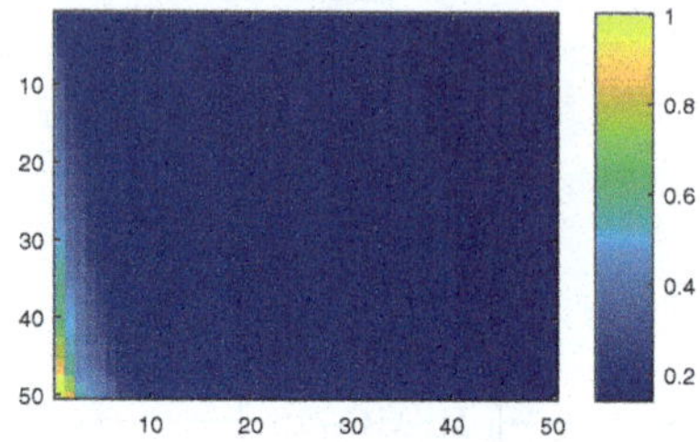

(b) Sorted cost distribution obtained from fitness function

Fig. 3 **a** Controller parameters distribution, **b** sorted cost distribution obtained from fitness function

4 Genetic Algorithm Based Tuning of Controller Parameters

In this paper, three of the controller parameters, which provide for a guaranteed convergence to the equilibrium are tuned using the above mentioned algorithm. The parameters to be tuned are k_1, k_2 and k_3, as observed from Eq. (22).

For achieving this goal, the following GA parameters are defined as: number of generations: 50; population size: 50; number of elite children: $0.05 \times population$ size; cross over fraction: 0.8; number of sites of crossover: random; mutation: gaussian mutation.

Fitness function is taken as:

$$J = \sqrt{\int_0^t (x^T x - {x_d}^T x_d) \mathrm{d}t} \tag{24}$$

After the algorithm runs for the given number of generations, the following distribution of the three parameters is obtained, as shown in Fig. 3a. Similarly, the sorted cost distribution across the generations is shown in Fig. 3b.

5 Stability Analysis

Let $\ddot{s}(t) = \psi(t)$

For the stability analysis, the candidate Lyapunov function is chosen as:

$$V(t) = \frac{\psi^2}{2} \tag{25}$$

Differentiating (25) with respect to time t, we obtain following equation:

$$\dot{V}(t) = \psi\dot{\psi} \tag{26}$$

From Eqs. (21) and (22), $\dot{\psi}(t)$ can be written as:

$$\dot{\psi} = -k\psi - \eta|\psi| \tag{27}$$

Substituting (27) into the Eq. (26), following is obtained:

$$\dot{V}_i = \psi\dot{\psi} \tag{28}$$
$$= \psi(-k\psi - \eta sgn\psi) \tag{29}$$
$$= -k\psi^2 - \eta|\psi| \tag{30}$$

Calculating the energy of the fuction by integrating Eq. (30) on both sides, following is obtained:

$$\int_0^t \dot{V}\mathrm{d}\tau = \int_0^t (-k\psi^2 - \eta|\psi|)\mathrm{d}\tau \tag{31}$$

$$V(t) - V(0) = -\int_0^t (k\psi^2 + \eta|\psi|)\mathrm{d}\tau \tag{32}$$

$$V(0) = V(t) + \int_0^t (k\psi^2 + \eta|\psi|)\mathrm{d}\tau \tag{33}$$

$$\geq \int_0^t (k\psi^2 + \eta|\psi|)\mathrm{d}\tau \tag{34}$$

Applying limit to the integrated term in above equation:

$$\lim_{t\to\infty} \int_0^t (k\psi^2 + \eta|\psi|)\mathrm{d}\tau \leq V(0) < \infty \tag{35}$$

Barbalat's lemma states that, supposing $f(t)\epsilon C^1(a\infty)$ and $\lim_{x\to\infty} f(t) = \alpha$ where, $\alpha < \infty$. If $\dot{f}(t)$ is uniformly continuous, then $\lim_{x\to\infty} \dot{f}(t) = 0$.

According to Barbalat's lemma:

$$\lim_{t\to\infty}(k\psi^2 + \eta|\psi|) = 0 \tag{36}$$

According to (36) $\lim_{t\to\infty}\psi = 0$, which implies that the sliding layer is asymptotically stable.

6 Simulation Results

The performance of the proposed control law is compared with the conventional PID controller, whose parameters are tuned using the GA algorithm. The accuracy, efficiency and robustness of the controllers are compared after plotting their responses for various cases discussed below:
Matched disturbance:

$$z = .0002 \times \text{rand}(1, 1) * \sin(0.5) \tag{37}$$

Viscous friction uncertainty:

$$B' = B \times (1 + \text{rand}(1, 1)) \tag{38}$$

Moment of inertia uncertainty:

$$J' = J(1 \pm 20\%) \tag{39}$$

HOSMC parameters found by trial and error are $\epsilon_1 = 1$, $\epsilon_2 = 1$, $\epsilon_3 = 1$ and $k_4 = 300$. HOSMC parameters obtained using GA are $k_1 = 150$, $k_2 = 85$ and $k_3 = 23$.

PID controller structure used:

$$C(s) = K_p\left(1 + \frac{1}{T_i s} + \frac{T_d s}{0.001s + 1}\right) \tag{40}$$

PID controller parameters obtained using GA are $K_p = 0.8906$, $T_i = 59.5$ and $T_d = 0.0992$.

The following instances are defined for the analysis of proposed controller performance:
Instance 1: only the condition (37) is considered.
Instance 2: the conditions (37) and (38) are considered.
Instance 3: the conditions (37), (38) and (39) are considered.

The three instances are represented by the following legends in the plots:

Instance 1:

Instance 2:

Instance 3:

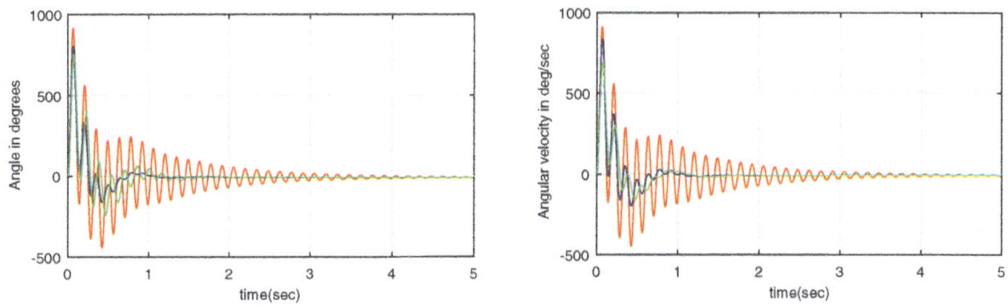

(a) Angular response (θ_3) of PID (b) Angular velocity ($\dot{\theta_3}$) PID

Fig. 4 **a** Angular response (θ_3) of PID, **b** angular velocity ($\dot{\theta_3}$) PID

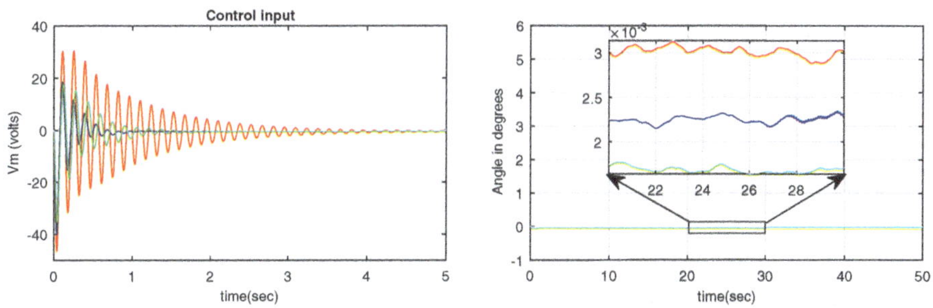

(a) Control input in PID (b) Angular response (θ) of HOHSMC

Fig. 5 **a** Control input in PID, **b** angular response (θ_3) of HOHSMC

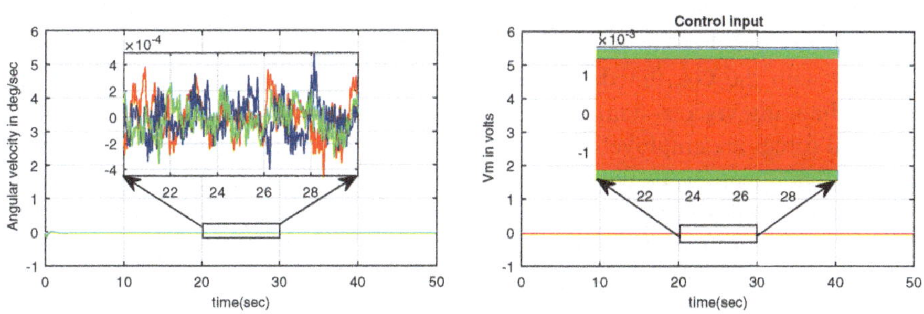

(a) Angular velocity ($\dot{\theta_3}$) of HOHSMC (b) Control input in HOHSMC

Fig. 6 **a** Angular velocity ($\dot{\theta_3}$) of HOHSMC, **b** control input in HOHSMC

Fig. 7 Sliding surface in HOHSMC

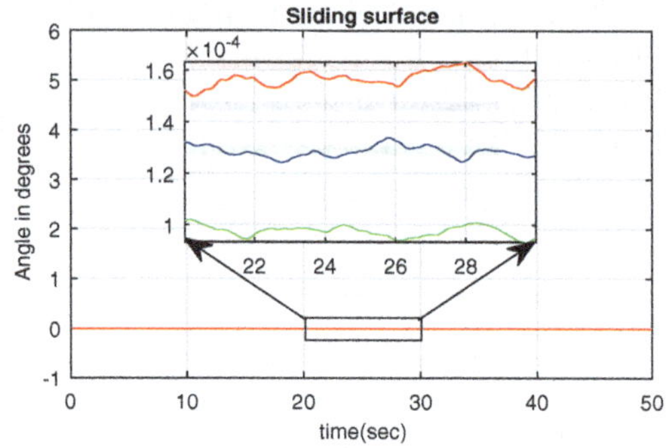

(a) Sliding surface in HOHSMC

The state responses of the proposed controller and the GA tuned PID controller can be seen in Figs. 4a and 5b. Since it is a regulation problem, the state of the system is assumed to be an error. It is observed that the states asymptotically die down to zero. It is also seen that some minute disturbances are observed because of the Gaussian matched disturbances and parametric uncertainties, but the response is chattering free because of now continuous control obtained by applying higher-order SMC (Figs. 6 and 7).

7 Conclusion

An effort has been made in this paper to enhance the performance of the 2-DoF torsion system in the presence of various disturbances and uncertainties. The HOSMC based on hierarchical sliding mode is put forward for obtaining the desired control design. The parameters of the proposed controller are tuned using the Genetic Algorithm scheme. Lyapunov theory and Barbalat's theorem are used for establishing the stability of the proposed controller. The 2-DoF torsion system is stabilized with the aforementioned controller and the plots obtained are compared with the plots obtained with the GA tuned PID controller. Gaussian matched disturbance, randomly changing viscous damping coefficient, parameter uncertainties, and varying frictional forces are employed to convey the robustness of the proposed controller. It is found that the proposed controller could robustly stabilize the 2-DoF system. One of the future scopes of the paper is the introduction of Kalman Filter in the control loop so that the effect of measurement noise and parametric uncertainties can be reduced on the output.

References

1. Fuchs G, Šátek V, Vopěnka V, Kunovský J, Kozek M (2013) Application of the modern taylor series method to a multi-torsion chain. Simul Modell Pract Theory 33(Complete):89–101
2. Elumalai VK, Subramanian RG (2017) A new algebraic lqr weight selection algorithm for tracking control of 2 dof torsion system. Arch Electr Eng 66(1)(Complete):55–75
3. Rojas-Moreno A, Salazar-Aquino O, Pampamallco-Jara J (2018) Control of the angular position of a rotary torsion plant using a 2dof fo pid controller. In: 2018 IEEE 38th central America and panama convention (CONCAPAN XXXVIII), pp 1–5
4. Navarro-López EM, Cortés D (2007) Sliding-mode control of a multi-dof oilwell drillstring with stick-slip oscillations. In: 2007 American control conference. IEEE, pp 3837–3842
5. Young KD, Utkin VI, Ozguner U (1999) A control engineer's guide to sliding mode control. IEEE Trans Control Syst Technol 7(3):328–342
6. Shtessel YB, Fridman L, Zinober A (2008) Higher order sliding modes. Int J Robust Nonlinear Control IFAC-Affil J 18(4–5):381–384
7. Levant A (1987) Higher order sliding modes and their application for controlling uncertain processes. Inst Syst Stud USSR Acad Sci Moskau Diss
8. Laghrouche S, Plestan F, Glumineau A (2007) Higher order sliding mode control based on integral sliding mode. Automatica 43(3):531–537
9. Defoort M, Floquet T, Kokosy A, Perruquetti W (2009) A novel higher order sliding mode control scheme. Syst Control Lett 58(2):102–108
10. Laghrouche S, Harmouche M, Ahmed FS, Chitour Y (2014) Control of pemfc air-feed system using lyapunov-based robust and adaptive higher order sliding mode control. IEEE Trans Control Syst Technol 23(4):1594–1601
11. Wang W (2004) Design of a stable sliding-mode controller for a class of second-order underactuated systems. IEE Proc Control Theor Appl 151(7):683–690 (2004). https://digital-library.theiet.org/content/journals/10.1049/ip-cta-20040902
12. Vadali SR, Kim ES (1991) Feedback control of tethered satellites using lyapunov stability theory. J Guid Control Dyn 14(4):729–735. https://doi.org/10.2514/3.20706
13. Dong H, Li T, Ding R, Sun J (2018) A novel hybrid genetic algorithm with granular information for feature selection and optimization. Appl Soft Comput 65:33–46
14. Li H, Yuan D, Ma X, Cui D, Cao L (2017) Genetic algorithm for the optimization of features and neural networks in ecg signals classification. Sci Rep 7:41011
15. Askarzadeh A (2017) A memory-based genetic algorithm for optimization of power generation in a microgrid. IEEE Trans Sustain Energy 9(3):1081–1089
16. Khazaii J (2016) Genetic algorithm optimization. In: Advanced decision making for HVAC engineers. Springer, pp 87–97
17. Kramer O (2017) Genetic algorithm essentials, vol 679. Springer, Berlin
18. Zahir A, Alhady S, Othman W, Ahmad M (2018) Genetic algorithm optimization of pid controller for brushed dc motor. In: Intelligent manufacturing and mechatronics. Springer, pp 427–437

Drowsiness Detection System in Real Time

Sheersh Kaushik, Purnima Gupta, Vineet Kumar Singh, Sonal Maan, and Saquib Faraz

Abstract The paper presents the study of the project on which we are working, to enhance the safety measures of the drivers. It will detect the drowsiness of the driver and will ensure their safety. As we know many drivers lose their lives every year due to tragic road accidents and the main reason behind that is the fatigue or the sleepiness of the drivers. The conventional methods that are used to detect the fatigue of the driver are the behavioural features while others need extravagant devices. For that reason, we developed a drowsiness detection system in real time. This system analyses the facial features and detects the drowsiness of the driver in real time by using an image processing technique CONVOLUTIONAL NEURAL NETWORK (CNN). In this eye aspect ratio and closure ratio is computed to detect the fatigue of the driver. PYTHON is used to test the intended approach competence.

Keywords Drowsy · CNN · Image processing · OpenCV · Safety

1 Introduction

In this project, we used both the hardware and software to create a system that will take care of driver's safety as one of the major causes behind the road accidents is driver's drowsiness. Studies have stated that majority of accidents occur due to driver's fatigue. The most important challenge is to map out a system that detects driver's drowsiness. Many heavy vehicles such as loaded trucks are mostly driven at night and the drivers of such vehicles who drive continuously for long times becomes more prone to these kinds of situations.

The methods that are used to identify driver's fatigue are mostly behavioural-based vehicle-based and psychological-based. The behavioural pattern includes yawning, closure and blinking of eyes etc. The psychological methods that are used are heart-beat, pulse rate etc. The vehicle-based methods include steering movements, accelerator patterns and acceleration. In this report, we will study the proposed approach

S. Kaushik (✉) · P. Gupta · V. K. Singh · S. Maan · S. Faraz
ADGITM GGSIPU, New Delhi, India

S. Suhag et al. (eds.), *Control and Measurement Applications for Smart Grid*,
Lecture Notes in Electrical Engineering 822,
https://doi.org/10.1007/978-981-16-7664-2_26

Fig. 1 Proposed module prototype

to identify driver's fatigue. The system is installed in driver's vehicle. The process starts when the image is taken as input from the camera. Then the images are fed to the CNN classifier which will speculate the condition of the eyes according to the score that whether the eyes are open or closed [1].

This project is made at an affordable price range without compromising its efficiency. The leading characteristics of our project are its compact size and low economic nature. This project is made under a budget of INR 1000. The main components that we used in this project are camera sensor, mount, thin electronic wires alongwith the microprocessor. As it is very compact in nature it can be used in any vehicle and also due to its lower economic value most of the people could reap benefits from it (Fig. 1).

2 Literature Survey

A survey that is done by CENTRAL ROAD REASEARCH INSTITUTE (CRRI) on Agra—Lucknow expressway which is 300 km long states that the fatigued drivers who doze off while driving are accountable for about 40% of the total road accidents [2].

The study related to heavy eyed drivers shows that in our country 3 life taking accidents occurs every minute, i.e. 3 people lose their lives every minute due to these fatal road accidents. The study has also emphasized the need for educating the highway auto-mobilists about the significance of taking proper sleep and regular breaks in between driving for safety purposes (Fig. 2).

In accordance with the road safety experts a person should not drive continuously for more than 3 h without taking a break of 15–30 min. To make sure that the driver stays alert he should not drive for more than 8 h in a day. According to the statistics that are revealed by Nitin Gadkari (Road Transport and Highway Minister) 147,913 people lost their lives in 2017 in road accidents having UP shows up the highest no. of deaths 20,124.

Approximately 4.5 lakh fatal road accidents take place in India every year. 70.7% of the fatal road accidents happens in 4 or more than 4-wheeled vehicles. 40% of these are due to drowsiness of the driver, which counts of 1.35 lakh deaths. Our system is capable of reducing this sadly high number of deaths [4] (Table 1).

Figure 3 shows that as the number of sleeping hours reduces, the chances of fatal accidents increases swiftly. It shows that when the number of sleeping hours is less than 4, at that time the vehicle is more prone to crash, while on the contrary when number of sleeping hours is more than 6 the vehicle is less prone to crash. It shows that being drowsy plays a crucial role when it comes to road accidents [5].

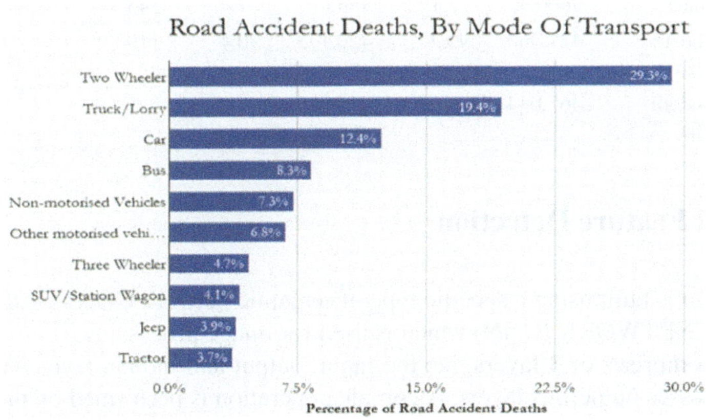

Fig. 2 Road accident statistics [3]

Table 1 Number of road accidents in top 15 states and their respective percentage shares in total accidents during 2017 and 2018

S. No	State	2018		2017		Increase/decrease over 2017	
		No. of accidents	% share in total	No. of accidents	% share in total	In numbers	% age terms
1	Tamil Nadu	63,920	13.7	65.562	14.1	− 1642	2.6
2	Madhya Pradesh	51,397	11.0	53,399	11.5	− 2002	3.9
3	Uttar Pradesh	42,560	9.1	38,783	8.3	3777	− 8.9
4	Karnataka	41,707	8.9	42,542	9.2	− 835	2.0
5	Kerala	40,181	8.6	38,470	8.3	1711	− 4.3
6	Maharashtra	35,717	7.6	35,853	7.7	− 136	0.4
7	Andhra Pradesh	24,475	5.2	25,727	5.5	− 1252	5.1
8	Telangana	22,230	4.8	22,484	4.8	− 254	1.1
9	Rajasthan	21,743	4.7	22,112	4.8	− 369	1.7
10	Gujarat	18,769	4.0	19,081	4.1	− 312	1.7
11	Chhattisgarh	13,864	3.0	13,563	2.9	301	− 2.2
12	West Bengal	12,705	2.7	11,631	2.5	1074	− 8.5
13	Haryana	11,262	2.4	11,258	2.4	4	0.0
14	Odisha	11,238	2.4	10,855	2.3	383	− 3.4
15	Bihar	9600	2.1	8855	1.9	745	− 7.8
	Total 15 states	421,368	90.2	420,175	90.4	1193	− 0.3
	Total all India	467,044		464,910			

3 CNN Feature Detection

This system is built using a specific type of semantic network CONVOLUTIONAL NEURAL NETWORK (CNN) which is used for image processing.

CNN comprises of 3 layers, i.e. the input, output and hidden layer. The hidden layer possesses numerous layers. A complex operation is performed on these layers employing a filter that carries out 2D matrix multiplication on the layers and filter [6].

Our algorithm works as follows.

- The first step is to take the images as input using device's webcam. To gain access to the webcam an unbounded loop is made that will capture each frame.
- The image is then converted to grayscale as the algorithm which we are using, i.e. the OpenCV algorithm takes only grey images as input. Then the detection is performed.

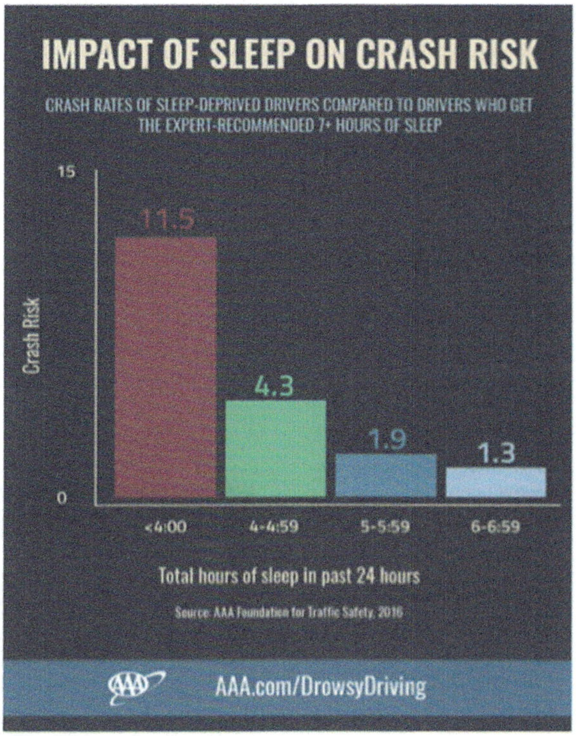

Fig. 3 Statistics of impact of sleep on crash risk [5]

- The approach that is used to detect the faces and is used to detect the eyes. Now we need to take only the eyes data from the captured image. This can be done by extracting the outer limit box of the eye, then we can take the eye image from the frame.
- The image data of the eye is contained by L_eye. The data is received by CNN classifier which predicts if the eyes are open or closed. Similarly, we will extract the data of right eye using R_eye.
- Here, we are using CNN classifier which predicts the status of the eye. As our model demands correct dimension to start with, firstly the colour images are converted to grayscale.
- Images are resized into 24 * 24 pixels. Our model predicts each eye. If the value of lpred[0] = 1, it represents that the eyes are open but if the value of lpred[0] = 0, then it represents that the eyes are closed.
- Here, the value of score will be determining for how long the eyes have been closed by the driver. So, if eyes are closed, the score will keep on increasing and if eyes are open the score will decrease. Result will be displayed on the screen which display status of the person in real time.
- A threshold value will be defined for the score which will predict if the eyes are open or closed. For example, if we have set the threshold value to 15 and if the

score becomes greater than 15 that will indicate that the person's eyes are closed for a longer period of time. And hence the alarm will start beeping.

4 Experimental results and analysis

As discussed earlier, the system created is used to detect facial expressions of the driver and thus stating the driver's condition whether he/she is drowsy or not. For this experiment camera and speaker raspberry pi modules were used connected to a screen for visuals [6] (Fig. 4).

In this picture, as we can see that the eyes of the operator are open, camera is still detecting but there will be no indication of the driver being drowsy (Fig. 5).

Fig. 4 Image when eyes are detected open

Fig. 5 Image when eyes are closed

Fig. 6 Alarming image when driver detected sleeping

In this picture, as we can see that the eyes are closed, camera is still detecting but in first 5 s there will be no light, no sound beep as we have set the threshold value to 5 s (Fig. 6).

In this picture, the eyes of driver are closed for more than 5 s and score exceeded the threshold value signifying that the driver is drowsy, hence we can see the red light which indicates the driver being drowsy and we can hear a beep sound as well along with the red light. Here, red light indicates that the indicating light of the vehicle has been turned on and beep sound indicates that the music system inside the vehicle has also been turned on.

5 Conclusion

In this work, we have proposed a system that monitors and detects the loss of attention of drivers of vehicles in real time. The face of the driver has been detected by capturing facial landmarks and warning is given to the driver by the system to avoid real time crashes. The proposed approach uses the aspect ratio and closure ratio of eyes with adaptive thresholding to detect the drowsiness of the driver in real time.

Acknowledgements We would like to offer sincere gratitude to our mentor Mr. Saquib Faraz, (Prof. Dept. of Electrical and Electronics) and our classmate Manjeet Singh for helping us with the Project. I learnt many new things I am really filled with gratitude by them. I would also like to thank everyone who assisted me in this project and make it work within the limited time frame.

References

1. IOP Conf Ser Mater Sci Eng 252:012097 (2017). https://doi.org/10.1088/1757-899X/252/1/012097
2. https://www.financialexpress.com/india-news/40-of-highway-accidents-occur-due-to-drivers-dozing-off/1659901/
3. Scroll.in; https://scroll.in/article/826264/three-killed-every-10-minutes-road-accident-deaths-in-india-up-9-in-4-years
4. Autocars Professionals; https://www.autocarpro.in/news-national/road-accidents-in-india-claimed-415-lives--injured-1-286-people-each-day-in-2018-44694
5. https://www.forensicdjs.com/blog/industry-update-acute-sleep-deprivation-crash-risk/
6. Real-Time driver drowsiness detection system using eye aspect ratio and eye closure ratio. 2019SUSCOM

An Improved Stabilization Criteria for Linear Systems with Time-Varying Delay Using a New Lyapunov–Krasovskii Functional

K. C. Rajendra Prasad, N. K. Arun, and M. Venkatesh

Abstract This study aims to design an improved stabilization controller for linear systems with time-varying delay. First, a proper augmented LKF has been constructed and then a tighter bound of the quadratic integral function in the LKF derivative is established by applying Wirtinger-based integral inequality (WBII) and extended reciprocally convex matrix inequality (RCMI). Congruence transformation has been performed to obtain the negative-definite condition for LKF derivative such that the closed-loop system becomes asymptotically stable in the sense of Lyapunov. Maximum permissible delay upper bound (MPDUB) and the corresponding control effort is computed using linear matrix inequality (LMI) convex optimization approach. A numerical example has been solved to verify the effectiveness of the proposed method using LMI control toolbox of MATLAB.

Keywords Time-varying delay · Stabilization control · Lyapunov–Krasovskii functional · Extended reciprocally convex matrix inequality · Linear matrix inequality

1 Introduction

Time-delay occurs in most of the practical systems such as power system, process control, cyber-physical systems, biological systems. Generally, networked control systems have time-varying delays due to the need of open communication channels between interconnected dynamical systems. Typical examples of such cases include networked load frequency control problem in deregulated power system [1], networked state feedback control of a DC servo position system [2], etc.

K. C. Rajendra Prasad (✉) · N. K. Arun
National Institute of Technology, Calicut, India

N. K. Arun
e-mail: arunkm@nitc.ac.in

M. Venkatesh
Indian Institute of Technology, Kharagpur, India

S. Suhag et al. (eds.), *Control and Measurement Applications for Smart Grid*,
Lecture Notes in Electrical Engineering 822,
https://doi.org/10.1007/978-981-16-7664-2_27

Time-delay system (TDS) belongs to a class of infinite-dimensional systems and the presence of time-delay may lead to inappropriate oscillations, degradation in performance of the system, etc. [3]. Delay–dependent stability conditions of TDS depend on the maximum delay margin. A large number of literatures have been published to reduce the gap between estimated and accurate delay value using Lyapunov–Krasovskii functional (LKF) approach whose feasibility is validated using linear matrix inequality (LMI) description technique.

In LKF approach, the first step is to select an appropriate LKF candidate. There are many techniques available in the literatures to construct a suitable LKF such as augmented, multiple-integral terms and delay-fractioning approach [4]. The second step is to estimate the quadratic integral function of LKF derivative using an efficient integral inequality approach. In this approach, quadratic integral function is first estimated by using Jensen [5], auxiliary-function [6] or Wirtinger-based integral inequality (WBII) [7] and then the reciprocally related quadratic term is estimated by using reciprocally convex matrix inequality (RCMI) [8], extended RCMI [9] or generalized RCMI [10]. The third step is to acquire the negative-definite condition of LKF derivative and it is generally guaranteed by the first two steps. However, for LKF derivative with second-order polynomial, the convexity or concavity is unknown and for such cases several techniques are available to acquire the negative-definite condition [11, 12].

In [13], by combining augmented LKF and WBII delay-dependent stability conditions for linear TDS is obtained. Though WBII is tighter than Jensen inequality its effectiveness can only be guaranteed by choosing an appropriate LKF, which is shown in [14]. A double integral WBII using single integral WBII has been proposed in [15]. Also, recently using double integral Jensen inequality, a double integral WBII has been obtained in [16]. In [17], an improved stability criterion is obtained for linear TDS by constructing an appropriate LKF and its derivative is estimated by WBII and extended RCMI. In [18], WBII has been used to construct a LKF for sampled-data systems.

When the TDS is unstable under open-loop condition, a stabilization controller is used to stabilize the closed-loop system in the sense of Lyapunov. In [19], a static state feedback controller for linear TDS is designed using a simple LKF and congruence transformation is done to obtain LMI conditions by invoking Finsler's lemma. In [20], for stabilization of uncertain linear systems with time-varying delay, a static state feedback controller is designed using augmented LKF and its derivative is tightly bounded by WBII and RCMI. In [21], an improved stability and stabilization result is obtained for Takaki-Sugeno fuzzy systems with time-varying delays by combining WBII and extended RCMI via delay-product-type functional approach.

This paper focuses on designing a delay-dependent static state feedback controller for linear systems with interval time-varying delay using a new LKF. First, an appropriate augmented LKF is constructed and then, a tighter bound of quadratic integral term in the LKF derivative is established using WBII and extended RCMI, which reduces the control effort with larger maximum permissible delay upper bound (MPDUB). The controller gain is found by performing congruence transformation and the controller design parameters with constraints are formulated in LMI framework. A numerical example has been solved to validate the superiority of the proposed delay-dependent stabilization criterion over the existing results.

Notations: $\Re^n$ represents the n-dimensional Euclidean space; $X > 0$ represents the positive-definite symmetric matrix; $X < 0$ represents the negative-definite symmetric matrix; I represents an identity matrix of appropriate dimension; 0 represents a zero matrix of appropriate dimension; A^T represents the transpose of matrix A; A^{-1} represents the inverse of matrix A;$\text{Sym}\{X\} = X + X^T$; $\text{col}\{x_1, x_2, \ldots, x_n\} = \left[x_1^T, x_2^T, \ldots, x_n^T\right]^T$;diag{.} represents a block-diagonal matrix; $(*)$ represents the symmetric elements in a symmetric matrix; and $\alpha(t)$ is represented as α for simplicity.

2 Problem Formulation and Preliminaries

2.1 Problem Formulation

Consider a class of linear time-varying delay system described by the following functional differential equation as shown below

$$\begin{aligned} \dot{x}(t) &= Ax(t) + A_d x(t - d(t)) + Bu(t), \ \forall t \geq 0, \\ y(t) &= Cx(t) + C_d x(t - d(t)), \ \forall t \geq 0, \\ x(t) &= \phi(t), \ \forall t \in [-h_2, 0], \end{aligned} \tag{1}$$

where $x(t) \in \Re^n$ is the state vector, $u(t) \in \Re^m$ is the control input vector, $y(t) \in \Re^p$ is the output vector, A, A_d, B, C and C_d are known real constant matrices of appropriate dimensions. $\phi(t)$ is a continuous-time differentiable vector-valued initial function defined on $[-h_2, 0]$ and $d(t)$ is an interval time-varying state delay which satisfies the below given conditions

$$0 \leq h_1 \leq d(t) \leq h_2 < \infty, \ \ \mu_1 \leq \dot{d}(t) \leq \mu_2, \tag{2}$$

where h_1 and h_2 are known positive real values, which represents the lower and upper bounds of $d(t)$, respectively, and μ_1 and μ_2 are known positive real values, which represents the minimum and maximum value of $\dot{d}(t)$, respectively.

For the TDS (1) with conditions satisfying (2), a static state feedback controller has to be designed as shown below

$$u(t) = Kx(t), \ K \in \Re^{m \times n}, \tag{3}$$

such that the closed-loop system

$$\dot{x}(t) = (A + BK)x(t) + A_d x(t - d(t)), \ \forall t \geq 0. \quad (4)$$

is asymptotically stable.

The following assumptions are made while deriving the delay-dependent stabilization controller (i) $(A + A_d, B)$ is stabilizable and (ii) (A, C) is detectable.

2.2 Preliminaries

The integral inequalities [7, 9, 20] are recalled to derive stabilization criterion for given system (1).

Lemma 1 *For a given $n \times n$ real matrix $Y > 0$, two scalars m_1 and m_2 with $m_2 > m_1$, and a continuously differentiable vector function z: $[m_1, m_2] \rightarrow \Re^n$, the below given quadratic integral inequality holds.*

$$\int_{m_1}^{m_2} \dot{z}^T(u) Y \dot{z}(u) du \geq \frac{1}{m_2 - m_1} (z(m_2) - z(m_1))^T Y (z(m_2) - z(m_1)) + \frac{3}{m_2 - m_1} \Pi^T Y \Pi, \quad (5)$$

where $\Pi = z(m_2) + z(m_1) - \frac{2}{m_2 - m_1} \int_{m_1}^{m_2} z(u) du.$

Lemma 2 *For a real constant $\alpha \in (0, 1)$, matrices $Y_1 > 0$ and $Y_2 > 0$, and any matrices S_1 and S_2, the below given inequality holds*

$$\begin{bmatrix} \frac{1}{\alpha} Y_1 & 0 \\ (*) & \frac{1}{1-\alpha} Y_2 \end{bmatrix} \geq \begin{bmatrix} Y_1 + (1-\alpha) M_1 & (1-\alpha) S_1 + \alpha S_2 \\ (*) & Y_2 + \alpha M_2 \end{bmatrix}, \quad (6)$$

where $M_1 = Y_1 - S_2 Y_2^{-1} S_2^T$ *and* $M_2 = Y_2 - S_1^T Y_1^{-1} S_1$.

Lemma 3 *For any given matrix X and $Y = Y^T > 0$ of appropriate dimensions, the below given inequality holds.*

$$X^T Y^{-1} X \geq X + X^T - Y. \quad (7)$$

The inequality (7) is obtained from $(Y - X)^T Y^{-1} (Y - X) \geq 0$.

3 Main Results

In this section, a stabilization controller is designed using an appropriate LKF.

Theorem 1 *For known positive constants h_1, h_2 and delay-derivative satisfying the condition (2), the TDS (1) is asymptotically stable by using the static state feedback controller (3), if there exist a $3n \times 3n$ matrix $\bar{P} > 0$, $n \times n$ matrices $\bar{Q}_i > 0$, $i = 1, 2, 3, 4$, $n \times n$ matrices $\bar{R}_j > 0$, $j = 1, 2$, a $n \times n$ matrix $U > 0$, a $m \times n$ any matrix V and $2n \times 2n$ any matrices $\bar{S}_1$, $\bar{S}_2$ such that the following LMIs hold:*

$$\begin{bmatrix} \Phi_{1,[h_1,\mu_1]} - E_1^T \widehat{\bar{R}}_1 E_1 - \Upsilon_1 & h_1 Y_1^T & h_D Y_1^T & E_2^T \bar{S}_2 \\ (*) & \bar{R}_1 - 2U & 0 & 0 \\ (*) & (*) & \bar{R}_2 - 2U & 0 \\ (*) & (*) & (*) & -\widehat{\bar{R}}_2 \end{bmatrix} < 0, \tag{8}$$

$$\begin{bmatrix} \Phi_{1,[h_1,\mu_2]} - E_1^T \widehat{\bar{R}}_1 E_1 - \Upsilon_1 & h_1 Y_1^T & h_D Y_1^T & E_2^T \bar{S}_2 \\ (*) & \bar{R}_1 - 2U & 0 & 0 \\ (*) & (*) & \bar{R}_2 - 2U & 0 \\ (*) & (*) & (*) & -\widehat{\bar{R}}_2 \end{bmatrix} < 0, \tag{9}$$

$$\begin{bmatrix} \Phi_{1,[h_2,\mu_1]} - E_1^T \widehat{\bar{R}}_1 E_1 - \Upsilon_2 & h_1 Y_1^T & h_D Y_1^T & E_3^T \bar{S}_1^T \\ (*) & \bar{R}_1 - 2U & 0 & 0 \\ (*) & (*) & \bar{R}_2 - 2U & 0 \\ (*) & (*) & (*) & -\widehat{\bar{R}}_2 \end{bmatrix} < 0, \tag{10}$$

$$\begin{bmatrix} \Phi_{1,[h_2,\mu_2]} - E_1^T \widehat{\bar{R}}_1 E_1 - \Upsilon_2 & h_1 Y_1^T & h_D Y_1^T & E_3^T \bar{S}_1^T \\ (*) & \bar{R}_1 - 2U & 0 & 0 \\ (*) & (*) & \bar{R}_2 - 2U & 0 \\ (*) & (*) & (*) & -\widehat{\bar{R}}_2 \end{bmatrix} < 0, \tag{11}$$

where

$$\begin{aligned}
\Phi_{1,[d(t),\dot{d}(t)]} &= \mathrm{Sym}\{Y_1^T e_1 + \Gamma_1^T P \Gamma_2\} + \widehat{\bar{Q}},\ h_D = h_2 - h_1, \\
Y_1 &= \left[AU + BV \ 0 \ A_d U \ 0\,0\,0\,0 \right], \\
\widehat{\bar{R}}_i &= \mathrm{diag}\{\bar{R}_i, 3\bar{R}_i\}, i = 1, 2 \\
\Gamma_1 &= \mathrm{col}\{h_1 e_5, (d(t) - h_1)e_6, (h_2 - d(t))e_7\}, \\
\Gamma_2 &= \mathrm{col}\{e_1 - e_2, e_2 - (1 - \dot{d}(t))e_3, (1 - \dot{d}(t))e_3 - e_4\}, \\
E_j &= \mathrm{col}\{e_j - e_{j+1}, e_j + e_{j+1} - 2e_{j+4}\},\ j = 1, 2, 3; \\
\widehat{\bar{Q}} &= \mathrm{diag}\{\bar{Q}_1 + \bar{Q}_2 + \bar{Q}_3, -\bar{Q}_2 + \bar{Q}_4, -(1 - \dot{d}(t))(\bar{Q}_1 + \bar{Q}_4), -\bar{Q}_3, 0, 0, 0\},
\end{aligned}$$

$$\Upsilon_1 = \begin{bmatrix} E_2 \\ E_3 \end{bmatrix}^T \begin{bmatrix} 2\widehat{\bar{R}}_2 & \bar{S}_1 \\ (*) & \widehat{\bar{R}}_2 \end{bmatrix} \begin{bmatrix} E_2 \\ E_3 \end{bmatrix}, \Upsilon_2 = \begin{bmatrix} E_2 \\ E_3 \end{bmatrix}^T \begin{bmatrix} \widehat{\bar{R}}_2 & \bar{S}_2 \\ (*) & 2\widehat{\bar{R}}_2 \end{bmatrix} \begin{bmatrix} E_2 \\ E_3 \end{bmatrix},$$

$$e_j = \left[0_{n\times(j-1)n}, I, 0_{n\times(7-j)n}\right], \; j = 1, 2, \ldots, 7, \quad \text{and} \quad K = VU^{-1}$$

Proof Let the LKF be

$$\begin{aligned} V(t) &= x^T(t)P_{11}x(t) + \tilde{x}^T(t)P\tilde{x}(t) \\ &+ \int_{t-d(t)}^{t} x^T(s)Q_1x(s)ds + \int_{t-h_1}^{t} x^T(s)Q_2x(s)\mathrm{d}s \\ &+ \int_{t-h_2}^{t} x^T(s)Q_3x(s)ds + \int_{t-d(t)}^{t-h_1} x^T(s)Q_4x(s)\mathrm{d}s \\ &+ h_1 \int_{t-h_1}^{t}\int_{\theta}^{t} \dot{x}^T(s)R_1\dot{x}(s)\mathrm{d}s\mathrm{d}\theta \\ &+ h_D \int_{t-h_2}^{t-h_1}\int_{\theta}^{t} \dot{x}^T(s)R_2\dot{x}(s)\mathrm{d}s\mathrm{d}\theta, \end{aligned} \tag{12}$$

where $\tilde{x}(t) = \mathrm{col}\left\{\int_{t-h_1}^{t} x(s)\mathrm{d}s, \int_{t-d(t)}^{t-h_1} x(s)\mathrm{d}s, \int_{t-h_2}^{t-d(t)} x(s)\mathrm{d}s\right\}$, $h_D = h_2 - h_1$, and $P_{11}, P, Q_1, Q_2, Q_3, Q_4, R_1, R_2$ are unknown weighted matrices which is real and symmetric.

The derivative of (12) becomes

$$\begin{aligned} \dot{V}(t) &= \dot{x}^T(t)P_{11}x(t) + x^T(t)P_{11}\dot{x}(t) \\ &+ \tilde{x}^T(t)P\dot{\tilde{x}}(t) + \dot{\tilde{x}}^T(t)P\tilde{x}(t) + x^T(t)Q_1x(t) \\ &- \left(1 - \dot{d}(t)\right)x^T(t - d(t))Q_1x(t - d(t)) \\ &+ x^T(t)Q_2x(t) - x^T(t - h_1)Q_2x(t - h_1) + x^T(t)Q_3x(t) \\ &- x^T(t - h_2)Q_3x(t - h_2) + x^T(t - h_1)Q_4x(t - h_1) \\ &- \left(1 - \dot{d}(t)\right)x^T(t - d(t))Q_4x(t - d(t)) \\ &+ h_1^2\dot{x}^T(t)R_1\dot{x}(t) - h_1 \int_{t-h_1}^{t} \dot{x}^T(s)R_1\dot{x}(s)\mathrm{d}s \\ &+ h_D^2\dot{x}^T(t)R_2\dot{x}(t) - h_D \int_{t-h_2}^{t-h_1} \dot{x}^T(s)R_2\dot{x}(s)\mathrm{d}s. \end{aligned} \tag{13}$$

By replacing $\dot{x}(t)$ and $\tilde{x}(t)$ from (1) and (12), respectively, then with Γ_1, Γ_2, e_1 defined in Theorem 1 and also by defining

$$\xi(t) = \text{col}\left\{\begin{bmatrix} x(t) \\ x(t-h_1) \\ x(t-d(t)) \\ x(t-h_2) \end{bmatrix}, \begin{bmatrix} \frac{1}{h_1}\int\limits_{t-h_1}^{t} x(s)\mathrm{d}s \\ \frac{1}{d(t)-h_1}\int\limits_{t-d(t)}^{t-h_1} x(s)\mathrm{d}s \\ \frac{1}{h_2-d(t)}\int\limits_{t-h_2}^{t-d(t)} x(s)\mathrm{d}s \end{bmatrix}\right\},$$

$$\hat{Q}_1 = \text{diag}(Q_1, 0, -(1-\dot{d}(t))Q_1, 0, 0, 0, 0),$$
$$\hat{Q}_2 = \text{diag}(Q_2, -Q_2, 0, 0, 0, 0, 0),$$
$$\hat{Q}_3 = \text{diag}(Q_3, 0, 0, -Q_3, 0, 0, 0),$$
$$\hat{Q}_4 = \text{diag}(0, Q_4, -(1-\dot{d}(t))Q_4, 0, 0, 0, 0),$$

$e_s = \begin{bmatrix} A + BK & 0 & A_d & 0 & 0 & 0 & 0 \end{bmatrix}$, (13) can be written as

$$\dot{V}(t) = \xi^T(t)\Big(\Lambda_{1,[d(t),\dot{d}(t)]} + h_1^2 e_s^T R_1 e_s + h_D^2 e_s^T R_2 e_s\Big)\xi(t) - h_1\int\limits_{t-h_1}^{t} \dot{x}^T(s)R_1\dot{x}(s)\mathrm{d}s - h_D\int\limits_{t-h_2}^{t-h_1} \dot{x}^T(s)R_2\dot{x}(s)\mathrm{d}s, \tag{14}$$

where $\Lambda_{1,[d(t),\dot{d}(t)]} = \text{Sym}\{e_s^T P_{11} e_1 + \Gamma_1^T P \Gamma_2\} + \hat{Q}$ and $\hat{Q} = \hat{Q}_1 + \hat{Q}_2 + \hat{Q}_3 + \hat{Q}_4$.

With E_1 defined in Theorem 1 and also by defining $\hat{R}_1 = \text{diag}\{R_1, 3R_1\}$, second term of (14) using WBII (5) can be written as

$$-h_1\int\limits_{t-h_1}^{t} \dot{x}^T(s)R_1\dot{x}(s)\mathrm{d}s \le -\xi^T(t)E_1^T\hat{R}_1E_1\xi(t). \tag{15}$$

With E_2 and E_3 defined in Theorem 1 and also by defining $\hat{R}_2 = \text{diag}\{R_2, 3R_2\}$, third term of (14) using WBII (5) can be written as

$$-h_D\int\limits_{t-h_2}^{t-h_1} \dot{x}^T(s)R_2\dot{x}(s)ds \le -\frac{h_D}{h_2-d(t)}\xi^T(t)E_3^T\hat{R}_2E_3\xi(t) - \frac{h_D}{d(t)-h_1}\xi^T(t)E_2^T\hat{R}_2E_2\xi(t). \tag{16}$$

By defining $d(t) = (1-\alpha)h_1 + \alpha h_2$, $d(t)-h_1 = \alpha h_D$, $h_2-d(t) = (1-\alpha)h_D$ where $\alpha \in (0, 1)$ and using extended RCMI (6), (16) can be written for any two matrices S_1 and S_2 as

$$-\xi^T(t)\begin{bmatrix} E_2 \\ E_3 \end{bmatrix}^T \begin{bmatrix} \frac{h_D}{d(t)-h_1}\hat{R}_2 & 0 \\ (*) & \frac{h_D}{h_2-d(t)}\hat{R}_2 \end{bmatrix}\begin{bmatrix} E_2 \\ E_3 \end{bmatrix}\xi(t) \le -\xi^T(t)\Lambda_{2,[d(t)]}\xi(t), \tag{17}$$

where

$$\Lambda_{2,[d(t)]} = \begin{bmatrix} E_2 \\ E_3 \end{bmatrix}^T \begin{bmatrix} \hat{R}_2 + \frac{h_2-d(t)}{h_D}\left(\hat{R}_2 - S_2\hat{R}_2^{-1}S_2^T\right) & \frac{h_2-d(t)}{h_D}S_1 + \frac{d(t)-h_1}{h_D}S_2 \\ (*) & \hat{R}_2 + \frac{d(t)-h_1}{h_D}\left(\hat{R}_2 - S_1^T\hat{R}_2^{-1}S_1\right) \end{bmatrix} \begin{bmatrix} E_2 \\ E_3 \end{bmatrix}$$

Using (15) and (17), (14) can be written as

$$\dot{V}(t) \leq \xi^T(t) \left(\Lambda_{1,[d(t),\dot{d}(t)]} - E_1^T\hat{R}_1E_1 - \Lambda_{2,[d(t)]} + h_1^2e_s^TR_1e_s + h_D^2e_s^TR_2e_s\right)\xi(t). \quad (18)$$

Define $\Lambda_{3,[d(t),\dot{d}(t)]} = \Lambda_{1,[d(t),\dot{d}(t)]} - E_1^T\hat{R}_1E_1 - \Lambda_{2,[d(t)]} + h_1^2e_s^TR_1e_s + h_D^2e_s^TR_2e_s$, and using Schur complement $\Lambda_{3,[d(t),\dot{d}(t)]}$ is equivalent to LMIs (19)–(22) as follows

$$\begin{bmatrix} \Lambda_{1,[h_1,\mu_1]} - E_1^T\hat{R}_1E_1 - \Lambda_4 & h_1e_s^TR_1 & h_De_s^TR_2 & E_2^TS_2 \\ (*) & -R_1 & 0 & 0 \\ (*) & (*) & -R_2 & 0 \\ (*) & (*) & (*) & -\hat{R}_2 \end{bmatrix} < 0, \quad (19)$$

$$\begin{bmatrix} \Lambda_{1,[h_1,\mu_2]} - E_1^T\hat{R}_1E_1 - \Lambda_4 & h_1e_s^TR_1 & h_De_s^TR_2 & E_2^TS_2 \\ (*) & -R_1 & 0 & 0 \\ (*) & (*) & -R_2 & 0 \\ (*) & (*) & (*) & -\hat{R}_2 \end{bmatrix} < 0, \quad (20)$$

$$\begin{bmatrix} \Lambda_{1,[h_2,\mu_1]} - E_1^T\hat{R}_1E_1 - \Lambda_5 & h_1e_s^TR_1 & h_De_s^TR_2 & E_3^TS_1^T \\ (*) & -R_1 & 0 & 0 \\ (*) & (*) & -R_2 & 0 \\ (*) & (*) & (*) & -\hat{R}_2 \end{bmatrix} < 0, \quad (21)$$

$$\begin{bmatrix} \Lambda_{1,[h_2,\mu_2]} - E_1^T\hat{R}_1E_1 - \Lambda_5 & h_1e_s^TR_1 & h_De_s^TR_2 & E_3^TS_1^T \\ (*) & -R_1 & 0 & 0 \\ (*) & (*) & -R_2 & 0 \\ (*) & (*) & (*) & -\hat{R}_2 \end{bmatrix} < 0, \quad (22)$$

where $\Lambda_4 = \begin{bmatrix} E_2 \\ E_3 \end{bmatrix}^T \begin{bmatrix} 2\hat{R}_2 & S_1 \\ (*) & \hat{R}_2 \end{bmatrix} \begin{bmatrix} E_2 \\ E_3 \end{bmatrix}$, $\Lambda_5 = \begin{bmatrix} E_2 \\ E_3 \end{bmatrix}^T \begin{bmatrix} \hat{R}_2 & S_2 \\ (*) & 2\hat{R}_2 \end{bmatrix} \begin{bmatrix} E_2 \\ E_3 \end{bmatrix}$.

Pre- and post-multiplying Eqs. (19)–(22) with $\text{diag}(\bar{U}, R_1^{-1}, R_2^{-1}, \bar{U}_1)$, where $\bar{U} = \text{diag}(U, U, U, U, U, U, U)$, $U = P_{11}^{-1}$ and $\bar{U}_1 = \text{diag}(U, U)$, and by substituting $\bar{Q}_i = UQ_iU$, $i = 1, 2, 3, 4$; $\bar{P} = \bar{U}_2 P \bar{U}_2$, $\bar{R}_1 = UR_1U$, $\bar{R}_2 = UR_2U$, $\bar{S}_1 = \bar{U}_1 S_1 \bar{U}_1$ and $\bar{S}_2 = \bar{U}_1 S_2 \bar{U}_1$ where $\bar{U}_2 = \text{diag}(U, U, U)$, we get the LMIs with inverse terms in (2, 2) and (3, 3) blocks. By invoking lemma (7) in these two blocks of obtained LMIs, we get LMIs (8)–(11). This concludes the proof.

Remark 1 Theorem 1 leads to reduced conservatism than [20] due to (i) two additional states $\int_{t-d(t)}^{t-h_1} x(s)\mathrm{d}s$, $\int_{t-h_2}^{t-d(t)} x(s)\mathrm{d}s$ are augmented in $\tilde{x}(t)$ of LKF (12) compared to [20] and (ii) in the LKF derivative a tighter bound of the quadratic integral term $h_D \int_{t-h_2}^{t-h_1} \dot{x}^T(s) R_2 \dot{x}(s)\mathrm{d}s$ is established using WBII and extended RCMI.

4 A Numerical Example

In this section, the proposed stabilization criterion is validated using a numerical example.

Example 1 Consider a linear time-varying delay system (23) described as follows.

$$A = \begin{bmatrix} 0 & 0 \\ 0 & 1 \end{bmatrix}, A_d = \begin{bmatrix} -1 & -1 \\ 0 & -0.9 \end{bmatrix}, B = \begin{bmatrix} 0 \\ 1 \end{bmatrix} \tag{23}$$

Tables 1 and 2 give the numerical results obtained for system (23). From Tables 1 and 2, the following observations are made (i) MPDUB h_2 is larger in Theorem 1 compared to [19, 20] which results in reduced conservatism, (ii) control effort, which

Table 1 MPDUB h_2, controller gain K and control effort for $h_1 = 0.5$, $\mu_1 = 0$, $\mu_2 = 0.5$

Method	h_2	K	Control effort
[19]	0.967	$\begin{bmatrix} -0.5290 & -2.7166 \end{bmatrix}$	7.6598
[20]	1.535	$\begin{bmatrix} -0.2172 & -2.0351 \end{bmatrix}$	4.1888
Theorem 1	1.798	$\begin{bmatrix} -0.0513 & -1.4228 \end{bmatrix}$	2.0270

Table 2 MPDUB h_2, controller gain K and control effort for $h_1 = 0.5$, $\mu_1 = 0$, $\mu_2 = 0.9$

Method	h_2	K	Control effort
[19]	0.776	$\begin{bmatrix} -0.2628 & -2.4470 \end{bmatrix}$	6.0569
[20]	1.476	$\begin{bmatrix} -0.1930 & -1.8068 \end{bmatrix}$	3.3018
Theorem 1	1.781	$\begin{bmatrix} -0.0573 & -1.4149 \end{bmatrix}$	2.0052

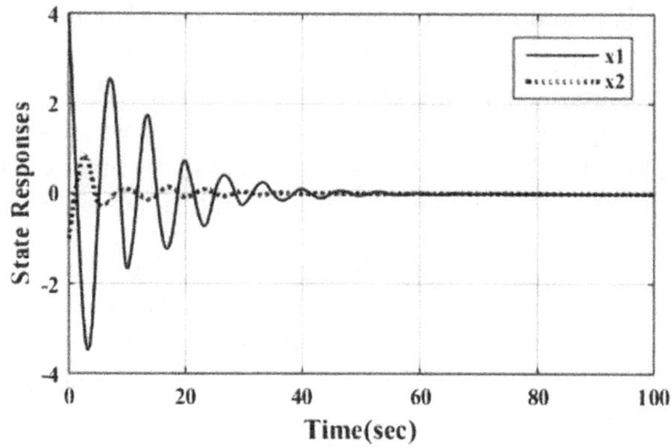

Fig. 1 State trajectories

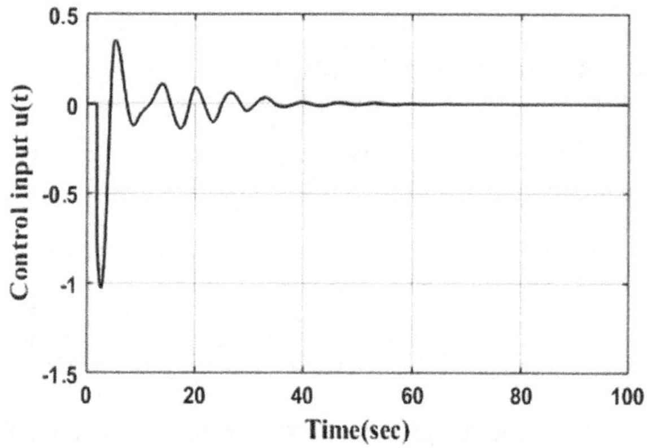

Fig. 2 Control input

is calculated as the maximum eigenvalue of $K^T K$, is reduced in Theorem 1 compared to [19, 20] and (iii) MPDUB h_2 decreases with increase in the maximum value of delay-derivative for the same delay lower bound h_1.

Also, the state trajectories and the corresponding control input has been plotted as shown in Figs. 1 and 2 with $h_1 = 0.5$, $h_2 = 1.798$, $\mu_1 = 0$, $\mu_2 = 0.5$, $x(t) = \phi(t) = [4, -1]^T$, $\forall t \in [-h_2, 0]$,$d(t) = 1 + 0.798|\sin(0.6265t)|\mu_1 = 0$ and $K = \begin{bmatrix} -0.0513 & -1.4228 \end{bmatrix}$. It is clear from Fig. 1 that the closed-loop system state trajectories are asymptotically converging to the origin.

5 Conclusions

An improved delay-dependent stabilization controller is designed by using a new LKF. From the numerical results it is verified that the proposed stabilization controller is less conservative and also control effort has been reduced with the compared techniques. Practical control problems can be solved using the proposed work by considering unmeasurable states, parametric uncertainties and external disturbances in the system. Further development in stabilization of practical systems using the proposed method will be performed in the future.

References

1. Jiang L, Yao W, Wu QH (2012) Delay-dependent stability for load frequency control with constant and time-varying delays. IEEE Trans Power Sys 27(4):932–941
2. Ramakrishnan K, Ray G (2016) Stability criterion for networked control systems with additive time-varying state-delays and bounded nonlinearity. Int J Syst Control Commun 7(1):68–82
3. Gu K, Kharitonov VL, Chen J, Shao H (2003) Stability analysis of time-delay system. Birkhauser, Boston
4. Han QL, Zhang XM, Seuret A, Gouaisbaut F (2019) An overview of recent advances in stability of linear systems with time-varying delays. IET Control Theor Appl 13(1):1–16
5. Gu K (2000) An integral inequality in the stability problem of time-delay systems. In: Proceedings of the 39th IEEE conference on decision and control, IEEE Press, Sydney, pp 2805–2810
6. Park P, Lee W, Lee SY (2015) Auxiliary function-based integral inequalities for quadratic functions and their applications to time-delay systems. J Franklin Inst 352(4):1378–1396
7. Seuret A, Gouaisbaut F (2013) Wirtinger based integral inequality: application to time-delay systems. Automatica 49:2860–2866
8. Park P, Jeong K, Jeong C (2011) Reciprocally convex approach to the stability of systems with time-varying delays. Automatica 47(1):235–238
9. Zhang CK, He Y, Jiang L, Wu M, Wang QG (2017) An extended reciprocally convex matrix inequality for stability analysis of systems with time-varying delay. Automatica 85:481–485
10. Seuret A, liu K, Gouaisbaut E (2018) Generalized reciprocally convex combination lemmas and its application to time-delay systems. Automatica 95:488–493
11. Long F, Lin WJ, He Y, Jiang L, Wu M (2020) Stability analysis of linear systems with time-varying delay via a quadratic function negative-definiteness determination method. IET Control Theor Appl 14(11):1478–1485
12. Mahto SC, Elavarasan RM, Ghosh S, Saket RK, Hossain E, Nagar SK (2020) Improved stability criteria for time-varying delay system using second and first order polynomials. IEEE Access 8(21):961–969
13. Kwon OM, Park MJ, Park JH, Lee SM, Cha EJ (2014) Improved results on stability of linear systems with time-varying delays via Wirtinger-based integral inequality. J Franklin Inst 351(12):5386–5398
14. Zhang CK, He Y, Jiang L, Wu M (2017) Notes on stability of time-delay systems: Bounding inequalities and augmented Lyapunov–Krasovskii functionals. IEEE Tans Auto Control 62(10):2768–2772
15. Mohajerpoor R, Shanmugam L, Abdi H, Rakkiyappan R, Nahavandi S, Shi P (2018) New delay range-dependent stability criteria for interval time-varying delay systems via Wirtinger-based inequalities. Int J Robust Control 28(2):661–677
16. Zhang L, Wang S (2018) Refined Wirtinger-type integral inequality. J Inequal Appl 109.https://doi.org/10.1186/s13660-018-1700-4
17. Venkatesh M, Patra S, Ramakrishnan K, Ray G (2018) An improved stability result for linear time-delay system using a new Lyapunov–Krasovskii functional and extended reciprocally convex inequality. Int J Syst Sci 49(12):2586–2600
18. Liu K, Fridman E (2012) Wirtinger's inequality and Lyapunov-based sampled data-stabilization. Automatica 48:102–108
19. Zhang J, Xia Y, Shi P, Mahmoud MS (2011) New results on stability and stabilisation of systems with interval time-varying delay. IET Control Theor Appl 5(3):429–436

20. Venkatesh M, Patra S, Ray G (2018) Stabilization of uncertain linear system with time-varying delay using a new Lyapunov–Krasovskii functional. In: Proceedings of TENCON 2018–2018 IEEE region 10 conference, IEEE Press, Jeju, pp 205–210
21. Lian Z, He Y, Zhang CK, Wu M (2019) Stability and stabilization of T-S fuzzy systems with time-varying delays via delay-product-type functional method. IEEE Trans Cyber 50(6):2580–2589

Load Frequency Control of Multi-area Thermal Power System Using Grey Wolf Optimization

Narender Saini and Jyoti Ohri

Abstract Load frequency control (LFC) immensely plays a serious role in supplying quality power in an interconnected power system. So in this regard, for better control of frequency in a multi-area power system, an effort has been made by designing a proportional-integral-derivative (PID) controller using the grey wolf optimization (GWO) algorithm. The controller is designed for a two area power system consisting of non-reheated turbines with the generation rate constraint (GRC) effect. The GWO-based PID controller gives superior performance in improving the system's parameters whilst comparing with the results of other optimization techniques such as genetic algorithm (GA), Firefly algorithm (FA), conventional Ziegler Nichols (ZN), teaching learning-based optimization (TLBO) and bacteria foraging optimization algorithm (BFOA) for the same integrated power system. From the simulation results, it can be seen that the GWO approach's performance provides better dynamic responses in steady-state error, settling time and integral time absolute error (ITAE).

Keywords Load frequency control (LFC) · PID controller · Grey wolf optimization (GWO) · Frequency deviation

1 Introduction

The power system is one of the vast networks with several integrated systems. An integrated electrical power station aims to generate, transport and distribute electrical energy at the desired system's frequency (50 Hz in India) and voltage. Power system control theory reveals that generated real power and reactive power control the system's frequency and terminal voltage. The mismatch between load demand and generated power leads to disturbing the power system stability. This deviates the frequency from the desired value. The LFC or automatic generation control

N. Saini (✉) · J. Ohri
National Institute of Technology Kurukshetra, Kurukshetra 136119, India

J. Ohri
e-mail: ohrijyoti@nitkkr.ac.in

S. Suhag et al. (eds.), *Control and Measurement Applications for Smart Grid*,
Lecture Notes in Electrical Engineering 822,
https://doi.org/10.1007/978-981-16-7664-2_28

(AGC)'s primary purpose is to retain a zero frequency deviations by maintaining proper symmetry between the power generated and power required by regulating two control loops. The speed governor acts as a primary control and the conventional PID, integral (I), and the proportional plus integral (PI) controllers are generally employed as secondary/supplementary control in the AGC to reduce the system frequency variations. Tuning or selection of gains of controller is uphill task as the functioning of controller largely depends on the gains. Metaheuristic optimization techniques like bat algorithm (BA) [1], FA [2], GA [3], GWO [4], artificial bee colony (ABC) optimization [5], ZN technique [6], fuzzy logic [7], etc., are employed in the literature for the optimal selection of gains or controller.. Because of simplicity and local optima avoidance of the metaheuristic techniques, these algorithms are mostly used by the researchers.

Literature survey reveals that various researchers have successfully designed numerous control approaches over the past decade for the LFC problem. The first modern optimal control concept was presented by Elgerd and Fosha [8] for AGC of multi-area power system. Kothari et al. [9] introduced the frequency control of two area single unit thermal power station consisting of reheat turbine and GRC by using the classical controllers. To reduce the frequency error in two area integrated thermal power system, Çam and Kocaarslan [7] designed a fuzzy gain scheduled PI (FGPI) controller. Wang et al. [10] presented AGC of a single area thermal power system considering GRC effect. Demirören and Zeynelgil [3] tried to optimize integral gains by GA for a modified three-area single unit interconnected power system. Padhan and Majhi [11] presented the PID controller for the LFC of the single and multi-area power system. By expanding the transfer function of controller with the Laurent series, author finds out the optimal control parameters. According to Gozde and Taplamacioglu [12], the PI controller optimized by the use of craziness-based particle swarm optimization (CRAZYPSO) improves the system's performance. Das et al. [1] designed a proportional-derivative–PID (PD–PID) cascade controller by using BA optimization approach for LFC of an interconnected power system. According to Ismayil et al. [13], GA optimized fractional order PID (FOPID) controller provides better performances compared with integer-order control. Sharma et al. [14] had shown the effectiveness of solar thermal power plant (STPP) in AGC of a multi-area system provided with reheated turbine and GRC. Out of the three areas, one area is considered a STPP. PID, integral and PI controller performance are measured in the system. TLBO technique is applied by R. Sahu et al. for the LFC of multi-area power system [5]. Jagatheesan et al. [2] presented five areas reheated thermal power station with PID as a secondary controller. The PID gain values are optimized with nature bio-inspired FA. The multi-objective optimization problem (MOP) was proposed by Daneshfar and Bervani, and GA is used to design and analyze a well-tuned PI controllers in three-area single unit system. The proposed control method was implemented in the three-area system, so that the LFC issue can be resolved [15]. The method of ABC optimization was used by Gozde et al. [16] to evaluate the efficiency of AGC of the two area power station.

From the above discussion, it is clear that the most frequent controller used by the researchers is the PID controller to control the error and improve the system's performance. The only difficult task is to determine optimal gain values. So gain value selection is considered an optimization problem, and optimization techniques are employed to determine optimal gain. The present study aimed to reduce the frequency deviations of the tested system and find out the gain values by using the GWO algorithm. The effectiveness of GWO is recognized by contrasting the responses with other metaheuristic and non-metaheuristic optimization approaches described in literature for the same system. The organization of this paper is as follows, Sect. 2 represents the modelling of the tested system. In Sect. 3, brief information about GWO algorithm is given. Simulation results obtained by GWO are provided in next Section, and lastly, in Sect. 5, the conclusion of this paper is given.

2 Power System Model

As shown in Fig. 1, the current work considers a two area single unit thermal power system model provided with non-reheated turbine and GRC effect. The literature survey reveals that the structure is commonly used for the design and testing of AGC. Symbols used in this system are defined as below.

B_1 and B_2	frequency bias parameters of area-1 and area-2.
T_{12}	synchronizing coefficient.
R_1 and R_2	speed regulation parameters (p.u.).
T_{t1} and T_{t2}	time constants of non-reheat turbine (s).
ACE_1 and ACE_2	area control errors.
ΔP_{D1} and ΔP_{D2}	change in load demand.
ΔP_{tie}	incremental tie-line power change (p.u.).

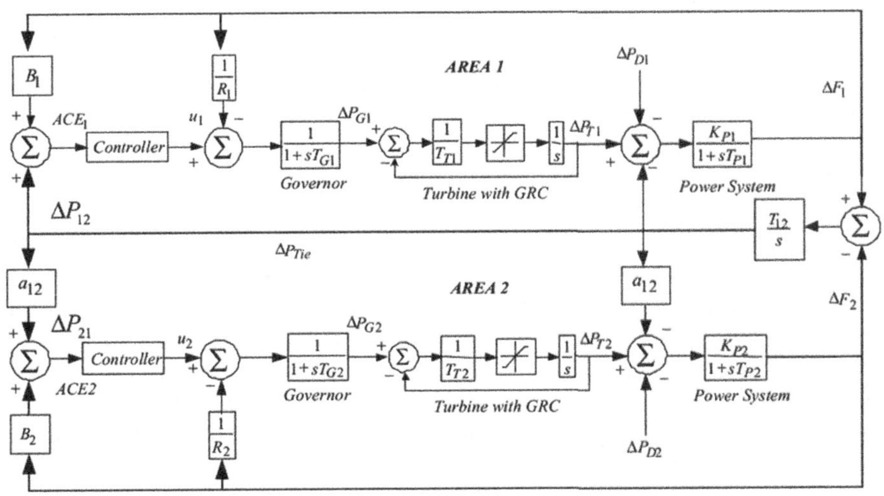

Fig. 1 Two area thermal power system model with non-reheat turbine and GRC

Table 1 Parameters value for test system

$T_{T1} = 0.3$ s	$T_{T2} = 0.3$ s
$R_1 = 2.4$ p.u	$R_2 = 2.4$ p.u
$K_{P1} = 120$	$K_{P2} = 120$
$a_{12} = -1$	$T_{12} = 0.545$
$T_{G1} = 0.08$ s	$T_{G2} = 0.08$ s
$T_{P1} = 20$ s	$T_{P2} = 20$ s
$B_1 = 0.425$ p.u	$B_2 = 0.425$ p.u

T_{g1} and T_{g2} speed governor time constants (s).
K_{p1} and K_{p2} power system gain.
u_1 and u_2 control inputs derived from the control outputs.
T_{P1} and T_{P2} power system time constants (s).
Δf_1 and Δf_2 change in system frequency (Hz).

The control input u_1 and u_2 are given as follow;

$$u_1 = K_{p1}\left(\text{ACE}_1 + \frac{1}{sT_{i1}}\text{ACE}_1 + sT_{d1}\text{ACE}_1\right) \tag{1}$$

$$u_2 = K_{p2}\left(\text{ACE}_2 + \frac{1}{sT_{i2}}\text{ACE}_2 + sT_{d2}\text{ACE}_2\right) \tag{2}$$

The parameters values for test system [5] are stated in Table 1.

The ACE signal is comprised of the incremental tie-line power change and the associated control area frequency error and is given by

$$\text{ACE}_1 = B_1\Delta f_1 + P_{\text{tie1}} \tag{3}$$

$$\text{ACE}_2 = B_2\Delta f_2 + P_{\text{tie2}} \tag{4}$$

A PID controller which is utilized as secondary source aims to reduce the tie-line power variations/imbalances and frequency error to zero in no time after the unexpected change in load demand. Moreover, its optimal parameters values are determined with GWO approach. Whenever any sudden load change interrupted the controlled areas, the controller movement is called by the required ACE in each area such that ACE can be set back to zero. Designing of good controllers primarily based totally on a suitable choice of the objective function. Integral absolute error (IAE), integral squared error (ISE), ITAE and integral time multiplied squared error (ITSE) are performance criterion commonly employed in the control design. ITAE objective function is selected in this work [5]. The expression for the ITAE incorporating the frequency change and tie-line power change is given as below.

$$\text{Obj.} = \text{ITAE} = \int_0^{t_{\text{sim}}} (|\Delta f_i| + |\Delta P_{\text{tie}-i-k}|) \cdot t\text{d}t \tag{5}$$

where

t_{sim} is the simulation time.
$\Delta P_{\text{tie}-i-k}$ is change in the tie-line power connecting i_{th} and k_{th} area.
Δf_i is change in frequency in i_{th} area.

3 Grey Wolf Optimization

GWO [4] is a metaheuristic approach that shows the leadership order and hunting behaviour of grey wolves in nature planned by Mirjalili et al. [4]. They are considered top predators, meaning they are at the food chain's top level. Grey wolves generally wish to survive in groups (packs). On average, each pack contains 6–14 members. Every member of the group has a strict social leading hierarchy. The social order includes four levels of wolves. Alpha (α) is the first level. They are the head of the pack and responsible for making all the decisions. The next level is known as beta (β). They are the subordinate wolves. The next level is delta (δ), and they have to follow α and β, but they control the omega. Omega (ω) is the fourth or the lowest level. They have to follow the alpha, beta and delta [4]. The hunting process by the grey wolf is tracking the target, chasing and approaching it, and finally, encircle and harassed it until prey stops moving then attacks towards the target. α, β and δ guess the location of the target, and then the remaining wolves renovate their place with respect to the positions of the finest search agents.

3.1 Mathematical Model of GWO Algorithm

The fittest solution in the GWO is considered the alpha; the next two most appropriate solutions are, respectively, called beta and delta, and the remaining solutions are named as omega. Hunting is generally guided by the α, β and δ in the GWO algorithm. These three are followed by the ω. For the period of the hunting, wolves encircle the target. The mathematical equations for encircling behaviour are

$$D_j^{(t)} = \left| C_i.P_{ij}^{*(t)} - P_{ij}^{(t)} \right| \tag{6}$$

$$P_{ij}^{(t+1)} = P_{ij}^{*(t)} - A_i.D_j^{(t)} \tag{7}$$

where

$i = 1, 2, 3, 4, 5 \ldots N_p$ and $j = 1, 2, 3, 4 \ldots N$.

t	is current iteration.
A_i and C_i	are the coefficient vectors.
P_{ij}^*	is the finest result obtained so far.

Vectors A_i and vector C_i are considered as follow

$$A_i = 2a.r_i - a \tag{8}$$

$$C_i = 2.r_i \tag{9}$$

where

r is the random vectors in [0,1].

The β and δ wolves might occasionally take part in hunting. α, β and δ have a more robust knowledge of the prey's location [4]. The top first three solutions, such as alpha, beta and delta, have saved, and the further agents have to renovate their place as per the location of the finest search agents;

$$D_{\alpha j}^t = \left|C_1 \cdot P_{\alpha j}^t - P_{ij}^t\right| \tag{10}$$

$$P_{1j}^t = P_{\alpha j}^t - A_1 D_{\alpha j}^t \tag{11}$$

$$D_{\beta j}^t = \left|C_1 \cdot P_{\beta j}^t - P_{ij}^t\right| \tag{12}$$

$$P_{2j}^t = P_{\beta j}^t - A_2 D_{\beta j}^t \tag{13}$$

$$D_{\delta j}^t = \left|C_1 \cdot P_{\delta j}^t - P_{ij}^t\right| \tag{14}$$

$$P_{3j}^t = P_{\delta j}^t - A_3 D_{\delta j}^t \tag{15}$$

$$P_{ij}{}^{t+1} = \frac{P_{1j}^t + P_{2j}^t + P_{3j}^t}{3} \tag{16}$$

When the movement of the target has stopped, the grey wolves attack it and terminate the search. When the magnitude of A is less than 1, wolves hit the target, representing an exploitation process. To look for the prey, they diverge from each other and converge to hit the target. When magnitude of A is more than 1, the wolves have to move away from the target to find another prey, representing an exploration process. Vector C also assists the exploration process. This can be seen from Eq. (9)

that vector C contains the random value in [0,2]. This allows GWO to show more random actions during optimization in favour of exploration and avoid local optima.

The conclusion is that by randomly creating the population of the grey wolves, the search process begins. α, β and δ approximate the expected location of the target throughout the iteration. Each solution renovates its remoteness from the prey. The parameter 'a' emphasized the exploitation and exploration. The solutions tend to move away from the target when magnitude of A is more than 1, and come together to the prey when magnitude of A is less than 1. At last, the GWO end by the satisfaction of an end condition [4].

4 Simulation Results and Discussion

The two area thermal power system model is designed by using MATLAB (Simulink) platform, and the technique to optimize the controller's gain, i.e. GWO algorithm code is written in a separate file. The load change in area-1 is considered as 5%. The ITAE-based objective function is considered in the optimization process. The population size and number of iterations are taken as 30 and 100, respectively, in this problem. The GWO algorithm's flow chart is shown in Fig. 2.

PID controller is provided with each area separately, and in the each case, PID controller's parameters are optimized at the same time by using the GWO algorithm. The process of optimization is repeated 20 times, and the finest result acquired amongst the 20 runs is selected. The final optimal gain values obtained are as follow;

For area-1 $K_{p1} = 1.44$, $K_{i1} = 1.998$, $K_{d1} = 0.608$;

For area-2 $K_{p2} = 0.2324$, $K_{i2} = 0.0005$, $K_{d2} = 0.251$.

The frequency change in both areas, i.e. Δf_1 and Δf_2 and tie-line power obtained from this simulation experiment using above obtained PID gains are displayed from Figs. 4, 5 and 6, respectively.

The convergence graph of the objective function (ITAE) by using the GWO technique is revealed in Fig. 3. The settling time (T_s) and ITAE values are obtained from Figs. 3, 4, 5 and 6, respectively. Figure 7 show the comparison based on objective i.e. Integral Time Absolute Error and settling time. From the bar graph shown below, one can observe that the GWO algorithm optimized controller provides much better response in respect of tie-line power deviations and settling times in frequency as compared with existing methods.

5 Conclusion

In this paper, GWO is applied to find optimal gain of the PID controller for LFC of two area single unit power system under GRC for 5% load change of area-1. An ITAE of the variations in frequency in both areas and tie-line power change is taken as the objective function to improve the system's performance. The effectiveness of

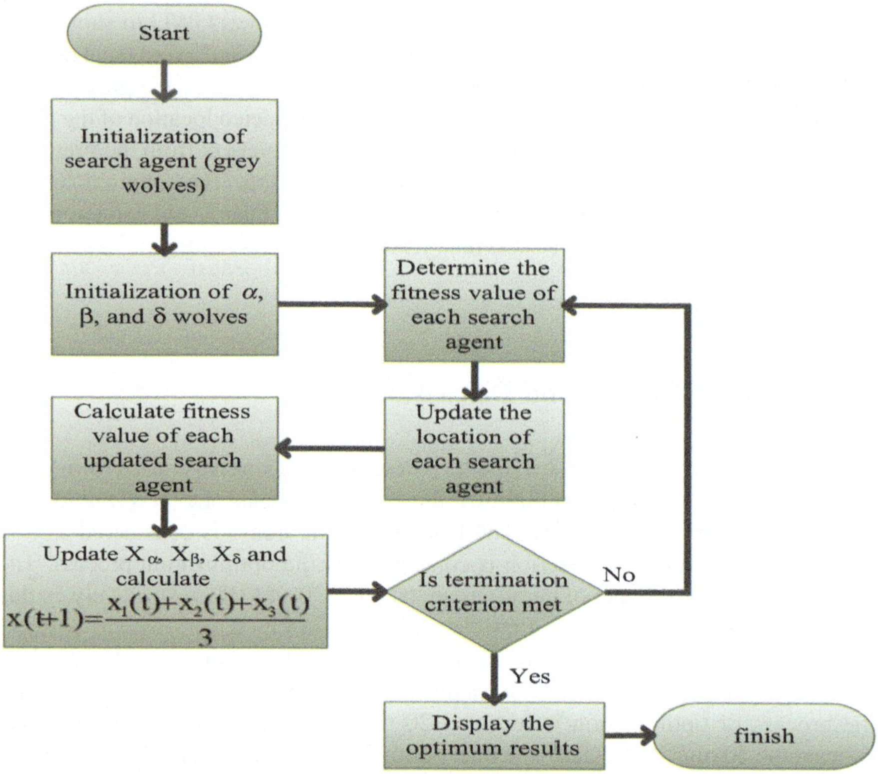

Fig. 2 Flow chart of GWO

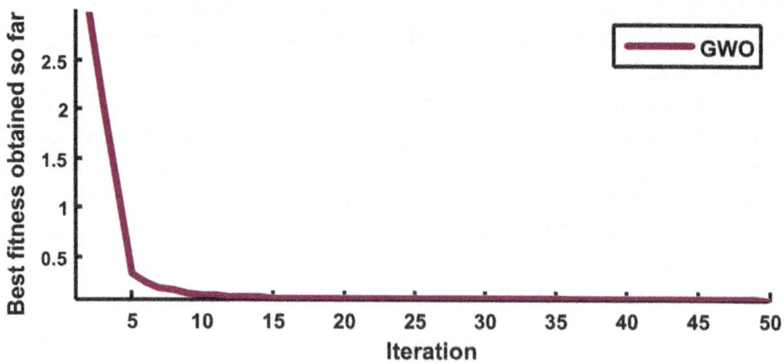

Fig. 3 Convergence of cost curve

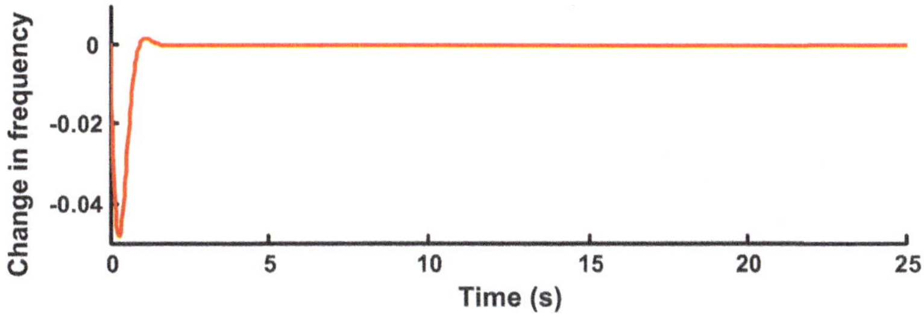

Fig. 4 Frequency change (Δf_1) of area-1

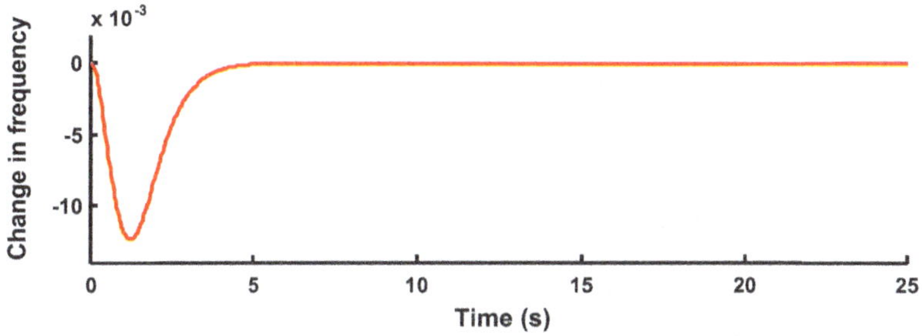

Fig. 5 Frequency change Δf_2) of area-2

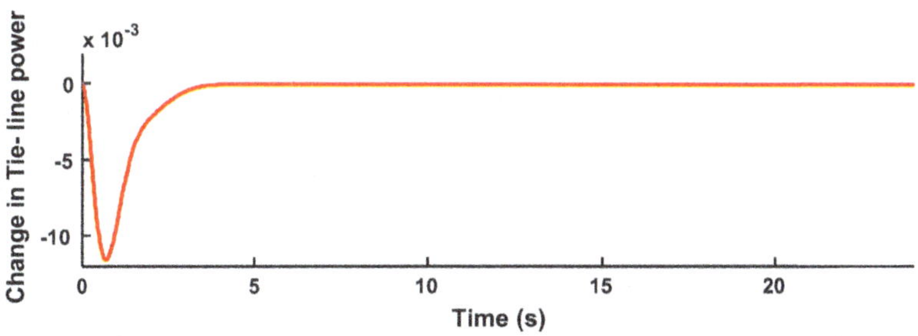

Fig. 6 Tie-line power change (ΔP_{tie}) transmit from area-1

the method in order to minimize the ITAE and settling time is compared with other metaheuristic and non-metaheuristic methods like ZN, GA, BFOA, FA and TLBO and clearly presented in Fig. 7 which proved that the presented work is fruitful than other.

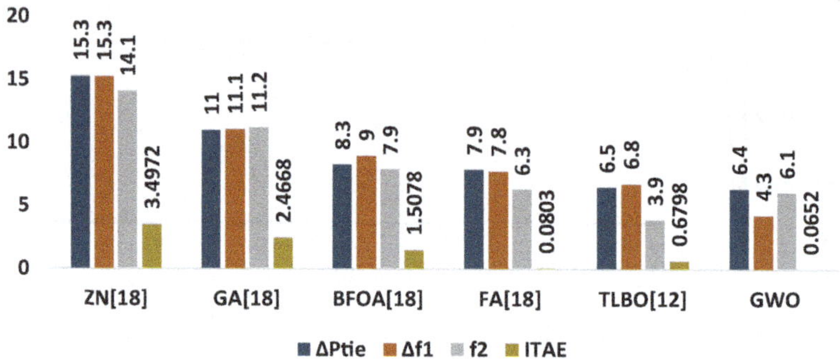

Fig. 7 Comparison of GWO performance based on error and settling time

References

1. Dash P, Chandra SL, Nidul S (2015) Automatic generation control of multi area thermal system using Bat algorithm optimized PD–PID cascade controller. Int J Electr Power Energy Syst 68:364–78
2. Jagatheesan K, Anad B, Samanta S, Dey N, Ashour AS, Balas VE (2019) Design of a proportional integral derivative controller for an automatic generation control of multi-area power thermal system using firefly algorithm, IEEE/CAA J Autom Sin 6(2)
3. Demirören A, Zeynelgil HL (2007) GA application to optimization of AGC in three-area power system after deregulation. Int J Electr Power Energy Syst 29:230–240
4. Mirjalili S, Mirjalili SM, Lewis A (2014) Grey wolf optimization. Adv Eng Softw 69:46–61
5. Sahu RK, Sekhar GT, Panda S (2016) Automatic generation control of multi-area power systems with diverse energy sources using teaching learning based optimization algorithm. Int J Electr Power Energy Syst 19:113–134
6. Ali ES, Abd Elazim SM (2013) BFOA based design PID controller for two-area load frequency control with nonlinearities. Int J Electr Power Energy Syst 51:224–231
7. Çam E, Kocaarslan I (2005) A fuzzy gain scheduling PI controller application for an interconnected electrical power system. Electr Power Syst Res 73:267–274
8. Elgerd OI, Fosha CE (1970) Optimum megawatt-frequency control of multi-area electric energy systems. IEEE Trans Power Appar Syst 89(4):556–563
9. Kothari ML, Satasngi PS, Nanda J (1981) Sampled-data automatic generation control of interconnected reheat thermal systems considering generation rate constraints. IEEE Trans Power Appar Syst 100(5):2334–2342
10. Wang Y, Zhou R, Wen C (1994) New robust adaptive load-frequency control with system parametric uncertainties. IEEE Trans Gener Transmission Distrib 141(3):184–190
11. Padhan DG, Majhi S (2013) A new control scheme for PID load frequency controller of single-area and multi-area power systems. ISA Trans 52(2):242–251
12. Gozde H, Taplamacioglu MC (2011) Automatic generation control application with craziness based particle swarm optimization in a thermal power system. Electr Power Energy Syst 33(1):8–16
13. Ismayil C, Sreerama Kumar R, Sindhu TK (2014) Automatic generation control of single area thermal power system with fractional order PID controllers. In: 3rd international conference on advances in control and optimization of dynamical systems, Kanpur
14. Sharma Y, Saikia LC (2015) Automatic generation control of a multi-area ST–thermal power system using grey wolf optimizer algorithm based classical controllers. Int J Electr Power Energy Syst 73:853–862

15. Daneshfar F, Bervani H (2012) Multi objective design of load frequency control using genetic algorithms. Electr Power Energy Syst 42:257–263
16. Gozde H, Taplamacioglu MC, Kocaarslan I (2012) Comparative performance analysis of artificial bee colony algorithm in automatic generation control for interconnected reheat thermal power system. Electr Power Energy Syst 42:167–178
17. Nagrath and Gopal (2008) Control system engineering. Anshan

Load Frequency Control of Two Area Deregulated Power System Using Hybrid PSOGSA-TVAC Algorithm

Sushrut Brahmachari and Saurabh Chanana

Abstract In power system, the load frequency control (LFC) is necessary to achieve a good quality and reliability of power supply. LFC main purpose is to preserve the frequency and tie-line power exchange within a specified limit. It is done through controlling the output of the generators for the consequent demand variation. The power industry is, nowadays, an open market configuration. This is known as power system deregulation. In deregulated environment, there are several GENCOs, TRANSCOs, and DISCOs, supervised under independent service operator (ISO). This manuscript presents the investigation of load frequency control (LFC) of a two area deregulated power system using the combination of particle swarm optimization with time-varying acceleration coefficient (PSO-TVAC) and gravitational search algorithm (GSA). The model consisting of GENCOs and DISCOs under deregulated market structure based on DISCO participation matrix (DPM) is designed and examined. The PID controller is used for tuning, and system's dynamic response is enhanced by minimizing the integral time squared error (ITSE). The simulation/programming is done in MATLAB.

Keywords Load frequency control (LFC) · Deregulation · Particle swarm optimization (PSO-TVAC) · Gravitational search algorithm (GSA) · Area control error (ACE)

1 Introduction

Power system generally experiences a power imbalance in between load demand and scheduled generation, leading to frequency fluctuations in the system. The changes of loads create frequencies of the par value to decrease/increase. The

S. Brahmachari (✉) · S. Chanana
Department of Electrical Engineering, NIT Kurukshetra, Kurukshetra, Haryana 136119, India
e-mail: sushrut_31904204@nitkkr.ac.in

S. Chanana
e-mail: saurabh@nitkkr.ac.in

S. Suhag et al. (eds.), *Control and Measurement Applications for Smart Grid*,
Lecture Notes in Electrical Engineering 822,
https://doi.org/10.1007/978-981-16-7664-2_29

frequency fluctuation is negligible and governable under normal conditions, however, frequency fluctuation undergoes above the limit, it adversely affects the operation of the system and reliability of the system. The interconnected power system when load varies, it results in fluctuation of both frequency and tie-line power flow. Therefore, the issue of LFC in an interconnected structure turn out more complicated. In this load frequency control, it is strictly linked to frequencies stabilization and condemnation of the error of unite power interchange in between different areas at fixed value so the frequency divergence is controlled in its limit [1].

As conventional electrical utilities have shifted from an interconnected vertical structure toward a free trade power system, which consisting GENCOs, TRANSCOs and DISCOs companies, and autonomous structure operators (ASO). The regulated power system structure, the power is transferred between GENCOS, DISCOs, and a power sharing agreements connecting them are signed. Those commitment connecting GENCO and DISCO may be consensual, poolco, or violations of these contracts may occur sometimes. In an interconnected free traded power structure with multiple power source and multiple-areas, an immediate shift in the demands for load in any area results in a changes in the frequency and the contractual sharing of power among different areas. The exchange of frequency and contractual of power in between unlike areas should be constant for the victorious functioning of the power system. During contract violations, load frequency control adjourns the creations of each GENCO to reopen the frequency and decrease the accidental flows of power [2]. Thus, the foremost purpose of load frequency control is to regulate the structure repetition at theoretical value and constitutional power interchange between dissimilar areas.

Due to its accuracy and speed, most of the researchers prefer PID controller in order to address LFC problem [3]. Since, the presentation of PID is straight proportional to its variables tuning. That is why, authors utilized artificial intelligence formed process such as fuzzy [4], GA [5], PSO [6], honey bee [7], DE [8], and GSA [9] to tune the parameters of PID in order to optimize the controller gain.

In this paper, a two area multisource system is implemented undergoing isolated marketing structure. Simulation is carried out under several normal operating circumstances and contract violation for tuning PID controller using TVAC-PSO, GSA, and combination of the both [10, 11]. The result shows that the hybrid approach to tune PID controller's parameters is better.

The following is the order for rest of the paper. Section 2 explains power system under deregulation in details. The formulation of the objective function is described in Sect. 3. Section 4 explains the proposed optimization algorithm to tune PID parameters along with the flowchart. The formulation of the objective function is described in Sect. 4. Sections 5 and 6 present the simulation results and the comparison between the results of the techniques. And at the last, Sect. 7 contains the conclusion of the work.

2 System Modeling Under Deregulated Environment

The system under deregulation consists of several GENCOs and DISCOs. Any DISCO can deal an agreement with any GENCO for power transaction to fulfill load demand. If a DISCO deals an agreement with a GENCO of the matching control area, it is known as “poolco” transaction and if both companies are of different control area, then this transaction type is known as “bilateral” transaction, which is administered by the ISO. In this study, two areas are considered under deregulated structure. Both the areas are identical and consist of a thermal non-reheat generation and a hydro generation unit.

DISCO participation matrix (DPM) shows the agreements made between DISCOs and GENCOs. In a DPM, rows signify the entire generation companies (GENCOs) and columns signify the entire distribution companies (DISCOs) contributing in the system. The entries in the DPM matrix are known as participation factor, i.e., portion of the overall agreement by a DISCO with a GENCO. Equation (1) demonstrates the DPM for a deregulated power system, where *ij*th element of the DPM constitutes the portion of the overall agreement by the DISCO “*j*” with the GENCO “” and summation of all the element along any column should be unity as shown by Eq. (2). A sample model for study is taken, as shown in Fig. 1. The assumption is taken that all the areas taken are identical and each consists of two GENCOs and two DISCOs.

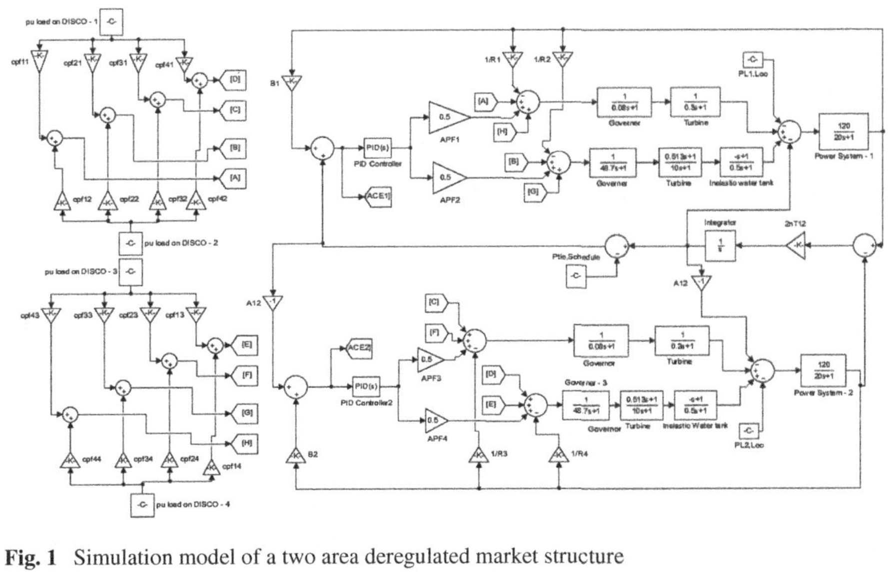

Fig. 1 Simulation model of a two area deregulated market structure

$$\begin{bmatrix} cpf_{11} & cpf_{12} & cpf_{13} & cpf_{14} \\ cpf_{21} & cpf_{22} & cpf_{23} & cpf_{24} \\ cpf_{31} & cpf_{32} & cpf_{33} & cpf_{34} \\ cpf_{41} & cpf_{42} & cpf_{43} & cpf_{44} \end{bmatrix} \tag{1}$$

$$\sum cpf_{ij} = 1 \tag{2}$$

The power generated as well as power intended for each generating unit for power exchange between various areas can be found with the help of the DPM. Power requirement of a DISCO "j" can be represented as ΔP_{Lj}. The overall power required of a DISCO in area "i" can be given as the totality of contracted power and contract violation.

$$\Delta P_{Di} = \Delta P_{L1} + \Delta P_{L2} + \cdots + \Delta P_{Ln} + \Delta P_{\text{Lviolation}} \tag{3}$$

Total power generation is given as

$$\Delta P_{gi} = \sum_{j=1}^{n} cpf_{ij} \Delta P_{Lj} \tag{4}$$

Power committed between area-1 and area-2 is given as

$$\Delta P^{sh}_{\text{tie}_{1,2}} = \sum_{i=1}^{2} \sum_{j=3}^{4} cpf_{ij} \Delta P_{ij} - \sum_{i=3}^{4} \sum_{j=1}^{2} cpf_{ij} \Delta P_{ij} \tag{5}$$

To govern the power interchange between area-1 and area-2, the error in tie-line at any instant can be calculated as

$$\Delta P^{\text{error}}_{\text{tie}_{1,2}} = \Delta P^{\text{act}}_{\text{tie}_{1,2}} - \Delta P^{sh}_{\text{tie}_{1,2}} \tag{6}$$

In Eq. (6), $\Delta P^{\text{act}}_{\text{tie}_{1,2}}$ is the quantity of actual power transmitted between area-1 and area-2 and is given as

$$\Delta P^{\text{act}}_{\text{tie}_{1,2}} = \frac{2\pi T_2}{s}(\Delta f_1 - \Delta f_2) \tag{7}$$

In order to regulate the frequency and scheduled power among the two areas, area control error (ACE) signal is obtained. It is stated for each areas and directed to the controller of that area. The signals can be denoted as

$$\text{ACE}_1 = B_1 \Delta f_1 + \Delta P^{\text{error}}_{\text{tie}_{1,2}} \tag{8}$$

$$\text{ACE}_2 = B_2 \Delta f_2 + a_{12} \Delta P^{\text{error}}_{\text{tie}_{1,2}} \tag{9}$$

3 Problem Formulation and Objective Function

The area control error (ACE) of the two area bilateral market structure is the input to the PID controller. Equation (10) represents the output of the PID controller U_{PID}.

$$U_{\mathrm{PID}} = K_{\mathrm{p}}\mathrm{ACE}_i + K_i \int_0^t \mathrm{ACE}_i \mathrm{d}t + K_{\mathrm{d}} \frac{\mathrm{d}(\mathrm{ACE}_i)}{\mathrm{d}t} \tag{10}$$

where $K_{\mathrm{p},}$ K_i, and K_d are the proportional gain, integral gain, and derivative gain, respectively.

The performance index taken for a system is a quantitative measure of the system and is selected so that the gains of the PID controller can be tuned to get the desired specification optimally. Equation (11) represents the performance index (integral time squared error) considered for optimal tuning of the PID controller parameters (error minimization).

$$\mathrm{ITSE} = \oint_0^{\infty} t e^2(t)\mathrm{d}t \tag{11}$$

4 Hybrid PSOGSA-TVAC Approach

PSO is a popular technique based on particle's position and population [6]. Gravitational search algorithm [9] is related to Newton's theory. A hybrid approach of both the algorithms (PSOGSA) [11] empowers the algorithm to be more robust. In this hybrid approach, time-varying acceleration coefficients ($\emptyset_1$, $\emptyset_2$) and time-varying inertial weight (w) of PSO are combined with the acceleration coefficient (a_i^k) of GSA. In PSO, the equation of position of particles is defined as

$$\begin{aligned} v_{\mathrm{p}}^k(t_{t+1}) = w.v_{\mathrm{p}}^k(t_t) + \emptyset_1.r_1.\left(p_{\mathrm{b}} - x_{\mathrm{p}}^k(t_t)\right) \\ + \emptyset_2.r_2.\left(g_{\mathrm{b}} - x_{\mathrm{p}}^k(t_t)\right) \end{aligned} \tag{12}$$

$$x_{\mathrm{p}}^k(t_{t+1}) = x_{\mathrm{p}}^k(t_t) + v_{\mathrm{p}}^k(t_{t+1}) \tag{13}$$

where $\emptyset_1$, $\emptyset_2$ are coefficient of acceleration, w is the inertial weight, r_1 and r_2 are random numbers between 0 and 1. Individual best position of the particle is represented as p_{b}, and globally best position is represented as g_{b}.

Instead of using a constant value for $Ø_1$, $Ø_2$, and w, the optimal value can be obtained by using other values also. Thus, the value of these constants is chosen as mentioned in (14), (15), and (16) [10].

$$w' = (w_f - w_i).\frac{t_{\max} - t_{\text{current}}}{t_{\max}} + w_i \tag{14}$$

$$Ø'_{1_{t+1}} = (c_{1\max} - c_{1\min}).\frac{t_{\text{current}}}{t_{\max}} + c_{1\min} \tag{15}$$

$$Ø'_{2_{t+1}} = (c_{2\max} - c_{2\min}).\frac{t_{\text{current}}}{t_{\max}} + c_{2\min} \tag{16}$$

Here, $c_{1\min}$ and $c_{2\min}$ are the initial values and $c_{1\max}$ and $c_{2\max}$ are the final values of $Ø'_{1_{t+1}}$ and $Ø'_{2_{t+1}}$, respectively. With the help of (14), (15), and (16) in (12), the governing equation of PSO-TVAC can be written as

$$v_{\text{p}}^k(t_{t+1}) = \mathsf{C}\left\{w'.v_{\text{p}}^k(t_t) + Ø'_{1_{t+1}}.r_1.\left(p_{\text{b}} - x_{\text{p}}^k(t_t)\right) + Ø'_{2_{t+1}}.r_2.\left(g_{\text{b}} - x_{\text{p}}^k(t_t)\right)\right\} \tag{17}$$

where C is known as constriction factor and its equation is as follows:

$$\mathsf{C} = \frac{2}{2 - \mu - \sqrt{\mu^2 - 4\mu}} \quad 4.1 \le \mu \le 4.2 \tag{18}$$

Equation (17) is once again modified by a new acceleration coefficient inspired from GSA [9]. The new equation is formed by exchanging $p_{\text{b}} - x_{\text{p}}^k(t_t)$ with a_{p}^k in (17).

The acceleration coefficient (a_{p}^k) is given by

$$a_{\text{p}}^k(t_{\text{t}}) = \frac{F_{\text{p}}^k(t_{\text{t}})}{M_{\text{p}}(t_{\text{t}})} \tag{19}$$

Here, $F_{\text{p}}^k(t_{\text{t}})$ is the overall external force in dimension (k) on the particle in current iteration and can be calculated by using Eq. (20), $M_{\text{p}}(t_{\text{t}})$ is the particle mass in the current iteration and $a_{\text{p}}^k(t_{\text{t}})$ is the acceleration at current iteration.

$$F_{\text{p}}^k(t_{\text{t}}) = \sum_{q=1, j \ne i}^{N} \text{rand}_q F_{pq}^k(t_{\text{t}}) \tag{20}$$

$$F_{pq}^k(t_{\text{t}}) = G(t_{\text{t}}).\frac{M_{pp}(t_{\text{t}}).M_{aq}(t_{\text{t}})}{R_{pq}(t_{\text{t}}) + \epsilon}\left(x_q^k(t_{\text{t}}) - x_{\text{p}}^k(t_{\text{t}})\right) \tag{21}$$

$$G(t_{\text{t}}) = G_0.e^{-\alpha.\frac{t_{\text{current}}}{t_{\max}}} \tag{22}$$

Here, rand_q is a random constant between 0 and 1, $M_{pp}(t_t)$ and $M_{aq}(t_t)$ are the passive and active mass p and mass q, respectively, $R_{pq}(t_t)$ is the recent location of particle "p" and "q".

Therefore, the equation of velocity of a particle is given by Eq. (23) and the equation of position is given by Eq. (13). The flow chart of the proposed algorithm is shown in the figure below (Fig. 2).

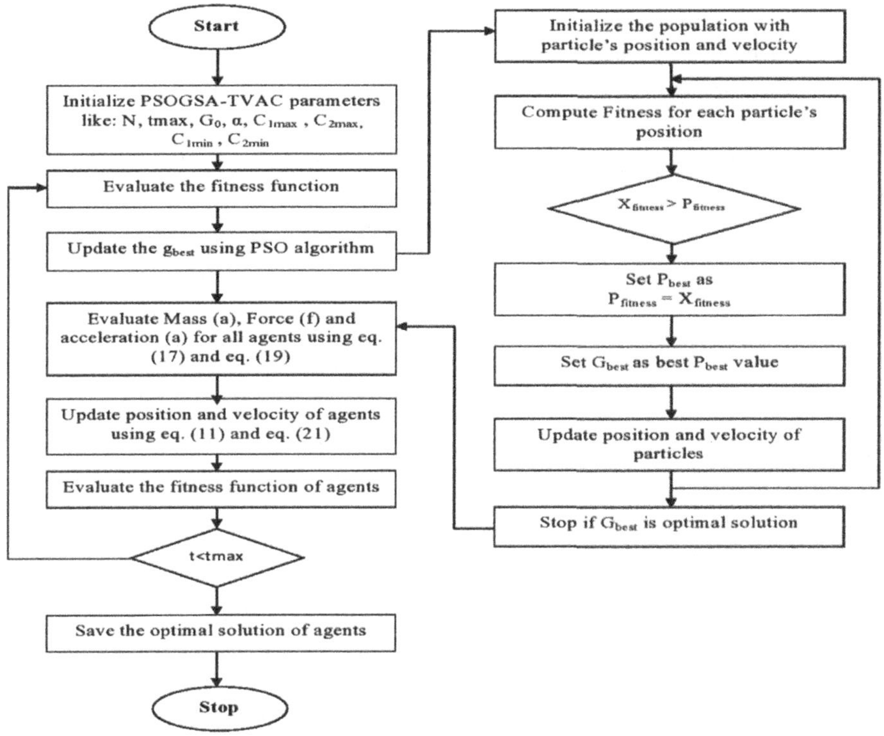

Fig. 2 Flowchart of hybrid PSOGSA-TVAC algorithm

$$v_{\mathrm{p}}^{k}(t_{t+1}) = \mathrm{C}\left\{w'.v_{\mathrm{p}}^{k}(t_{\mathrm{t}}) + \emptyset'_{1_{t+1}}.r_1.a_{\mathrm{p}}^{k}(t_{\mathrm{t}}) + \emptyset'_{2_{t+1}}.r_2.\left(g_{\mathrm{b}} - x_{\mathrm{p}}^{k}(t_{\mathrm{t}})\right)\right\} \tag{23}$$

5 Simulation

The model simulation, as shown in Fig. 1, was done by using MATLAB Simulink software. Both the areas are taken to be identical. For optimal tuning the gains of the PID controller in the above aforementioned system model, the algorithms were written in

MATLAB (.m file) and the simulation was implemented in a personal computer (PC) with the subsequent specifications: 64-bit OS with × 64-based processor (Intel(R) Core(TM) i5-7200U CPU @ 2.50 GHz) and a RAM of 4 GB. The minimum and maximum boundary value for the controller gain parameters are chosen as − 5.0 and 0, respectively. Three different scenarios are considered during this study. They are as follows:

5.1 Scenario 1: Poolco-Based Transaction in Deregulated Environment

In this case, DISCOs meet its demanded power only from the GENCOs of its local area. A load change is assumed to occur in area-1 and the GENCOs of this area participate equally (poolco transaction). Therefore, $APF_1 = 0.5$, $APF_2 = 0.5$, $APF_3 = 0.5$, $APF_4 = 0.5$.

The DPM obtained for this case is shown as follows:

$$\text{DPM} = \begin{bmatrix} 0.5 & 0.5 & 0 & 0 \\ 0.5 & 0.5 & 0 & 0 \\ 0 & 0 & 0 & 0 \\ 0 & 0 & 0 & 0 \end{bmatrix}$$

5.2 Scenario 2: Bilateral Transactions in Deregulated Environment

In this type of contracts, all the DISCOs can have a contract with the GENCOs of their own or with the other interconnected area. The APFs are considered as $APF_1 = 0.75$, $APF_2 = 0.25$, $APF_3 = 0.5$, $APF_4 = 0.5$.

The DPM obtained for this case is shown as follows:

$$\text{DPM} = \begin{bmatrix} 0.5 & 0.25 & 0 & 0.3 \\ 0.2 & 0.25 & 0 & 0 \\ 0 & 0.25 & 1 & 0.7 \\ 0.3 & 0.25 & 0 & 0 \end{bmatrix}$$

5.3 Scenario 3: Contract Violation in Deregulated Environment

A DISCO in area-2 is considered that violates the contract and has demanded 0.1 p.u. MW of surplus power which should be reflected as the load of the local area-2 and not as a contract demand.

The DPM will be same as it was in scenario-2. This un-contracted load is shown in the power generation of GENCOs of area-2. GENCOs of area-1 will be unaffected by this surplus load, which is under supervision of ISO. It is assumed that at-least one GENCO from each area is participating in LFC.

$\Delta P_{\text{tie}_{1,2}}^{\text{sch}} = -0.05$ p.u. MW as per Eq. (5) which is same as in 5.2.

6 Results

Figure 3 represents the dynamic response plot of both the areas for the first scenario as mentioned in 5.1 by tuning PID controller with PSO, GSA, and PSOGSA-TVAC algorithm. Figure 3a–c shows the frequency change and tie-line power exchange between area-1 and area-2. By investigating and comparing the frequency responses of both the areas and the tie-line power flow obtained from PSO, GSA, and PSOGSA-TVAC, it is clear that frequency deviance in each area settles to zero in steady state faster in PSOGSA-TVAC than the PSO and GSA algorithm. Also, it can be analyzed that the peak undershoot and peak overshoot are less in hybrid PSOGSA-TVAC controlled PID controller than in PSO and GSA controlled PID controller. The detailed transient response for all the three algorithms is mentioned in Table 1.

Figure 4 represents the dynamic response plot of both the areas for the second scenario as mentioned in 5.2 by tuning PID controller with PSO, GSA, and PSOGSA-TVAC algorithm. Figure 4a–c shows the frequency change and tie-line power exchange between area-1 and area-2. By investigating and comparing the frequency responses of both the areas and the tie-line power flow obtained from PSO, GSA, and PSOGSA-TVAC, it is clear that frequency deviance in each area settles to zero in steady state faster in PSOGSA-TVAC than the PSO and GSA algorithm. Also, it can be analyzed that the peak undershoot and peak overshoot are less in hybrid PSOGSA-TVAC controlled PID controller than in PSO and GSA controlled PID controller. The detailed transient response for all the three algorithms is mentioned in Table 1.

Figure 5 represents the dynamic response plot of both the areas for the third scenario as mentioned in 5.3 by tuning PID controller with PSO, GSA, and PSOGSA-TVAC algorithm. Figure 5a–c shows the frequency change and tie-line power exchange between area-1 and area-2. By investigating and comparing the frequency responses of both the areas and the tie-line power flow obtained from PSO, GSA, and PSOGSA-TVAC, it is clear that frequency deviation in each area settles to zero in steady state faster in PSOGSA-TVAC than the PSO and GSA algorithm. Also,

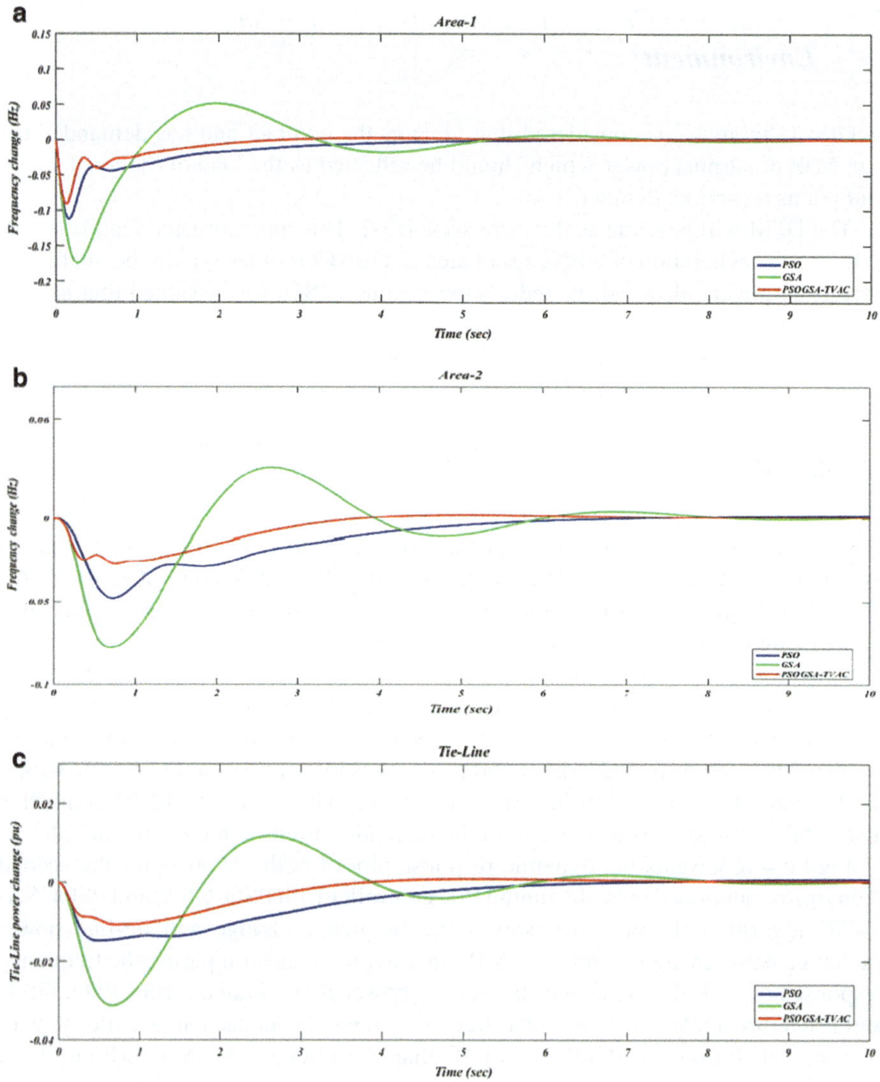

Fig. 3 **a** Frequency change of area-1 in scenario-1, **b** frequency change of area-2 in scenario-1, **c** tie-line power change in scenario-1

it can be analyzed that the peak undershoot and peak overshoot are less in hybrid PSOGSA-TVAC controlled PID controller than in PSO and GSA controlled PID controller. The detailed transient response for all the three algorithms is mentioned in Table 1.

A comparison chart is displayed in Table 1 which compares the frequency responses of both the areas and tie-line power exchange with respect to settling time, peak overshoot, and peak undershoot obtained from PSO, GSA, and hybrid

Table 1 Comparison table for frequency change and tie-line power flow for PSO, GSA, and PSOGSA-TVAC tuned PID controller

Cases	Controller	Layout	Settling time (s)	Peak overshoot	Peak undershoot
Case-1	PSO	Area-1	5.65	0 Hz	– 0.11 Hz
		Area-2	7.05	0 Hz	– 0.04 Hz
		Tie-line	7.5	0 p.u.	– 0.017 p.u.
	GSA	Area-1	6.22	0.051 Hz	– 0.0155 Hz
		Area-2	7.92	0.03 Hz	– 0.075 Hz
		Tie-line	8.46	0.015 p.u.	– 0.03 p.u.
	PSOGSA-TVAC	Area-1	4.72	0 Hz	– 0.008 Hz
		Area-2	5.15	0 Hz	– 0.025 Hz
		Tie-line	5.26	0 p.u.	– 0.01 p.u.
Case-2	PSO	Area-1	7.12	0.044 Hz	– 0.06 Hz
		Area-2	7.14	0.028 Hz	– 0.21 Hz
		Tie-line	5.8	0 p.u.	– 0.02 p.u.
	GSA	Area-1	6.72	0.018 Hz	– 0.116 Hz
		Area-2	7.28	0.008 Hz	– 0.21 Hz
		Tie-line	5.2	0 p.u.	– 0.019 p.u.
	PSOGSA-TVAC	Area-1	6.31	0.08 Hz	– 0.049 Hz
		Area-2	5.16	0.005 Hz	– 0.215 Hz
		Tie-line	4.88	0 p.u.	– 0.018 p.u.
Case-3	PSO	Area-1	9.19	0.058 Hz	– 0.122 Hz
		Area-2	6.95	0.015 Hz	– 0.325 Hz
		Tie-line	7.25	0 p.u.	– 0.015 p.u.
	GSA	Area-1	9.05	0 Hz	– 0.074 Hz
		Area-2	7.63	0 Hz	– 0.255 Hz
		Tie-line	7.19	0 p.u.	– 0.02 p.u.
	PSOGSA-TVAC	Area-1	5.86	0.082 Hz	– 0.044 Hz
		Area-2	4.52	0.005 Hz	– 0.245 Hz
		Tie-line	4.69	0 p.u.	– 0.025 p.u.

PSOGSA-TVAC tuned PID controller for poolco-based transaction, bilateral transaction, and contract violation case under power system deregulation.

7 Conclusion

This paper work explores the functioning of a PID-based controller and parameters are tuned using hybrid PSOGSA-TVAC optimization algorithm for the purpose of

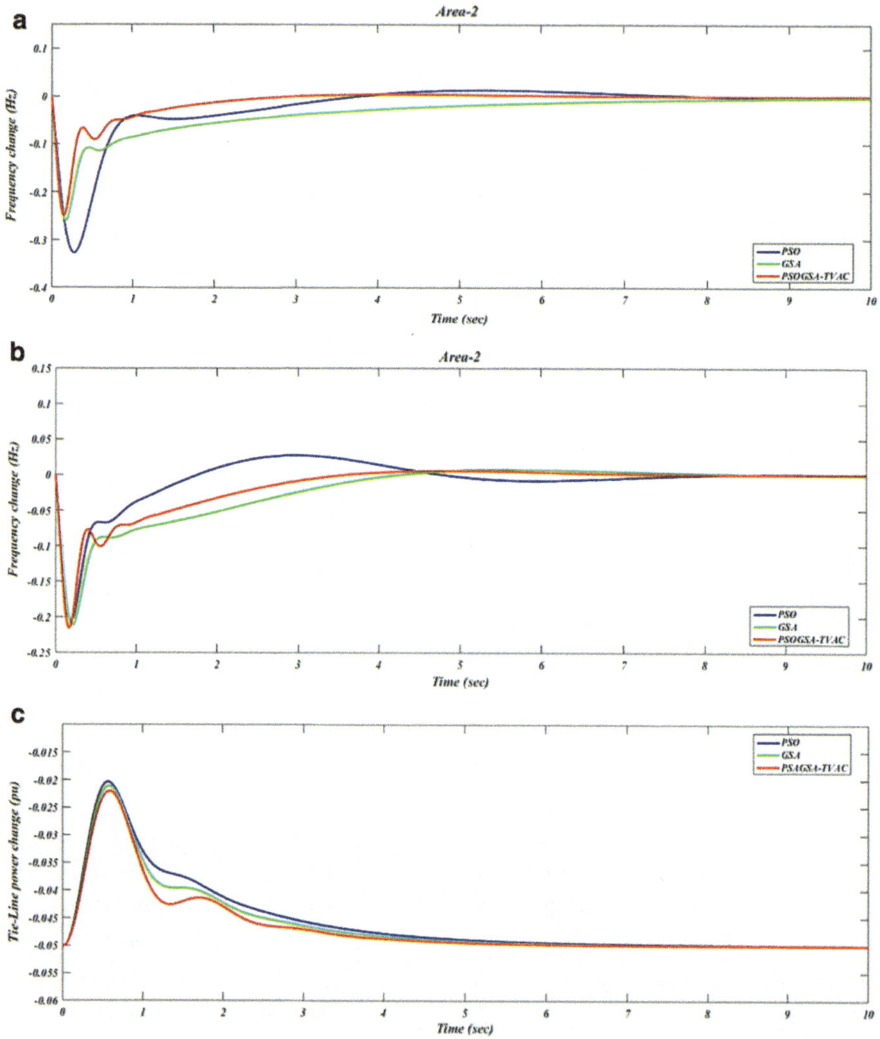

Fig. 4 **a** Frequency change of area-1 in scenario-2, **b** frequency change of area-2 in scenario-2, **c** tie-line power change in scenario-2

LFC under deregulating market structure for various transactions and violation. A comparison is performed between PSO, GSA, and PSOGSA-TVAC algorithm, and it is visualized that PSOGSA-TVAC tuned PID controller performs superior than PSO and GSA controlled PID controller in terms of time required for settling, peak overshoot, and peak undershoot. The performance characteristics under integral time squared error demonstrate that hybrid approach gives promising control scheme for LFC and can be utilized to get reliable and optimal power in deregulated system.

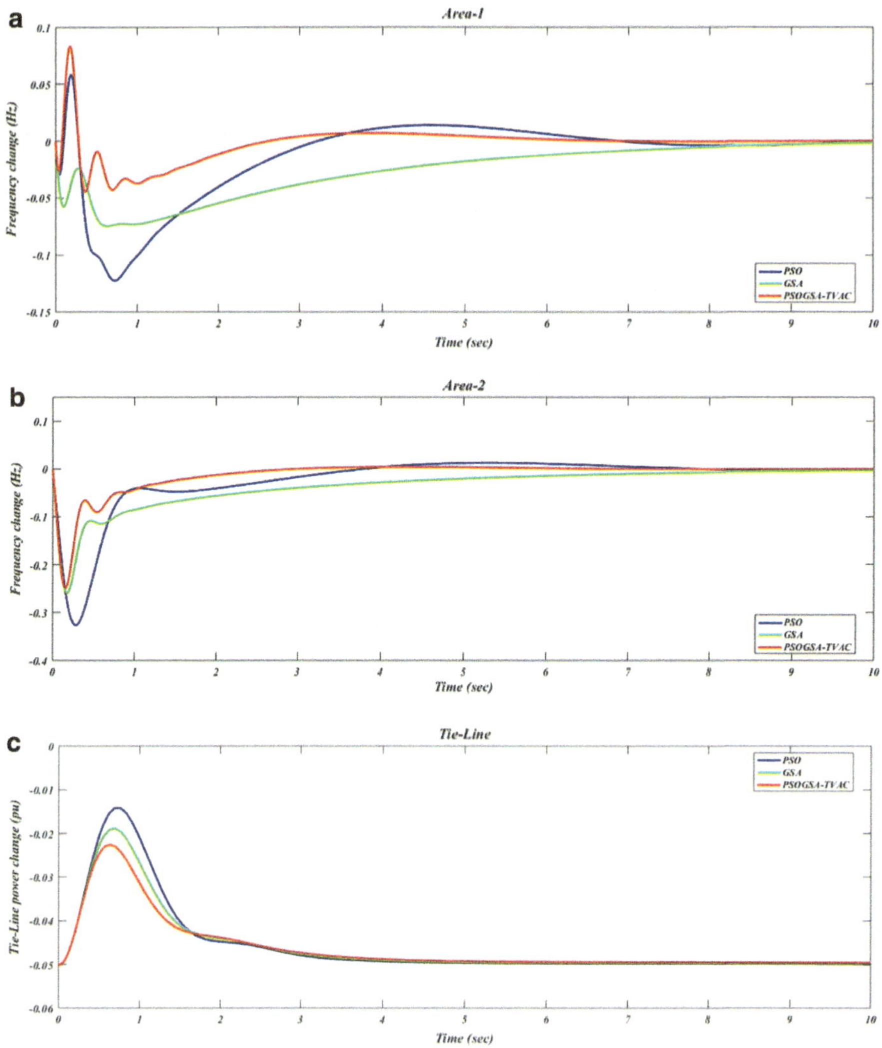

Fig. 5 **a** Frequency change of area-1 in scenario-3, **b** frequency change of area-2 in scenario-3, **c** tie-line power change in scenario-3

References

1. Davidson RA, Ushakumari S (2015) Optimal load frequency controller for a deregulated non-reheat thermal power system. Int Conf Control Commun Comput India (ICCC) 2015:319–324
2. Hobbs BE (2001) Linear complementarity models of Nash-Cournot competition in bilateral and POOLCO power markets. IEEE Trans Power Syst 16(2):194–202
3. Alhelou HH, Hamedani-Golshan M-E, Zamani R, Heydarian-Forushani E, Siano P (2018) Challenges and opportunities of load frequency control in conventional, modern and future smart power systems: a comprehensive review. Energies 11:2497

4. Bhateshvar YK, Mathur H (2014) Frequency stabilization for thermal-hydro power system with fuzzy logic controlled SMES unit in deregulated environment, advanced computing and communication technologies (ACCT). In: 2014 fourth international conference on IEEE, 2014, pp 536–540
5. Deepak M, Abraham RJ (2015) Load following in a deregulated power system with thyristor controlled series compensator. Int J Electr Power Energy Syst 65:136–145
6. Lakshmi D, Fathima A, Muthu R (2016) Simulation of the two-area deregulated power system using particle swarm optimization. Int J Electr Eng Inform. 8:93–107
7. Abedinia O, Amjady N, Naderi MS (2012) Multi-stage fuzzy PID load frequency control via SPHBMO in deregulated environment. In: 2012 11th international conference on environment and electrical engineering, Venice, Italy, 2012, pp 473–478
8. Hota P, Mohanty B (2016) Automatic generation control of multisource power generation under deregulated environment. Int J Electr Power Energy Syst 75(205–214):10
9. Kumar M, Sharma D, Singh A (2017) Improved PI using GSA algorithm for LFC of two area thermal power system. Int J Electron Electr Comput Syst (IJEECS) 6:39–44
10. Sharma V, Walde P, Siddiqui AS (2019) A new hybrid PSOGSA-TVAC algorithm for transmission line congestion management in deregulated environment. In: 2019 6th international conference on signal processing and integrated networks (SPIN), Noida, India, 2019, pp 1116–1121
11. Veerasamy V, Wahab A, Izzri N, Ramachandran R, Arangarajan V, Othman ML, Hizam H, Kumar J (2019) Automatic load frequency control of a multi-area dynamic interconnected power system using a hybrid PSO-GSA-tuned PID controller. Sustainability 11

Performance Comparison of GA, PSO and WCA for Three Area Interconnected Load Frequency Control System

Kirtiman Godara, Narendra Kumar, and Karni Pratap Palawat

Abstract Automatic generation control (AGC) and load frequency control (LFC) are very important parts of the power systems in order to provide reliable and quality power supply to the end users. Load variations can lead to the deviation generated power frequency which ultimately can destabilize the entire power system. Here comes the LFC for rescue by maintain the power frequency at desired value. In LFC, area controllers are used to perform this task. In this work, three area interconnected LFC system has been used for simulation purpose. Fractional order proportional integral (FOPI) controllers were used as an area controller in order to mitigate the effect of load variation in power system. In order to optimize the parameters of FOPI controllers, three optimization algorithms namely genetic algorithm (GA), particle swarm optimization (PSO) and water cycle algorithm (WCA) have been used. Performance comparison between optimized FOPI controllers using these algorithms has also been performed. Comparative analysis shows that the WCA optimization performed better than the other two GA and PSO algorithms. LFC system was simulated on MATLAB with the load disturbance of 0.01 p.u. in all areas at the same time.

Keywords Load frequency control · Optimization · ITAE · Power systems · Water cycle algorithm

1 Introduction

In this emergent power scenario, it is absolutely a challenge for electric suppliers to deliver safe and sound electric power to meet the demands of consumers. In the enormous generating stations, exchange of bulk amount of power has been done by the help of tie-lines. It is a very hard-hitting job of power system to stay within stability by balancing between whole produced and demanded power simultaneously [1]. Automatic generation control is answerable for supporting the basic

K. Godara (✉) · N. Kumar · K. P. Palawat
Department of Electrical Engineering, Delhi Technological University, New Delhi, Delhi, India

S. Suhag et al. (eds.), *Control and Measurement Applications for Smart Grid*,
Lecture Notes in Electrical Engineering 822,
https://doi.org/10.1007/978-981-16-7664-2_30

constraints in the accepted scope by following the active power and reactive power stability mechanism. Conventional technologies alone are not enough to react rapidly against the intolerable power fluctuations. Area controller error (ACE) conveys to zero by appraising net variation in the required generation to protect and restore the equilibrium state of the system with an economical manner [2, 3].

Many researchers give emphasis towards AGC due to its regulation against the instability regarding nonlinearities of unified power system. Mechanism of AGC in closed loop determines the generations to fulfil the prescheduled system constraints in order to keep protect the system from abnormalities related governor wind-up. In Refs. [4, 5], suitable modelling of AGC is presented to achieve the stimulating tasks concerning fluctuations regarding faults, load demand etc. Recently, researchers improve many relevant researches about AGC that described in below section. In present research to obtain the fruitful operation of load frequency control (LFC), a multiple combination of reheat based thermal, hydro and gas sources are deliberated to keep an eye on LFC complications [6]. In [7], an effort has been done along with hybrid optimization technique named hybrid firefly algorithm and pattern search over a multi-area system with Proportional-integral-derivative (PID) controller mechanism. Dated back proportional-integral (PI) controller caught the eye of researches to keep on progress on AGC problems. A modified objective function integral time absolute error (ITAE) along with suitable weight coefficients of settling time of frequency and interline power alternations are further developed to acquire the minimal values of differential evolution (DE) tuned PI controller [8]. Release of carbon and rapidly lack of conventional energy resources paid attention towards renewable energy sources. To fulfil the energy demands, solar and wind energy sources were employed as alternate sources and implemented over a hybrid system. Addition of fossil fuel which has a great impact on global warming along with solar, wind and diesel have been imposed for research work [9–11]. Wind energy is considered as the utmost promising resources not only for quick reduction in the cost of wind generators, but also it is eco-friendly and eye-catching for wind provinces. In 2015, Lu Chia-Liang gave emphasis to solve the wind–thermal coordination dispatch problem tuned for AGC [12]. The controllers are implemented to find out the appropriate instructions to acquire the desired state with less fluctuations. Dated back proportional-integral-derivative controller is introduced to make a control over stability and dynamic responses of a hydro-power system [13]. A population-based artificial bee colony algorithm is imposed for optimization process of PI and PID controller for AGC [14]. Fuzzy logic has quickly become the utmost methodology for the designing of classy control system. It resembles the human thinking with the capability of producing exactness solutions. Multi-stage fuzzy proportional-integral-derivative controller [15], proportional with integral controller with crazy PSO [16] have employed to meet the challenges of AGC. Later on in [17, 18], the multi-area power system is verified with the presence of generation rate constraint and thyristor control phase shifter (TCPS) along with fuzzy logic-based PI and PID controller. From the above discussion, it can be conferred that LFC for multi-area system is a challenging task and lot to be done in order to improve the overall quality of the power systems. Keeping this in mind, in this work, a three area interconnected thermal power system has been designed. In order to mitigate the effect of load variations in any area on the generated power frequency, fractional order proportional integral (FOPI) controllers have been used in each area as an area controller.

Three optimization algorithms named as: genetic algorithm (GA), particle swarm optimization (PSO) and water cycle algorithm (WCA) have been used to tune the parameters of the three FOPI controllers. Finally, comparison have been done among the optimization algorithms in order to show the effectiveness of WCA algorithm for tuning the FOPI controllers for highly nonlinear systems.

Further, organization of rest of the paper is as follows: Sect. 2 describes the FOPI controller structure, Sect. 3 presents the description of three area interconnected LFC system, Sect. 4 presents the results obtained in this study and Sect. 5 concludes the paper with possible direction for future work.

2 FOPI Controller

In recent times, fractional order calculus has gained lot of attention due to its improved performance over integer powered calculus. Based on this motivation, a new kind of controllers have been developed named as fractional order controllers. In this section, a brief about FOPI controller has been presented.

At first, researcher Podunk introduced the methodologies of the robust FOPID controller into the research work. The integral part and derivative part of the FOPID controller do not belongs to the integer, which makes this subjected controller very efficient and dynamic as compare to PID controller. Presence of additional optimization algorithms enhances the beauty of FOPI controller. Therefore, traditional proportional, integral gains are tuned along with the integral order $\{\lambda\}$. In fractional order control system, researchers mostly preferred the Caputo's fractional differentiation expression. The explanation of fractional order derivative controller is usually employed to find out transfer function models of fractional calculus out of fractional order usual differential equations having zero initial situations.

Mathematical equation of the FOPI controller can be given by (1).

$$G_C(s) = K_p + \frac{K_i}{S^\lambda} \tag{1}$$

Depending upon the value of the λ, the FOPI controller can also work as a PI controller. To obtain PI controller, the value of λ must be 1. In this work, FOMCON toolbox has been used in order to implement the FOPI controller in SIMULINK.

3 Plant Description

A multi-area network comprises of more than one interconnected area. A basic three equal areas non-reheat based power system with its transfer functions are modelled for investigation. In this section, description of various components of the plant have been provided.

3.1 Plant Model

Each area involves of a governor, a generator, a turbine and speed governing system. We have to find out the ratio of the output of Laplace transform to the inputs of Laplace transform in order to find out the transfer function.

The Laplace transfer function of governor output response ΔP_G is defined as

$$\Delta P_{\mathrm{G}} = \Delta P_{\mathrm{ref}}(s) - \frac{1}{R}\Delta F(\mathrm{s}) \tag{2}$$

The Laplace transfer function for governor is given by Eq. (3) as follows:

$$G_{\mathrm{G}}(\mathrm{s}) = \frac{\Delta P_{\mathrm{V}}(\mathrm{s})}{\Delta P_{\mathrm{G}}(\mathrm{s})} = \frac{1}{1 + T_G(\mathrm{s})} \tag{3}$$

The Laplace transfer function for non-reheated turbine is given by Eq. (4)

$$G_{\mathrm{T}}(s) = \frac{\Delta P_{\mathrm{T}}(s)}{\Delta P_{\mathrm{V}}(s)} = \frac{1}{1 + T_{\mathrm{T}}(s)} \tag{4}$$

In generator, variation within the increments in power demand and generated turbine power assists as input and the output signal is formed by using the following transfer function

$$G_{\mathrm{p}}(s) = \frac{K_{\mathrm{ps}}}{T_{\mathrm{p}}s + 1} \tag{5}$$

A non-reheat three multi area interconnected networks is modelled to investigate the load frequency control. The appropriate parameters with their corresponding values of all the components are presented in Table1.

Figure 1 shows the SIMULINK model of three area interconnected thermal power system with non-reheated turbines. Step load disturbance (ΔP_{L}) has been added into each area. For the simulation purpose, load disturbance was provided at the same time in each area.

Table 1 Notations used in the model

Parameter symbol	Parameter name	Value
B	Frequency bias parameter	0.425
K_{pm}	Gain of electric power system (p.u, MW)	120
T_G	Time constant of governor (s)	0.08
T_T	Time constant of non-reheat turbine (s)	0.3
T_p	Time constant of electric power system (s)	20
R	Speed regulation	2.4
ΔP_L	Disturbance in load	
ΔP_g	Incremental variation in governor input	
ΔF	Incremental fluctuations in system frequency (Hz)	
ΔP_v	Incremental variation in governor valve location	
ΔP_m	Incremental variation in turbine output (p.u, MW)	

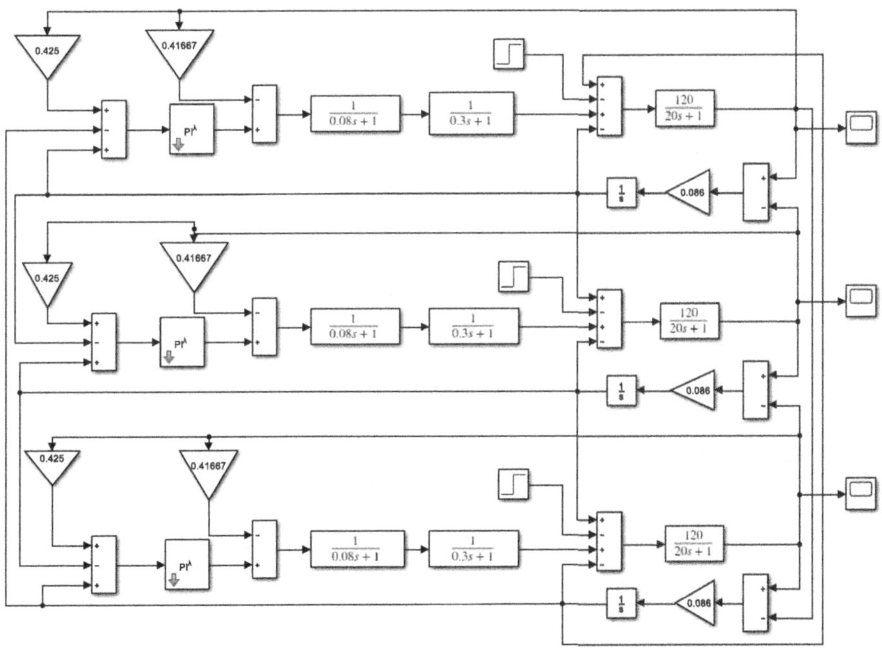

Fig. 1 SIMULINK model of three area interconnected thermal power system

3.2 *Objective Function*

Optimization algorithms use objective or cost function in order to tune the parameters of the underline controller. They tried to minimize the value of objective function and then in return, response of the system become better. In this work, ITAE objective function is used which incorporates controller error, area frequency deviation and tie power. Mathematical equation of the objective function is given by Eq. (6)

$$\mathrm{ITAE} = \int_{0}^{t} \left(\sum_{i=1}^{n} \left(|e_i| + |\Delta F_i| + |\Delta P_{\mathrm{tie}_i}| \right) \right) \cdot t \cdot \mathrm{d}t \quad (6)$$

where n is the number of areas, e_i is the error input to the controller, ΔF_i is the area frequency deviation and $\Delta P_{\mathrm{tie}_i}$ is the change in tie power for each area.

4 Result and Discussion

In this section, results obtained from simulation study of the three area interconnected LFC system has been presented. For this study, MATLAB/SIMULINK© was used. ODE 45 solver was used in SIMULINK for the simulation purpose.

4.1 Optimization Algorithm Result

GA, PSO and WCA algorithms were run with same parameters in order to do comparison among them. Details of the various optimization parameters used in this study are given in Table 2. All optimization algorithms were run with the population size of 50 and for 50 iterations. As discussed in earlier section, there are three FOPI controllers for each area and hence, there are total six gain parameters named as K_{p1}, K_{i1}, K_{p2}, K_{i2}, K_{p3}, K_{i3} and three integrator order parameters named as λ_1, λ_2, λ_3. Lower and upper bound (UB) for gain parameters were chosen as −2 and +2, respectively. LB and UB for integrator order were chosen as 0 and 2, respectively.

Figure 2 shows the optimization convergence curve for all three GA, PSO and WCA optimization algorithms. From the Fig. 2, it is very much clear that WCA gives better performance compared to GA and PSO in terms of convergence. WCA took

Table 2 Optimization parameters

S. No.	Optimization parameter	Value
1	Population size	50
2	Iterations	50
3	Lower bound	−2 for gain values 0 for integrator order
4	Upper bound	2 for gain values 2 for integrator order

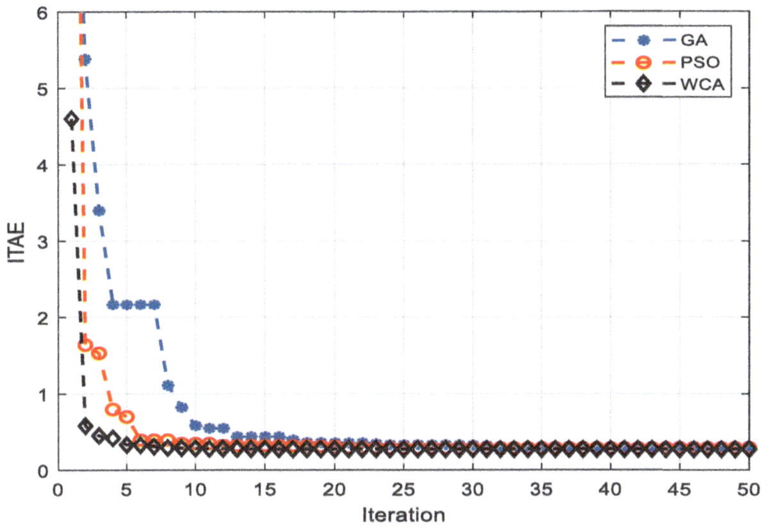

Fig. 2 Optimization convergence curve for GA, PSO and WCA

Table 3 FOPI controllers parameters value

Parameter name	GA	PSO	WCA
K_{p1}	−1.595	−1.684	−1.231
K_{i1}	−1.862	−2.000	−2.000
K_{p2}	−1.443	−1.681	−1.009
K_{i2}	−1.308	−1.995	−1.297
K_{p3}	−1.533	−1.683	−1.231
K_{i3}	−1.898	−2.000	−2.000
λ_1	1.457	1.420	1.388
λ_2	1.007	1.421	0.982
λ_3	1.416	1.420	1.388
ITAE	**0.3046**	**0.3003**	**0.2706**

around five iterations only to get minimum ITAE value, while GA took around 25 iterations to reach up to that level. PSO performed in between GA and WCA as it took around 15 iterations to converge at minimum value of ITAE objective function. Table 3 presents the optimized values of three FOPI controller parameters along with minimum ITAE obtained using optimization algorithms. From Table 3, it is very much clear that WCA algorithm gives minimum ITAE value as 0.2706.

4.2 LFC Results for Each Area

The three area LFC system was simulated with step load disturbance of 0.01p.u. in each area at the same time. By this, performance of the proposed FOPI controller can be judged effectively. Optimized FOPI controller parameters were used to obtain the frequency deviation result for each area. Figures 3, 4 and 5 show the frequency deviation curve for area 1, area 2 and area 3, respectively. All graphs have been colour coded in order to distinguish between the responses between optimization algorithms. From these figures, it is clear that WCA-optimized parameters performs better than GA and PSO.

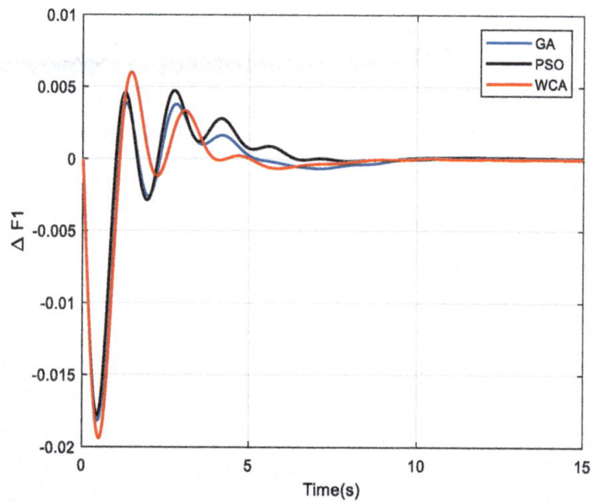

Fig. 3 Frequency deviation for area 3

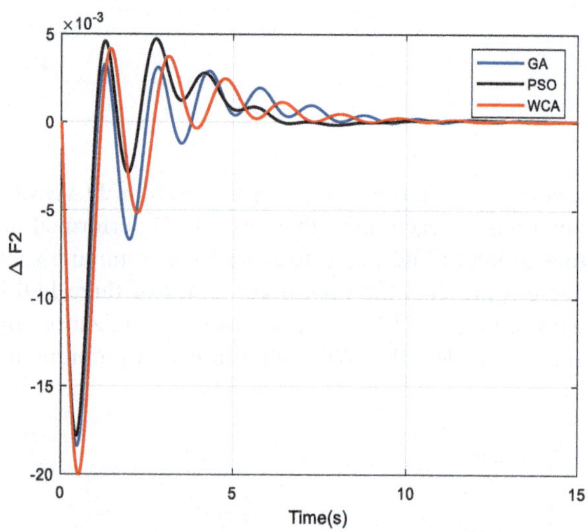

Fig. 4 Frequency deviation for area 2

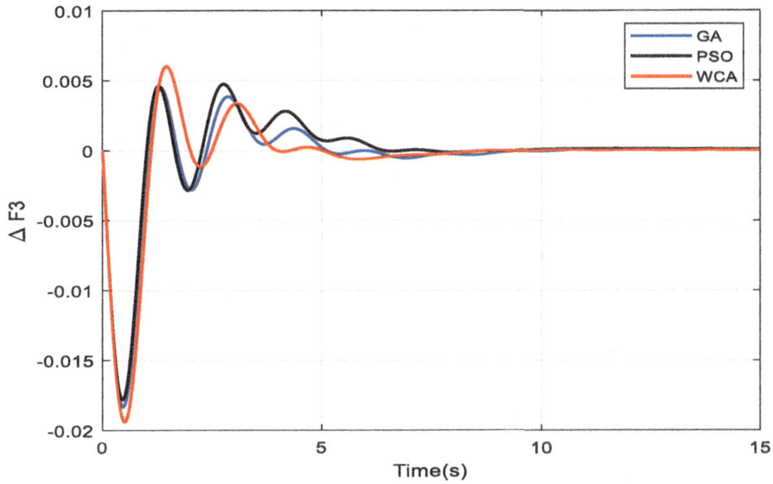

Fig. 5 Frequency deviation for area 3

5 Conclusion and Future Scope

In this paper, three area interconnected thermal power system has been simulated with non-reheated turbine. Fractional order PI (FOPI) controllers were used as an area controller in order to minimize the effect of load variation in any area on the generated power frequency. Three optimization algorithms, namely GA, PSO and WCA were used in order to optimize the parameters of FOPI controllers. Further, performance comparison between these three algorithms shows that WCA gives better optimized parameters and that too in lesser number of iterations. In future work, authors will try to do the similar comparison between optimization algorithms for power system with different kind of energy generation systems.

References

1. Kundur P, Balu NJ, Lauby MG (1994) Power system stability and control, vol 7. McGraw-Hill, New York
2. Bevrani H, Hiyama T (2016) Intelligent automatic generation control. CRC Press
3. Wood AJ, Wollenberg BF (2012) Power generation, operation, and control. Wiley, New York
4. Patel MR (2005) Wind and solar power systems: design, analysis, and operation. CRC Press
5. Glavitsch H, Stoffel J (1980) Automatic generation control. Int J Electr Power Energy Syst 2(1):21–28
6. Parmar KPS, Majhi S, Kothari DP (2012) Load frequency control of a realistic power system with multi-source power generation. In J Electr Power Energy Syst 42(1):426–433
7. Sahu RK, Panda S, Padhan S (2015) A hybrid firefly algorithm and pattern search technique for automatic generation control of multi area power systems. Int J Electr Power Energy Syst 64:9–23

8. Rout UK, Sahu RK, Panda S (2013) Design and analysis of differential evolution algorithm based automatic generation control for interconnected power system. Ain Shams Eng J 4(3):409–421
9. Rahman A, Saikia LC, Sinha N (2017) Automatic generation control of an interconnected two-area hybrid thermal system considering dish-stirling solar thermal and wind turbine system. Renew Energy 105:41–54
10. Das DC, Roy AK, Sinha N (2012) GA based frequency controller for solar thermal–diesel–wind hybrid energy generation/energy storage system. Int J Electr Power Energy Syst 43(1):262–279
11. Kurokami S, Fukae K, Takehara N (1999) Power control apparatus for solar power generation system. U.S. Patent No. 5923158, 13 July1999
12. Lu C-L et al (2008) Effects of wind energy supplied by independent power producers on the generation dispatch of electric power utilities. Int J Electr Power Energy Syst 30(9):553–561
13. Khodabakhshian A, Hooshmand R (2010) A new PID controller design for automatic generation control of hydro power systems. Int J Electr Power Energy Syst 32(5):375–382
14. Gozde H, Cengiz Taplamacioglu M, Kocaarslan I (2012) Comparative performance analysis of Artificial Bee Colony algorithm in automatic generation control for interconnected reheat thermal power system. Int J Electr Power Energy Syst 42(1):167–178
15. Shayeghi H, Shayanfar HA, Jalili A (2006) Multi-stage fuzzy PID power system automatic generation controller in deregulated environments. Energy Convers Manage 47(18–19):2829–2845
16. Gozde H, Cengiz Taplamacioglu M (2011) Automatic generation control application with craziness based particle swarm optimization in a thermal power system. Int J Electr Power Energy Syst 33(1):8–16
17. Behera A et al (2018) Hybrid ITLBO-DE Optimized fuzzy PI controller for multi-area automatic generation control with generation rate constraint. In: Smart computing and informatics. Springer, Singapore, pp 713–722
18. Lal DK, Barisal AK, Tripathy M (2016) Grey wolf optimizer algorithm based Fuzzy PID controller for AGC of multi-area power system with TCPS. Procedia Comput Sci 92:99–105

A Reliable Numerical Approach for Liouville-Caputo Time-fractional Belousov-Zhabotinsky Equation Arising in Chemical Oscillatory Reaction

Amit Prakash, Hardish Kaur, Rahul, and Manish Goyal

Abstract This paper suggests a computationally effective homotopy perturbation Sumudu transform technique (HPSTT) to investigate various time-fractional nonlinear models in Liouville-Caputo sense arising in mathematical physics. The nonlinear terms are presented in terms of He's polynomials. The error analysis of HPSTT is discussed. The numerical simulation results are illustrated graphically to study effects of the arbitrary order of fractional derivative on the behavior of obtained solution.

Keywords Time-fractional Belousov-Zhabotinsky (BZ) equation · Homotopy perturbation Sumudu transform technique (HPSTT) · He's polynomials · Liouville-Caputo fractional derivative

1 Introduction

Fractional derivatives are used in modeling of numerous important models in significant areas such as signal and systems, control theory, mechanics, chemical engineering, biological sciences, fluid dynamic traffic, acoustics, neurophysiology, plasma physics, and many engineering sciences. The universe is full of nonlinear fractional-order models, and it is not possible to find their exact solution due to their nonlinearity in nature. So, we have to choose some numerical methods for their convergent solution. In literature, there are few techniques for finding approximate solution to these models, e.g., homotopy perturbation [1–4] scheme, q-homotopy

A. Prakash (✉) · H. Kaur · Rahul
National Institute of Technology, Kurukshetra, Haryana 136119, India

Rahul
e-mail: rahul_61900119@nitkkr.ac.in

M. Goyal
Institute of Applied Sciences and Humanities, GLA University, Mathura, UP 281406, India
e-mail: manish.goyal@gla.ac.in

S. Suhag et al. (eds.), *Control and Measurement Applications for Smart Grid*, Lecture Notes in Electrical Engineering 822,
https://doi.org/10.1007/978-981-16-7664-2_31

analysis [5] via the Laplace transform, iterative method via the Sumudu transform [6], variational iteration [7] scheme. For more methods, (see, for details, [8–53]).

This paper studies the merit of scheme HPSTT to get the numerical solution of nonlinear time-fractional Belousov-Zhabotinsky (BZ) equation. HPSTT is a smooth combination of homotopy perturbation scheme and Sumudu transform. This technique offers the numerical solution simply in a convergent series. It is evident that some semi-analytical methods, when combined with the Sumudu transform, take less CPU runtime in calculation than other techniques. Sumudu transform [54, 55] has an advantage of the "unity" feature over other transforms. Watugala [56] proposed the Sumudu transform, and Asiru [57] proved its properties. Weerakoon [58, 59] applied it in finding solution to wave equation.

Singh et al. [60] presented homotopy perturbation method via the Sumudu transform (HPSTM). It is largely due to the works of Ghorbani and Saberi-Nadjafi [61]. Ghorbani [62] used He's polynomials in nonlinear term. The benefit of HPSTT is its power of embracing two robust computational schemes for tackling a fractional differential equation. These approaches can reduce the time and computation work more in comparison with other existing schemes, preserving efficiency of the results.

We examine the time-dependent nonlinear Belousov-Zhabotinsky (BZ) equation with fractional derivative in Liouville-Caputo sense in this paper. The Belousov-Zhabotinsky reaction is an experimental model for pattern formation and chemical oscillatory reaction. It is a famous example of the self-organizing chemical system. It is given as

$$
\begin{aligned}
D_t^{\alpha} u(x,t) &= u(x,t)(1-u(x,t)-rv(x,t)) + u_{xx}(x,t),\\
D_t^{\alpha} v(x,t) &= -au(x,t)v(x,t) + v_{xx}(x,t),
\end{aligned} \tag{1}
$$

where $0 \leq \alpha \leq 1, 0 < t < R < 1$.

The fractional model given by Eq. (1) is studied for the first time by Ali Jaradat et al. [63] with generalized Taylor series and discussed the effects of arbitrary order on its solution. Our paper is prepared as follows.

After introduction in Sect. 1, we give few definitions and some properties of Liouville-Caputo arbitrary-order derivative and the Sumudu transform in next section. In Sect. 3, analysis of HPSTT is presented. Next section has the error analysis, and in Sect. 5, HPSTT is implemented on the time-fractional nonlinear BZ equation. In Sect. 6, we discuss the results and their importance using figures. Also, in last Sect. 7, we recap outcomes and find a conclusion.

2 Preliminaries

We write some definitions of fractional-order derivatives and integrals in Liouville-Caputo sense along with few properties of the Sumudu transform (see, for details, [64, 65]).

Definition 2.1 A real function $g(\mu)$, $\mu > 0$, lies in spaces:

(a) $C_\zeta, \zeta \in \mathbb{R}$ if there exists a real number $p > \zeta$ in such a manner that

$$g(\mu) = \mu^p g_1(\mu), \quad \text{while} \quad g_1(\mu) \in C[0, \infty).$$

(b) C_ζ^l if $g^l \in C_\zeta, l \in N$.

Definition 2.2 Liouville-Caputo arbitrary-order derivative of $g(\mu), g \in C_{-1}^m, m \in$ N, $m > 0$, is stated as

$$D^\beta g(\mu) = I^{m-\beta} D^m g(\mu) = \frac{1}{r(m-\beta)} \int\limits_0^\mu (\mu - \eta)^{m-\beta-1} g^{(m)}(\eta) \mathrm{d}\eta,$$

where $m - 1 < \beta \leq m$.

The operator D^β has following basic properties:

1. $D^\beta I^\beta g(\mu) = g(\mu)$,
2. $I^\beta D^\beta g(\mu) = g(\mu) - \sum\limits_{k=0}^{m-1} g^{(k)}(0^+) \frac{\mu}{r(k+1)}, m > 0.$

Definition 2.3 Sumudu transform [66, 67] is stated over a set of function.

$$A = \{ f(t) | \exists M, t_1, t_2 > 0, |f(t)| < M e^{\frac{|t|}{t_j}} \quad \text{if} \ t \in (-1)^j \times [0, \infty)$$

by the following formula,

$$S[f(t)] = \int\limits_0^\infty f(ut) e^{-t} \mathrm{d}t, \quad u \in (-t_1, t_2).$$

Definition 2.4 Sumudu transform [66, 67] of Liouville-Caputo derivative is

$$S\left[D_x^{m\alpha} u(x,t)\right] = s^{-m\alpha} S[u(x,t)] - \sum_{k=0}^{m-1} s^{(-m\alpha+k)} u^k(0,t), m - 1 < m\alpha \leq m.$$

3 Analysis of Homotopy Perturbation Sumudu Transform Technique (HPSTT)

Ponder over a nonlinear time-fractional differential equation of arbitrary order is

$$D_t^{\beta}u(x,t) + Ru(x,t) + Nu(x,t) = f(x,t), l-1 < \beta \leq l, \tag{2}$$

with the condition

$$u^{m}(x,0) = f_m(x), m = 0, 1, 2, \ldots, l-1, \tag{3}$$

where $D_t^{\beta}u(x,t)$ is β-order derivative of $u(x,t)$ in Liouville-Caputo sense and R and N are used for linear and nonlinear differential operators, respectively. $f(x,t)$ is actually the source term.

By Sumudu transform on Eq. (2),

$$\begin{aligned} u^{-\alpha}S[u(x,t)] - \sum_{k=0}^{m-1} u^{-(\alpha-k)}u^{k}(x,0) \\ = -S(Ru(x,t) + Nu(x,t) - f(x,t)). \end{aligned}$$

On simplification,

$$\begin{aligned} S[u(x,t)] = \sum_{k=0}^{m-1} u^{k}u^{k}(x,0) \\ - u^{\alpha}\{S[Ru(x,t)] + S[Nu(x,t)] - S[f(x,t)]\} = 0. \end{aligned} \tag{4}$$

Taking inverse Sumudu transform

$$\begin{aligned} u(x,t) = S^{-1}\left[\sum_{k=0}^{m-1} u^{k}u^{k}(x,0)\right] \\ - S^{-1}\{u^{\alpha}[S(Ru(x,t) + Nu(x,t) - f(x,t))]\} = 0 \end{aligned} \tag{5}$$

By homotopy perturbation method, we have

$$u(x,t) = \sum_{i=0}^{\infty} p^{i}u_i(x,t). \tag{6}$$

The nonlinear term is expressed in He's polynomials as

$$Nu(x,t) = \sum_{i=0}^{\infty} p^{i}H_i(u), \tag{7}$$

where He's polynomial $H_i(w)$ is given as:

$$H_i(w) = \frac{1}{i!}\frac{\partial^i}{\partial p^i}\left[N\left(\sum_{j=0}^{\infty} p^j w_j\right)\right]_{p=0}, \quad i = 0, 1, 2, 3, \ldots \tag{8}$$

Putting Eqs. (6) and (7) in Eq. (5), we write

$$\begin{aligned}\sum_{i=0}^{\infty} p^i u_i(x,t) &= S^{-1}\left(\sum_{k=0}^{m-1} u^k u^k(x,0) + S[f(x,t)]\right) \\ &- p\left[S^{-1}\left\{u^\alpha S\left\{\left[R\sum_{i=0}^{\infty} p^i u_i(x,t)\right] + \sum_{i=0}^{\infty} p^i H_i(u)\right\}\right\}\right]\end{aligned} \tag{9}$$

Equating on each side the coefficients of identical powers of p, we find

$$\begin{aligned}p^0 &: u_0(x,t) = S^{-1}\left(\sum_{k=0}^{m-1} u^k u^k(x,0) + S[f(x,t)]\right), \\ p^1 &: u_1(x,t) = -S^{-1}\{u^\alpha S\{[Ru_0(x,t)] + H_0(u)\}\}, \\ p^2 &: u_2(x,t) = -S^{-1}\{u^\alpha S\{[Ru_1(x,t)] + H_1(u)\}\}, \\ p^3 &: u_3(x,t) = -S^{-1}\{u^\alpha S\{[Ru_2(x,t)] + H_2(u)\}\},\end{aligned}$$

In the same way, the next iterates can be found.
Hence, the solution is

$$u(x,t) = \lim_{p\to 1}\lim_{N\to\infty}\sum_{i=0}^{N} p^i u_i(x,t) = \lim_{N\to\infty}\sum_{i=0}^{N} u_i(x,t). \tag{10}$$

4 Error Analysis

Now, we give the error analysis of presented approach found with HPSTT.

Theorem 4.1 If there exists $0 < \gamma < 1$, such that

$$\|u_{i+1}(\tau,\xi)\| \leq \gamma\|u_i(\tau,\xi)\|, \forall i \in N,$$

then, the maximum value of absolute truncation error in HPSTT solution given by Eq. (10) of time-dependent nonlinear Belousov-Zhabotinsky equation is estimated as

$$\left| u(\tau, \xi) - \sum_{i=0}^{j} u_i(\tau, \xi) \right| \le \frac{\gamma^{j+1}}{(1-\gamma)} u_0(\tau, \xi).$$

5 Applications

In this section, HPSTT is implemented on the fractional model of BZ equation.

Test Example. Consider the following fractional model of Belousov-Zhabotinsky equation

$$\begin{aligned} D_t^\alpha u(x,t) &= u(x,t)(1 - u(x,t) - rv(x,t)) + u_{xx}(x,t), \\ D_t^\alpha v(x,t) &= -au(x,t)v(x,t) + v_{xx}(x,t), \end{aligned} \tag{11}$$

where $0 \le \alpha \le 1, 0 < t < R < 1$. The initial conditions are

$$\begin{aligned} u(x,0) &= -\frac{1}{2}\left(1 - \tanh^2\left(\frac{x}{2}\right)\right), \\ v(x,0) &= -\frac{1}{2} + \tanh\left(\frac{x}{2}\right) + \frac{1}{2}\tanh^2\left(\frac{x}{2}\right). \end{aligned} \tag{12}$$

The exact solution of Eq. (11) when $\alpha = 1$, for $r = 2, a = 3$ is given as

$$\begin{aligned} u(x,t) &= -\frac{1}{2}\left(1 - \tanh^2\left(\frac{x}{2} + t\right)\right), \\ v(x,t) &= -\frac{1}{2} + \tanh\left(\frac{x}{2} + t\right) + \frac{1}{2}\tanh^2\left(\frac{x}{2} + t\right) \end{aligned}$$

Firstly, exerting Sumudu transform operator on Eq. (11)

$$\begin{aligned} S\left[D_t^\alpha u(x,t)\right] &= S[u(x,t)(1 - u(x,t) - rv(x,t)) + u_{xx}(x,t)], \\ S\left[D_t^\alpha v(x,t)\right] &= S[-au(x,t)v(x,t) + v_{xx}(x,t)]. \end{aligned}$$

Applying the inverse Sumudu transform

$$\begin{aligned} u(x,t) &= u(x,0) - S^{-1}\left\{u^\alpha S\left[-u + u^2 + ruv - u_{xx}\right]\right\}, \\ v(x,t) &= v(x,0) - S^{-1}\{u^\alpha S[auv - v_{xx}]\}, \end{aligned}$$

By applying the homotopy perturbation method, using

$$u(x,t) = \sum_{i=0}^{\infty} p^i u_i(x,t) \quad \text{and} \quad v(x,t) = \sum_{i=0}^{\infty} p^i v_i(x,t),$$

And, the nonlinear term is decomposed as

$$Nu(x,t)=\sum_{i=0}^{\infty}p^{i}H_{i}(u)\quad\text{and}\quad Nv(x,t)=\sum_{i=0}^{\infty}p^{i}H_{i}(v).$$

where $H_i(u)$, $H_i(v)$ are homotopy polynomials that are representing the nonlinear terms. So, Eq. (11) becomes

$$\begin{aligned}\sum_{i=0}^{\infty}p^{i}u_{i}(x,t)&=u(x,0)\\&\quad-pS^{-1}\left\{u^{\alpha}S\left[\begin{array}{c}-\sum_{i=0}^{\infty}p^{i}u_{i}(x,t)\\+\sum_{i=0}^{\infty}p^{i}H_{i}(u)-\left(\sum_{i=0}^{\infty}p^{i}u_{i}(x,t)\right)_{xx}\end{array}\right]\right\},\\\sum_{i=0}^{\infty}p^{i}v_{i}(x,t)&=v(x,0)-pS^{-1}\left\{u^{\alpha}S\left[\sum_{i=0}^{\infty}p^{i}H_{i}(v)-\left(\sum_{i=0}^{\infty}p^{i}v_{i}(x,t)\right)_{xx}\right]\right\},\end{aligned}\tag{13}$$

where

$$\sum_{i=0}^{\infty}p^{i}H_{i}(u)=u^{2}+ruv\quad\text{and}\quad\sum_{i=0}^{\infty}p^{i}H_{i}(v)=auv.$$

$$H_0(u)=u_0^2+ru_0v_0,\quad H_0(v)=au_0v_0,$$

$$H_1(u)=2u_0u_1+2u_0v_1+2v_0u_1,\quad H_1(v)=a(u_0v_1+v_0u_1),$$

Solving above equations, we get

$$u_0(x,t)=-\frac{1}{2}\left(1-\tanh^2\left(\frac{x}{2}\right)\right),\quad v_0(x,t)=-\frac{1}{2}+\tanh\left(\frac{x}{2}\right)+\frac{1}{2}\tanh^2\left(\frac{x}{2}\right),$$

$$\begin{aligned}&u_1(x,t)\\&=-\frac{t^{\alpha}}{\Gamma(\alpha+1)}\left\{\frac{5}{4}-2\tanh^2\left(\frac{x}{2}\right)+\frac{3}{4}\tanh^4\left(\frac{x}{2}\right)-\tanh\left(\frac{x}{2}\right)\mathrm{sech}^2\left(\frac{x}{2}\right)\right.\\&\left.-\frac{1}{4}\mathrm{sech}^4\left(\frac{x}{2}\right)+\frac{1}{2}\tanh^2\left(\frac{x}{2}\right)\mathrm{sech}^2\left(\frac{x}{2}\right)\right\},\end{aligned}$$

$$v_1(x,t)=-\frac{t^{\alpha}}{\Gamma(\alpha+1)}\left\{\frac{3}{4}+\frac{3}{2}\tanh^2\left(\frac{x}{2}\right)+\frac{3}{4}\tanh^4\left(\frac{x}{2}\right)-\frac{3}{2}\tanh\left(\frac{x}{2}\right)\mathrm{sech}^2\left(\frac{x}{2}\right)\right.$$

$$-\frac{1}{4}\mathrm{sech}^4\left(\frac{x}{2}\right)+\frac{1}{2}\tanh^2\left(\frac{x}{2}\right)\mathrm{sech}^2\left(\frac{x}{2}\right)+\frac{1}{2}\tanh\left(\frac{x}{2}\right)\Bigg\},$$

$$\begin{aligned}u_2(x,t) &= \frac{t^{2\alpha}}{\Gamma(2\alpha+1)}\Bigg\{\frac{9}{2}+\frac{43}{4}\tanh^2-\frac{34}{4}\tanh^4\left(\frac{x}{2}\right)\\
&+\frac{9}{2}\tanh\left(\frac{x}{2}\right)\mathrm{sech}^2\left(\frac{x}{2}\right)\\
&+2\mathrm{sech}^4\left(\frac{x}{2}\right)-6\tanh^2\left(\frac{x}{2}\right)\mathrm{sech}^2\left(\frac{x}{2}\right)+\frac{9}{4}\tanh^6\left(\frac{x}{2}\right)\\
&-\frac{3}{2}\tanh^3\left(\frac{x}{2}\right)\mathrm{sech}^2\left(\frac{x}{2}\right)-\frac{1}{4}\tanh^2\left(\frac{x}{2}\right)\mathrm{sech}^4\left(\frac{x}{2}\right)\\
&+\frac{5}{2}\tanh^4\left(\frac{x}{2}\right)\mathrm{sech}^2\left(\frac{x}{2}\right)\\
&+2\tanh\left(\frac{x}{2}\right)-\frac{7}{2}\tanh^3\left(\frac{x}{2}\right)+\frac{3}{2}\tanh^5\left(\frac{x}{2}\right)-\frac{5}{2}\tanh\left(\frac{x}{2}\right)\mathrm{sech}^4\left(\frac{x}{2}\right)\\
&-\frac{1}{2}\mathrm{sech}^6\left(\frac{x}{2}\right)\Bigg\},\end{aligned}$$

$$\begin{aligned}v_2(x,t) &= \frac{t^{2\alpha}}{\Gamma(2\alpha+1)}\Bigg\{-\frac{3}{8}+\frac{43}{4}\tanh^2\left(\frac{x}{2}\right)-\frac{3}{8}\tanh^4\left(\frac{x}{2}\right)\\
&+\frac{9}{2}\tanh\left(\frac{x}{2}\right)\mathrm{sech}^2\left(\frac{x}{2}\right)+\frac{1}{4}\mathrm{sech}^4\left(\frac{x}{2}\right)\\
&-6\tanh^2\left(\frac{x}{2}\right)\mathrm{sech}^2\left(\frac{x}{2}\right)+\frac{9}{4}\tanh^6\left(\frac{x}{2}\right)\\
&-\frac{3}{2}\tanh^3\left(\frac{x}{2}\right)\mathrm{sech}^2\left(\frac{x}{2}\right)-\frac{9}{4}\tanh^2\left(\frac{x}{2}\right)\mathrm{sech}^4\left(\frac{x}{2}\right)\\
&+\frac{5}{2}\tanh^4\left(\frac{x}{2}\right)\mathrm{sech}^2\left(\frac{x}{2}\right)\\
&+2\tanh\left(\frac{x}{2}\right)-\frac{7}{2}\tanh^3\left(\frac{x}{2}\right)+\frac{3}{2}\tanh^5\left(\frac{x}{2}\right)-\frac{5}{2}\tanh\left(\frac{x}{2}\right)\mathrm{sech}^4\left(\frac{x}{2}\right)\\
&-\frac{1}{4}\mathrm{sech}^6\left(\frac{x}{2}\right)\Bigg\}+\cdots,\end{aligned}$$

Hence, the next iterates of series solution can be calculated.
The HPSTT solution is given as

$$u(x,t)=\lim_{N\to\infty}\sum_{i=0}^{N}u_i(x,t),\quad v(x,t)=\lim_{N\to\infty}\sum_{i=0}^{N}v_i(x,t).$$

6 Results and Discussion

Figures 1a–f represent the behavior of solution at $\alpha = 1$, the estimated solution for Eq. (11), and the absolute error. We can clearly see from the plots of solution

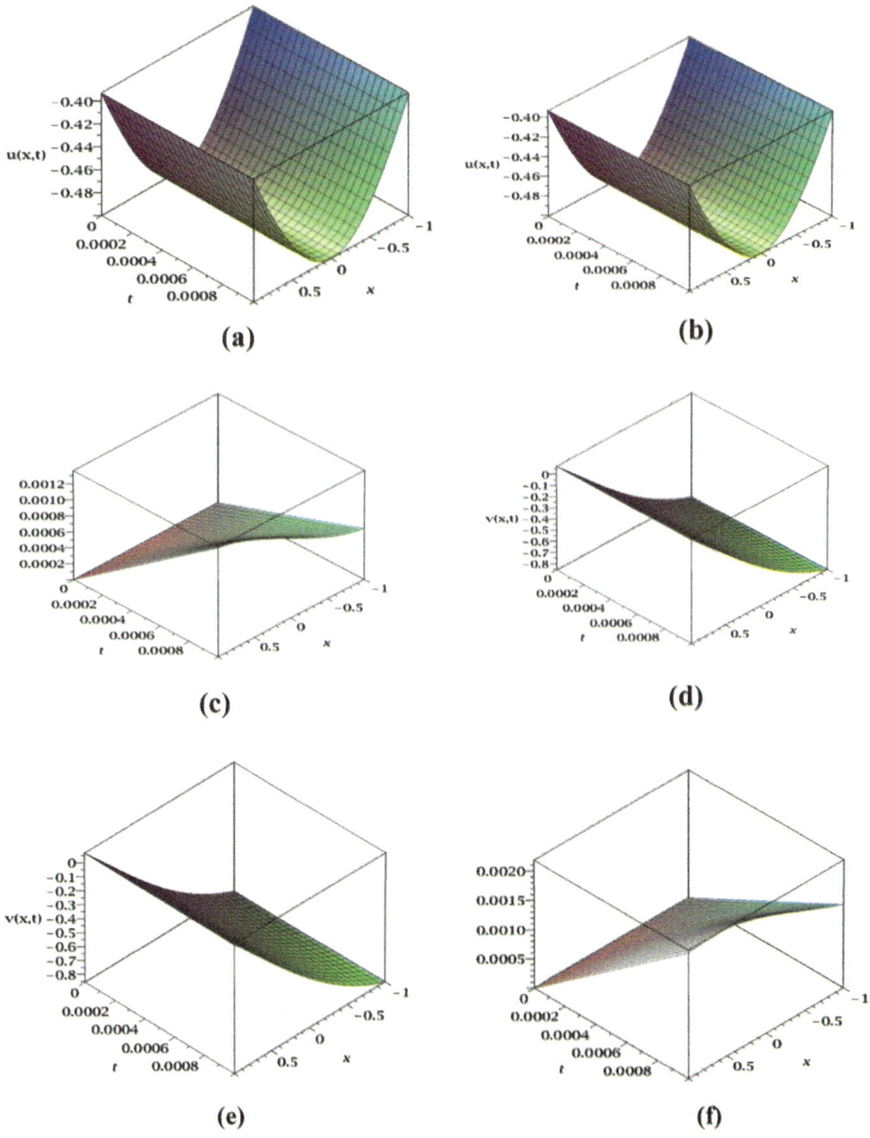

Fig. 1 **a** $u_{\text{exact}}(x,t)$ when $\alpha = 1$, **b** $u_{\text{approx.}}(x,t)$ when $\alpha = 1$, **c** absolute error $= |u_{\text{exact}} - u_{\text{approx.}}|$, when $\alpha = 1$, **d** $v_{\text{exact}}(x,t)$, when $\alpha = 1$, **e** $v_{\text{approx.}}(x,t)$, when $\alpha = 1$, **f** absolute error $= |v_{\text{exact}} - v_{\text{approx.}}|$ when $\alpha = 1$

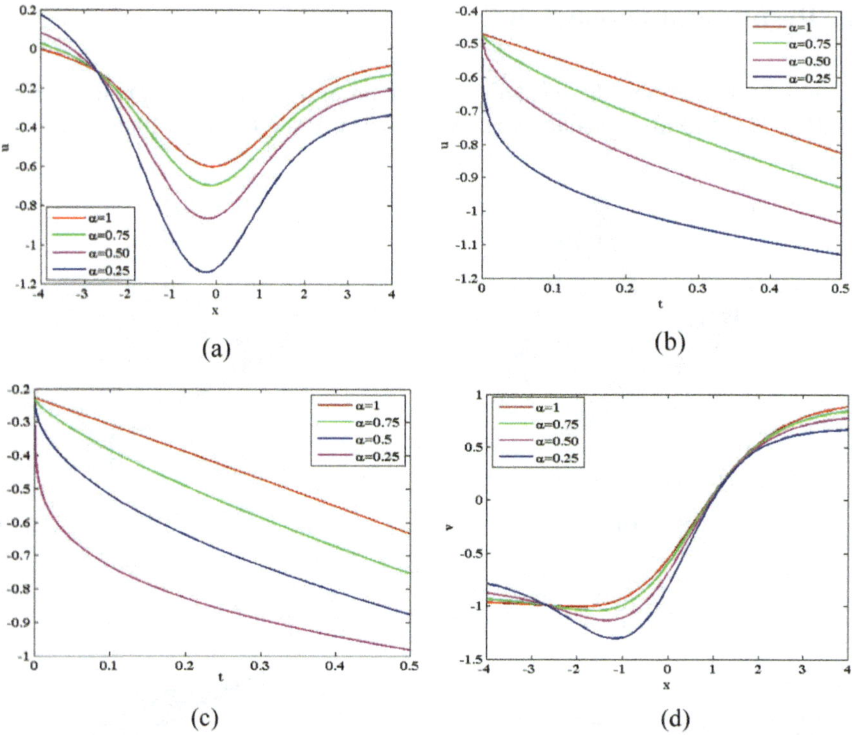

Fig. 2 For different values of α, the plots of **a** $u_{\text{approx.}}(x, t)$ with x when $t = 0.1$, **b** $u_{\text{approx.}}(x, t)$ with t when $x = 0.5$, **c** $v_{\text{approx.}}(x, t)$ with t when $x = 0.5$, and **d** $v_{\text{approx.}}(x, t)$ with x, when $t = 0.1$

that the proposed numerical approach is accurate. Figures 2a–d give 2D graphical representations of the numerical solution for varying fractional order, and we clearly observe from Figs. 2a, d that the oscillatory behavior of u and v with x as the order α increases. Figures 2b, c show decreasing behavior of numerical solution with time for order.

7 Conclusion

In this work, we used a numerical scheme HPSTT to investigate the nonlinear time-dependent Belousov-Zhabotinsky system and to check effects of the fractional order on obtained solution. The simulation results are illustrated graphically. The results are derived using the third-order iterates, and the accuracy of the proposed technique can be enhanced by calculating further approximations. The supremacy of HPSTT is its quality of joining two strong methods for the possible solution of nonlinear differential equations. We conclude that this effective scheme reduces time and computation compared to the standard scheme simultaneously conserving the high accuracy of results.

References

1. Yildrim A (2009) An algorithm for solving the Fractional Nonlinear Schrödinger equation by means of the Homotopy perturbation method. Int J Nonlinear Sci Numer Simul 10(4):445–450
2. Gupta PK (2011) Approximate analytical solutions of fractional Benney-Lin equation by reduced differential transform method and the homotopy perturbation method. Comput Math Appl 58:2829–2842
3. Sakar MG, Uludag F, Erdogan F (2016) Numerical solution of time-fractional nonlinear PDEs with proportional delays by homotopy perturbation method. Appl Math Model 40:6639–6649
4. Prakash A (2016) Analytical method for space-fractional telegraph equation by Homotopy perturbation transform method. Nonlinear Eng. 5(2):123–128
5. Prakash A, Kaur H (2017) Numerical solution for fractional model of Fokker-Planck equation by using q-HATM. Chaos Soliton Fract 105:99–110
6. Prakash A, Kumar M, Baleanu D (2018) A new iterative technique for a fractional model of nonlinear Zakharov-Kuznetsov equations via Sumudu transform. Appl Math Comput 334:30–40
7. Prakash A, Kumar M (2016) He's Variational iteration method for the solution of nonlinear Newell-Whitehead-Segel equation. J Appl Anal Comput 3:738–748
8. Prakash A, Goyal M, Gupta S (2019) A reliable algorithm for fractional Bloch model arising in magnetic resonance imaging. Pramana 92(2):1–10
9. Prakash A, Goyal M, Gupta S (2019) Fractional variational iteration method for solving time-fractional Newell-Whitehead-Segel equation. Nonlinear Eng. 8(1):164–171
10. Prakash A, Verma V, Kumar D, Singh J (2018) Analytic study for fractional coupled Burger's equations via Sumudu transform method. Nonlinear Eng 7(4):323–332
11. Prakash A, Kumar M (2019) Numerical method for time-fractional Gas dynamic equations. Proc Natl Acad Sci India Sect A Phys Sci 89(3):559–570
12. Prakash A, Kumar M, Sharma KK (2015) Numerical method for solving coupled Burgers equation. Appl Math Comput 260:314–320
13. Prakash A, Verma V (2019) Numerical method for Fractional model of Newell-Whitehead-Segel equation. Front Phys 7(15):1–10
14. Prakash A, Kaur H (2018) An efficient hybrid computational technique for solving nonlinear local fractional partial differential equations arising in fractal media. Nonlinear Eng 7(3):229–235
15. Prakash A, Kaur H (2018) q-homotopy analysis transform method for space and time- fractional KdV-Burgers equation. Nonlinear Sci Lett A 9(1):44–61
16. Prakash A, Kumar M (2017) Numerical method for fractional dispersive partial differential equations. Commun Numer Anal 2017(1):1–18
17. Prakash A, Veeresha P, Prakasha DG, Goyal M (2019) A homotopy technique for a fractional order multi-dimensional telegraph equation via the Laplace transform. Eur Phys J Plus 134(19):1–18
18. Goyal M, Prakash A, Gupta S (2019) Numerical simulation for time-fractional nonlinear coupled dynamical model of romantic and interpersonal relationships. Pramana 92(82):1–14
19. Prakash A, Goyal M, Baskonus HM, Gupta S (2020) A reliable hybrid numerical method for a time dependent vibration model of arbitrary order. AIMS Math 5(2):979–1000
20. Prakash A, Goyal M, Gupta S (2019) Numerical simulation of space-fractional Helmholtz equation arising in seismic wave propagation, imaging and inversion. Pramana 93(28):1–15
21. Prakash A, Goyal M, Gupta S (2020) q-homotopy analysis method for fractional Bloch model arising in nuclear magnetic resonance via the Laplace transform. Ind J Phys 94:507–520

22. Goyal M, Baskonus HM, Prakash A (2019) An efficient technique for a time fractional model of lassa hemorrhagic fever spreading in pregnant women. Eur Phys J Plus 134(10):482:1–10
23. Prakash A, Veeresha P, Prakasha DG, Goyal M (2019) A new efficient technique for solving fractional coupled Navier–Stokes equations using q-homotopy analysis transform method. Pramana 93(1): 6, 1–10
24. Goyal M, Baskonus HM, Prakash A (2020) Regarding new positive, bounded and convergent numerical solution of nonlinear time fractional HIV/AIDS transmission model. Chaos Soliton Fract 139(110096):1–12
25. Goyal M, Gupta S (2017) Variational iteration method for solving advection-dispersion equations. Acta Ciencia Indica, XLIII M, pp 215–219
26. Goyal M, Gupta S (2017) On some systems of partial differential equations. Acta Ciencia Indica, XLIII M, pp 103–107
27. Gupta S, Goyal M, Prakash A (2020) Numerical treatment of Newell-Whitehead-Segel equation. TWMS J App Eng Math 10:312–320
28. Goyal M, Sharma U (2010) Solution of linear and non-linear higher order BVPs using MDM. Acta Ciencia Indica XXXVI M:691–698
29. Goyal M, Prakash A, Gupta S (2020) Numerical Analysis of coupled time-fractional differential equations arising in epidemiological models. In: Mathematical Modelling and soft computing in epidemiology. CRC Press, Boca Raton, pp 173–198
30. Goyal M (2009) Numerical solution of tenth order BVP using modified decomposition method. Acta Ciencia Indica XXXV M:33–37
31. Goyal M, Bhardwaj VK, Prakash A (2021) Investigating new positive, bounded, and convergent numerical solution for the nonlinear time-dependent breast cancer dynamic competition model. Math Meth Appl Sci 1–18. https://doi.org/10.1002/mma.7057
32. Prakash A, Kumar M (2016) Numerical solution of two-dimensional time fractional order biological population model. Open Phys 14:177–186
33. Goyal M (2009) Solution of two dimensional heat flow equation by Adomian decomposition method. Acta Ciencia Indica XXXVM(1):1–4
34. Prakash A, Kumar M (2017) Numerical method for solving time-fractional multi-dimensional diffusion equations. Int J Comput Sci Math 8(3):257–267
35. Goyal M (2008) A new approach of solving tenth order boundary value problems. Acta Ciencia Indica XXXIVM(4):1941–1945
36. Prakash A, Kumar A, Baskonus HM, Kumar A (2021) Numerical analysis of nonlinear fractional Klein–Fock–Gordon equation arising in quantum field theory via Caputo-Fabrizio fractional operator. Math Sci. https://doi.org/10.1007/s40096-020-00365-2
37. Goyal M (2008) Solution of fluid dynamics problems by Adomian decomposition method. Acta Ciencia Indica XXXIVM 3:1417–1424
38. Kumar D, Singh J, Prakash A, Swaroop R (2019) Numerical simulation for system of time-fractional linear and nonlinear differential equations. Prog Fract Differ Appl 5(1):65–77
39. Goyal M, Gupta GD, Kishan H (2008) Solution of wave equations of two and three dimensions by Adomian decomposition method. Acta Ciencia Indica XXXIVM (1):415–420
40. Prakash A, Kaur H (2021) Analysis and numerical simulation of fractional Biswas-Milovic model. Math Comput Simul 181:298–315
41. Verma V, Prakash A, Kumar D, Singh J (2019) Numerical study of fractional model of multi-dimensional dispersive partial differential equation, Journal of Ocean. Eng Sci 4:338–351
42. Goyal M, Gupta GD (2008) Solution of various one dimensional wave equations using Adomian decomposition method. Acta Ciencia Indica XXXIVM(1):383–388
43. Prakash A, Kaur H (2019) Analysis and numerical simulation of fractional order Cahn-Allen model with Atangana-Baleanu derivative, Chaos, Soliton. Fractals 124:134–142
44. Goyal M, Gupta GD (2007) Solution of special three-dimensional inhomogeneous differential equations using Adomian decomposition method. Acta Ciencia Indica XXXIIIM(4):1569–1574
45. Goyal M, Watkins CD (2007) Computer based numerical and statistical techniques. Infinity Science Press, Massachusetts, pp 1–500

46. Prakash A, Verma V (2019) Numerical solution of nonlinear fractional Zakharov-Kuznetsov equation arising in ion-acoustic waves. Pramana 93(66):1–19
47. Padmavathi V, Prakash A, Alagesan K, Magesh N (2021) Analysis and numerical simulation of novel coronavirus (COVID-19) model with Mittag-Leffler Kernel, Math Meth Appl Sci 1–15. https://doi.org/10.1002/mma.6886
48. Bali NP, Goyal M (2019) A text book of engineering mathematics, vol I, II, 10th edn. Laxmi Publications, New Delhi, pp 1–1584
49. Prakash A, Verma V (2020) Two efficient computational technique for fractional nonlinear Hirota–Satsuma coupled KdV equations. Eng Comput 1–28. https://doi.org/10.1108/EC-02-2020-0091
50. Prakash A, Kumar M (2019) Numerical method for space- and time-fractional telegraph equation with generalized lagrange multipliers. Prog Fract Differ Appl 5(2):111–123
51. Prakash A, Veeresha P, Prakasha DG (2019) A reliable algorithm for time-fractional Navier-Stokes equations via Laplace transform. Nonlinear Eng 8:695–701
52. Prakash A, Kaur H (2019) A reliable numerical algorithm for fractional model of Fitzhugh-Nagumo equation arising in the transmission of nerve impulses. Nonlinear Eng 8:719–727
53. Prakash A, Kumar M (2019) Numerical solution of time-fractional order Fokker-Planck equation. TWMS J App Eng Math 9(3):446–454
54. Belgacem FBM, Silambarasan R, Hammouch Z (2017) New and extended applications of the natural and sumudu transforms: fractional diffusion and Stokes fluid flow realms. In: Advances in real and complex analysis with applications. Birkhauser, Singapore, pp 107–120
55. Belgacem FBM, Karaballi AA (2006) Sumudu transform fundamental properties investigations and applications. J Appl Math Stoch Anal 91083, 23 p
56. Watugala GK (2002) The Sumudu transform for functions of two variables. Mat Eng Ind 8:293–302
57. Asiru MA (2002) Further properties of the Sumudu transform and its applications. Int J Math Educ Sci Technol 33:441–449
58. Weerakoon S (1994) Applications of Sumudu transform to partial differential equations. Int J Math Educ Sci Technol 25:277–283
59. Weerakoon S (1998) Complex inversion formula for Sumudu transforms. Int J Math Educ Sci Technol 29:618–621
60. Singh J, Kumar D (2011) Sushila: homotopy perturbation Sumudu transform method for nonlinear equations. Adv Theor Appl Mech 4:165–175
61. Ghorbani A, Saberi-Nadjafi J (2007) He's Homotopy perturbation method for calculating. Adomian polynomials. Int J Nonlinear Sci Numer Simul 8:229–232
62. Ghorbani A (2009) Beyond Adomian polynomials: he polynomials. Chaos Soliton Fract 39:1486–1492
63. Jaradat A, Noorani MSM, Alquran M, Jaradat HM (2018) Numerical investigations for time-fractional nonlinear model arise in physics. Res Phys 8:1034–1037
64. Podlubny I (1999) Fractional differential equations. Academic Press, San Diego
65. Kilbas AA, Srivastava HM, Trujillo JJ (2006) Theory and applications of fractional differential equations, 1st edn. Elsevier Science, New York, pp 1–540
66. Watugala GK (1993) Sumudu transform: a new integral transform to solve differential equation and control engineering problems. Int J Math Educ Sci Technol 24(1):35–43
67. Bulut H, Baskonus HM, Belgacem FBM (2013) The analytical solutions of some fractional ordinary differential equations by Sumudu transform method. Abstr Appl Anal 2013(203875):1–6

Detection of High Impedance Fault Using Advanced ELM-based Neuro-fuzzy Inference System

K. V. Shihabudheen

Abstract In power systems, protective systems like circuit breakers and relays are used to detect abnormal conditions. High impedance fault (HIF) is a category of fault in power systems where energized conductors in contact with high impedance material, which cannot be detected by conventional protective measures. In this paper, an advanced neuro-fuzzy system called online sequential fuzzy extreme learning machine (OS-fuzzy-ELM) is utilized for efficient detection of HIF. OS-fuzzy-ELM is an adaptive strategy that combines the learning capability of the extreme learning machine (ELM) and the cognitive capability of the Takgi-Sugeno-Kang (TSK) fuzzy inference system. Discrete wavelet transform (DWT) and its energy signals are used for feature extraction. The simulation has been carried out using IEEE-9 bus standard distribution system, and the high impedance detection performance of OS-fuzzy-ELM is compared with ELM and SVM.

Keywords Extreme learning machine · Online sequential ELM · Discrete wavelet transform · Classification

1 Introduction

High impedance fault (HIF) is a type of fault happens in the power system where energized conductor comes in contact with high impedance surfaces like tree limb, broken pole, water flood, etc. HIF exhibits arcing and flashing at the high impedance point due to high the potential difference between the two contact points. HIF cannot be detected by a conventional protection scheme because of the low fault current. Failure of HIF detection may lead to fire hazards and electric shocks. The HIF currents are very chaotic in nature characterized by low magnitude, asymmetry, instability, non-linearity, non-stationarity, intermittence, buildup and shoulder [1].

K. V. Shihabudheen (✉)
Department of Electrical Engineering, Indian National Institute of Technology Calicut, Calicut 673601, India
e-mail: shihabudheen@nitc.ac.in

S. Suhag et al. (eds.), *Control and Measurement Applications for Smart Grid*, Lecture Notes in Electrical Engineering 822,
https://doi.org/10.1007/978-981-16-7664-2_32

In the past few decades, many researchers have shown interest in detecting HIFs and several work have been done on it so far. These works include some conventional schemes such as mechanical and electrical methods [2] for which installation and maintenance costs are high, as six units need to be mounted on each pole for bi-directional coverage. It also leads to false tripping and needs more operating time. The harmonics-based HIF identification scheme provides the detection by current harmonic content in different fault situations [3]. Finding threshold value itself is a big task and detection accuracy also gets affected. It can be overcome through intelligent techniques which avoid the concept of threshold value.

Artificial neural networks (NNs) are widely used for the detection of HIF in distribution networks [4]. Sarlak [5] introduced the detection of HIF using multi-layer perceptron. Neural network and S-transform (ST) strategy are used detection of HIF in radial distribution system [6]. In [7], a polynomial function-based voltage model and the online trained neural network have been designed for competent HIF detection. Most of the feedforward neural networks use gradient-based backproro-gation algorithm to find out the optimal parameters. But learning using the gradient decent method may produce the local minima trapping and slow convergence. The kernel-based method called support vector machine (SVM) and its variants provide a revolutionary strategy for finding optimal decision boundary from given nonlinear data [8, 9].

Recently, a non-iterative technique for the learning of single layer feedforward neural networks (SLFFNNs) was proposed and named as extreme learning machine (ELM) [10]. In ELM, parameters connecting between the hidden layer and the input layer are found randomly and parameters of the output layer calculated by linear least square algorithm. Introduction of extreme learning in SLFFNN produces better generalization accuracy with minimum computational effort compared to backprop-agation algorithm. ELM is successfully applied for smart grid protection with an HIF detection scheme [11]. In spite of these advantages, ELM may suffer inherent randomness in the hidden parameter which produces redundancy in hidden layers [12].

The learning capability of ELM and the reasoning ability of Takagi–Sugeno-Kang (TSK) inference system can be combined to get better approximation capability. In fuzzy extreme learning machine (FELM), the membership functions and fuzzy if–then rules are embedded into the hidden layers of the ELM [13]. In most of the FELM systems, membership parameters are calculated randomly and antecedent parameters are found by extreme learning [14, 15]. In [16], an online sequential extreme learning-based TSK fuzzy inference system (OS-fuzzy-ELM) is proposed for sequential training of the network. It uses sequential concept of learning where data are coming in chunk by chunk, so that algorithm is more suitable for online application [17, 18].

This paper proposes the detection of HIF using an online sequential fuzzy extreme learning machine (OS-fuzzy-ELM). Discrete wavelet transform (DWT) is used for efficient feature extraction. Initially, currents from monitoring points in the power system network are captured using CT and then meaningful features are extracted using discrete wavelet transform (DWT). And finally, classification of HIFs from

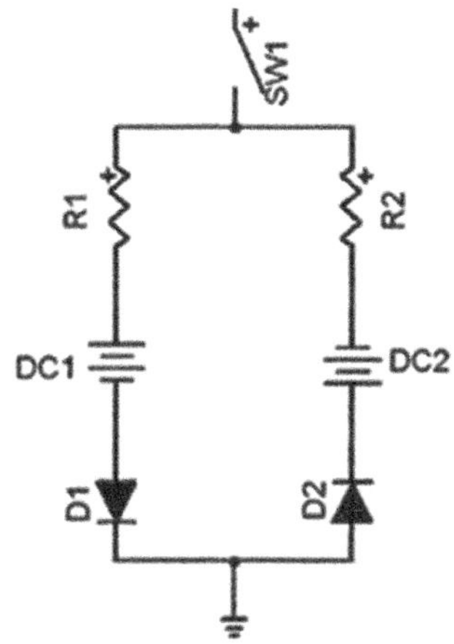

Fig. 1 HIF fault model

LIFs and normal switching events (SEs) like capacitor switching using OS-fuzzy-ELM techniques. Chapter 2 shows the details of the HIF model. Proposed mythology is described in Chap. 3. Chapter 4 depicts simulation study, and Chap. 5 gives the conclusion.

2 HIF Model

In order to simulate the effect of HIF on the distribution system, model of HIF is considered here. A perfect model should consider the effect of arcing, linear, nonlinear and asymmetric characteristics of HIF currents. Emmanuel model (Fig. 1) is one of the promising model of HIF used by the research community [19]. It is simple and considered the nonlinear characteristics of HIF behaviour. The model consists of two diodes $D1$ and $D2$ and two unequal resistances $R1$ and $R2$. Two unequal voltages DC1 and DC2 are incorporated in series with diode and resistors. The detailed analysis of the model is provided in [19].

3 Methodology

The proposed HIF technique essentially pattern recognition problem consists of three steps as shown in Fig. 2. Initially, currents from monitoring points in the power system network are captured using CT. The sufficient features from the current signal are extracted in the second stage. Discrete wavelet transform (DWT) is used for extracting t from the current signals. The energy of the decomposed high frequency signal is chosen as a feature for the classifier. The classifier is used to distinguish between the HIF and non-HIF signals. In this paper, OS-fuzzy-ELM classifier is used to

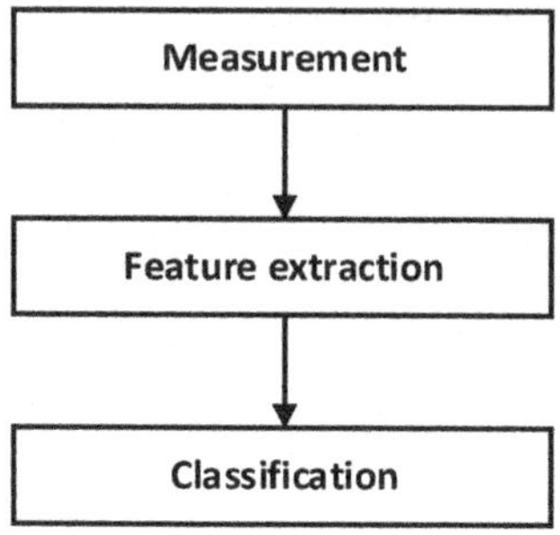

Fig. 2 Basic methodology of HIF detection

differentiate between HIF and non-HIFs. For comparison, ELM and SVM are also used for the classification.

3.1 Discrete Wavelet Transform (DWT)

Discrete wavelet transform (DWT) is one of the signal decomposition techniques which convert the signals into different frequency time domain signals by down sampling and up sampling. DWT exhibits multiresolution in time and frequency. In DWT, the input signal is decomposed into approximation and detailed coefficient using filters. The detailed coefficient is again decomposed into components. The pictorial representation of three-level decomposition is shown in Fig. 3.

Let input signal is represented as $a[n]$, detail coefficients are indicated as $x[n]$ and $y[n]$ are the approximation coefficients. Equations (2) and (3) represent detailed and approximated coefficients obtained from the output of low pass filter and high pass filter, respectively.

$$\mathrm{DWT}_{\nabla}(i, j) = \langle p(t), \lambda_{i.j} \rangle = q^{\frac{i}{2}} \int_{\infty}^{\infty} p(t) \nabla^{*}(q^{i} t - j) \mathrm{d}t \tag{1}$$

$$\mathrm{Det}_{k}[n] = \sum_{i} a[n] * x[2n - i] \tag{2}$$

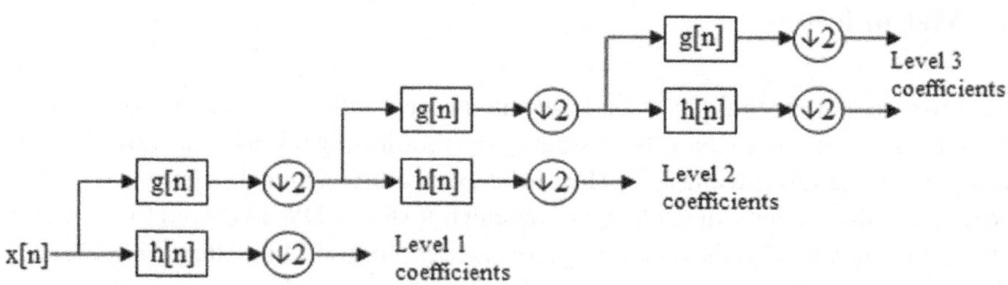

Fig. 3 DWT decomposition

$$\mathrm{App}_k[n] = \sum_i a[n] * y[2n - i] \tag{3}$$

3.2 Support Vector Regression (SVR)

In machine learning, support vector machines (SVM) are the most popular algorithm based on kernel methods. SVM is gaining popularity compared to other soft computing techniques such as artificial neural networks (ANNs) as it overcomes the bottleneck issues of ANN such as overfitting and local minima. Support vector regression (SVR) [20] is an extended version of SVM where it can effectively use for regression problems. All the input combinations are transferred to the higher dimensional plane using special function called kernel function. The input–output relation of SVR is written as

$$\hat{y} = w^T \phi(x) + b \tag{4}$$

where w is the weight connecting between input and output, b indicates the bias in output node and $\phi(x)$ is transferred function in higher dimensional space. Using methods of quadratic optimization, the optimum value of weight and base can be found. The output equation after simplification can be written as

$$\hat{y} = \sum_i \alpha_i K(x_i, x) + b \tag{5}$$

where α_i is Lagrangian function to be calculated by quadratic optimization and $K(x_i, x)$ indicates kernel function.

3.3 Extreme Learning Machine (ELM)

Extreme learning machine (ELM) is a learning strategy applied for SLFFNN. Consider SLFFNN with distinct S samples (P_i, Q_i) as training data, where $P_i = [p_{i1}, p_{i2}, \ldots, p_{in}]^{\mathrm{T}}$ and $Q_i = [q_{i1}, q_{i2}, \ldots, q_{im}]^{\mathrm{T}}$. The output of the network with $\tilde{S}$ hidden neuron is represented as

$$O_j = \sum_{i=1}^{\tilde{S}} \varphi_i f(X_i P_j + b_i); \; j = 1, 2, \ldots, S; \tag{6}$$

where $X_i = [x_{i1}, x_{i2}, ..., x_{in}]^{\mathrm{T}}$ is a weight vector connecting between input layer and ith hidden layer, $\phi_i = [\phi_{i1}, \phi_{i2}, ..., \phi_{im}]^{\mathrm{T}}$ is weight vector connecting between ith hidden layer and output neuron, b_i is bias on hidden layer and $f(.)$ is activation function in the hidden neuron.

Under zero training error, the network can be represented for all N training samples in matrix form,

$$\begin{bmatrix} f(X_1 P_1 + b_1) & \cdots & f(X_{\tilde{S}} P_1 + b_{\tilde{S}}) \\ \vdots & \vdots & \vdots \\ f(X_1 P_S + b_1) & \cdots & f(X_{\tilde{S}} P_S + b_{\tilde{S}}) \end{bmatrix}_{S \times \tilde{S}} \begin{bmatrix} \phi_1^T \\ \vdots \\ \phi_{\tilde{S}}^T \end{bmatrix}_{\tilde{S} \times m} = \begin{bmatrix} Q_1^T \\ \vdots \\ Q_N^T \end{bmatrix}_{S \times n} \quad (7)$$

which is represented in the form, $H\phi = Q$.

Training algorithm: For distinct S samples (P_i, Q_i), do following steps:

Step 1: Randomly calculate X_i and b_i based on continues random distribution.

Step 2: Find out the H matrix using Eq. (7)

Step 3: Find the output parameter $\phi = H^{\otimes} Q$, where $H^{\otimes}$ is Moore–Penrose generalized inverse of H.

3.4 OS-Fuzzy-ELM

Online sequential fuzzy ELM (OS-fuzzy-ELM) introduces a sequential modification of TSK-based fuzzy ELM using recursive least square algorithm. It can lean the data which are coming in one by one or block by block. In OS-fuzzy-ELM, premise parameters are randomly selected and consequent parameters are determined analytically based on arriving of data in sequential. The output of the OS-fuzzy-ELM can be represented in the form, $G\beta = D$, where G is matrix indicating normalized fuzzy rule which containing premise parameters, β is array containing consequent parameters and D is the target matrix. The training algorithm of OS-fuzzy-ELM is summarized below.

1. **Initialization phase**:
   (a) Initialize a block of training data $N_0 = \{(x_i, d_i)\}_{i=1}^{N_0}$ from the given set N
   (b) Randomly, select premise parameters
   (c) Compute initial matrix G_0 by using N_0
   (d) For $j = 0$, find out the initial consequent parameter matrix $\beta_0 = T_0 G_0^{\mathrm{T}} D_0$, where $T_0 = \left(G_0^{\mathrm{T}} G_0\right)^{-1}$
2. **Sequential learning phase**:
   For the $(j+1)$th block of observation $N_{j+1} = \{(x_i, d_i)\}_{i=jN_0+1}^{(j+1)N_0}$
   (a) Calculate the matrix G_{j+1} by using N_{j+1}

(b) Calculate consequent parameter matrix

$$
\begin{aligned}
T_{j+1} &= T_j - T_j G_{j+1}^{\mathrm{T}} (I + G_{j+1} T_j G_{j+1}^{T})^{-1} G_{j+1} T_j \\
\beta_{j+1} &= \beta_j + T_{j+1} G_{j+1}^{\mathrm{T}} (D_{j+1} - G_{j+1} \beta_j)
\end{aligned} \tag{8}
$$

(c) set $j = j + 1$, go to step 2.

4 Simulation Study

In this section, the testing of the proposed HIF detection scheme is performed using various faults at different locations of the power system. The simulation is performed on IEEE-9 bus distribution system with Immanual model for creating HIF using MATLAB 2018. Figure 4 shows a single line diagram for a 22 kV IEEE-9 bus system. IEEE-9 bus system includes three generators, 9 buses, 6 transmission line, 3 transformer and 3 constant impedance load [21]. Buses 2, 4 and 9 are considered as monitoring points, hence CTs are connected at these points. The HIF fault model (Immanual model) is included in buses 4, 7 and 3, where these buses are chosen as fault points. The conventional faults like line ground (LG), line to line (LL), line to line ground (LLG) and three phase fault (LLLG) are also incorporated at the fault points. Extensive simulations of IEEE-9 bus system are performed under HIF and other fault conditions to collect the training data. Total of 750 values of currents for HIF conditions and 750 values for non-HIF conditions are taken. Non-HIF data consist of normal conditions and conventional faults like LG, LL, LLG, LLLG. Data with capacitance switching are also taken by connecting capacitance bank at bus point 9. Signal decomposition of the measured current signal is performed

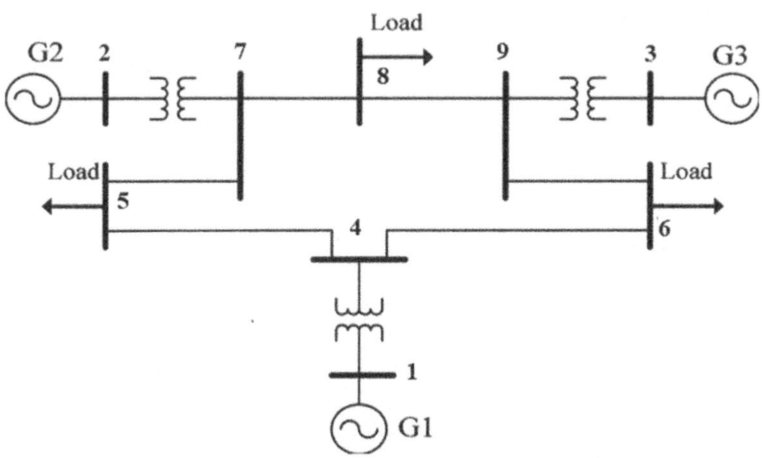

Fig. 4 Single line diagram of IEEE-9 bus system

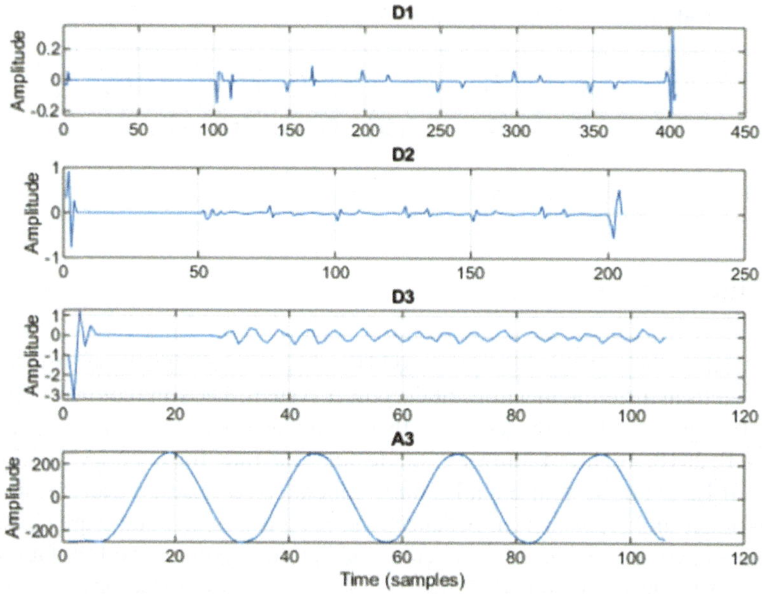

Fig. 5 DWT coefficients of high impedance fault (HIF)

using three-level DWT using Daubechies (D-4) mother wavelet. It comprises three detail coefficients ($D1$, $D2$, $D3$) and one approximate coefficient ($A1$). The sampling frequency is chosen as 10 kHz. According to three-level DWT algorithm, $D1$ has a frequency range from 2.5 to 5 kHz, $D2$ has a frequency range from 1.25 to 2.5 kHz, $D3$ has a frequency range from 0.625 to 1.25 kHz and A1 has frequency range from 0 to 0.625 kHz. Three-level DWT for HIF condition simulated in the bus is shown in Fig. 5.

After three-level DWT decomposition, the following features of the detailed coefficient are considered for classification purpose, (i) energy, (ii) standard deviation, (iii) RMS value, (iv) entropy and (v) power. These features are given to the input of the classification algorithm.

4.1 Optimal Parameter Selection of Classifier

Out of the total of 1500 data collected, 1000 data were used for training and validation and the remaining 500 data were used for testing of the classifier. The hyperparameters of SVM, ELM and OS-fuzzy-ELM are selected by validation. In SVM, the hyperparameter such as regularization parameter (C) and kernel parameter (σ) is calculated in the range $\{2^{-10}, 2^{-9}, ..., 2^{0}, ..., 2^{20}\}$ and optimum value calculated by validation set. Hidden neurons of ELM are varied from 1 to 100, and the best number of the hidden neuron is found by validation. In OS-fuzzy-ELM, rules are varied from

Table 1 User defined parameters of different classifiers

Classifier	Parameters	Optimal value
ELM	#hidden neuron	22
SVM	Regularization parameter (C)	2^{-1}
	Kernel parameter (σ)	2^3
OS-fuzzy-ELM	#rules	55

1 to 100 is selected and the best value is obtained by validation set. Block size of OS-fuzzy-ELM is chosen as 1. The optimal values of the parameters selected for different classifiers are shown in Table 1.

4.2 Results for HIF Detection

After the training and optimal parameter selection, the classifier is tested with testing data. The performance of the HIF detection is tested by calculating the confusion matrix, classification accuracy, sensitivity and the specificity. The confusion matrix of the classifiers such as ELM, SVM and OS-fuzzy-ELM for the testing data is shown in Tables 2, 3 and 4. It can see that ELM-based neuro-fuzzy inference systems provide better classification between the HIF and non-HIF dataset. The classification accuracy, sensitivity and specificity of different classifiers are shown in Table 5. It is shown that OS-fuzzy-ELM obtained higher accuracy (90.2%) compared to ELM and SVM. Sensitivity and specificity also greater in OS-fuzzy-ELM compared to other methods. This indicates the capability of OS-fuzzy-ELM to detect the HIF from the data collected in power distribution systems.

Table 2 Confusion matrix for ELM

	Predicted: no HIF	Predicted: HIF
Actual: no HIF	226	34
Actual: HIF	37	205

Table 3 Confusion matrix for SVM

	Predicted: no HIF	Predicted: HIF
Actual: no HIF	225	33
Actual: HIF	33	209

Table 4 Confusion matrix for OS-fuzzy-ELM

	Predicted: no HIF	Predicted: HIF
Actual: no HIF	239	19
Actual: HIF	30	212

Table 5 Comparison of HIF detection using different classifiers

	Accuracy (%)	Sensitivity (%)	Specificity (%)
ELM	85.8	86.92	84.71
SVM	86.8	87.20	86.36
OS-Fuzzy-ELM	**90.2**	**92.26**	**87.6**

Bold values corresponding to higher accuracy

Table 6 Confusion matrix for OS-fuzzy-ELM

	Predicted: HIF	Predicted: LG	Predicted: 3ϕ	Predicted: CS	Predicted: NF
Actual: HIF	77	5	3	10	5
Actual: LG	4	82	9	3	2
Actual: 3ϕ	3	10	80	3	4
Actual: CS	13	4	3	76	4
Actual: NF	5	4	3	6	82

4.3 Classification of HIF with Other Faults

In order to test the capability of OS-fuzzy-ELM to detect the different types of faults including HIF, extensive simulations on IEEE-9 system are conducted for data collection. Total 1500 data are collected with 300 cases in each fault category, which includes HIF, LG, three phase fault (3ϕ), capacitor switching (CS) and no fault (NF) condition. 1000 data are used for training and validation, and the remaining 500 data are used for testing. The ELM, SVM and OS-fuzzy-ELM are used for this multi-class classification. The optimum number of rules in OS-fuzzy-ELM is obtained as 66. After the training and validation, the network is tested with testing data. The classification accuracy of SVM, ELM and OS-fuzzy-ELM is obtained as 75%, 76.2% and 79.4%, respectively. The confusion matrix for OS-fuzzy-ELM is shown in Table 6. The more accuracy of OS-fuzzy-ELM is due to the incorporation of minimum norm solution using least square method for calculating consequent parameters.

5 Conclusion

The automated high impedance fault detection scheme is very essential in the current scenario, especially in smart gird environment. In this work, a machine learning scheme is proposed for HIF detection in distribution networks. One of the ELM-based neuro-fuzzy systems, OS-fuzzy-ELM is used for classification. The meaningful features are extracted from discrete wavelet transform-based decomposed signal. Better accuracy and precision are obtained in OS-fuzzy-ELM for HIF detection compared with ELM and SVM classifier. The more accuracy of OS-fuzzy-ELM is due to the incorporation of minimum norm solution using least square method for calculating consequent parameters.

References

1. Ghaderi A, Ginn HL, Mohammadpour HA (2017) High impedance fault detection: a review. Electr Power Syst Res 143:376–388
2. Wester CG (1998) High impedance fault detection on distribution systems. In: 1998 Rural electric power conference presented at 42nd annual conference, pp c1–5. St. Louis, MO
3. Wei M, Shi F, Zhang H, Jin Z, Terzija V, Zhou J, Bao H (2020) High impedance arc fault detection based on the harmonic randomness and waveform distortion in the distribution system. IEEE Trans Power Deliv 35:837–850
4. Ledesma JJG, do Nascimento KB, de Araujo LR, Penido DRR (2020) A two-level ANN-based method using synchronized measurements to locate high-impedance fault in distribution systems. Electr Power Syst Res 188:106576
5. Sarlak M, Shahrtash SM (2011) High impedance fault detection using combination of multi-layer perceptron neural networks based on multi-resolution morphological gradient features of current waveform. IET Gener Transm Distrib 5:588–595
6. Routray P, Mishra M, Rout PK (2015) High impedance fault detection in radial distribution system using S-transform and neural network. In: 2015 IEEE power, communication and information technology conference (PCITC), pp 545–551
7. Farias PE, de Morais AP, Rossini JP, Cardoso G (2018) Non-linear high impedance fault distance estimation in power distribution systems: a continually online-trained neural network approach. Electr Power Syst Res 157:20–28
8. Tong S, Koller D (2001) Support vector machine active learning with applications to text classification. J Mach Learn Res 2:45–66
9. Suykens JAK, Vandewalle J (1999) Least squares support vector machine classifiers. Neural Process Lett 9:293–300
10. Huang GB, Zhou H, Ding X, Zhang R (2012) Extreme learning machine for regression and multiclass classification. IEEE Trans Syst Man Cybern Part B Cybern 42:513–29
11. AsghariGovar S, Pourghasem P, Seyedi H (2019) High impedance fault protection scheme for smart grids based on WPT and ELM considering evolving and cross-country faults. Int J Electr Power Energy Syst 107:412–421
12. Olatunji SO, Selamat A, Abdulraheem A (2014) A hybrid model through the fusion of type-2 fuzzy logic systems and extreme learning machines for modelling permeability prediction. Inf Fusion 16:29–45
13. Wong SY, Yap KS, Yap HJ, Tan SC, Chang SW (2015) On equivalence of FIS and ELM for interpretable rule-based knowledge representation. IEEE Trans Neural Netw Learn Syst 26:1417–1430
14. Cao W, Gao J, Ming Z, Cai S, Shan Z (2018) Fuzziness-based online sequential extreme learning machine for classification problems. Soft Comput 22:3487–3494
15. Chu Y, Lin H, Yang L, Zhang D, Zhang S, Diao Y, Yan D (2020) Fuzzy ELM for classification based on feature space. Multimed Tools Appl 79:27439–27464
16. Rong HJ, Bin HG, Sundararajan N, Saratchandran P, Rong HJ (2009) Online sequential fuzzy extreme learning machine for function approximation and classification problems. IEEE Trans Syst Man Cybern Part B Cybern 39:1067–1072
17. Wang X, Han M (2015) Improved extreme learning machine for multivariate time series online sequential prediction. Eng Appl Artif Intell 40:28–36
18. Liu Z, Loo CK, Pasupa K, Seera M (2020) Meta-cognitive recurrent kernel online sequential extreme learning machine with kernel adaptive filter for concept drift handling. Eng Appl Artif Intell 88:103327

19. Emanuel AE, Cyganski D, Orr JA, Shiller S, Gulachenski EM (1990) High impedance fault arcing on sandy soil in 15kV distribution feeders: contributions to the evaluation of the low frequency spectrum. IEEE Trans Power Deliv 5:676–686
20. Maldonado S, González A, Crone S (2019) Automatic time series analysis for electric load forecasting via support vector regression. Appl Soft Comput J 83:105616
21. Shihabudheen KV, Kunju B, Ahammed I, Guruvarurappan A, Jose J, Keerthana D, Revathi PB (2019) Detection of high impedance fault using machine learning techniques. In: TENCON 2019—2019 IEEE Region 10 conference (TENCON), pp 2117–2122

Implementation of Neural Network-based PID Controller for Speed Control of an IC Engine

V. N. Siva Praneeth, V. Bharath Kumar, Dasa Sampath, Y. V. Pavan Kumar, D. John Pradeep, Ch. Pradeep Reddy, and Ramani Kannan

Abstract In the present day, transportation plays an important role in any country's economy and sustenance. Even though electric vehicles have started market intrusion, at present, main commuting vehicles such as cars, ships and planes work on internal combustion engines (ICEs). In line with any complex system, an ICE exhibits poor time domain characteristics when not controlled properly. Generally, PID controller is used to control the ICE to give better time domain characteristics. There are various conventional methods available to tune the PID controller such as OLTR methods and ultimate cycle methods. Generally, these offline controller tuning methods cannot address non-linear disturbances effectively. So, to overcome these drawbacks, there is a need for using artificial intelligence-based tuning methods. Hence, this paper implements an artificial neural network-based PID controller and compares it with a conventional method and track the rate of change of PID parameters with the injection of disturbances. This paper concludes that the response of the ICE system tuned with

V. N. Siva Praneeth · V. Bharath Kumar · D. Sampath · Y. V. Pavan Kumar (✉) · D. John Pradeep
School of Electronics Engineering, VIT-AP University, Amaravati, AP 522237, India
e-mail: pavankumar.yv@vitap.ac.in

V. N. Siva Praneeth
e-mail: praneeth.18bev7022@vitap.ac.in

V. Bharath Kumar
e-mail: bharath.18bec7093@vitap.ac.in

D. Sampath
e-mail: sampath.18bec7101@vitap.ac.in

D. John Pradeep
e-mail: john.darsy@vitap.ac.in

Ch. Pradeep Reddy
School of Computer Science and Engineering, VIT-AP University, Amaravati, AP, India
e-mail: pradeep.ch@vitap.ac.in

R. Kannan
Electrical and Electronics Engineering, Universiti Teknologi PETRONAS (UTP), Seri Iskandar, 32610 Perak, Malaysia
e-mail: ramani.kannan@utp.edu.my

S. Suhag et al. (eds.), *Control and Measurement Applications for Smart Grid*,
Lecture Notes in Electrical Engineering 822,
https://doi.org/10.1007/978-981-16-7664-2_33

ANN-PID gives a better output when compared to the key conventional PID tuning methods.

Keywords Speed control · IC engine · PID controller · ANN-PID controller

1 Introduction

Due to the rapid increase of population, the requirement for transportation is heavily increased. So, the vehicles should be designed in such a way that most of the space should be used for seating and very less place should be used for the engine and other parts without compromising on the performance of the vehicle. In such a case, combustion engine is introduced as it occupies less space and gives high torque. The combustion engine works by burning air–fuel mixture converting it into mechanical energy. Generally, there are two types of combustion engines, viz. external combustion engine (ECE) and the internal combustion engine (ICE).

In external combustion engines, heat is produced by the combustion of air and fuel and then the mixture is transferred to another liquid which is the actual working fluid inside the cylinder. In ICEs, the mixture of air and fuel is directly used as a working fluid inside the cylinder for combustion. ICE is widely used as it gives better performance in terms of efficiency. Authors of [1] describe the working of an ICE, and in [2, 3], the mathematical model for an ICE is derived. In [4], implementation of a PI-based controller for an ICE is presented and [5] deals with the control and design of an intelligent ICE cooling method. In [6], adaptive PID controller implementation for speed regulation of BLDCM is presented. In [7], a comparison of various PID algorithms for Cuk converter is presented, and in [8], implementation of ZN and CHR methods for a thermal plant is presented. Implementation of fractional order PID is given in [9] and [10]. In [11], the implementation of a controller for robot application based on fuzzy theory is proposed. Further, in [12], a qualitative comparison of fuzzy PID methods with other offline methods for magnetic levitation system is presented. A detailed comparison of PID methods with genetic algorithm-based tuning is given in [13], and detailed comparison and implementation of various algorithms in ANN-PID for a system were presented in [14, 15]. IC engine models are widely discussed in [16, 17]. This paper implements the ANN-PID controller for ICE with a six-cylinder configuration [18], having a transfer function (1).

$$\mathrm{TF} = \frac{760s + 3578}{0.83s^2 + 766s + 3578} \tag{1}$$

System control plays a crucial role to get better output by resisting the system from disturbances. For tuning the PID controller, the foremost algorithm is Ziegler-Nichols, which was introduced in the year 1942. Since then, a variety of algorithms has been evolved for the tuning of controllers. These tuning techniques are majorly classified into open-loop methods (OLTR) including error performance index (EPI) methods and ultimate cycle (UC) or closed-loop methods. In most cases, UC methods give a better response than OLTR methods. But, this cannot cope up with practical disturbances of the system. This paper implements the Ziegler-Nichols UC method as the conventional best method to test the usefulness of the proposed ANN-PID controller.

2 Description of the Proposed ANN-PID Controller

Generally, artificial neural networks possess a connection between points or nodes in a pattern. This pattern resembles the structure of neurons in the human brain, thus, these are called artificial neural networks. Perceptron is a fundamental type of neural network. Some of the widely used neural network types are 'feed-forward neural network (FFNN), radial basis function neural network (RBNN) and recurrent neural network (RNN)', which are discussed below [14].

FFNN is one of the simplest architectures in the neural network, in which only one-way data flow is present. It is also called a front propagation network since no recursion of data takes place. The output is given as the sum of products of input and weights at nodes. According to the number of hidden layers, it is further classified into two types, namely single-layer FFNN and multi-layer FFNN. The advantage of this architecture is to have great disturbance rejection. A typical FFNN is shown in Fig. 1a. The applications which use this architecture is pattern recognition, speech recognition, computer vision, etc. RBNN shown in Fig. 1b is similar to that of FFNN architecture, but the main difference lies in taking the approximation. The radial basis function is used in the hidden layer, which acts as the activation function for calculating the approximations. Each neuron will classify the input by calculating the Euclidian distance between the input and its prototype. Because of this classification property, this works best with continuous input. RBNN has a fast learning rate. It is generally used for function prediction and time-series prediction. RNN shown in Fig. 1c is also similar to FFNN architecture, but the difference lies in having a memory element present in the hidden layer of RNN architecture. Initially, RNN gives output using the sum of products of inputs and weights which is the same as FFNN. Later, this information is stored in memory and RNN predicts the output of the next layers. If the prediction is incorrect, an adjustment is made to the learning rate to match the predictions. Finally, makes it a better output predictor by keeping track of the previous values. It is also known as backpropagation architecture. This architecture is generally used in the language translation, grammar auto checking, etc.

3 Implementation of ANN-PID Controller

In this proposed method, the training data to compute ANN parameters are obtained using the process flow shown in Fig. 2. Initially, the ICE system is coupled with the

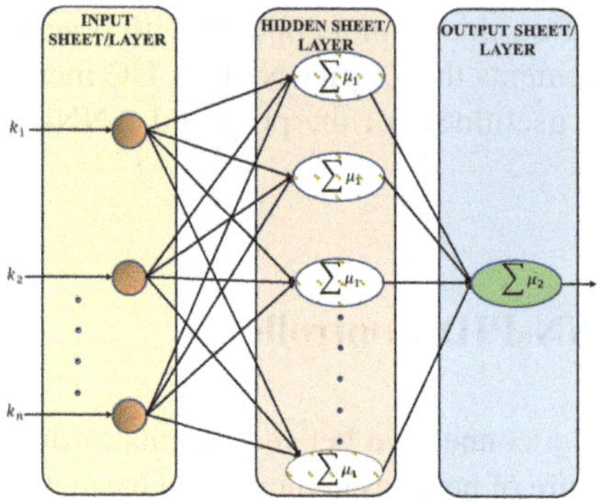

(a) Structure of FFNN

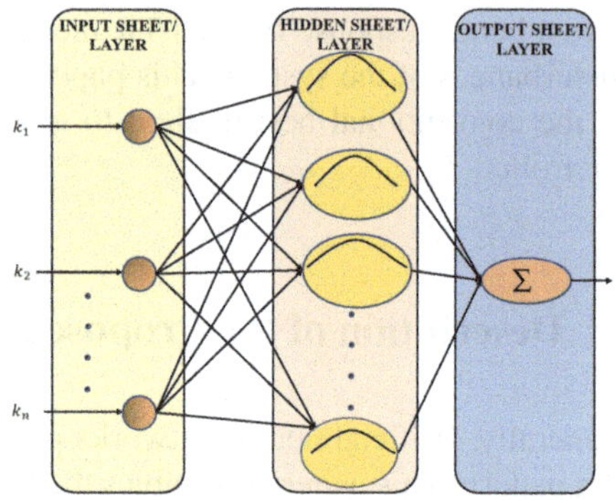

(b) Structure of RBNN

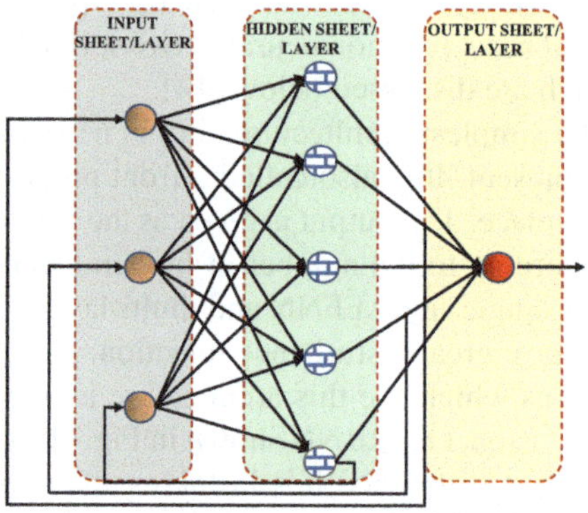

(c) Structure of RNN

Fig. 1 Different architectures of artificial neural networks

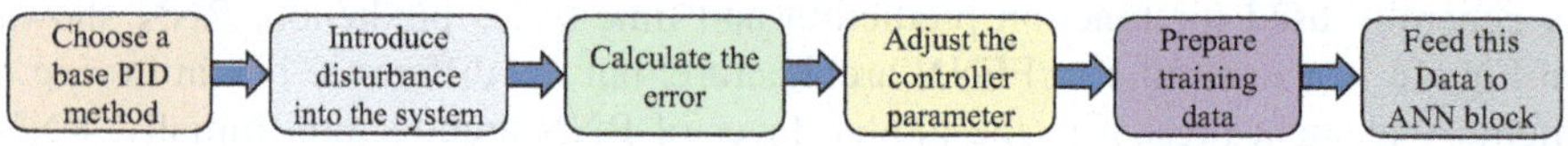

Fig. 2 Flowchart represents the implementation of ANN-PID controller

PID controller that is tuned with the ZN-2 method. In the second step, a disturbance is introduced into the system and the corresponding error in the response is calculated. This is used to update the controller parameters. The controller parameters are computed under different disturbances and are noted. This acts as training data for the proposed controller. The overall system model is shown in Fig. 3. The FFNN architecture with ten hidden layers is considered for the ANN implementation. For processing the training data, Levenberg–Marquardt algorithm is used, where the division of data for training, testing and validation is used as 75%, 10% and 15%, respectively. The regression coefficient obtained is 0.974.

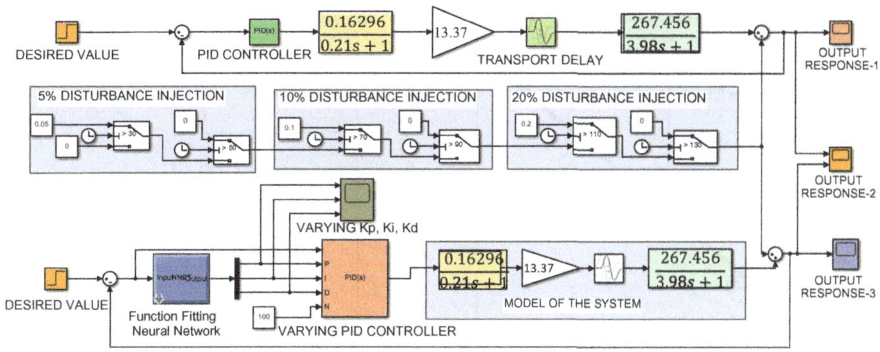

Fig. 3 MATLAB/Simulink model of ICE with PID and ANN-PID controllers

4 Result and Analysis

The proposed neural network structure has one input parameter (error) and three output parameters (K_p, K_i, K_d). The accuracy of training is admitted by observing the regression coefficient value. This determines the strength of the relation between input and output data. The more, it is nearer to 1, the stronger the relation. As shown in Fig. 4, for the proposed system, the regression coefficient obtained is 0.974, which indicates satisfactory training. The efficacy of this proposed ANN-PID controller is assessed by observing time domain and frequency domain responses given as follows.

4.1 Time Domain Analysis

For safe operation of the system, it is preferred to have a smooth (non-oscillatory) response during the transient operation including the disturbance conditions. To probe the stability of the system, three different disturbances (5, 10, 20% of the input step) are given to the system and the corresponding response is observed. The system, when tuned with the ZN-2 method, produces oscillatory response both in the initial transient region and the disturbance condition as shown in Fig. 5. But, the system with the proposed ANN-PID controller gives a smooth response both in the initial transient region and the disturbance condition as shown in Fig. 6. Comparative response of the plant with traditional PID controller and proposed ANN-PID controller is presented in Fig. 7. The overall performance parameters computed for these two responses are compared in Table 1. PID controller is a mixture of proportional constant, integral constant, derivative constant. The optimal combination of these constants sets the system response to the desired value. So, when the system undergoes different disturbances, the change in PID gains will compensate the disturbance to the maximum extent. In Fig. 8, the dotted circles represent a change in the proportional gain value for various disturbances injected into the system. The corresponding changes can

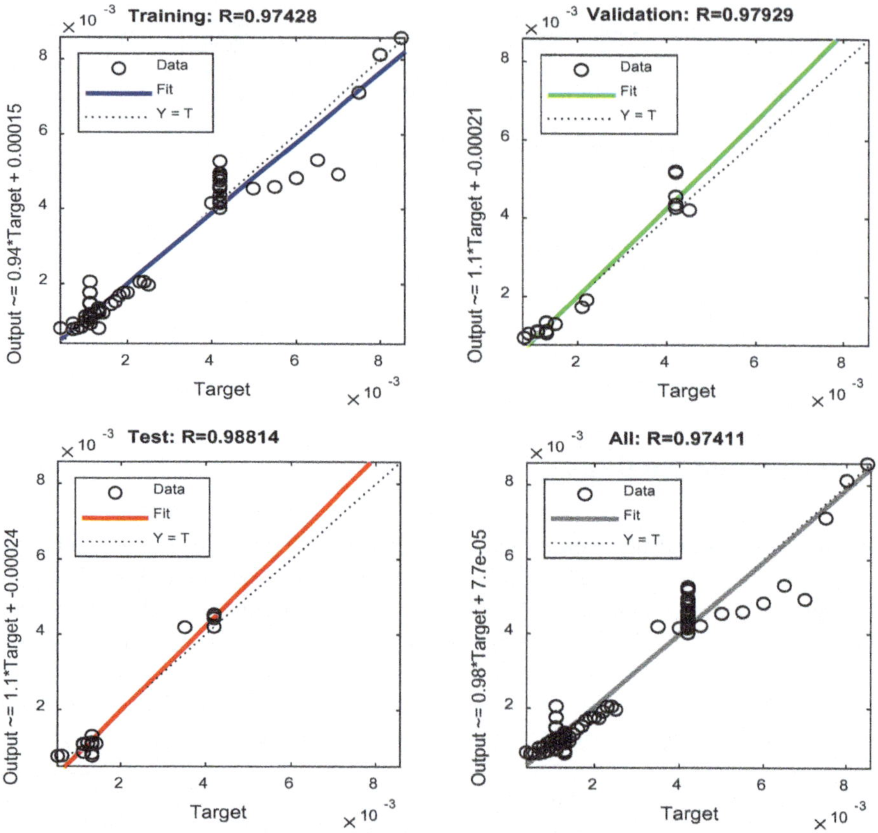

Fig. 4 Plot between the target and the output for various regression coefficients

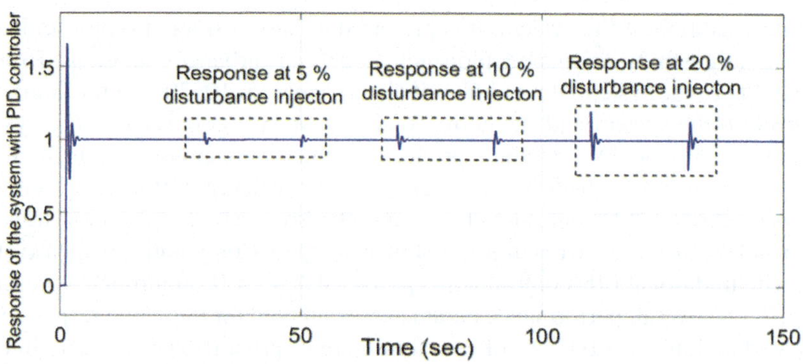

Fig. 5 System response with the conventional PID controller

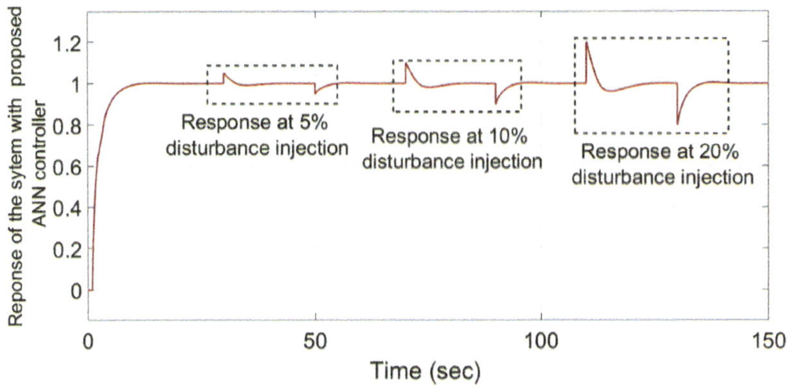

Fig. 6 System response with proposed ANN-PID controller

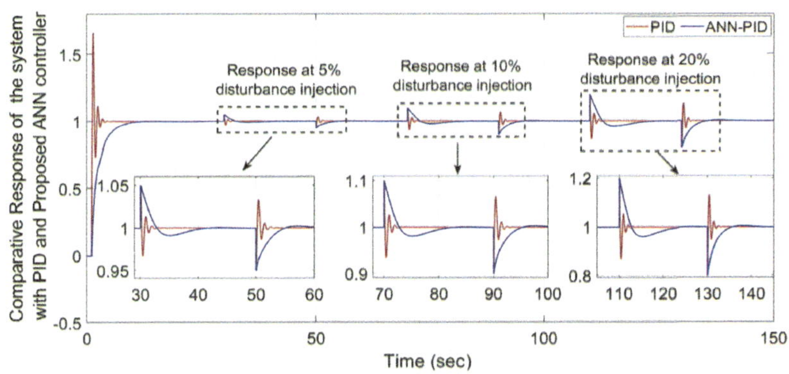

Fig. 7 System response with Ziegler-Nichols-2 and proposed ANN-PID methods

Table 1 Comparison of key time domain index for conventional PID and proposed ANN-PID

S. no.	Controller type	Peak overshoot	Settling-time	Rise-time	Delay-time	Transient behaviour
1	Conventional PID controller	39.75%	3.2	1.224	1.163	Oscillatory
2	Proposed ANN-PID controller	0	8.5	5.0	1.82	Smooth
	Superior controller	ANN-PID	PID	PID	PID	ANN-PID

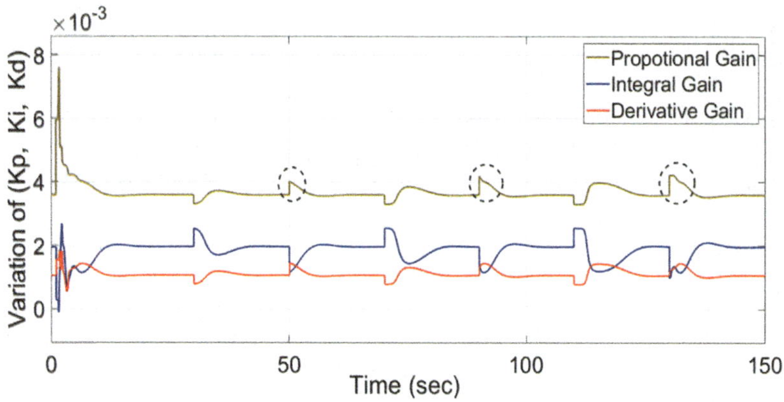

Fig. 8 Variation of (K_p, K_i, K_d) gains during various disturbances

also be observed in the integral and derivative gains. These changes in the controller gains can effectively address various transient issues in the system response.

4.2 Frequency Domain Analysis

In the frequency domain, the judgement of system stability can be done using gain (G_m) and phase margin (P_m) values. The system that has high G_m or P_m values are considered a more stable system. The system response with the conventional method has $P_m = 99.9$ and $G_m = 0$ as shown in Fig. 9, and the system response with the proposed ANN-PID has $P_m = 135$ and $G_m = 0$ as shown in Fig. 10. This concludes that the proposed ANN-PID system is more stable.

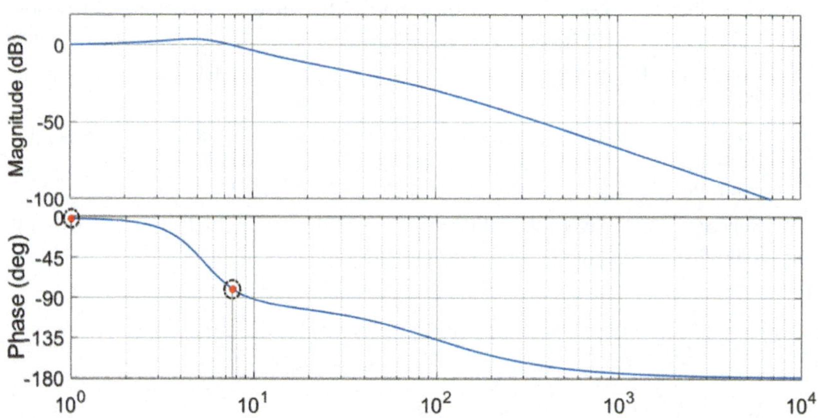

Fig. 9 Bode plot response of the system with PID controller

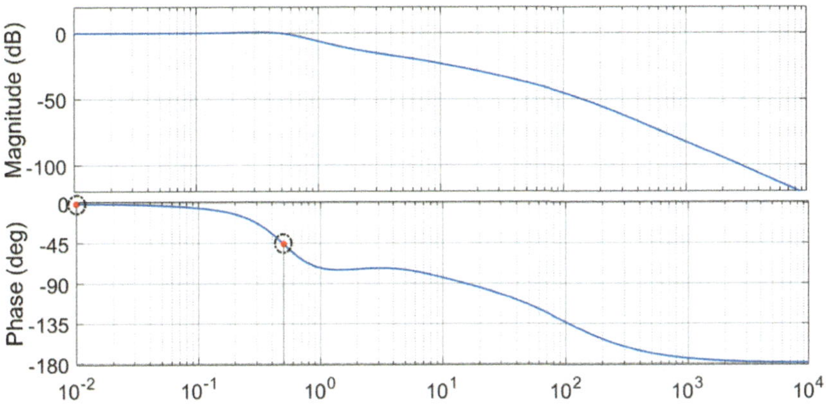

Fig. 10 Bode plot response of the system with ANN-PID controller

5 Conclusion

This paper mainly reports the issue with the classical PID controller and proposes a modified PID controller to provide a good time domain and frequency response characteristics for an ICE. By comparing different transient parameters and stability parameters, the following points are concluded.

- Conventional PID has more peak overshoot and oscillations which may harm the ICE system.
- The proposed ANN-PID control method has no peak overshoot and no oscillations and it can address the external disturbances robustly.

In conclusion, an ANN-PID-based controller provides the best output for an ICE when compared to a PID controller tuned with conventional tuning methods.

References

1. Dovgyallo AI, Kudinov VA, Shestakova DA (2017) Working cycle analysis of the internal combustion engine with heat regeneration. In: International conference on mechanical, system and control engineering (ICMSC). St. Petersburg, Russia, pp 36–39
2. Simanenkov AL, Rozhkov SA (2018) marine internal combustion engine mathematical model. In: IEEE 5th international conference on methods and systems of navigation and motion control (MSNMC), Kiev, Ukraine, pp 238–241
3. Nguyen HH, Horak V, Cornak S (2019) Theory of the internal combustion engine cycle with the thermochemical model of combustion. In: International conference on military technologies (ICMT), Brno, Czech Republic
4. Busher V, Horoshko V (2019) Dual electromagnetic retarder control system for tuning internal combustion engines. In: IEEE International conference on modern electrical and energy systems (MEES), Kremenchuk, Ukraine, pp 26–29

5. Kalelı A, Kaltakkiran G, Dumlu A, Ayten KK (2017) Design and control of intelligent cooling system for IC engine. In: International conference on engineering and technology (ICET), Antalya, Turkey
6. Mahmud M, Motakabber SMA, Zahirul Alam AHM, Nordin AN (2020) Adaptive PID controller using for speed control of the BLDC motor. In: IEEE international conference on semiconductor electronics (ICSE), Kuala Lumpur, Malaysia, pp 168–171
7. Bharath Kumar V, Charan G, Pavan Kumar YV (2020) Design of robust PID controller for improving voltage response of a Cuk converter. In: Innovation in electrical and electronic engineering, vol 661, Delhi, India, pp 301–318
8. Torres WL, Araujo IBQ, Filho JBM, Junior AGC (2017) Mathematical modeling and PID controller parameter tuning in a didactic thermal plant. IEEE Lat Am Trans 15(7):1250–1256
9. Zhu D, Liu L, Liu C (2014) Optimal fractional-order PID control of chaos in the fractional-order BUCK converter. In: 9th IEEE conference on industrial electronics and applications, Hangzhou, China, pp 787–791
10. Paliwal S (2017) Stabilization of mobile inverted pendulum using fractional order PID controllers. In: International conference on innovations in control, communication and information systems, Greater Noida, India
11. Wang TY, Chang CD (2018) Hybrid fuzzy PID controller design for a mobile robot. In: IEEE international conference on applied system invention, Chiba, Japan, pp 650–653
12. Singh S, Mitra R (2014) Comparative analysis of robustness of optimally offline tuned PID controller and Fuzzy supervised PID controller. In: Recent advances in engineering and computational sciences (RAECS), Chandigarh, India
13. Korkmaz M, Aydogdu O, Dogan H (2012) Design and performance comparison of variable parameter nonlinear PID controller and genetic algorithm based PID controller. In: International symposium on innovations in intelligent systems and applications, Trabzon, Turkey
14. Sandeep Rao K, Siva Praneeth VN, Pavan Kumar YV, John Pradeep D (2020) Investigation on various training algorithms for robust ANN-PID controller design. Int J Sci Technol Res 9(2):5352–5360
15. Pavan Kumar YV, Kiran KMNS, Yugandhar S, Raju KP (2013) Online attitude controlling of longitudinal autopilot for general aviation aircraft using artificial neural networks. In: Nirma University international conference on engineering. IEEE, Ahmedabad, India
16. Caton JA (2018) The thermodynamics of internal combustion engines: examples of insights. Inventions 32(2):33
17. Roy SK, Mohanty AR (2020) Combustion detection in IC engine by analysis of instantaneous angular acceleration. In: Advances in Mechanical Engineering. Springer, Singapore, pp 1127–1137
18. Rao KS, Praneeth VNS, Kumar YVP (2020) Fuzzy logic-based intelligent PID controller for speed control of linear internal combustion engine. In: Innovations in electrical and electronic engineering, vol 661, Delhi, India, pp 505–521

Methods for Effective Speed Control of DC Shunt Motor—Part 1: Classical PID Controller Tuning Methods

Ramireddy Karthik, E. Tejendra, R. Gowrav, Harshit Harsh, and Y. V. Pavan Kumar

Abstract Motors are electromechanical devices that find numerous uses in everyday life. Electric motors are of both AC and DC types. Normally, DC-powered electric motors are widely used because of their advantages such as powerful starting torque and easy control structure. The main concern in the DC motors is that whenever the load is changed, the motor speed is also forced to change. Even though the shunt motors produce constant speed, it is not guaranteed in all cases. So, an efficient control scheme for robust speed control of shunt motor is of importance for research. To address this issue, in this paper, an analysis of different classical tuning techniques of the PID controller used in DC shunt motor control which alters the armature voltage to produce constant speed is presented. The analysis is performed on MATLAB/Simulink, and the comparative results are presented. From this comparative analysis of all the key tuning methods, the best method is suggested based on the computation of the time-domain performance index.

Keywords Armature voltage · DC motor · Frequency-domain analysis · PID controller · Shunt motor · Time-domain responses

---

R. Karthik · E. Tejendra · R. Gowrav · H. Harsh · Y. V. Pavan Kumar (✉)
School of Electronics Engineering, VIT-AP University, Amaravati, A. P 522237, India
e-mail: pavankumar.yv@vitap.ac.in

R. Karthik
e-mail: karthik.18bev7028@vitap.ac.in

E. Tejendra
e-mail: tejendra.18bec7059@vitap.ac.in

R. Gowrav
e-mail: gowrav.18bec7072@vitap.ac.in

H. Harsh
e-mail: harshit.18bev7002@vitap.ac.in

S. Suhag et al. (eds.), *Control and Measurement Applications for Smart Grid*,
Lecture Notes in Electrical Engineering 822,
https://doi.org/10.1007/978-981-16-7664-2_34

1 Introduction

DC shunt electric motors are widely used in many applications such as lathe machines, centrifugal and reciprocating pumps, blowers, fans, conveyors, lifts, weaving machine, spinning machines, machine tools, milling machines, drilling machines where constant speed is desired and starting torques are not severe. Normally, the DC motor's speed is controlled easily when compared to the AC motor. These are more efficient and can make use of lower input voltages. There are two control mechanisms available traditionally to regulate the speed of DC motors. They are armature control type with fixed field current and field control type with fixed armature current.

In the armature control process, the speed of the motor is varied by altering the voltage of armature by fixing the field flux as constant, while in the field control method, motor speed is varied by altering the field flux by making armature voltage as constant. Usually, the armature control mode is preferred because it is less complex in operation as compared to the field control mode. When the armature control mode is applied for both the shunt and series motors, the series motor gives the good starting torque and drops instantly when it is connected to the load, while the shunt motor rotates almost at a constant speed. Even though it has a low starting torque, it produces the same speed irrespective of load. Basic models of series motor and shunt motors are displayed in Figs. 1 and 2, respectively.

To control the DC shunt motor speed, many control techniques are proposed such as PID control mechanisms, fuzzy logic control mechanisms, and neural network-based methods [1]. The genetic algorithm-based PID control methods for speed regulation of the DC motor were discussed in [2]. A converter-based smooth starter for DC motor is given in [3], whereas different online tuning methods-based controller design was discussed in [4–6]. A nonlinear control method for motor control was discussed in [7]. Controlling the torque of the motor was dealt in [8]. A PID controller for different converters was presented in [9], and a rheostat-based control process for speed control of shunt motor was discussed in [10]. Further, some advanced PID controller design methods for motors were discussed in [11–15]. In line with

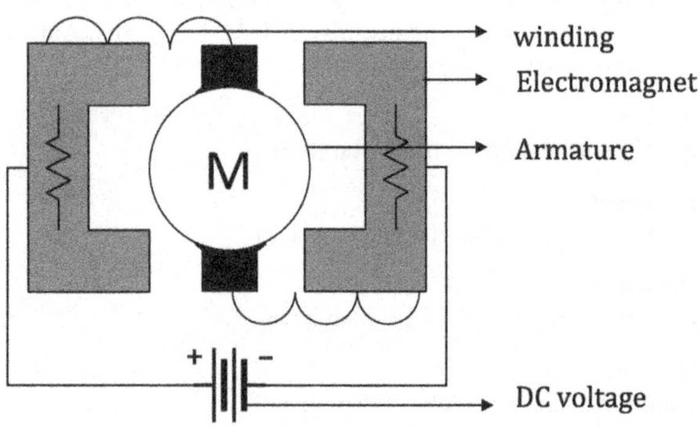

Fig. 1 Series DC motor schematic

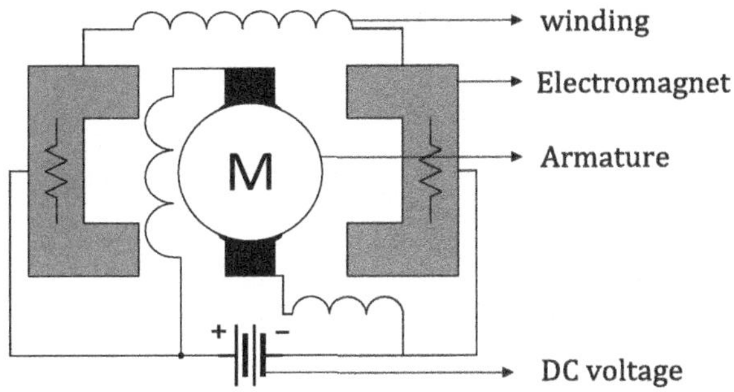

Fig. 2 Shunt DC motor schematic

these state-of-the-art works, this paper analyzes various important PID tuning procedures for the speed control of the DC shunt motor. Out of this analysis, the tuning method which gives the best transient response is suggested for the controller design. The effective analysis is done with the help of the computation of the time-domain performance index under different system conditions.

2 Modeling of DC Shunt Motor

DC motor modeling is viewed in terms of the electrical model and mechanical model separately. The armature side and the input source are considered for the electrical model, and the load side is considered for the mechanical model that is derived below.

2.1 Electrical Model

The DC motor's emf (E) is given as (1), where it is equivalent to the product of armature speed (N) and field flux (ϕ). The speed of the armature is represented by (2), where V is the terminal voltage, I_a is the current at the armature, R_a is the resistance of the armature, and k is the flux constant as given in (3). Figure 3 shows an armature-controlled DC shunt motor. The differential equation involving the quantities of armature circuit is given as (4), whose Laplace transform is given as (5).

$$E = \frac{ZNP\Phi}{60A} \tag{1}$$

$$N = \frac{V - I_a R_a}{k\Phi} \tag{2}$$

$$k = \frac{ZP}{60A} \tag{3}$$

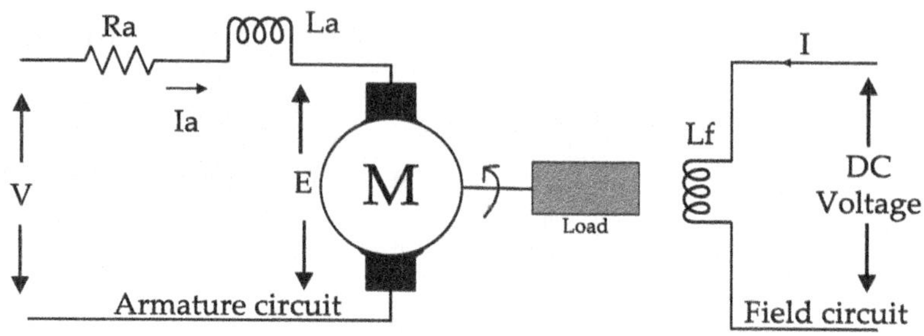

Fig. 3 Circuit of armature-controlled DC shunt motor

$$L_a \frac{dI_a}{dt} + R_a I_a + E = V \tag{4}$$

$$\left.\begin{aligned} &(L_a s + R_a) I_a(s) = V(s) - E(s) \\ &\Rightarrow I_a(s) = \frac{V(s) - E(s)}{L_a s + R_a} \end{aligned}\right\} \tag{5}$$

2.2 Mechanical Model

The motor is connected to the load with the shaft having a viscous friction coefficient (B) and inertia (J). The torque produced by the motor is T, which causes the rotation and thereby produces an angular velocity (ω) and angular acceleration (α). The torque generated by the motor is opposed by inertia torque and frictional torque. The differential equation representing the mechanical quantities of the motor is given by (6), whose Laplace transform equation is given by (7).

$$J \frac{d^2\theta}{dt^2} + B \frac{d\theta}{dt} = T_m = k_T I_a \tag{6}$$

$$\left.\begin{aligned} &(J s^2 + B s)\theta(s) = T_m(s) = k_T I_a(s) \\ &\Rightarrow \theta(s) = \frac{k_T}{J s^2 + B s} I_a(s) \end{aligned}\right\} \tag{7}$$

$$\frac{\theta(s)}{V(s)} = \frac{K_T}{s((Js + B)(L_a + R_a) + K_T K_b)} \tag{8}$$

From (1)–(7), the system mathematical block model is developed as given in Fig. 4, which is simulated with the parameters given in Table 1. The open-loop transfer function of the system is represented as (8), which is considered for the simulation.

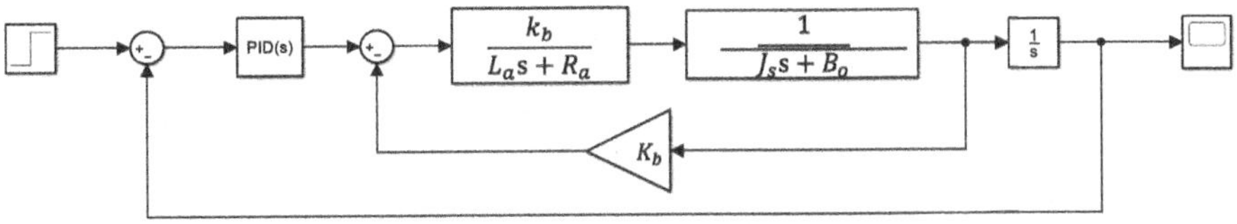

Fig. 4 Block diagram representation of the shunt motor control system

Table 1 Shunt motor simulation specifications

S. no.	Parameter	Specifications	Units
1	L_a	0.23	H
2	R_a	1.35	Ω
3	K_T	1.41	N m/A
4	J_s	0.000052	Kg m^2
5	B_o	0.01	N/(rad/s)
6	K_b	0.235	V/(rad/s)

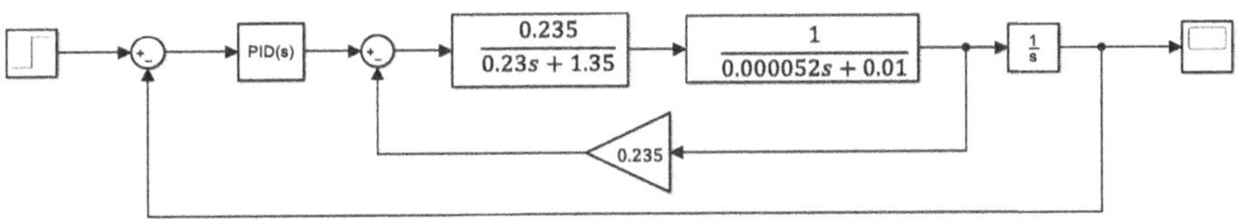

Fig. 5 Simulink model for the PID controller interfaced DC shunt motor

3 Controller Design

PID controller is interfaced with the shunt motor as shown in Fig. 5 to control the speed of the shaft as desired. The parameters of the controller are tuned by various traditional tuning methods like open-loop transfer function (OLTR) methods, ultimate cycles (UC) methods, and error performance indices (EPI) techniques as presented in Fig. 6. PID controller tuned by these methods gets different K_P, K_I, K_D values. Out of these methods, OLTR and EPI methods design PI configuration and UC methods design PID configuration. All these tuned values obtained by several techniques are tabulated in Table 2.

4 Simulation and Analysis

The simulation results of the shunt motor control system involving the PID controller that is tuned with OLTR, EPI, and UC tuning approaches are given in Figs. 7, 8, and 9, respectively. From these plots, it can be observed that the use of all these methods

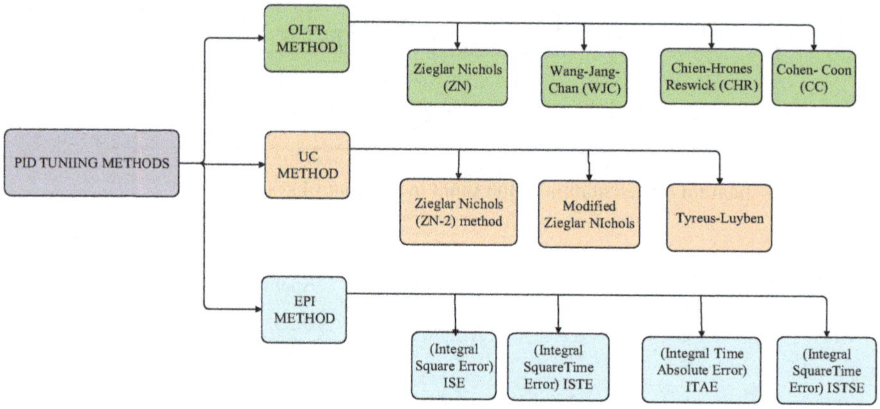

Fig. 6 Classification of traditional PID controller tuning methods

Table 2 Shunt motor controller specifications designed by various methods

Method	Algorithm	(K_P)	(K_I)	(K_D)
OLTR	ZN-1	0.27027	3.3	–
	WJC	0.5472	0.8	–
	CHR	0.10510	1.2	–
	CC	0.3768	1.764	–
EPI	ISE	0.3562	0.3627	–
	ISTE	0.31244	2.2303	–
	ISTSE	0.35418	1.5427	–
	ITAE	0.34709	0.9735	–
UC	ZN-2	1.2	30	0.012
	MZN	0.4	5	0.0106
	TL	0.9	5.113	0.01142

ensures the stability of the system. Further, to recognize the best method, various performance metrics are taken into consideration. All these metrics computed for each of the response are tabulated in Table 3. From these values, the efficient tuning method is suggested as the technique which obtains lesser values in all the time-domain metrics computed when compared to other methods. From Table 3, it can be observed that the WJC method leads to less performance index values in most of the cases when compared to other methods. Hence, the WJC method is considered the best tuning algorithm for controller design for a shunt motor application.

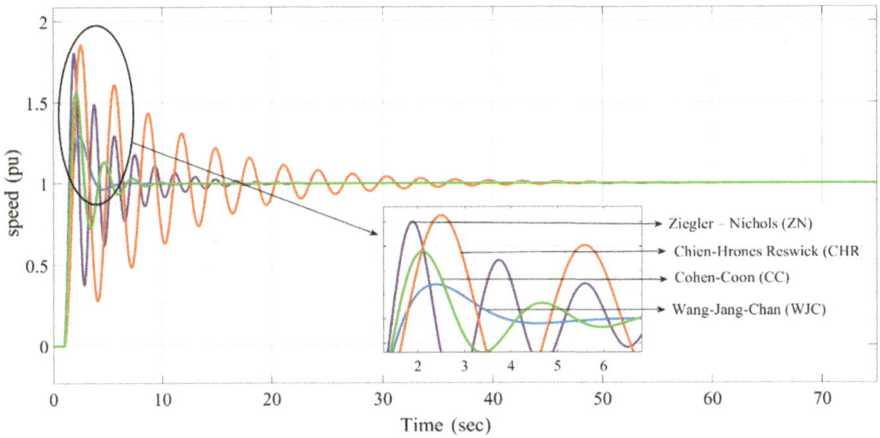

Fig. 7 System performance with PID controller tuned by OLTR approaches

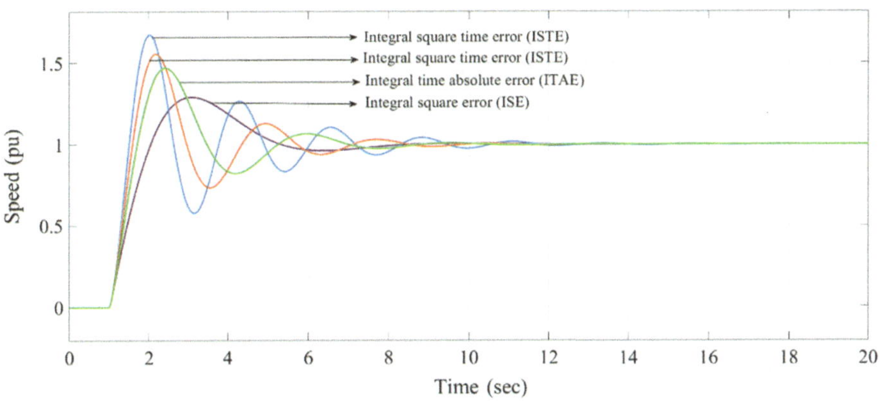

Fig. 8 System performance with PID controller tuned by EPI approaches

5 Conclusion

Thus, this paper suggests the best PID controller tuning method for the speed control of the DC shunt motor. This plays a vital role as the DC motors find numerous applications in day to day life. In the majority of all those applications, the PID controller is the most widely used device to regulate the motor speed with respect to the varying load on the system. Out of this view, this paper realizes all the traditional PID controller design procedures. The efficacy of all these methods is investigated by computing time-domain metrics. Results showed that Wang-Jang-Chan (WJC) method is the best method to set the PID factors for DC shunt motor application, especially if the peak overshoot and settling time are of major interest.

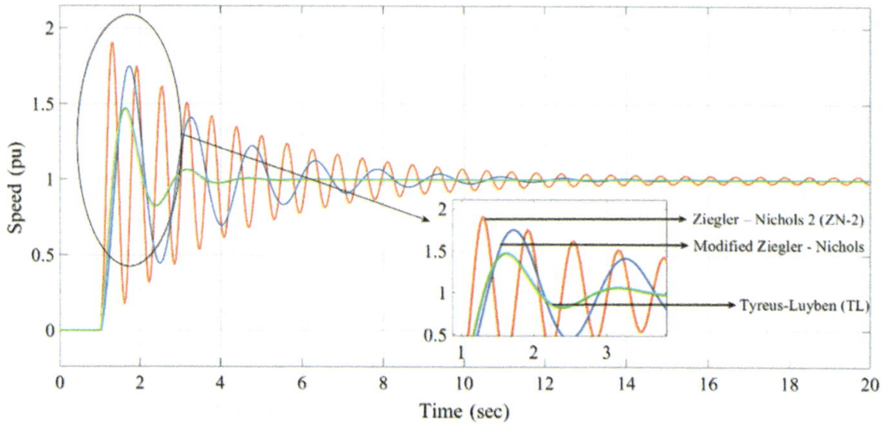

Fig. 9 System performance with PID controller tuned by UC approaches

Table 3 Time-domain performance index parameters

Method	Algorithm	Delay time (T_d)	Rise time (T_r)	Peak time (T_p)	Peak overshoot (M_p in %)	Settling time (T_s)	Steady-state error (e_{SS})
OLTR	ZN-1	1.2775	1.441	1.8975	80.37	28	0
	WJC	1.3087	1.6929	2.388	28.91	8	0
	CHR	1.4765	1.7472	2.4974	85.62	60	0
	CC	1.3105	1.5522	2.1035	56.2	14	0
EPI	ISE	1.4588	2.0466	3.0894	29.01	14	0
	ISTE	1.3066	1.5144	2.0399	67.52	20	0
	ISTSE	1.3299	1.5898	2.1745	55.64	15	0
	ITAE	1.3767	1.7151	2.4242	47.06	16	0
UC	ZN-2	1.1983	1.1518	1.3019	91.02	20	1.02
	MZN	1.2179	1.3589	1.7278	74.86	22	0
	TL	1.1648	1.3101	1.6265	47.31	9	0
Superior method		TL	ZN-2	ZN-2	WJC	WJC	All except ZN-2

References

1. Karthik R, Hari AS, Kumar YVP, Pradeep DJ (2020) Modelling and control design for variable speed wind turbine energy system. In: International conference on artificial intelligence and signal processing (AISP), Amaravati, India, pp 1–6
2. Tutkun N, Maden D (2010) Determination of the PID controller parameters for a DC shunt motor using the genetic algorithm method. In: National conference on electrical, electronics and computer engineering, Bursa, pp 330–333
3. Kumar R, Choudhary A, Shimi SL (2017) Smooth starter for DC shunt motor using buck-boost

power converter. In: 2017 International conference on innovations in control, communication and information systems (ICICCI), Greater Noida, India, pp 1–8
4. Derugo P (2017) Adaptive neuro fuzzy PID type II DC shunt motor speed controller with Petri transition layer. In: 2017 IEEE 26th international symposium on industrial electronics (ISIE), Edinburgh, pp 395–400
5. Ahmad MdA, Rai P, Mahato A, Mahapatra M (2017) Speed control of a DC motor using fuzzy logic application. Int J Res Eng Technol Sci 7:1–11
6. Tapia R, Rosas JC, Aguilar O, Templos A (2014) On-line speed control of the shunt-connected DC motor via a neurocontroller. In: 7th IET International conference on power electronics, machines and drives (PEMD), pp 1–6
7. Sedghisigarchi K, Hasanovic A, Feliachi A, Davari A (2001) Evaluation of three algorithms for nonlinear control of a DC shunt motor. In: Proceedings of the 33rd southeastern symposium on system theory, Athens, OH, USA, pp 407–411
8. Magsino JC, Magsino E (2009) Speed and torrque control of a DC shunt motor. In: 5th NOCEI research forum, pp. 1–11
9. Kumar VB, Charan G, Kumar YVP (2020) Design of robust PID controller for improving voltage response of a Cuk converter. Innov Electr Electron Eng 301–318
10. Dwivedi A (2013) Speed control of DC Shunt motor with field and armature Rheostat control simultaneously. Adv Electron Electr Eng 3(1):77–80
11. Shih F, Chen DY, Wu YP, Chen YT (1996) A procedure for designing EMI filters for AC line applications. Power Electron IEEE Trans 11(1):170–181
12. Barinov IA, Melnichenko OV (2019) Power IGBTs application in AC-wire DC-motor locomotive thyristor-based power circuit for regenerative Brake energy efficiency increase. In: 2019 International conference on industrial engineering, applications and manufacturing (ICIEAM), pp 1–5, Sochi, Russia
13. Mahmud K, Tao L (2013) Vehicle speed control through fuzzy logic. In: 2013 IEEE global high-tech congress on electronics, Shenzhen, pp 30–35
14. Afrin A, Ahmed MR (2019) Design, improvement & analysis of solar based three-stage interleaved boost converter for BLDC motor. In: 2019 International conference on electrical, computer and communication engineering, Cox'sBazar, Bangladesh, pp 1–5
15. Rao KS, Praneeth VNS, Kumar YVP (2020) Fuzzy logic-based intelligent PID controller for speed control of linear internal combustion engine. Innov Electr Electron Eng 505–521

Modelling of Neural Network-based MPPT Controller for Wind Turbine Energy System

Ramireddy Karthik, Harshit Harsh, Y. V. Pavan Kumar, D. John Pradeep, Ch. Pradeep Reddy, and Ramani Kannan

Abstract Wind energy is one of the best renewable energy sources, used for energy generation in modern-day power generation system. Nowadays, wind energy is widely used to power up devices that consume huge power. As wind speed changes rapidly over time, its power generating capacity also varies, this gives rise to a need for a controller which controls the power harnessed from the wind energy system. The procedure to achieve maximum power from a renewable energy system is known as maximum power point tracking (MPPT). There are many methods to achieve maximum power from the wind turbine, and in this paper, a neural network-based controller for MPPT is proposed. Firstly, the mathematical model of a wind power turbine system is presented, followed by designing a neural network-based controller to achieve maximum power profile. The influence of the proposed controller on power point tracking is investigated, and the time domain parameters are presented. In this paper, MATLAB/Simulink software is used for the simulating the system and to verify the controller efficacy.

R. Karthik · H. Harsh · Y. V. Pavan Kumar (✉) · D. John Pradeep
School of Electronics Engineering, VIT-AP University, Amaravati, AP 522237, India
e-mail: pavankumar.yv@vitap.ac.in

R. Karthik
e-mail: karthik.18bev7028@vitap.ac.in

H. Harsh
e-mail: harshit.18bev7002@vitap.ac.in

D. John Pradeep
e-mail: john.darsy@vitap.ac.in

Ch. Pradeep Reddy
School of Computer Science and Engineering, VIT-AP University, Amaravati, AP, India
e-mail: pradeep.ch@vitap.ac.in

R. Kannan
Electrical and Electronics Engineering, Universiti Teknologi PETRONAS (UTP), Seri Iskandar, 32610 Perak, Malaysia
e-mail: ramani.kannan@utp.edu.my

S. Suhag et al. (eds.), *Control and Measurement Applications for Smart Grid*,
Lecture Notes in Electrical Engineering 822,
https://doi.org/10.1007/978-981-16-7664-2_35

1 Introduction

At present, renewable energy sources are emerging as the main source of energy because of the fossil fuel depletion and pollution issues. It has many advantages compared to conventional fuel-based sources of energies like less/no waste byproduct, environment friendly and renewable. Wind energy power system is one of the renewable energy sources which has got huge demand to produce power in far-field places. Normally, the energy production from wind energy system is dependent on wind speed which is a great matter of concern when facilitating high power requirement. To reduce its dependency on wind speed, several MPPT algorithms were proposed that would give high power at lower wind velocities. In this paper, transfer function-based model of neural network-based MPPT controller for wind energy power system [1, 2] is proposed and corresponding results were obtained by varying the wind speed as the input to the system. Tip speed ratio (TSR) control, optimum relationship-based (ORB) control, hybrid control, intelligent control methods with the use of fuzzy and neural networks are different types of MPPT control algorithms discussed in literature. In TSR [3, 4] control method, the speed of rotation of the generator is varied to maintain maximum power profile. It heavily depends on the wind speed measurement which varies unevenly across the surface of the turbine. Despite of low complexity and faster MPPT tracking, it suffers from its limitation of application to regions of lower wind velocities. In ORB [5] control method, the wind system parameters are probed for an optimal relationship which may involve two or more parameters and those values will be used for MPPT control. This method suffers from the fact that it requires system information Apriori.

As mentioned above, many MPPT algorithms were proposed and the most well-known one is perturb and observation method. There are generally two ways to achieve maximum power through perturb and observation method. The first way is by controlling the output from boost converter [6] which is an essential component in renewable energy system, and other way is through controlling the rotational speed of shaft by using flexible wind shaft whose length can be controlled. Artificial intelligence techniques provide a control mechanism to alter the shaft length for good power harvesting. The artificial neural network can be implemented to achieve MPPT in wind power systems. Recent research is being carried out in wind energy harvesting like sensor less control scheme for the wind turbine in 1.5 MW, doubly-fed induction generator (DFIG) application with maximum power point tracking (MPPT) facility is discussed in [7], an adaptive MPPT-based rotor speed control of DFIG wind turbine is discussed in [8], an intelligent speed control integrated with MPPT facility for medium power wind energy systems is proposed in [9]. An imperialist competitive algorithm-based induction motor's speed control provision that was supplied by wind turbine is discussed in [10], a review of conventional and advanced MPPT algorithms for wind energy systems is discussed in [11]. Modelling of smart airborne wind turbine system with neural network based on MPPT is proposed in [12], MPPT algorithms for Magnus effect wind turbine control system are proposed and discussed in [13–16]. This paper mainly focuses on achieving MPPT through the rotational speed of the shaft.

2 Description of Wind Turbine Power System

The wind turbine is an energy system which converts mechanical energy to electrical energy. It receives wind speed as input, which in turn rotate, flaps of wind turbine and its rotation generates voltage. The above phenomenon is governed by Faraday's law of electromagnetic induction which states that 'when a moving coil is placed in the magnetic field, it induces an EMF in the coil'. The flaps of the wind turbine rotate and voltage is produced. The relation between bending moment and dimension parameters is given in (1) and (2).

$$M = -\frac{1}{2}w(L - r)^2 \tag{1}$$

$$\frac{\sigma}{y} = \frac{M}{I} = \frac{E}{R} \tag{2}$$

where L = total blade length, M = bending moment, w = UDL, r = radial distance from the hub, σ = stress, y = distance from the neutral axis, M = bending moment, I = second moment of Inertia, E = modulus of elasticity, R = radius of curvature.

The wind turbine's mechanical torque generated is given in (3).

$$T_{\mathrm{m}}(t) = T_{\mathrm{e}}(t) + J_{\mathrm{t}}\frac{\mathrm{d}\omega_r(t)}{\mathrm{d}t} + F\omega_{\mathrm{r}}(t) \tag{3}$$

where F is the frictional effect, J_{t} is the inertial component of turbine system and ω_{r} is the wind turbine's speed.

It is a combination of three components such as mechanical model which consists of the gearbox and other mechanical parts, permanent magnet synchronous motor (PMSG) model and DC-DC boost converter model with a step-up transformer connected to the grid. The transfer function of the mechanical model is obtained from [1] as follows.

$$T_{\mathrm{M}} = \frac{1}{0.11s + 0.05} \tag{4}$$

The transfer function of the PMSG model is obtained from [1] as follows.

$$T_{\mathrm{P}} = \frac{0.56}{0.000174s + 1.22} \tag{5}$$

The transfer function of the DC-DC boost converter model is as follows [1].

$$T_B = \frac{9.904}{0.00005s + 9.6} \tag{6}$$

A boost converter boosts the voltage harvested from the wind turbine. The current through boost converter is the output current drawn from the wind turbine. The current from the boost converter is calculated in two steps. The first step is to calculate $V_{in(min)}$, which is dependent on the duty cycle of the converter. In this paper, the duty cycle is chosen to be 0.8.

$$D = 1 - \frac{V_{IN(min)} \times \eta}{V_{OUT}} \tag{7}$$

where $V_{IN(min)}$ = minimum input voltage, V_{OUT} = desired output voltage, η = efficiency of the converter.

The next step is to calculate inductor ripple current (ΔI_L), which is dependent on maximum switching frequency f_s and device parameters as given in (8).

$$\Delta I_L = \frac{V_{IN(min)} \times D}{f_s \times L} \tag{8}$$

To calculate the power output, I_{out}(max) is computed using (9)

$$\Delta I_L = (0.2 \text{ to } 0.4) \times I_{OUT(max)} \times \frac{V_{OUT}}{V_{IN}} \tag{9}$$

The algebraic multiplication of I_{out}(max) and V_{out} results in calculation of power. As mentioned above, V_{out} and V_{in} are presented in frequency domain. So, the current is also computed in s domain.

3 Design of Neural Network Controller

In the field of science and technology, artificial intelligence plays a vital role and has gained a lot of popularity in recent years due to the advancement of machine learning. The neural network is one of the artificial techniques which works in the same principle as humans' neurons and hence was named as neural network. A neural network is a system software or hardware which works similar to human neurons. The neural network consists of an input layer, output layer and hidden layer which help to get the desired output.

As it is an artificial intelligence technique, it works on data provided to the system. In online mode of deploying the system controller, the output changes according to input and this is done based on the data fed to the system offline. The input and corresponding output are fed to achieve the desired performance. Large samples of data are need for proper tuning of the controller to get desired system response. In wind energy system, for maximum power point tracking, output is the angular

Table 1 Input wind speed and angular velocity

Input wind speed (km/h)	Angular velocity (rad/s)
4	120
5	170
6	220
7	250
8	300
9	320

velocity of flaps whereas the input to the neural network is wind speed. The data for the input wind speed and angular velocity are considered from [17]. The data for the input wind speed and angular velocity are given in Table 1. But, that data are not sufficient for the effective training of neural network. So, a mathematical tool—'Newton's forward interpolation' is used to extract the data for unknown values of wind speeds other than the ones given in Table 1. The MATLAB code used to obtain the interpolated values of the wind speed is given in Table 2.

The wind speed and angular velocity of wind turbine data sets obtained from the Newton interpolation technique are fed to the neural network controller as input and output, respectively. Levenberg Marquardt algorithm is used to train the neural network controller. The error histogram graph, plot fit, training state analysis and regression plot analysis of the trained neural network are shown in Figs. 1, 2, 3 and 4, respectively. The complete Simulink model of neural network-based MPPT

Table 2 Code that is used to obtain the interpolated values of the function

Code implemented in MATLAB environment

```
x= [4:1:10]; % Write the values of independent variable x.
y = [120 170 200 220 270 300 320]; % Write the values of dependent variable y.
n=length(x); % Number of terms of X or Y
d=zeros(n-1);
h=x(2)-x(1); %step length
x0=input('Enter the value of x where we want to find the value of f(x): ');
u=(x0-x(1))/h;
for i=2:n %Calculation of first forward differences
        d(i-1,1)=y(i)-y(i-1);
end
for k=2:n-1 %Calculation of second and rest forward differences
        for i=1:n-k
            d(i,k)= d(i+1,k-1)-d(i,k-1);
        end
end
disp('The forward difference table is:')
d
s=y(1); t=u;
for k=1:n-1 %Calculation of result
s=s+t*d(1,k);
t=(u-k)/(k+1)*t;
end
fprintf('The required value is f(%f)= %5.5f',x0,s);
```

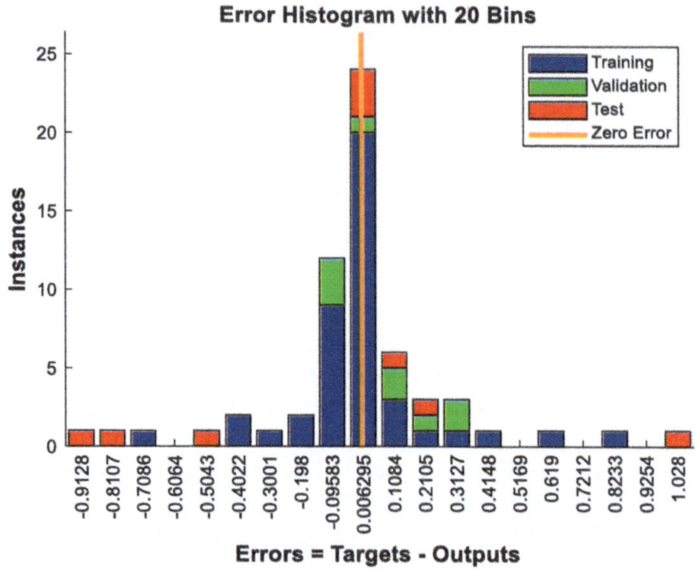

Fig. 1 Error histogram of the trained neural network controller

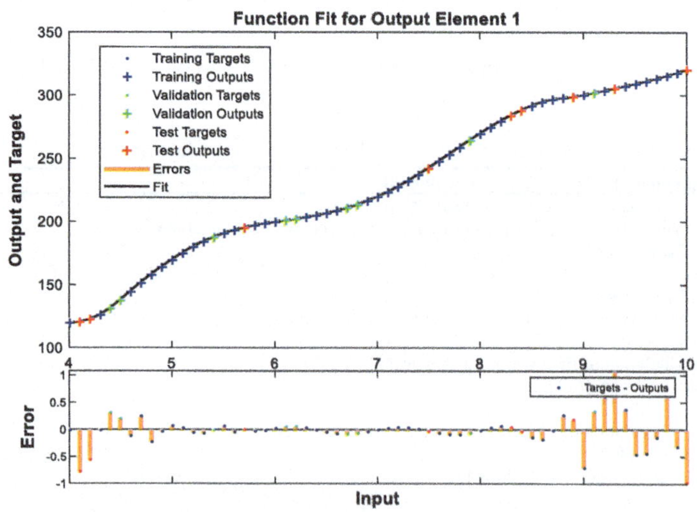

Fig. 2 Function fit graph of the trained neural network controller

controller interfaced wind turbine energy system is shown in Fig. 5. The elaborated design of current circuit diagram is shown in Fig. 6.

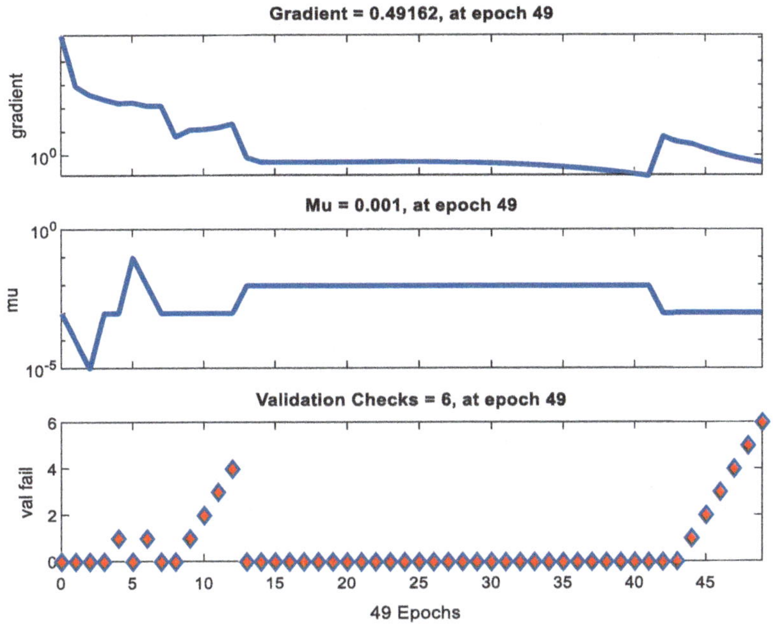

Fig. 3 Training state analysis of the neural network controller

4 Simulation Results and Analysis

The responses of the neural network trained MPPT algorithm-based wind turbine and conventional wind turbine systems are analyzed by comparing the transient responses of the systems in their stable conditions and under various disturbances. The table representing the comparative analysis of conventional wind turbine and proposed wind turbine systems under similar input conditions are represented in Table 3. The transient responses of the proposed wind turbine energy system and conventional wind turbine energy systems are represented in Figs. 7 and 8, respectively. Responses of output power produced by proposed and conventional wind turbine energy systems when random disturbances are injected are shown in Figs. 9 and 10, respectively.

5 Conclusion

From the simulation results of the proposed and conventional wind turbines power vs time in Figs. 7, 8, 9 and 10 show that there is a huge voltage boosting in the proposed system and also from Table 3, almost in all cases we can obtain maximum power for all rotational speeds. Hence, by using the proposed architecture of wind turbine, we can get the maximum power in almost all cases. Hence, it can be concluded that

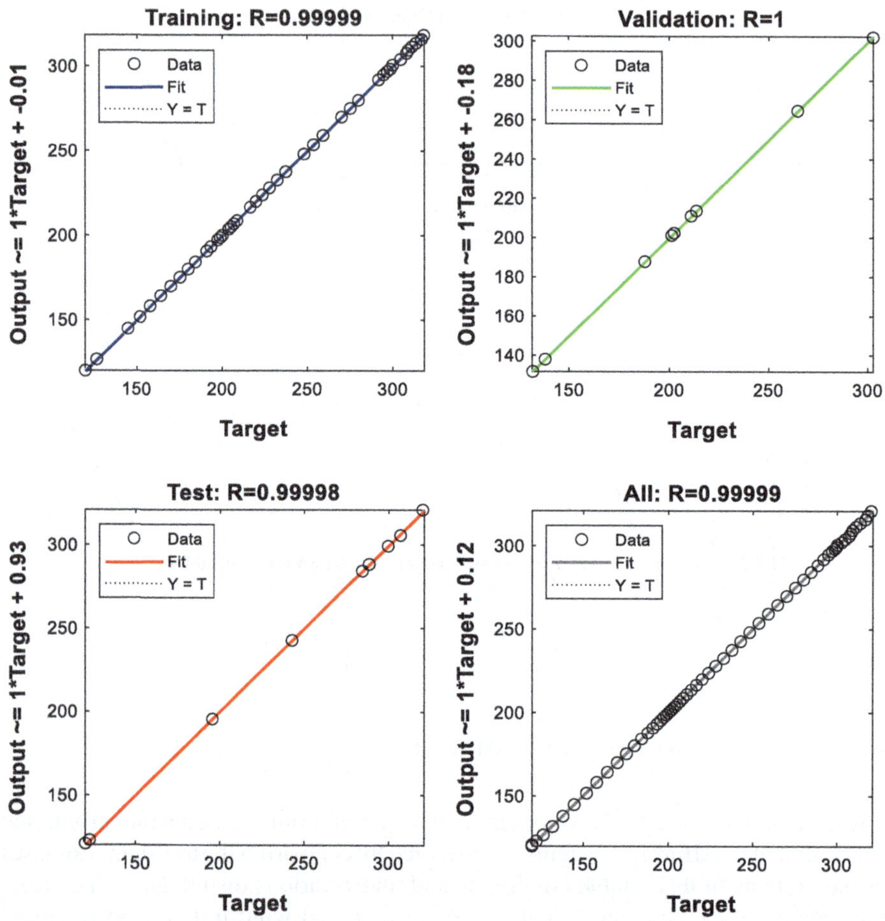

Fig. 4 Regression analysis of the trained neural network controller

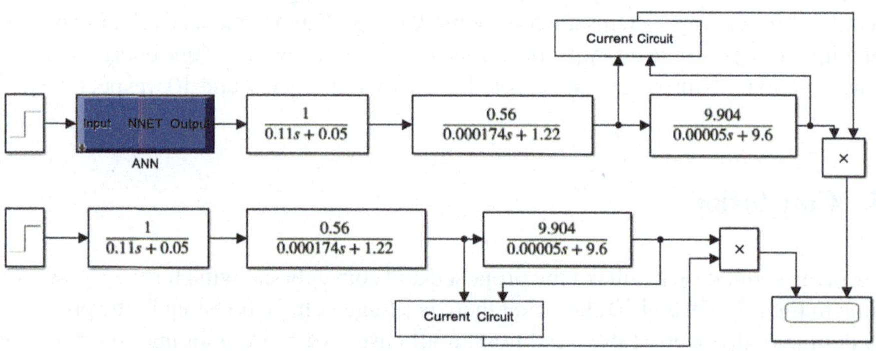

Fig. 5 Simulink model of neural network-based MPPT controller for wind turbine energy system

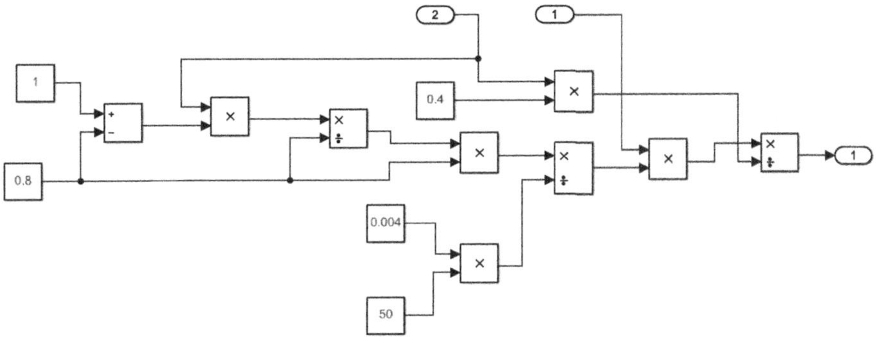

Fig. 6 Elaborated design of current circuit

Table 3 The power generated by conventional and proposed wind turbine

S. no.	Wind speed	Conventional system	Proposed system
1	0.1	2.174	1.195e+07
2	0.2	8.695	1.1951 e+07
3	0.3	19.56	1.1959 e+07
4	0.4	34.78	1.19523 e+07
5	0.5	54.34	1.19535 e+07

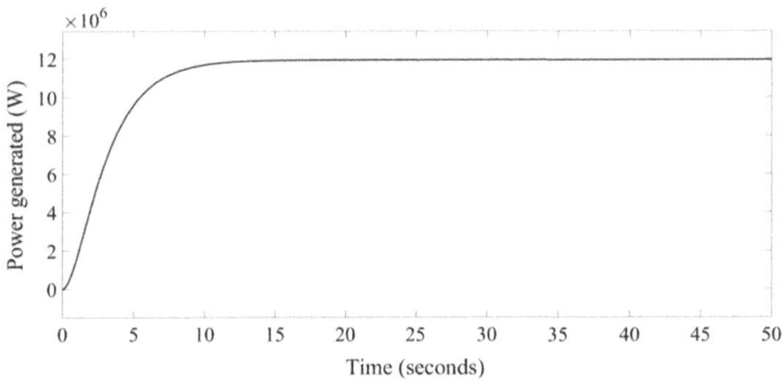

Fig. 7 Transient response of the proposed wind turbine system

the maximum power can be achieved by using the proposed neural network-based MPPT controller the wind turbine energy system.

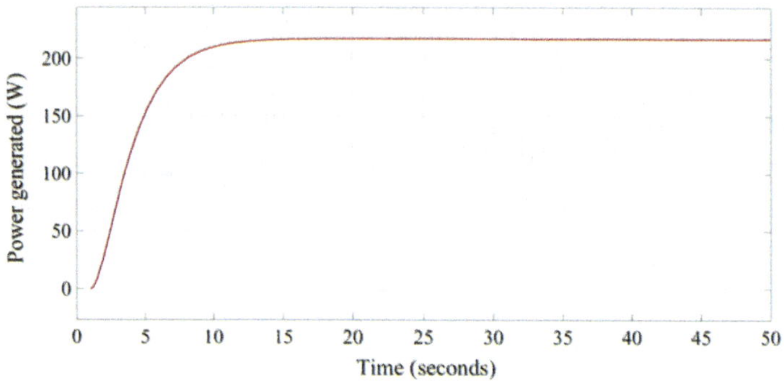

Fig. 8 Transient response of the conventional wind turbine system

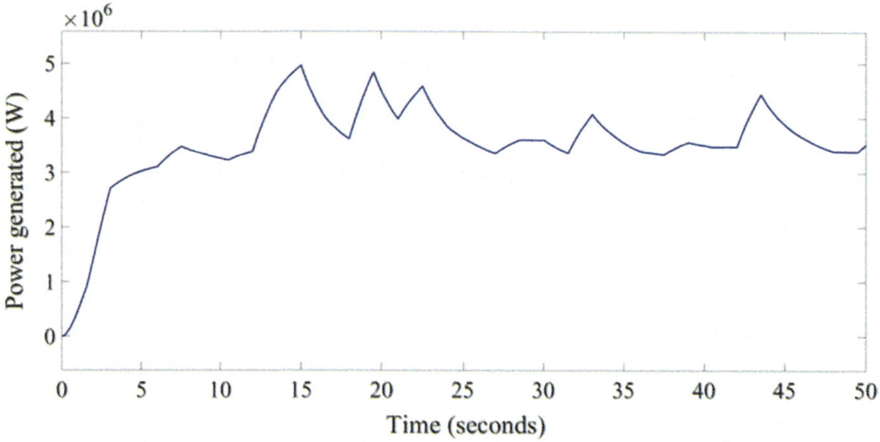

Fig. 9 The response of the wind turbine system with proposed MPPT controller under various disturbances

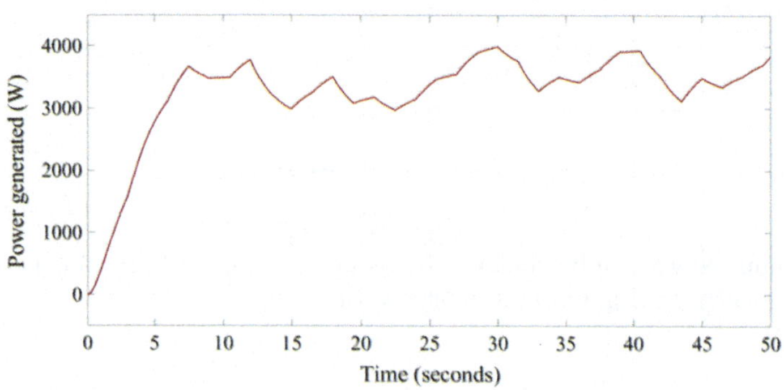

Fig. 10 The response of the conventional wind turbine system under various disturbances

References

1. Karthik R, Hari AS, Kumar YVP, Pradeep DJ (2020) Modelling and control design for variable speed wind turbine energy system. In: 2020 International conference on artificial intelligence and signal processing (AISP 2020), Amaravati, India, pp 1–6
2. Ramireddy K, Hari AS, Kumar YVP (2021) Artificial intelligence based control methods for speed control of wind turbine energy system. Intell Comput Control Commun 702:203–217
3. Nasiri M, Milimonfared J, Fathi SH (2014) Modeling, analysis and comparison of TSR and OTC methods for MPPT and power smoothing in permanent magnet synchronous generator-based wind turbines. J Energy Convers Manag 86:892–900
4. Balasundar C, Sudharshanan S, Elakkiyavendan R (2015) Design of an optimal tip speed ratio control MPPT algorithm for standalone WECS. Int J Res Appl Sci Eng Technol 3:442–450
5. Ibarra E, Kortabarria I, Andreu J, Martinez de Alegria I, Martin J, Ibanez P (2012) Improvement of the design process of matrix converter platforms using the switching state matrix averaging simulation method. IEEE Trans Ind Electron 59:220–234
6. Ramireddy K, Hirpara Y, Kumar YVP (2020) Transient performance analysis of buck boost converter using various PID gain tuning methods. In: 2020 12th international conference on computational intelligence and communication networks (CICN), Bhimtal, India, pp 321–326
7. Essam HA, Abdel RY, Salah K, Mohamed MA (2018) Sensorless wind speed control of 1.5 MW DFIG wind turbines for MPPT. In: Twentieth international middle east power systems conference (MEPCON), pp 700–704
8. Dinh CP, Shigeru Y (2016) Rotor speed control of doubly fed induction generator wind turbines using adaptive maximum power point tracking. Energy J 111(c):377–388
9. Vlaho P, Carlo LB (2017) Wind turbine envelope protection control over the full wind speed range. Renew Energy J 111:836–848
10. Andrea T, Christian C (2011) Speed control for medium power wind turbines: an integrated approach oriented to mppt. IFAC Proc Vol 44(1):544–550
11. Kumar D, Chatterjee K (2016) A review of conventional and advanced MPPT algorithms for wind energy systems. Renew Sustain Energy Rev 55:957–970
12. Karthikeyan S, Veeraraghavan AK, Karni U, Srininvasan SR (2018) Modelling of smart airborne wind turbine system with neural network based on MPPT. In: 2018 2nd Borneo international conference on applied mathematics and engineering (BICAME), Balikpapan, Indonesia, pp 1–6
13. Babayeva M, Abdullin A, Polyakov N, Aranovskiy S (2020) MPPT algorithms for magnus effect wind turbine control system. In: 2020 XI international conference on electrical power drive systems (ICEPDS), Saint Petersburg, Russia, pp 1–4
14. Anjaneyulu BU, Satish T, Kumar NY (2019) Comparative analysis of fuzzy/ANN techniques for compensation of unbalanced voltages in grid connected PMSG based wind turbine. Int J Res Adv Technol 7:285–289
15. Singh S, Bhatti TS, Kothari DP (2007) Wind power estimation using artificial neural network. J Energy Eng 133(1)
16. Marugan AP, Arquez FPG, Perez JMP, Ruiz HD (2018) A survey of artificial neural network in wind energy systems. Appl Energy 228:1822–1836
17. Mesemanolis A, Mademlis C (2012) A neural network based MPPT controller for variable speed wind energy conversion systems. In: 8th Mediterranean conference on power generation, transmission, distribution and energy conversion (MEDPOWER 2012), Cagliari, pp 1–6

Monitoring and Analyzing the Daily Routine of Senior Citizens in Old-Age Home Using IoT

MD. Saquib Faraz, Areeb Kamal, Mohd. Aaqil Rehman, Noman Ahmed, and Mohd Ali Taqvi

Abstract In this modern era of technology, everything is designed to be accessible from Internet, the idea and methodologies of smart products and services are much more easily accessible, and people's interest with respect to this is expanding at a significant rate. Operation of such systems and products can be easily monitored through Internet even from far distant places, the use of such facilities has been made possible only because of the progress and research in the department of Internet of Things (IoT). In the field of IoT, hard work and attempts are being made to improve the efficiency and reliability of urban infrastructure. In this paper, using various IoT modules, we are representing a cloud integrated system to monitor and analyze the daily routine of senior citizens in old-age home using various IoT modules. The proposed system consists of motion detection, movement detection, and facial recognition techniques which are implemented by different IoT modules. The paper also defines a high-level view of the design of the system. Finally, to illustrate the correctness of the proposed model, the paper describes the operation of the system using different examples.

Keywords IoT · Node MCU · ESP32 CAM module · Cloud computing

1 Introduction

The idea of smart products and services is playing an essential role to influence the lifestyle of people, also to improve the working methodologies. The Internet of Things (IoT) plays an important role in the modern world in solving different challenges and problems. IoT helps us to connect different devices through a common platform and store the data recorded in all the devices at a common database which can be easily accessed by the authorized person even from all over the globe. The data recorded in the database can be used for analysis and evaluation to achieve the desired result [1]. The aim of the project is to discuss about designing of a

MD. Saquib Faraz · A. Kamal (✉) · Mohd. Aaqil Rehman · N. Ahmed · M. Ali Taqvi
EEE Department, Dr. Akhilesh Das Gupta institute of Technology and Management (Formerly Niec), New Delhi, Delhi, India

S. Suhag et al. (eds.), *Control and Measurement Applications for Smart Grid*, Lecture Notes in Electrical Engineering 822,
https://doi.org/10.1007/978-981-16-7664-2_36

smart system to analyze and monitor the activities of the senior citizens residing in a particular building. As old people require special care, permanent resident or a caretaker is required to help all those people in all aspects, i.e., for medicine, for physical movement, or for any other problem. Because of reduced visibility, weak immune system, and other health problems, a permanent caretaker is required to assist them. To improve the care taking methodologies, as per our observation and knowledge, we have used three techniques for monitoring of movement, i.e., by using motion detection, movement detection, and facial recognition techniques. The output of all these three techniques are stored in a database, and using various algorithms, an alarm system has been designed to update the caretaker about the person's condition.

2 Components Used

A. *ESP8266 module*

ESP8266 is a chip which merges a 32-bit microcontroller, digital peripheral interface, antenna switch, power amplifier filters, and power management modules into a compact package. The processor core, called "L106 by Espressif Company, is established on Tensilica's Diamond Standard 106Micro 32-bit processor controller core and runs at 80 MHz. It dispenses proficiency for 2.4 GHz Wi-Fi (supporting WPA/WPA2), general-purpose input/output (16 GPIO), analog-to-digital conversion (10-bit ADC), I^2S interfaces with DMA (sharing pins with GPIO), UART (on dedicated pins), serial peripheral interface (SPI), and pulse-width modulation (PWM). It has a 64 KB boot ROM, 32 KB instruction RAM, and 80 KB user data RAM. External flash memory can be accessed through SPI (Fig. 1).

The ESP32 CAM has an inbuilt Wi-Fi and Bluetooth modules integrated with OV2640 camera of two megapixel which can be used for face recognition. The basic requirements for camera module are system with a frame of 40 × 27 mm and an

Fig. 1 ESP8266 module B. ESP32—camera module

Fig. 2 ESP32 CAM module

operating current of up to 6 mA. This module takes on a dual in-line package and can be directly thrusted into the backplane. It provides high-fidelity connection, which is convenient for appropriate in various IoT hardware terminals. ESP has Wi-Fi and BLE Beacon, with two modules of 32-bit LX6 CPUs. It has the main frequency adjustment range of 80–240 MHz, on-chip sensor, Hall sensor, temperature sensor, etc. It is sketched for smart devices in home, industrial wireless control, wireless monitoring, and lot of other IoT applications (Fig. 2).

B. *Raspberry Pi module*

The prerequisite of a processor is essential for a mechanized gadget coordinated with a memory and force supply. This chip has a solitary center with 700 MHz CPU and simply 256 MB RAM, and the most advanced model has a quad-core 1.4 GHz CPU with 1 GB RAM. The processor helps in the preparing of info sign to produce a relative yield sign to get the ideal outcome [2]. Everywhere on the world, individuals use Raspberry Pi to construct equipment projects on home automation, and even use them in mechanical applications. The Raspberry Pi is a modest PC that runs Linux; however, it additionally gives a bunch of GPIO (universally useful info/yield) sticks that permit you to control electronic segments for actual processing and investigate the Internet of Things (IoT) (Figs. 3, 4, and 5).

C. *Arduino UNO*

It is a microcontroller with 14 pins, 6 PWM pins, 6 analog pins, a 16 MHz resonator, a USB port with an ICSP header, and a reset push button. It is designed on an open-source platform. It comprises of a programmable circuit board as well as a portion of programmable memory that assists in interfacing it to PC which is utilized to create and transfer code to the actual board so as to get a desired output and performance.

Fig. 3 Raspberry Pi module

Fig. 4 PIR sensor

Fig. 5 Arduino UNO

D. *PIR sensor*

PIR sensors are utilized to detect motion. It is utilized to identify or check the presence, regardless of whether a human has been moved in or out inside the sensors capable range. These sensors are little in size and relatively less expensive with low-force utilization. PIRs are of pyroelectric sensor, which attempt to recognize levels of

Fig. 6 PIR sensor coupled with relay and Node MCU

infrared radiation. Because of such highlights, they are utilized in different machines and devices utilized in home and enterprises. Each body present around us has some degree of temperature and heat present in them either living or non-living. The sensor in a motion detector is actually break up in two parts. The purpose behind that will be that our rationale is to recognize movement not normal IR levels. The two parts are wired up with the goal that they offset one another. On the off chance that one half sees pretty much IR radiation than the other, the yield will move from high to low or the other way around (Fig. 6).

3 Description of Each Section

A. *Motion detection*

It is defined as the computing of detecting an abnormality with respect to an object relative to its environment or an exchange in the surroundings relative to an object. The aim is to detect the motion of the person which is fulfilled by various PIR sensors connected with ESP8266, i.e., Node MCU and relay setup. Such PIR sensors can be installed in the room or around the entryways of the washroom. The output of all the PIR sensors is stored in a real-time database and further analyzed to monitor their movement along various pathways of the residence or within a room. Such sensors can be further placed near drawers where medicines are being kept, chairs or almirah or cupboard where essential commodities are being kept.

B. *Movement detection*

To detect the person position, a Raspberry Pi module includes camera module along with libraries of OpenCV, time, and Pandas. A program has been designed to detect the presence and movement of a person over a particular area. It consists of Raspberry Pi module integrated with a camera which detects the presence of a person, and based on the code stored in the module, it compares the present output with the past output stored in it and creates an output in the form of digital signal which is stored in a real-time database linked with Google Spreadsheet. As per the designed program, if

Fig. 7 ESP32 module with Arduino

the camera does not observe any movement in the person for a specific duration, the system will deliver a notification to the guardian or caretaker about his/her status. So that, the guardian should immediately check that person whether he/she is having any problem in any manner or suffering from anything.

C. *Facial recognition*

A facial recognition system is a technology capable of matching a human face from a digital image or a video frame against a database of faces.

Researchers are currently developing multiple methods in which facial recognition systems work. One of the most advanced face recognition method, which is also employed to authenticate users through ID verification services, works by pin pointing and measuring facial features from a given image. Such techniques can be used in case of two or more people residing in the same residence, to identify whom among them has entered or left the room. In our case, we have placed an ESP32 camera module connected with an Arduino UNO module near the washroom to detect whom among all the people has entered the toilet. In such system, the camera is programmed to detect the face of the person and compare the facial data of that particular person with the facial data of all the person stored in the database, and output of the data compared is stored in a real-time database which would be used to analyze the physical activity of all the person (Figs. 7, 8, and 9).

4 Working

To detect the movement of senior citizens, we have written Python code using Python libraries like NumPy and Pandas. The code will be compiled and starts running, and using camera, our system will detect whether any movement is detected or not. Based on the input of movement, our camera will give its output which is whether it is an emergency situation or a normal one. At different times of day when our senior citizen goes to washroom, the ESP32 webcam module will do a facial recognition on our senior citizen and do what is required. Once facial recognition is done, the

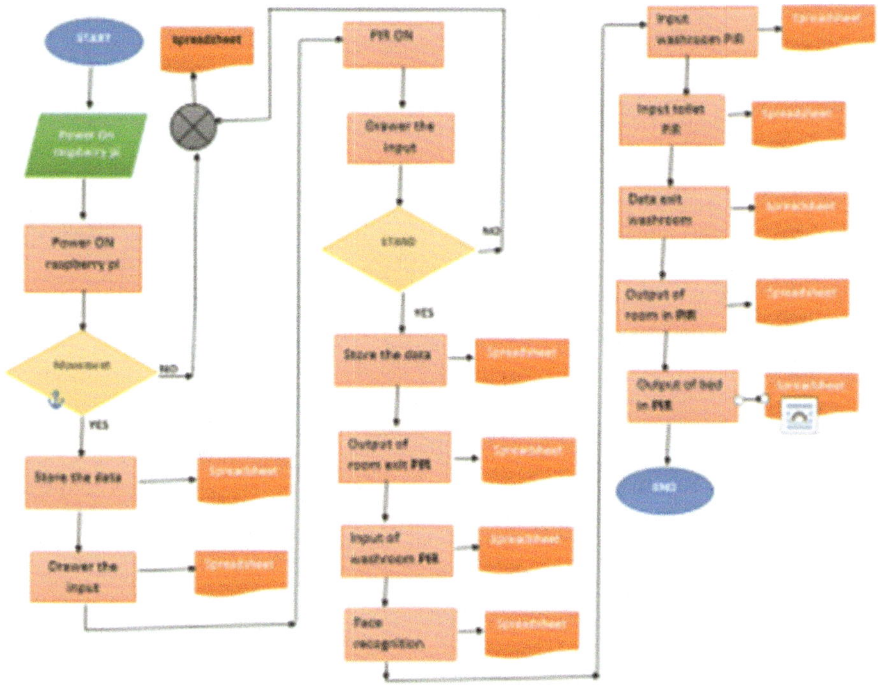

Fig. 8 Flowchart of sensor data

PIR sensor connected to Node MCU and placed in the entryways of washrooms will start collecting readings with date and time and store it in spreadsheets. Once we have enough data points to run a ML algorithm, we will use any regression algorithm which best fits our model and which can predict the time our senior citizen can be spending in washrooms, and if our algorithm detect any anomalous behavior, then it will raise an alarm (Figs. 10 and 11).

5 Output Data in Database

See Figs. 12 and 13.

6 Conclusion

We have proposed an IoT-based system to monitor the life of senior citizens from his rise from bed till he goes to bed. Our movement detection system will detect if our human has done any movement in the past 1 h, and if not, then it will raise an

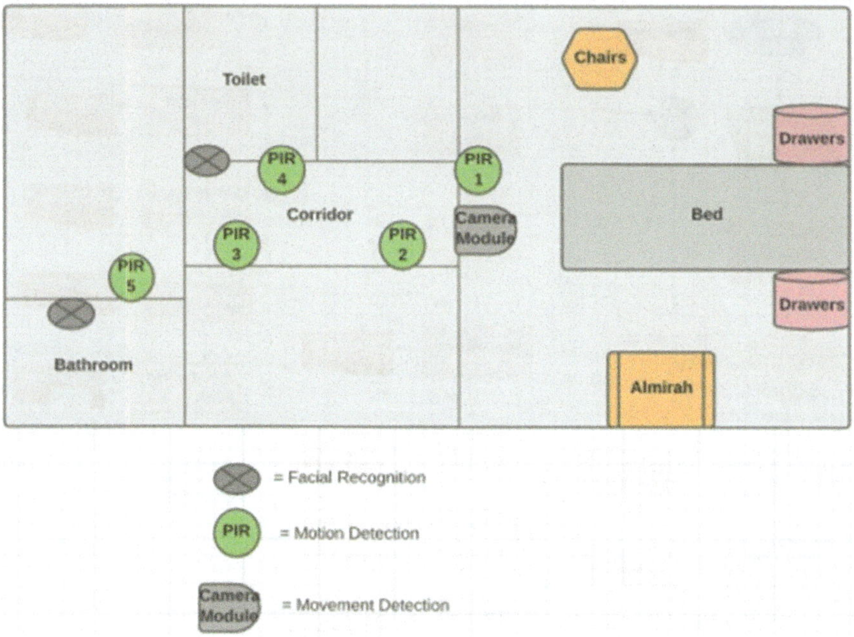

Fig. 9 Positioning of sensors

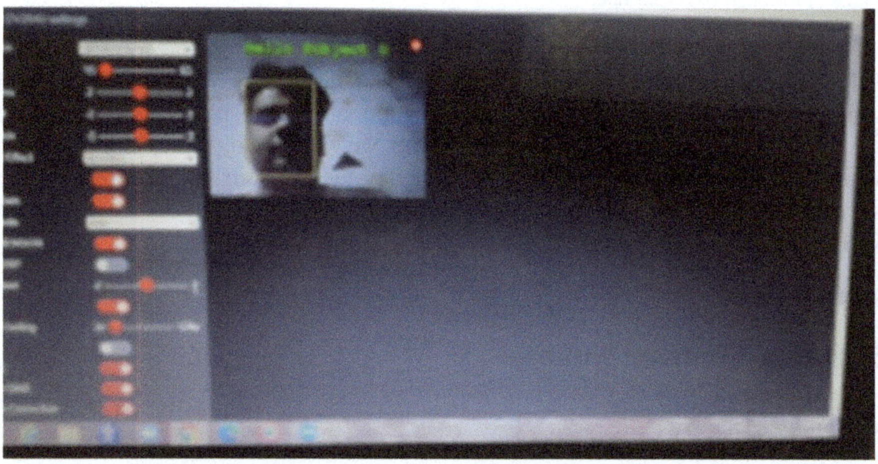

Fig. 10 Facial recognition interface

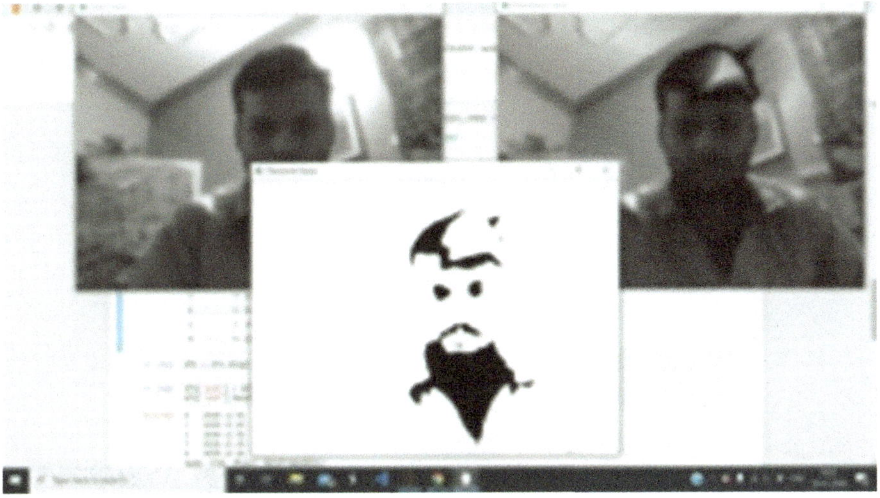

Fig. 11 Movement detection interface

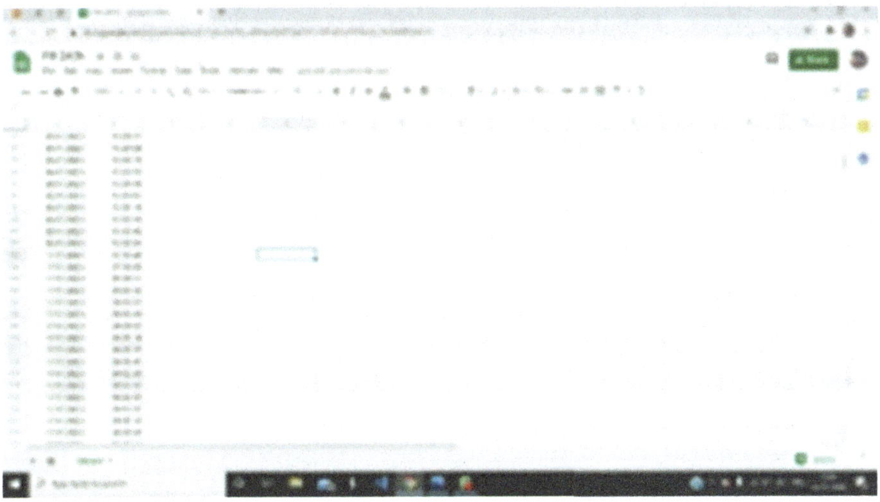

Fig.12 Output of Raspberry Pi module

emergency alarm. Our moment detection system will detect the entry and leaving time from washrooms. If the senior citizen takes very long in the washroom, then it can be an emergency situation and our system will raise an alarm. Our biometrics system will detect the face of senior citizens, and based on output, it will operate (Fig. 14).

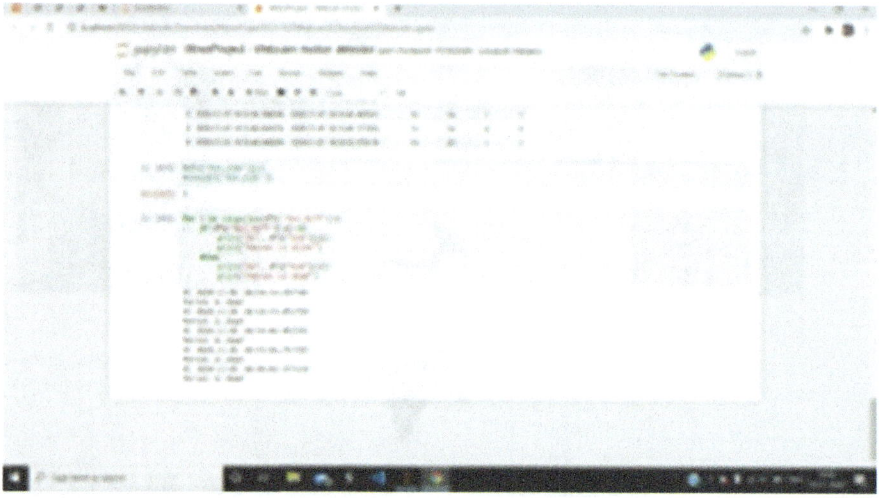

Fig. 13 PIR data in spreadsheet

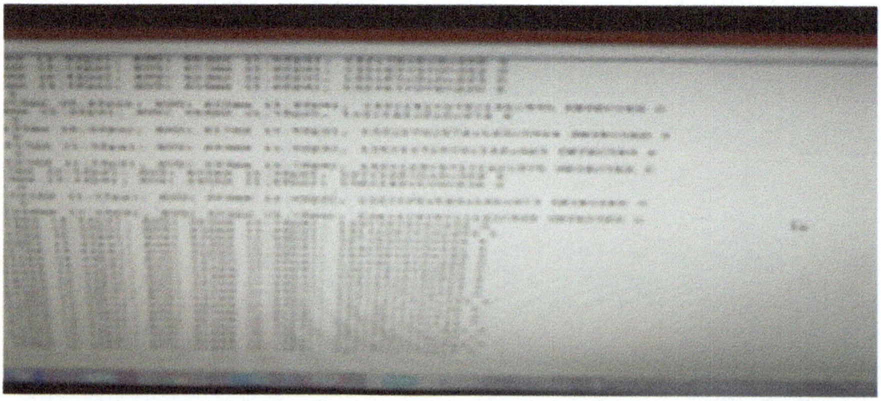

Fig. 14 Output of ESP32 module serial monitor

7 Application

The system and program described above are designed to install in all places where senior citizens reside. As due to the busy schedule of different people, sometimes, it becomes difficult to take care of our loved ones. This system can easily help us to monitor the activity of all the people residing with us. Or, this system can also be used when there is any sick person residing with us specially suffering from any infectious disease or COVID-19. Or, this system can also be used if there is any pet living with us and all the residents are out. It will easily help us to look after them. It will be really beneficial to monitor their activity.

Literature Survey

S. no.	Remarks	Reference
1. 978-1-5386-7159-7/18	Controlling the equipment used in the smart farm using a Blynk app	Smart farm monitoring via the blynk platform, *2018 Sixteenth International Conference on ICT and Knowledge Engineering*
2. 978-1-4799-4/15	Using Raspberry Pi as a microcontroller to store and access the data. And, getting the output as per the code programmed	Globally Accessible Machine Automation Using Raspberry Pi, *2015 International Conference on Advance in Computing, Communications and Informatics (ICACCI*
3. 749-75210-23,919/2018	The data from various devices used in the farm are stored in a Google Spreadsheet as a database	Implement smart farm with IoT technology, 2018 Twentieth International Conference on Advanced Communication Technology (ICACAT)
4. https://doi.org/10.1109/2018.00035	The data from various devices used in the farm are stored in a Google Spreadsheet as a database and can be easily accessed by the admin and authorized accounts to monitor and analyze the total consumption using machine learning	An IoT based remote monitoring system for Electrical Power Consumption via Web Application, 2018 International Conference on Informatic systems and Computer Science (INCISCOS)

References

1. Serikul P, Nakpong N, Nakjuatong N (2018) Smart farm monitoring via the blynk platform. In: 2018 Sixteenth international conference on ICT and knowledge engineering. 978-1-5386-7159-7/18
2. Sandeep V, Lalith K, Naveen S, Amudhan A, Kumar LS (2015) Globally accessible machine automation using Raspberry Pi. In: 2015 International conference on advance in computing, communications and informatics (ICACCI). 978-1-4799-4/15. IEEE
3. Yoon C, Huh M, Kang S, Park J, Lee C (2018) Implement smart farm with IoT technology. In: 20th International conference on advanced communication technology (ICACAT), Chuncheon-siGangwon-do, Korea (South), pp 749–752. https://doi.org/10.23919/ICACT.2018.8323908
4. Jain S, Vaibhav A, Goyal L (2014) Raspberry Pi based Interactive home automation system. In: 2014 International conference on reliability, optimization and information technology-ICROIT 2014, India, 6–8 Feb 2014
5. Alulema D, Zapata M, Zapata MA (2018) An IoT based remote monitoring system for electrical power consumption via web application. In: 2018 International conference on informatic systems and computer science (INCISCOS). https://doi.org/10.1109/INCISCOS.2018.00035

A Comparative Analysis of LQR and SMC for Quanser AERO

Saurabh Kumar and Lillie Dewan

Abstract In this paper, mathematical modeling of multi-input multi-output Quanser AERO system is obtained using Euler–Lagrange equation. Model obtained is nonlinear, and there exits cross-coupling. This nonlinearity and cross-coupling are challenging tasks for designing the controller for the Quanser AERO system. LQR controller has been widely used in literature, but it is not able to meet the desired performance specifications. To overcome this, SMC has been implemented in addition to LQR, and their performance has been compared.

Keywords Quanser AERO · Linear quadratic regulator (LQR) · Sliding mode control (SMC) · Multi-input multi-output (MIMO)

1 Introduction

Quanser AERO is a nonlinear, cross-coupled, multi-input multi-output (MIMO) system [1]. The modeling of the system dynamics is a highly challenging task due to the presence of high interactions among the various process variables and the non-accessibility of certain states [2]. It is an experimental setup that provides replication of flight dynamics. This prototype model has obtained high popularity among the control system community because of the difficulties involved in performing direct experiment to the flight vehicles [3]. Aerodynamically, Quanser AERO consists of two rotors, viz. main rotor and tail rotor at both ends of the beam, and both of the rotors are driven by the DC motor [4]. This system can move freely in horizontal as well as in vertical plane. The state of beam is defined by four process variables, viz. yaw, pitch angles and their respective velocities [5]. The yaw and pitch angles are

S. Kumar · L. Dewan (✉)
Department of Electrical Engineering, National Institute of Technology Kurukshetra, Kurukshetra, Haryana 136119, India
e-mail: l_dewan@nitkkr.ac.in

S. Kumar
e-mail: saurabh.k@aero.iitb.ac.in

S. Suhag et al. (eds.), *Control and Measurement Applications for Smart Grid*, Lecture Notes in Electrical Engineering 822,
https://doi.org/10.1007/978-981-16-7664-2_37

being measured by the encoders which are fixed at the pivot. For the measurement of these velocities, speed sensors coupled with the DC motors are used. In the literature, commonly used controller for control of these variables is LQR. It is observed that it does not satisfy the desired performance specifications [6]. The Quanser AERO is a prototype model of the helicopter, but there are some significant difference in controlling the Quanser AERO and the helicopter. In the helicopter, controlling is done by varying the angles of the rotors, while in Quanser AERO, controlling is done by changing the speed of rotors [7].

In this paper, mathematical model of nonlinear cross-coupled multi-input multi-output Quanser AERO system is obtained. All of the four process variables such as yaw, pitch angles and their respective velocities have been controlled using sliding mode controller (SMC) and linear quadratic regulator (LQR), and their performance has been compared.

In the first section, the dynamic modeling of Quanser AERO is obtained using the Euler–Lagrange equation, and from there, nonlinear dynamics are being obtained. Dynamics are linearized using Taylor series expansion, and from there, linear equations of motion as well as state-space matrices are obtained. In the next section, the various steps involved in the controller design (LQR and SMC) are given, and in the further upcoming section, both the controllers will be implemented on the Quanser AERO system. In the fourth section, results of both controllers are presented, and performance is compared followed by conclusion and references.

2 Modeling of Quanser AERO System

In this section, the mathematical model of the Quanser AERO helicopter configuration using Euler–Lagrange equation is presented. Quanser AERO model is shown in Fig. 1.

2.1 Dynamical Model of AERO 2-DoF Configuration

The dynamics of the AERO model are represented in Fig. 2. This system is configured with two high-efficiency rotors coupled with two DC motors, which produces the thrust forces, F_P and F_Y at distances r_p and r_y from the Z-axis along the X-axis. The propeller which is horizontal to the ground (front) generates a torque along Y-axis which yields pitch motion, while the back (or tail) propeller handles the yaw motion [8].

The gravitational force, F_g, is acting on the body at a distance of l_{cm} from the Z-axis. The exact position of center of mass with respect to fixed frame (pivot) of the Quanser AERO system is found using the homogeneous transformation matrices [9], and the Cartesian position of center of mass is obtained from this.

Fig. 1 Experimental model of Quanser AERO [7]

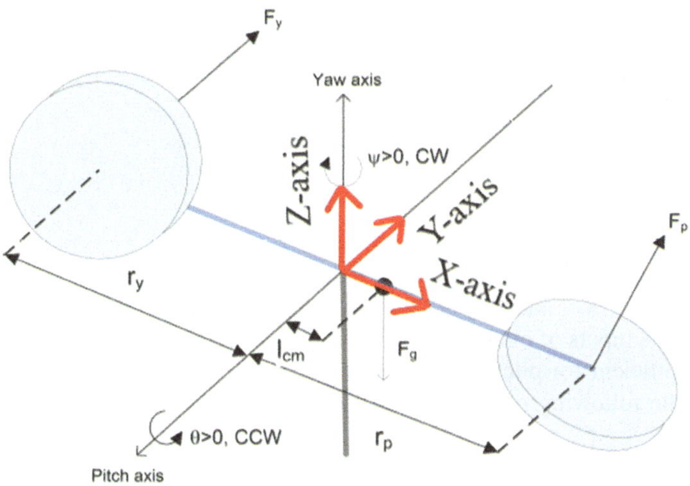

Fig. 2 Free body diagram of Quanser AERO [4]

$$X_{\mathrm{cm}} = l\cos\theta\cos\psi,\ Y_{\mathrm{cm}} = -l\cos\theta\sin\psi,\ Z_{\mathrm{cm}} = l\cos\theta \tag{1}$$

where l is the distance of center of mass from Z-axis
The total potential energy will be due to gravity only, and the total kinetic energy will be the sum of rotational kinetic energy and transnational kinetic energy.

$$P_{\mathrm{total}} = \mathrm{mgl}\cos\theta \tag{2}$$

$$T_{\mathrm{total}} = T_{\mathrm{rot},p} + T_{\mathrm{rot},y} + T_{\mathrm{trans}} \tag{3}$$

$$T_{\text{rot},p} = \frac{1}{2}J_{eq,p}\dot{\theta}^2,\ T_{\text{rot},y} = \frac{1}{2}J_{eq,y}\dot{\psi}^2 \tag{4}$$

$$T_{\text{trans}} = \frac{1}{2}m(\dot{X}_{\text{cm}}^2 + \dot{Y}_{\text{cm}}^2 + \dot{Z}_{\text{cm}}^2) \tag{5}$$

Substituting the values from Eqs. (4) and (5) into Eq. (3), the total kinetic energy is obtained as

$$T_{\text{total}} = \frac{1}{2}J_{eq,p}\dot{\theta}^2 + \frac{1}{2}J_{eq,y}\dot{\psi}^2 + \frac{1}{2}ml^2(\cos^2\theta\dot{\psi}^2 + \dot{\theta}^2) \tag{6}$$

Euler–Lagrange equations are defined as

$$\frac{\partial L}{\partial \theta} - \frac{\partial}{\partial t}\left(\frac{\partial L}{\partial \dot{\theta}}\right) = -Q_1 \quad \text{and} \quad \frac{\partial L}{\partial \psi} - \frac{\partial}{\partial t}\left(\frac{\partial L}{\partial \dot{\psi}}\right) = -Q_2 \tag{7}$$

where L is Lagrange variable which is defines as $L = T_{\text{total}} - P_{\text{total}}$

$$L = \frac{1}{2}J_{eq,p}\dot{\theta}^2 + \frac{1}{2}J_{eq,y}\dot{\psi}^2 + \frac{1}{2}ml^2(\cos^2\theta\dot{\psi}^2 + \dot{\theta}^2) - \text{mgl}\sin\theta \tag{8}$$

Q_1 and Q_2 are the generalized forces which are defined as

$$\begin{aligned} Q_1 &= K_{pp}V_p + K_{py}V_y - B_p\dot{\theta} \\ Q_2 &= K_{yp}V_p + K_{yy}V_y - B_p\dot{\psi} \end{aligned} \tag{9}$$

where K_{pp}, K_{py}, K_{yy} and K_{yp} are the thrust torque gain for pitch and yaw axis. V_p and V_y are the inputs of pitch and yaw axis, respectively. B_P and B_Y are the viscous friction coefficient for pitch and yaw axis, respectively. By using Euler–Lagrange equation, the following nonlinear equations of motion for Quanser AERO system are obtained.

$$\begin{aligned} \ddot{\theta} &= \frac{1}{a}(K_{pp}V_p + K_{py}V_y - B_p\dot{\theta} - \text{mgl}\cos\theta - ml^2\sin\theta\cos\theta\dot{\psi}^2) \\ \ddot{\psi} &= \frac{1}{b}(K_{yp}V_p + K_{yy}V_y - B_y\dot{\psi} + 2ml^2\sin\theta\cos\theta\dot{\theta}\dot{\psi}) \end{aligned} \tag{10}$$

where $a = J_{eq,p} + ml^2$ and $b = Jeq,\ y + ml^2\cos^2\theta$.

2.2 State-Space Representation

On linearizing of the nonlinear equations, the linear equations of motion are obtained, and from there, state-space matrices are obtained given below. Taking the state vector as $x^T(t)$ as $[\theta(t)\ \psi(t)\ \dot{\theta}(t)\ \dot{\psi}(t)]$ and input vector as $u^T(t) = [V_p\ V_y]$. The linear state-space model is given as

$$\dot{x}(t) = Ax(t) + Bu(t)$$
$$y(t) = Cx(t) + Du(t)$$

where $y(t)$ is output vector, and A, B, C, D are known as system matrix, input matrix, output matrix and feed-forward matrix, respectively.

$$A = \begin{bmatrix} 0\,1 & 0 & 0 \\ 0\,0 & 0 & 1 \\ 0\,0 & -B_p/a & 0 \\ 0\,0 & 0 & -B_y/b_1 \end{bmatrix}$$
$$B = \begin{bmatrix} 0 & 0 \\ 0 & 0 \\ K_{pp}/a & K py/a \\ K_{yp}/b_1 & K_{yy}/b_1 \end{bmatrix}$$
$$C = \begin{bmatrix} 1\,0\,0\,0 \\ 0\,1\,0\,0 \end{bmatrix}$$
$$D = \begin{bmatrix} 0\,0 \\ 0\,0 \end{bmatrix}$$

where $a = J_{eq,p} + ml^2$ and $b_1 = Jeq, y + ml^2$
The controllability and observability analysis is performed on the system, and it is observed that system is completely controllable and observable.

3 Controller Design

In the present section, two control strategies will be designed for the Quanser AERO system. In the present work, LQR and SMC proposed in the literature [10, 11] have been used, and the main steps involved in design are given.

3.1 Linear Quadratic Regulator (LQR)

LQR is a type of optimal control which minimizes the cost function [10]:

$$J = \int \frac{1}{2} x^T(t) Q x(t) + \frac{1}{2} u^T(t) R u(t) \mathrm{d}t \tag{11}$$

where Q and R are known as weight matrices of states and inputs, respectively [12, 13]. Q will be symmetric positive semi-definite or positive definite, whereas R will always be symmetric positive definite matrix [14].
The state feedback law $u(t) = -Kx(t)$ [14] will be used for designing the controller. Where

$$K = R^{-1}B^T P$$

And P is the solution of algebraic Riccati equation:

$$PA + A^T P - PBR^{-1}B^T P + Q = 0 \tag{12}$$

3.2 Sliding Mode Controller (SMC)

As a control system engineer to design any control scheme for the any system, we consider the mathematical model of that system, but there is always some discrepancy between the actual plant and the mathematical model. These discrepancies may be due to the presence of external disturbances or parameter variations of the plant or due to unmodelled dynamics [15]. Due to these reasons, designing a control law becomes a challenging task. Sliding mode control can take care of these disturbances and parameter variation. It is a nonlinear robust control design technique [16].

SMC design involves two steps, and first step is to design the control input which will drive the states to the sliding surface in finite time and will keep it there. The second step is to design a sliding surface [17].

The nonlinear equations of motion are given as

$$\begin{aligned} \ddot{\theta} &= A_1 + B_{11}V_p + B_{12}V_y \\ \ddot{\psi} &= A_2 + B_{21}V_p + B_{22}V_y \end{aligned} \tag{13}$$

where $B_{11} = \frac{K_{pp}}{a}, B_{12} = \frac{K_{py}}{a}, B_{21} = \frac{K_{yp}}{b}, B_{22} = \frac{K_{yy}}{b}$ and

$$A_1 = \frac{-\text{mgl}\cos\theta - B_p\dot{\theta} - ml^2\sin\theta\cos\theta\dot{\psi}^2}{a}, \quad A_2 = \frac{2ml^2\sin\theta\cos\theta\dot{\theta}\dot{\psi} - B_y\dot{\psi}}{b} \tag{14}$$

Taking the sliding surface for the set-point tracking as [11]:

$$[s] = \begin{bmatrix} s_1 \\ s_2 \end{bmatrix} = \begin{bmatrix} \dot{e}_1 + c_1 e_1 \\ \dot{e}_2 + c_2 e_2 \end{bmatrix}$$

where c_1, c_2 will satisfy Hurwitz condition, and $c_1 > 0$ and $c_2 > 0$ and e_1 and e_2 are the errors for pitch and yaw, respectively, and defined as

$$\begin{aligned} e_1 &= \theta - \theta_{\text{desired}} \\ e_2 &= \psi - \psi_{\text{desired}} \end{aligned}$$

Taking the time derivative of sliding surface:

$$[\dot{s}] = \begin{bmatrix} \ddot{\theta} + c_1\dot{\theta} \\ \ddot{\psi} + c_2\dot{\psi} \end{bmatrix} \tag{15}$$

$$= \begin{bmatrix} A_1 + B_{11}V_p + B_{12}V_y + c_1\dot{\theta} \\ A_2 + B_{21}V_p + B_{22}V_y + c_2\dot{\psi} \end{bmatrix} \tag{16}$$

$$= \begin{bmatrix} A_1 \\ A_2 \end{bmatrix} + \begin{bmatrix} B_{11} & B_{12} \\ B_{21} & B_{22} \end{bmatrix} \begin{bmatrix} V_p \\ V_y \end{bmatrix} + \begin{bmatrix} c_1\dot{\theta} \\ c_2\dot{\psi} \end{bmatrix} \tag{17}$$

The exponential reaching law is used for designing the control law of SMC, and it is given by

$$[\dot{s}] = -\begin{bmatrix} k_1\text{sign}(s_1) + k_3 s_1 \\ k_2\text{sign}(s_2) + k_4 s_2 \end{bmatrix} \tag{18}$$

where the first term guarantees faster convergence speed when s is small. The second term, i.e., proportional term forces the state to reach the switching surface faster when s is large [18].

On solving the Eqs. (17) and (18), we get our control input as

$$\begin{bmatrix} V_p \\ V_y \end{bmatrix} = -\begin{bmatrix} B_{11} & B_{12} \\ B_{21} & B_{22} \end{bmatrix}^{-1} \begin{bmatrix} A_1 + c_1\dot{\theta} + k_1\text{sign}(s_1) + K_3 s_1 \\ A_2 + c_1\dot{\psi} + k_2\text{sign}(s_2) + k_4 s_2 \end{bmatrix} \tag{19}$$

To reduce the chattering and speedup, the response tanh has been used instead of sign function [19].

Now, in the next section, these controllers will be implemented on the system, and the results are given.

4 Results

In this section, both LQR and SMC controllers are being implemented on Quanser AERO using the mathematical model. The values of the parameters are given in Table 1. Based on the results, a comparative analysis is performed.

The values of c_1 and c_2 are taken as 10, and it follows the Hurwitz condition. The value of k_1, k_2 is taken as 1, and the value of k_3, k_4 is taken as 10.
These values are taken on the based of observation that bigger the values of proportional term coefficient (k_3 and k_4) in reaching law and smaller the value of first term coefficient (k_1 and k_2) causes less chattering. The desired values of pitch and yaw angles are taken as $\theta_{\text{desired}} = 1$ and $\psi_{\text{desired}} = 1$.
The initial conditions are taken as $\theta_0 = 0$ and $\psi_0 = 0$.

Table 1 System parameters of Quanser AERO [7]

Parameters	Value	Unit
$J_{eq,p}$	0.0219	kg m^2
$J_{eq,y}$	0.0220	kg m^2
B_p	0.00711	Vs/rad
B_y	0.0220	Nm/rad
K_{pp}	0.0011	Nm/V
K_{py}	0.0021	Nm/V
K_{yp}	-0.0027	Nm/V
K_{yy}	0.0022	Nm/V
l	0.0071	m
m	1.0750	kg

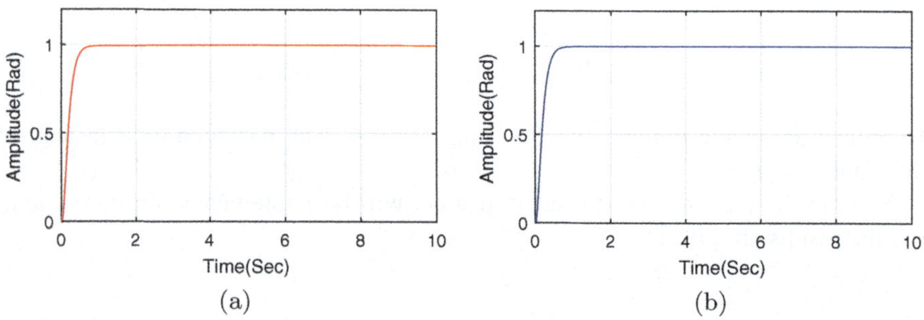

Fig. 3 Angular position of **a** pitch, **b** yaw on applying SMC

On applying this SMC controller on the Quanser AERO dynamics, the following results are obtained for pitch, yaw and their respective velocities shown in Figs. 3 and 4.

For designing LQR, weight matrices have always an impact on the system performance. These weight matrices are chosen in such a way that $Q = Q^T$ and $R = R^T$ [14]. For the implementation of LQR, weight matrices are chosen by trial and error method, and it is given by

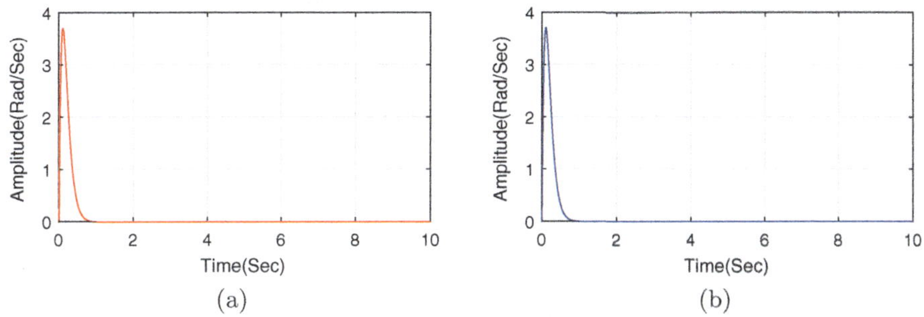

Fig. 4 Angular velocity of **a** pitch, **b** yaw on applying SMC

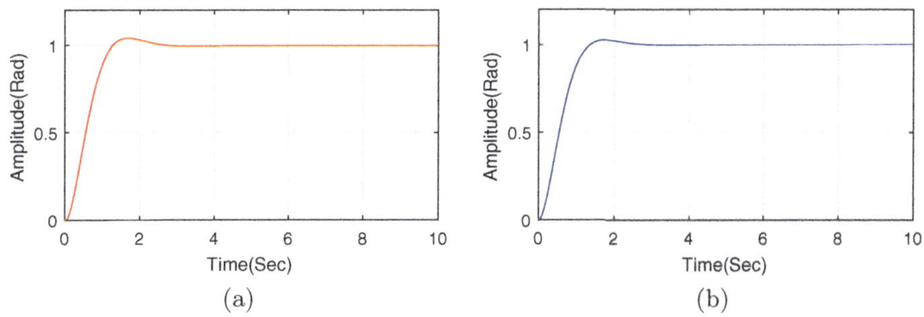

Fig. 5 Angular position of **a** pitch, **b** yaw on applying LQR

$$Q = \begin{bmatrix} 200 & 0 & 0 & 0 \\ 0 & 75 & 0 & 0 \\ 0 & 0 & 0 & 0 \\ 0 & 0 & 0 & 0 \end{bmatrix} \text{ and } R = 0.05 * \begin{bmatrix} 1 & 0 \\ 0 & 1 \end{bmatrix}$$

By using these weight matrices, the following gains are obtained shown in Eq. (19).

$$K = \begin{bmatrix} 33.6648 & -32.7873 & 17.9880 & -14.5220 \\ 53.5414 & 20.6154 & 25.9476 & 8.4231 \end{bmatrix} \tag{20}$$

These gains are applied on the system, and the following results are obtained shown in Figs. 5 and 6.

Based on the performances of both the controllers on the Quanser AERO system, observations made are given in Table 2.

On implementation of LQR on Quanser AERO, the peak overshoot percentage for pitch and yaw are 4.4 and 2.8, respectively, with settling time 2.412 and 2.230 s. While with SMC, there is no overshoot, and also, settling time is 0.5462 s for both pitch and yaw, which is lesser in comparison with LQR. With both the controllers, steady-state error is zero. Rise time and delay time are lesser for pitch and yaw with SMC in comparison with LQR.

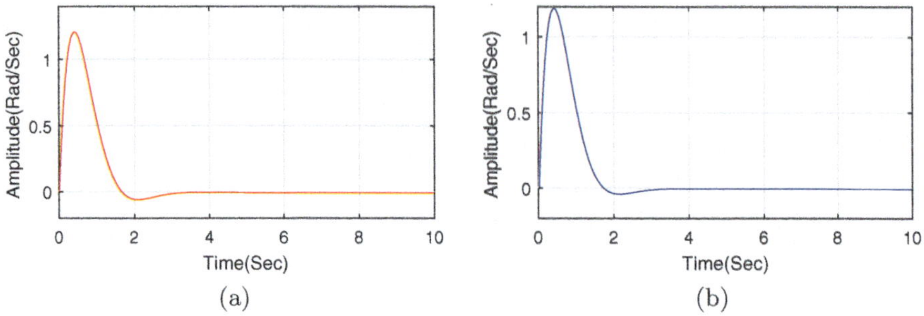

Fig. 6 Angular velocity of **a** pitch, **b** yaw on applying LQR

Table 2 Comparison of performances of LQR and SMC

Performance specifications	LQR (pitch)	LQR (yaw)	SMC (pitch)	SMC (yaw)
Peak value	1.044	1.028	1	1
Rise time (s)	1.476	1.571	1.223	1.223
Overshoot percentage	4.4	2.8	0	0
Steady-state error	0	0	0	0
Delay time (s)	0.721	0.731	0.1711	0.1711
Settling time (2%)	2.412	2.230	0.5899	0.5899

5 Conclusion

In this paper, the dynamics of Quanser AERO are presented. From there, it can be observed that it is a nonlinear, cross-coupled multiple input multiple output (MIMO) system. From controllability and observability analysis, it is observed that system is completely controllable and completely observable. Linear optimal control (LQR) and a nonlinear control scheme sliding mode control have been successfully implemented on the Quanser AERO. SMC controller settles down fast as compared to LQR controller with zero steady-state error. Work can be further extended in the direction of optimizing parameters of controllers and implementing on real-time system.

References

1. Ahmad SM, Chipperfield AJ, Tokhi O (2000) Dynamic modeling and optimal control of a twin rotor MIMO system. In: Proceedings of the IEEE 2000 national aerospace and electronics conference. NAECON 2000. Engineering Tomorrow (Cat. No. 00CH37093), pp 391–398

2. Rahideh A, Shaheed MH, Huijberts HJC (2008) Dynamic modelling of a TRMS using analytical and empirical approaches. Control Eng Practice 16(3):241–259
3. Luo B, Wu HN, Huang T (2017) Optimal output regulation for model-free Quanser helicopter with multistep Q-learning. IEEE Trans Ind Electron 65(6):4953–4961
4. Quanser (2006) Quanser 2-DOF helicopter user and control manual. Quanser Inc
5. Pratap B, Agrawal A, Purwar S (2012) Optimal control of twin rotor MIMO system using output feedback. In: 2012 2nd international conference on power, control and embedded systems, pp 1–6
6. Vonckx K, Janiak G, Miah MS (2019) Optimal tracking control experiments for 2-DOF helicopter: an open-implementation approach. In: 2019 7th international conference on mechatronics engineering (ICOM), pp 1–6
7. Inc Quanser (2016) (2016) Quanser AERO laboratory guide. Technical Report, Quanser
8. https://www.quanser.com/products/quanser-aero/ . Last accessed 30th Dec 2019
9. Spong MW, Hutchinson S, Vidyasagar M (2006) Forward and inverse kinematics. Robot modelling and control, pp 74–118
10. Shieh LS, Dib HM, Yates RE (1988) Sequential design of linear quadratic state regulators via the optimal root-locus techniques. IEE Proc D (Control Theory Appl) 135(4):289–294
11. Khalil H (2000) Nonlinear systems, 1st edn. Prentice Hall
12. Kadmiry B, Driankov D (2004) A fuzzy gain-scheduler for the attitude control of an unmanned helicopter. IEEE Trans Fuzzy Syst 12(4):502–515
13. Ghosh S, Gude S (2012) A genetic algorithm tuned optimal controller for glucose regulation in type 1 diabetic subjects. Int J Numer Methods Biomed Eng 28(8):877–889
14. Ogata K (2010) Modern control engineering. Prentice hall
15. Esfandiari F, Khalil HK (1991) Stability analysis of a continuous implementation of variable structure control. IEEE Trans Autom Control 36(5):616–620
16. Huang YJ, Kuo TC, Way HK (2003) Robust vertical takeoff and landing aircraft control via integral sliding mode. IEEE Proc-Control Theory Appl 150(4):383–388
17. Cao WJ, Xu JX (2004) Nonlinear integral-type sliding surface for both matched and unmatched uncertain systems. IEEE Trans Autom Control 49(8):1355–1360
18. Levant A (2010) Chattering analysis. IEEE Trans Autom Control 55(6):1380–1389
19. Eltayeb A, Rahmat MF, Basri MAM, Mahmoud MS (2020) An improved design of integral sliding mode controller for chattering attenuation and trajectory tracking of the quadrotor UAV. Arab J Sci Eng 45:6949–6961

Numerical Investigation of Fractional Model of Biswas–Milovic Equation via Laplace Transform

Amit Prakash and Hardish Kaur

Abstract In this paper, the fractional Biswas–Milovic equation with Kerr and parabolic law nonlinearities is studied with homotopy perturbation transform technique (HPTT) which is a compilation of homotopy perturbation method and Laplace transform. The graphical and numerical outcomes explicitly reveal the complete reliability of the proposed algorithms with high accuracy of the results.

Keywords Fractional Biswas–Milovic model · Homotopy perturbation technique · Caputo's fractional derivative · Laplace transform

1 Introduction

Fractional calculus (FC) has been used to model physical and engineering processes [1–6] which can be best described by fractional differential equations (FDEs). The reason is that a reasonable modeling of a physical phenomenon depends not only on the time instant but also on the prior time history. Numerous techniques have been suggested for solving FDEs numerically [7–10]. The theory of FC is an emerging topic of applied mathematics, and different fractional models were studied by many authors (see, e.g., [11–24]). In this manuscript, the space- and time-fractional Biswas–Milovic equation (BME) is investigated via HPTT. BME is a generalization of the nonlinear Schrodinger's equation describing the propagation of solitons through optical fibers for transcontinental and transocean distances and is given by [25]

$$iq^r + \lambda {q^r}_{\zeta\zeta} + \eta H\left(|q|^2\right)q^r = 0, \tag{1}$$

ζ and τ are the independent variables, $q(\zeta, \tau)$ is the wave profile, the parameter $r \geq 1$, and λ and η are the real-valued constants with the property $\lambda\eta > 0$. Here,

A. Prakash · H. Kaur (✉)
Department of Mathematics, National Institute of Technology, Kurukshetra 136119, India
e-mail: amitmath@nitkkr.ac.in

S. Suhag et al. (eds.), *Control and Measurement Applications for Smart Grid*, Lecture Notes in Electrical Engineering 822,
https://doi.org/10.1007/978-981-16-7664-2_38

H is real-valued algebraic function, and $H\left(|q|^2\right) : C \to C$ is a smooth complex function. The complex plane C is taken as 2D linear space R^2, and $H(|q|^2)$ is n times differentiable function, so

$$H\left(|q|^2\right) \in \bigcup\nolimits_{l,m=1}^{\infty} C^n\left((-m, m) \times (-l, l); R^2\right).$$

Here, parameter m is for the power law nonlinearity, and β is the coefficient of the nonlinear term. Various researchers have examined the BME with different semi-analytical, analytical and numerical techniques [26–32]. In this manuscript, we investigate the following BME when $r = 1$, expressed as

$$iD_\tau^\mu q(\zeta, \tau) + \lambda q_{\zeta\zeta}^{2\beta}(\zeta, \tau) + \eta|q|^2 q = 0. \tag{2}$$

The aim of this work is to investigate Eq. (2) numerically with the compilation of homotopy perturbation method (HPM) [33] and Laplace transform (LT). The applications of homotopy theory have become a powerful mathematical tool in the nonlinear problems due to the excellent work done, and also, the method is really capable of reducing the size of the computational work besides being convenient for solving nonlinear fractional equations.

2 Preliminaries

Definition 2.1 A real-valued function $f(t), t > 0$ is said to be in the space $C_\mu, \mu \in \mathbb{R}$ if $\exists$ [34] a real number $p(> \mu)$ so that $f(t) = t^p f_1(t),\ f_1 \in C[0, \infty)$, and in C_μ^m if $f^m \in C_\mu, M \in \mathrm{N}\bigcup\{0\}$.

Definition 2.2 The Caputo fractional derivative of $f, f \in C_{-1}^m, m \in \mathbb{N}\bigcup\{0\}$

$$D_t^\alpha f(t) = \begin{cases} \frac{\mathrm{d}^n f(t)}{\mathrm{d}t^n}, \alpha = n \in N, \\ \frac{1}{\Gamma(n-\alpha)} \int_0^t (t-\tau)^{n-\alpha-1} f^n(\tau)\mathrm{d}\tau, n-1 < \alpha < n, n \in \mathbb{N}. \end{cases}$$

Definition 2.3 The LT, $L[u(x, t)]$ of the Caputo fractional derivative is written as $L\left[D_t^{n\alpha} u(x, t)\right] = s^{n\alpha} L[u(x, t)] - \sum_{k=0}^{n-1} s^{n\alpha-k-1} u^k(x, 0), n-1 < n\alpha \le n.$

3 Basic Idea of Proposed Technique

Consider the nonlinear differential equation of arbitrary order

$$D_t^\mu u(x, t) + Ru(x, t) + Nu(x, t) = f(x, t), 0 < \mu \le 1, \tag{3}$$

with

$$u(x, 0) = f(x), \tag{4}$$

here $D_t^{\mu} u(x,t)$ is derivative of $u(x,t)$ of order μ in Caputo's sense, R and N are used for the linear and nonlinear differential operator, and $f(x,t)$ is the source term.

Exerting LT on Eq. (3), it yields

$$L[u(x,t)] - \sum_{k=0}^{l-1} s^{-k-1} u^k(x,0) + \frac{1}{s^{\mu}}\{L[Ru(x,t) + Nu(x,t)] - L[f(x,t)]\} = 0. \tag{5}$$

Take inverse LT on Eq. (5), then

$$u(x,t) = L^{-1}\left(\sum_{k=0}^{l-1} s^{-k-1} u^k(x,0) + L[f(x,t)]\right) - L^{-1}\left\{\frac{1}{s^{\mu}}\{L[Ru(x,t) + Nu(x,t)]\}\right\}.$$

HPM defines the solution in terms of infinite series given as

$$u(x,t) = \sum_{i=0}^{\infty} p^i u_i(x,t), \tag{6}$$

and the nonlinear term is decomposed as

$$Nu(x,t) = \sum_{i=0}^{\infty} p^i H_i(u), \tag{7}$$

here $H_i(u)$ is the homotopy polynomial, given by

$$H_i(u) = \frac{1}{i!}\frac{\partial^i}{\partial p^i}\left[N\left(\sum_{j=0}^{\infty} p^j u_j\right)\right]_{p=0}, i = 0, 1, 2, \ldots \tag{8}$$

Using Eqs. (6) and (7) in (5),

$$\sum_{i=0}^{\infty} p^i u_i(x,t) = L^{(-1)}\left(\sum_{k=0}^{l-1} s^{(-k-1)} u^k(x,0) + L[f(x,t)]\right) - p[L^{(-1)} 1/\left\{s^{\mu} L\left\{\left[R\sum_{i=0}^{\infty} p^i u_i(x,t)\right] + \sum_{i=0}^{\infty} p^i H_i(u)\right\}\right\}. \tag{9}$$

Equating the coefficients of like powers of p,

$$p^0: u_0(x,t) = L^{-1}\left(\sum_{k=0}^{l-1} s^{-k-1} u^k(x,0) + L[f(x,t)]\right),$$

$$p^1: u_1(x,t) = -L^{-1}\left\{\frac{1}{s^{\mu}} L\{[Ru_0(x,t)] + H_0(u)\}\right\},$$

$$p^2: u_2(x,t) = -L^{-1}\left\{\frac{1}{s^{\mu}} L\{[Ru_1(x,t)] + H_1(u)\}\right\},$$

$$p^3: u_3(x,t) = -L^{-1}\left\{\frac{1}{s^{\mu}} L\{[Ru_2(x,t)] + H_2(u)\}\right\},$$

Calculating in this way, the solution is expressed as

$$u(x,t) = \lim_{p\to 1}\lim_{N\to\infty} \sum_{i=0}^{N} p^i u_i(x,t) = \lim_{N\to\infty} \sum_{i=0}^{N} u_i(x,t). \tag{10}$$

4 Applications of the Proposed Iterative Approach on Nonlinear Fractional Model of BME

Example 4.1 Consider the following nonlinear fractional BME

$$iD_{\tau}^{\mu} q + \lambda q_{\zeta\zeta}^{2\nu} + \eta |q|^2 q = 0, \tag{11}$$

with

$$q(\zeta, 0) = e^{i\zeta}, \tag{12}$$

Applying LT,

$$L(q(\zeta,\tau)) = \frac{q(\zeta,0)}{s} + \frac{1}{s^{\mu}} L\{\lambda i q_{\zeta\zeta}^{2\nu} + \eta i |q|^2 q\}.$$

Taking inverse LT,

$$q(\zeta,\tau) = L^{-1}\left(\frac{q(\zeta,0)}{s}\right) + L^{-1}\left\{\frac{1}{s^{\mu}} L\left(\lambda i q_{\zeta\zeta}^{2\nu} + \eta i |q|^2 q\right)\right\}. \tag{13}$$

By HPM,

$$\sum_{n=0}^{\infty} p^n q_n(\zeta,\tau) = q(\zeta,0)$$

$$+\, pL^{-1}\left\{\frac{1}{s^{\mu}}L\left[\lambda i\left(\sum_{n=0}^{\infty} p^{n} q_{n}(\zeta,\tau)\right)_{\zeta\zeta}^{2\nu} + \eta i\left(\sum_{n=0}^{\infty} p^{n} H_{n}(q)\right)\right]\right\}, \tag{14}$$

where

$$\sum_{n=0}^{\infty} p^{n} H_{n}(q) = |q|^{2} q.$$

Simplifying, we get

$$\begin{aligned}
H_0(q) &= |q_0|^2 q_0,\\
H_1(q) &= \frac{1}{1!}\frac{\partial}{\partial p}\left[\left(|q_0 + pq_1|^2\right)(q_0 + pq_1)\right]_{p=0},\\
H_2(q) &= \frac{1}{2!}\frac{\partial^2}{\partial p^2}\left[\left(|q_0 + pq_1 + p^2 q_2|^2\right)\left(q_0 + pq_1 + p^2 q_2\right)\right]_{p=0}.
\end{aligned}$$

After simplification, we have the following iterates

$$\begin{aligned}
q_0(\zeta,\tau) &= e^{i\zeta},\\
q_1(\zeta,\tau) &= i\left(\lambda e^{i\pi\nu} + \eta\right)e^{i\zeta}\frac{t^{\mu}}{\Gamma(\mu+1)},\\
q_2(\zeta,\tau) &= \left[\eta\left(\lambda e^{-i\pi\nu} + \eta\right) - 2\eta\left(\lambda e^{i\pi\nu} + \eta\right) - \lambda\left(\lambda e^{2i\pi\nu} + \eta\right)\right]\\
&\quad e^{i\zeta}\frac{t^{2\mu}}{\Gamma(2\mu+1)},\\
q_3(\zeta,\tau) &= \left\{\frac{-i\lambda}{2}\left(\lambda e^{i\pi\nu} + \eta\right)e^{i\zeta} + \frac{3}{2}\eta i\left(\lambda e^{-i\pi\nu} + \eta\right)e^{i\zeta} + \eta i\left(\lambda e^{3i\pi\nu} + \eta\right)e^{i\zeta}\right\}\\
&\quad \frac{t^{3\mu}}{\Gamma(3\mu+1)},
\end{aligned}$$

in the same pattern, further iterations can be calculated Tables 1 and 2. Therefore,

$$q(\zeta,\tau) = q_0(\zeta,\tau) + q_1(\zeta,\tau) + q_2(\zeta,\tau) + q_3(\zeta,\tau) + \ldots$$

Example 4.2 Consider the following nonlinear fractional BME

$$iD_{\tau}^{\mu} q(\zeta,\tau) + \lambda q_{\zeta\zeta}(\zeta,\tau) + \eta|q|^2 q = 0, \tag{15}$$

with

Table 1 L^2 and L^∞ error norms when $\mu = 0.9$, $\nu = 0.9$, $h = 0.001$ **for** Example 4.1

ζ	L^2		L^∞	
	Real part	Imaginary part	Real part	Imaginary part
−0.6	2.7361×10^{-5}	2.9238×10^{-4}	8.6522×10^{-4}	0.0092
−0.4	8.4902×10^{-5}	2.8111×10^{-4}	0.0027	0.0089
−0.2	1.3906×10^{-4}	2.5864×10^{-4}	0.0044	0.0082
0.0	1.8767×10^{-4}	2.2586×10^{-4}	0.0059	0.0071
0.2	2.2880×10^{-4}	1.8407×10^{-4}	0.0072	0.0058
0.4	2.6081×10^{-4}	1.3495×10^{-4}	0.0082	0.0042
0.6	2.8242×10^{-4}	8.0444×10^{-5}	0.0089	0.0025

Table 2 L^2 and L^∞ error norms **when** $\mu = 1$, $h = 0.001$ for Example **4.2**

ζ	L^2		L^∞	
	Real part	Imaginary part	Real part	Imaginary part
−0.6	1.3936×10^{-11}	9.8519×10^{-12}	3.4059×10^{-10}	2.4008×10^{-10}
−0.4	1.5615×10^{-11}	6.8689×10^{-12}	3.8150×10^{-10}	1.6763×10^{-10}
−0.2	1.6672×10^{-11}	3.6475×10^{-12}	4.0720×10^{-10}	8.8496×10^{-11}
0.0	1.7065×10^{-11}	2.6519×10^{-13}	4.1667×10^{-10}	5.8333×10^{-12}
0.2	1.6776×10^{-11}	3.1334×10^{-12}	4.0952×10^{-10}	7.7062×10^{-11}
0.4	1.5820×10^{-11}	6.4038×10^{-12}	3.8605×10^{-10}	1.5688×10^{-10}
0.6	1.4232×10^{-11}	9.4189×10^{-12}	3.4718×10^{-10}	2.3045×10^{-10}

$$q(\zeta, 0) = e^{i\zeta}. \tag{16}$$

Applying LT, Eq. (15) becomes

$$L(q(\zeta, \tau)) = \frac{q(\zeta, 0)}{s} + \frac{1}{s^{\mu}} L\{\lambda i q_{\zeta\zeta} + \eta i |q|^2 q\}. \tag{17}$$

Exerting inverse LT, it yields

$$q(\zeta, \tau) = L^{-1}\left(\frac{q(\zeta, 0)}{s}\right) + L^{-1}\left\{\frac{1}{s^{\mu}} L\left(\lambda i q_{\zeta\zeta} + \eta i |q|^2 q\right)\right\}.$$

By HPM,

$$\sum_{n=0}^{\infty} p^n q_n(\zeta, \tau) = q(\zeta, 0) + pL^{-1}\left\{\frac{1}{s^{\mu}} L\left[\lambda i \left(\sum_{n=0}^{\infty} p^n q_n(\zeta, \tau)\right)_{\zeta\zeta} + \eta i \left(\sum_{n=0}^{\infty} p^n H_n(q)\right)\right]\right\}, \tag{18}$$

where

$$\sum_{n=0}^{\infty} p^n H_n(q) = |q|^2 q.$$

Simplifying, we get

$$H_0(q) = |q_0|^2 q_0,$$
$$H_1(q) = \frac{1}{1!}\frac{\partial}{\partial p}\left[\left(|q_0 + pq_1|^2\right)(q_0 + pq_1)\right]_{p=0},$$
$$H_2(q) = \frac{1}{2!}\frac{\partial^2}{\partial p^2}\left[\left(|q_0 + pq_1 + p^2 q_2|^2\right)\left(q_0 + pq_1 + p^2 q_2\right)\right]_{p=0}.$$

After simplification, we have the following iterates

$$q_0(\zeta, \tau) = e^{i\zeta},$$
$$q_1(\zeta, \tau) = i(\eta - \lambda)e^{i\zeta}\frac{t^{\mu}}{\Gamma(\mu + 1)},$$
$$q_2(\zeta, \tau) = (\eta - \lambda)(\lambda - \eta)e^{i\zeta}\frac{t^{2\mu}}{\Gamma(2\mu + 1)},$$
$$q_3(\zeta, \tau) = \frac{t^{3\mu}}{\Gamma(3\mu + 1)}\left\{\frac{-i\lambda}{2}(\eta - \lambda)(\lambda - \eta)e^{i\zeta} + \frac{3}{2}\eta i(\eta - \lambda)(\lambda - \eta)e^{i\zeta} + \eta i(\eta - \lambda)^2 e^{i\zeta}\right\},$$

in the same pattern, and further iterations can be calculated. Therefore,

$$q(\zeta, \tau) = q_0(\zeta, \tau) + q_1(\zeta, \tau) + q_2(\zeta, \tau) + q_3(\zeta, \tau) + \ldots,$$

5 Discussion and Concluding Remarks

The 2D and 3D graphs of $q(\zeta, \tau)$ are presented to capture the behavior of the nonlinear complex problems. The 2D plots of real and imaginary part of the numerical solution are shown in Fig. 2a, b for Example 4.1 and Fig. 3a, b for Example 4.2 when $\tau = 0.1, \lambda = 1, \eta = 2$. The 3D graphics are demonstrated in Fig. 1a, b for Example 4.1 and Fig. 4a, b for Example 4.2 when $\lambda = 1, \eta = 2$. The accuracy of the suggested method is checked through L^2 and L^{∞} error norms as the number of iterations increases the HPTT solutions converge to the exact solution. All the results are obtained with third iteration, and the accuracy can be improved by computing more terms of the approximate solution. In this paper, an efficient and simple technique is suggested by compiling LT with homotopy perturbation approach. It can be concluded that the proposed approach effectively analyzes the natural phenomena and investigates the nonlinear fractional order equations arising in mathematical physics and other scientific disciplines.

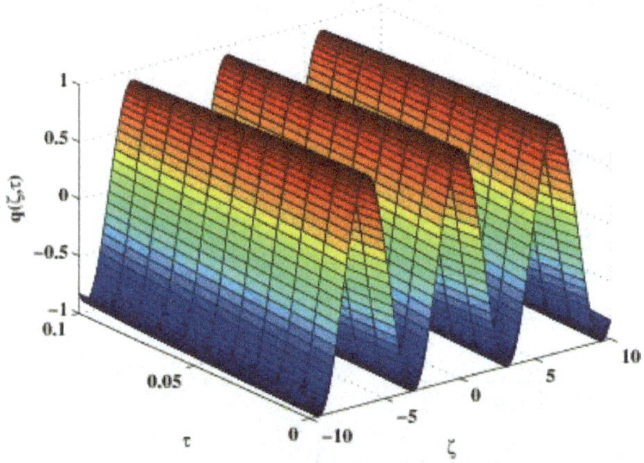

(a). Real division of $q(\zeta,\tau)$ when $\lambda = 1, \eta = 2, \mu = 0.9, \nu = 0.9$ for Ex. 4.1

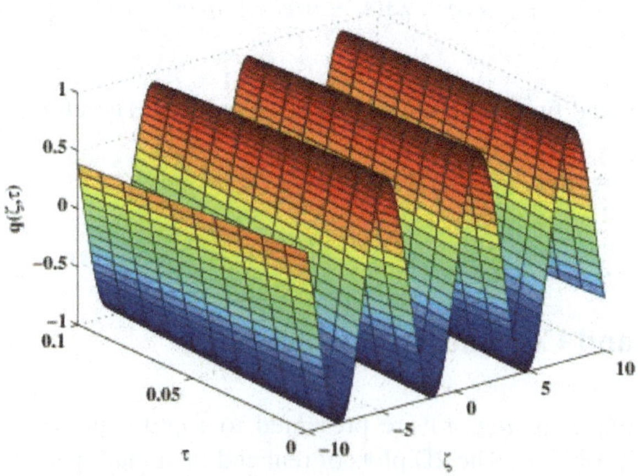

(b). Imaginary division of $q(\zeta,\tau)$ when $\lambda = 1, \eta = 2, \mu = 0.9, \nu = 0.9$ for Ex. 4.1

Fig. 1 **a** Real *division* of $q(\zeta, \tau)$ when $\lambda = 1, \eta = 2, \mu = 0.9, \nu = 0.9$ for Example 4.1. **b** Imaginary division of $q(\zeta, \tau)$ when $\lambda = 1, \eta = 2, \mu = 0.9, \nu = 0.9$ for Example 4.1

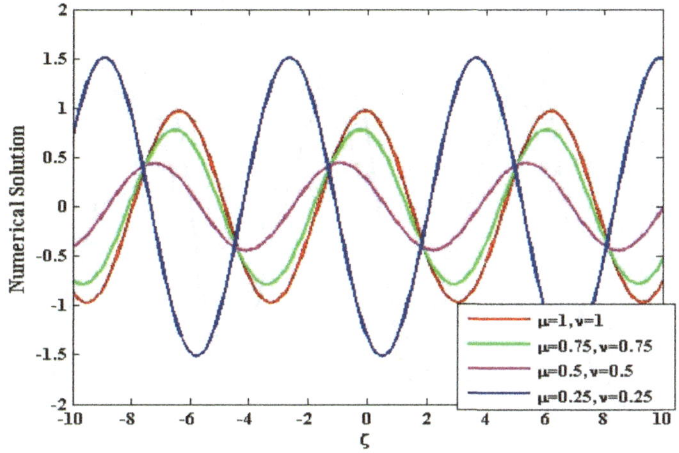

(a). Real part of $q(\zeta, \tau)$ in 2D when $\tau = 0.1, \lambda = 1, \eta = 2$ for Ex. 4.1

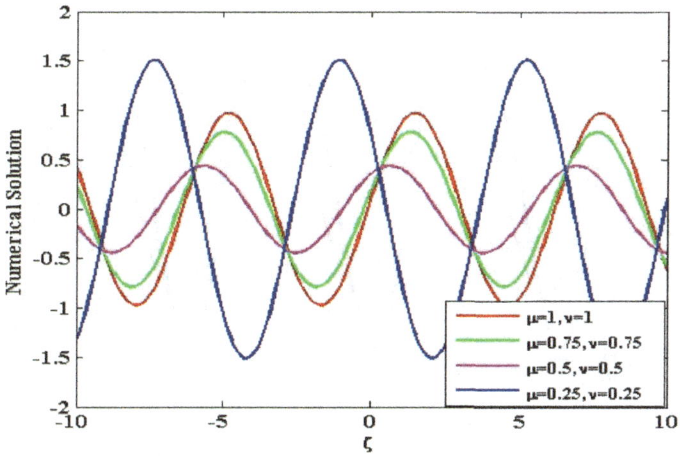

(b). Imaginary part of $q(\zeta, \tau)$ in 2D when $\tau = 0.1, \lambda = 1, \eta = 2$ for Ex. 4.1

Fig. 2 **a** Real part of $q(\zeta, \tau)$ in 2D when $\tau = 0.1, \lambda = 1, \eta = 2$ for Example 4.1. **b** Imaginary part of $q(\zeta, \tau)$ in 2D when $\tau = 0.1, \lambda = 1, \eta = 2$ for Example 4.1

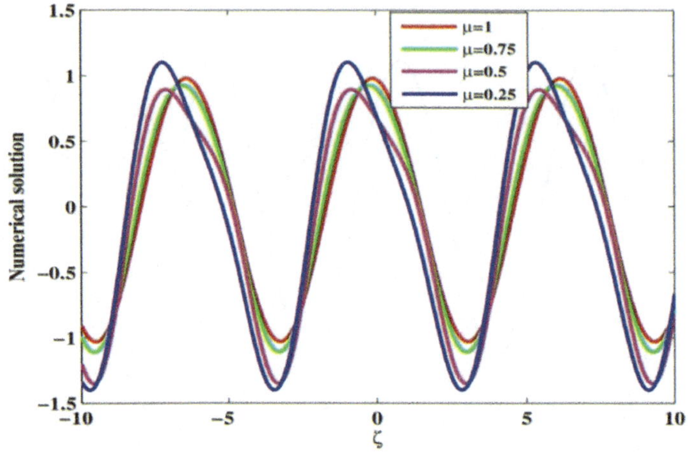

(a). Real part of $q(\zeta,\tau)$ in 2D when $\tau = 0.1, \lambda = 1, \eta = 2$ for Ex. 4.2.

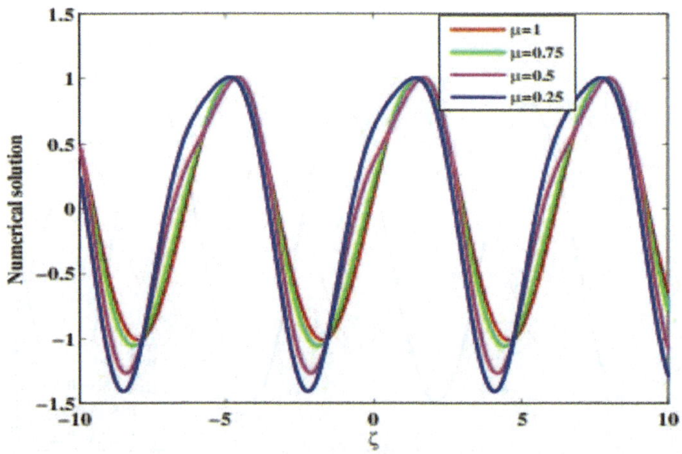

(b). Imaginary part of $q(\zeta,\tau)$ in 2D when $\tau = 0.1, \lambda = 1, \eta = 2$ for Ex. 4.2.

Fig. 3 **a** Real part of $q(\zeta, \tau)$ in 2D when $\tau = 0.1$, $\lambda = 1$, $\eta = 2$ for Example 4.2. **b** Imaginary part of $q(\zeta, \tau)$ in 2D when $\tau = 0.1$, $\lambda = 1$, $\eta = 2$ for Example 4.2

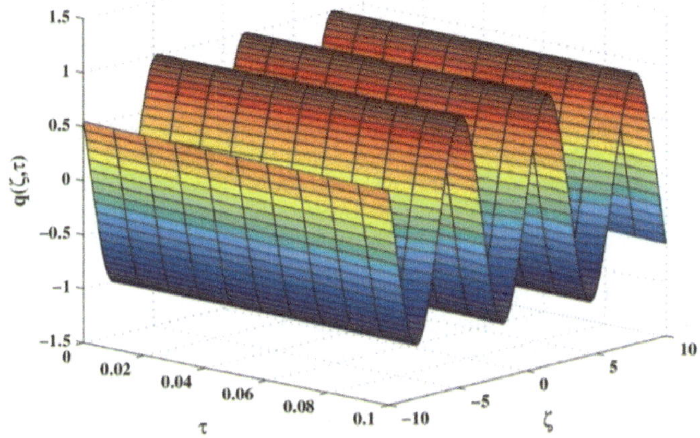

(a) Imaginary division of $q(\zeta,\tau)$ when $\lambda = 1$, $\eta = 2$, $\mu = 0.9$ for Ex. 4.2.

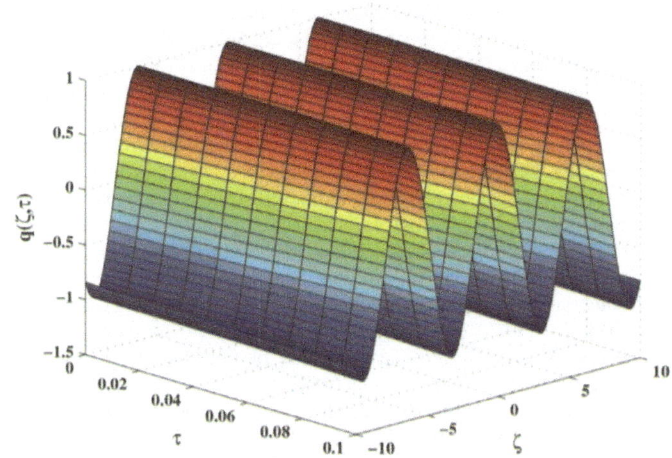

(b) Real division of $q(\zeta,\tau)$ when $\lambda = 1$, $\eta = 2$, $\mu = 0.9$ for Ex. 4.2.

Fig. 4 **a** Imaginary division of $q(\zeta, \tau)$ when $\lambda = 1, \eta = 2, \mu = 0.9$ for Example 4.2. **b** Real division of $q(\zeta, \tau)$ when $\lambda = 1, \eta = 2, \mu = 0.9$ for Example 4.2

References

1. Kumar D, Singh J, Baleanu D (2020) An efficient computational approach for a fractional order biological population model with carrying capacity. Chaos Solitons Fractals 138:109880
2. Prakash A, Kumar M, Baleanu D (2018) A new iterative technique for a fractional model of nonlinear Zakharov–Kuznetsov equations via Sumudu transform. Appl Math Comput 334:30–40

3. Prakash A, Kaur H (2019) A reliable numerical algorithm for fractional model of Fitzhugh–Nagumo equation arising in the transmission of nerve impulses. Nonlinear Eng-Model Appl 8:719–727
4. Prakash A, Veeresha P, Prakasha DG, Goyal M (2019) A homotopy technique for a fractional order multi-dimensional telegraph equation via the Laplace transform. Euro Phys J Plus 134(19):1–18
5. Prakash A, Goyal M, Baskonus HM, Gupta S (2020) A reliable hybrid numerical method for a time dependent vibration model of arbitrary order. AIMS Math 5(2):979–1000
6. Gao W, Veeresha P, Prakasha DG, Baskonus HM (2020) Novel dynamic structures of 2019-nCoV with nonlocal operator via powerful computational technique. Biology 9(5):107
7. Sakar MG, Uludag F, Erdogan F (2016) Numerical solution of time-fractional nonlinear PDEs with proportional delays by homotopy perturbation method. Appl Math Model 40:6639–6649
8. Prakash A, Kaur H (2017) Numerical solution for fractional model of Fokker–Planck equation by using q-HATM. Chaos Solitons Fractals 105:99–110
9. Goyal M, Baskonus HM, Prakash A (2019) An efficient technique for a time fractional model of lassa hemorrhagic fever spreading in pregnant women. Eur Phy J Plus 134(482):1–10
10. Jingtang M, Liu J, Zhou Z (2014) Convergence analysis of moving finite element methods for space fractional differential equations. J Comput Appl Math 255:661–670
11. Goyal M, Baskonus HM, Prakash A (2020) Regarding new positive, bounded and convergent numerical solution of nonlinear time fractional HIV/AIDS transmission model. Chaos Solitons Fractals 139:110096
12. Prakash A, Goyal M, Gupta S (2019) Numerical simulation of space-fractional Helmholtz equation arising in seismic wave propagation, imaging and inversion. Pramana J Phys 93:28
13. Prakash A, Goyal M, Gupta S (2020) q-homotopy analysis method for fractional Bloch model arising in nuclear magnetic resonance via the Laplace transform. Indian J Phys 94:507–520
14. Goyal M, Prakash A, Gupta S (2019) Numerical simulation for time-fractional nonlinear coupled dynamical model of romantic and interpersonal relationships. Pramana J Phys 92:82
15. Veeresha P, Prakasha DG, Magesh NM, Nandeppanavar M, John Christopher A, Sarwe DU (2021) Solution for fractional potential KdV and Benjamin equations using the novel technique. J Ocean Eng Sci. https://doi.org/10.1016/j.joes.2021.01.003
16. Veeresha P, Prakasha DG, Magesh NM, Nandeppanavar M, John Christopher A (2019) Numerical simulation for fractional Jaulent–Miodek equation associated with energy-dependent Schrödinger potential using two novel techniques. Waves Random Complex Media. https://doi.org/10.1080/17455030.2019.1651461
17. Prakash A, Kumar M (2019) Numerical solution of time-fractional order Fokker–Planck equation. TWMS J Appl Eng Math 9(3):446–454
18. Padmavathi V, Prakash A, Alagesan K, Magesh N. Analysis and numerical simulation of novel coronavirus (COVID-19) model with Mittag-Leffler Kernel. Math Methods Appl Sci:1–15. https://doi.org/10.1002/mma.6886
19. Prakash A, Verma V (2021) Two efficient computational technique for fractional nonlinear Hirota-Satsuma coupled KdV equations. Eng Comput 38(2):791–818
20. Verma V, Prakash A, Kumar D, Singh J (2019) Numerical study of fractional model of multi-dimensional dispersive partial differential equation. J Ocean Eng Sci 4:338–351
21. Prakash A, Verma V (2019) Numerical solution of nonlinear fractional Zakharov–Kuznetsov equation arising in ion-acoustic waves. Pramana J Phys 93(66):1–19
22. Prakash A, Veeresha P, Prakasha DG, Goyal M (2019) A new efficient technique for solving fractional coupled Navier–Stokes equations using q-homotopy analysis transform method. Pramana-J Phys 6 (93):1–10
23. Prakash A, Kaur H (2019) A reliable numerical algorithm for fractional model of Fitzhugh–Nagumo equation arising in the transmission of nerve impulses. Nonlinear Eng Model Appl 8:719–727
24. Prakash A, Veeresha P, Prakasha DG (2019) A reliable algorithm for time-fractional Navier–Stokes equations via laplace transform. Nonlinear Eng Model Appl 8:695–701

25. Biswas A, Milovic D (2010) Bright and dark solitons of the generalized nonlinear Schrodinger's equation. Commun Nonlinear Sci Numer Simul 15(5):1473–1484
26. Ahmed I, Mu C, Zhang F (2013) Exact solution of the Biswas–Milovic equation by Adomian decomposition method. Int J Appl Math Res 2(4):418–422
27. Mirzazadeh M, Arnous AH (2015) Exact solution of Biswas–Milovic equation using new efficient method. Electr J Math Anal Appl 3(2):139–146
28. Ahmadian S, Darvishi MT (2016) A new fractional Biswas–Milovic model with its periodic soliton solutions. Optik-Int J Light Electron Opt 127(19):7694–7703
29. Ahmadian S, Darvishi MT (2016) Fractional version of (1+1) dimensional Biswas–Milovic equation and its solutions. Optik 127(21):10135–10147
30. Zaidan LI, Darvishi MT (2017) Semi-analytical solutions of different kinds of fractional Biswas–Milovic equation. Optik 136:403–410
31. Singh J, Kumar D, Baleanu D (2019) New aspects of fractional Biswas–Milovic model with Mittag–Leffler law. Math Modell Natural Phenomena 14(3):303
32. Prakash A, Kaur H (2021) Analysis and numerical simulation of fractional Biswas–Milovic equation via the fractional complex transform. Maths Comput Simulation 181:298–315
33. He JH (1999) Homotopy perturbation technique. Comput Meth Appl Mech Eng 178:257–262
34. Podlubny I (1999) Fractional differential equations. Academic Press, New York

Performance Analysis of Cache Memory Architecture for Core Processor

Reeya Agrawal

Abstract A quantitative and yield analysis of single-bit cache memory architecture with different sense amplifiers such as voltage mode sense amplifier and charge-transfer sense amplifier has been done. Apart from that, power reduction techniques such as the sleep transistor technique, footer stack technique, and dual sleep technique also have been applied over different blocks of single-bit cache memory architecture to reduce the power consumption of the architecture.

Keywords Static random access memory cell (SRAMC) · Voltage mode sense amplifier (VMSA) · Charge-transfer sense amplifier (CTSA) · A sense amplifier (SA) · Write driver circuit (WDC)

1 Introduction

Improved density and efficiency of VLSI circuits have been obtained by scale-down transistors. Local connections are called wires and are used to connect transistors within integrated circuits. The clock and power routing wires within the chip are known as global links on the chip. The latency of global communication becomes essential for many applications, such as buses between cache memories and processors [1–3], as the VLSI technology scales down to the sub-micron level. With the length [4–6], the delay in this global interconnection on the chip raises four-way. Different signaling methods were suggested to transmit the signals by interconnection, and to reduce the delay, different kinds of transceivers were employed. In recent years, the design of low power storage has been powered by the exponential output of battery-operated computers. The leakage current has made the SRAM system a power-hungry block from both static and dynamic perspectives as the transistor count increases. The SRAM block is now also an important part of the SOC architecture.

R. Agrawal (✉)
GLA University, Mathura, India

S. Suhag et al. (eds.), *Control and Measurement Applications for Smart Grid*, Lecture Notes in Electrical Engineering 822,
https://doi.org/10.1007/978-981-16-7664-2_39

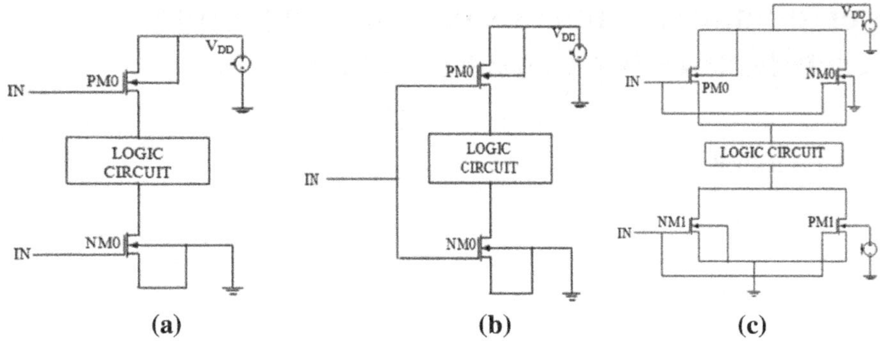

Fig. 1 a Sleep transistor technique. b Forced stack technique. c Dual sleep technique

1.1 Power Reduction Techniques

In the VLSI model, the designers suggested many strategies for reducing the power consumption of the circuit. The short descriptions of all the methods used in the circuits are given below [7].

1.1.1 Sleep Transistor Technique

The sleep transistor technique [8] is the most well-known method. A P_{MOS} transistor "Sleep" is situated between V_{DD} and network pull-up, and an N_{MOS} transistor is positioned between pull-down network and G_{ND} as shown in Fig. 1a.

1.1.2 Forced Stack Technique

Another power reduction technique is the stacking strategy, which pushes an individual transistor into two half-size transistors [9]. The result of the transistor stacking contributes to a reduction in the decrease in sub-threshold current. This strategy saves the actual state while the transistor is in the off state as shown in Fig. 1b.

1.1.3 Dual Sleep Technique

Both P_{MOS} and N_{MOS} types of transistors are used in this technique. Both P_{MOS} and N_{MOS} transistors are used in the header and footer. In on-state mode, one transistor is on and another transistor is switched off in the off-state mode. Both P_{MOS} and N_{MOS} are used in off-state mode to decrease leakage capacity as shown in Fig. 1c [10].

The single-bit cache memory architecture is clarified in Sect. 2 in depth, and SRAM architecture is explained. Experimental effects are shown in Sect. 3. Conclusions are obtained in Sect. 4.

2 Single-Bit Cache Memory Architecture

In this section, cache memory architecture has been described with their architectures as shown in Figs. 2 and 3. SRAM architecture comprises WDC, SRAMC, and SA. SA is of two types: (a) the voltage difference between the bit lines is amplified to determine the output voltage; voltage mode sense amplifier has five input pins (BL, BLB, Y_{sel}, Pch, and SA_{en}) and two output pins (V_3 and V_4); (b) charge-transfer sense amplifier, the higher bit line capacitance charge propagation can be used in narrower lines to achieve the output voltage; it has five input pins (BL, BLB, Y_{sel}, Pch, and SA_{en}) and two output pins (V_3 and V_4).

2.1 Write Driver Circuit

The write driver is responsible for rapidly discharging the bit lines to a level below the cell's write margin before or when the selected cell's word lines are involved. Two standard written drivers are seen in Fig. 4. The input of the data chooses which bit line is discharged. Just when the writing process is intended is the WE signal switched on.

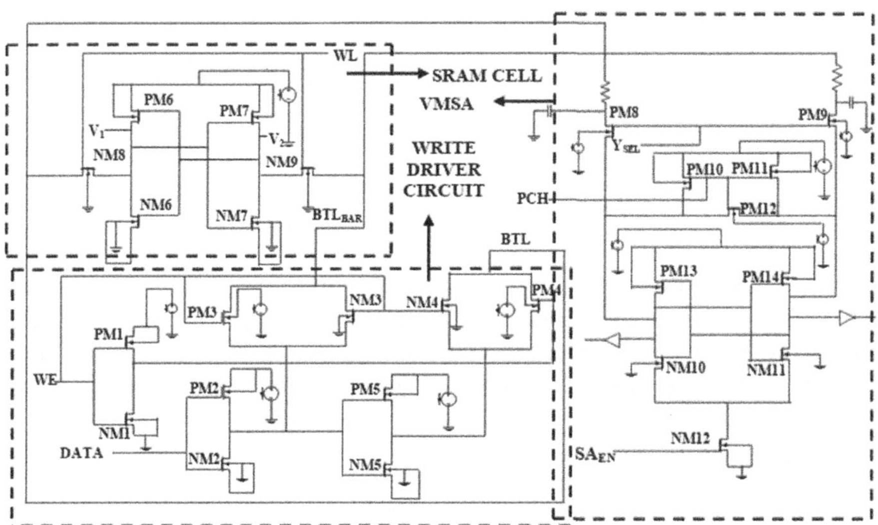

Fig. 2 Schematic of single-bit SRAM VMSA architecture

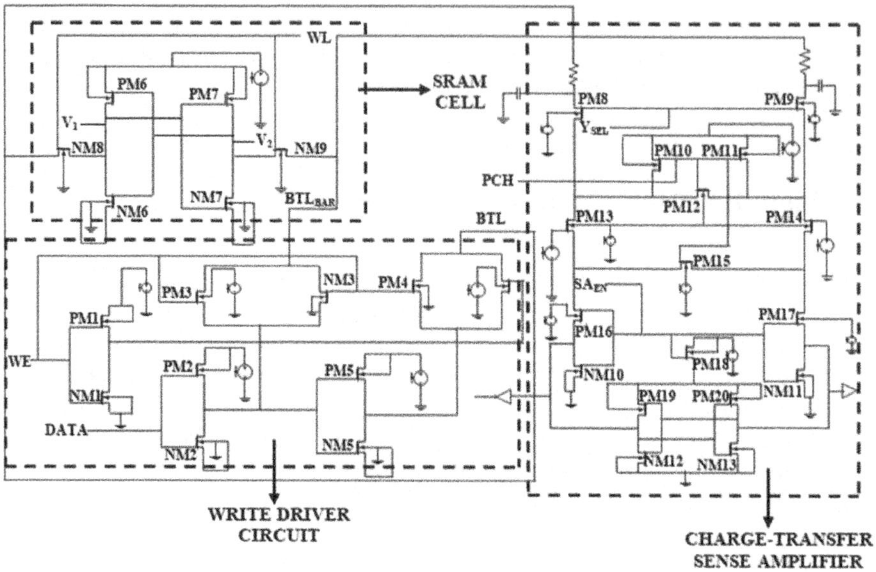

Fig. 3 Schematic of single-bit SRAM CTSA architecture

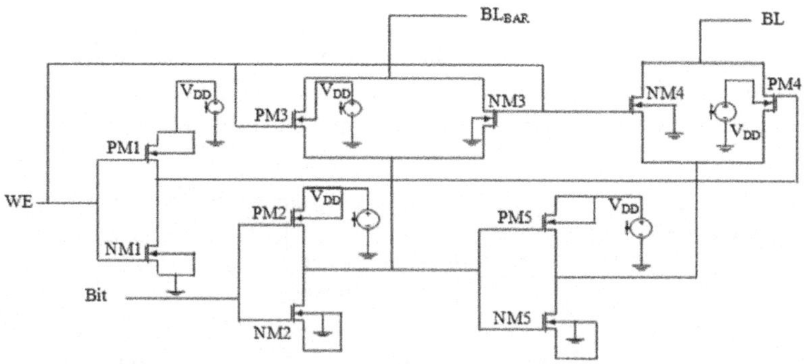

Fig. 4 Write driver circuit schematic

Otherwise, WE distinguishes the bit lines from the drivers of printing. It is quicker because, at the cost of sophistication, it has fewer stacked transistors in its discharge direction. Usually, the write operation is not a transaction that restricts speed, so the write driver is chosen for simplified setups that relax the layout specifications [11–13].

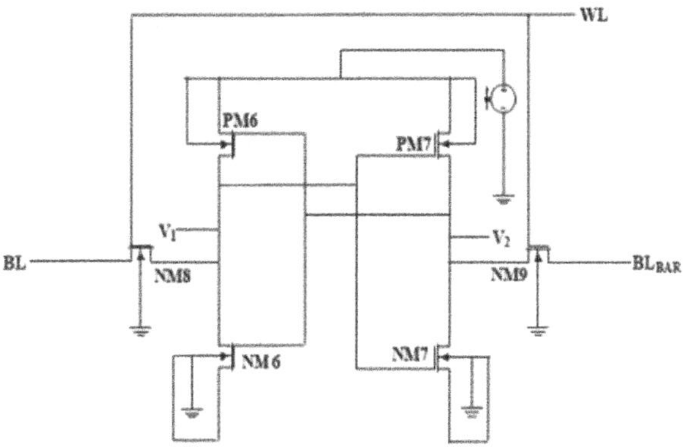

Fig. 5 SRAM cell **schematic**

2.2 *Conventional SRAM*

It is used for operations at low power, low voltage. Here, each bit is stored using bistable latching circuitry. Figure 5 shows the 6T SRAMC schematic, the pull-up transistors are M_1 and M_2 (P_{MOS}), while the driver transistors are M_3 and M_4 N_{MOS}. These bit lines enhance the margin of noise. The value of measurable output voltage swings is given by differential circuitry. Logic 0 or 1 is stored as long as the power is on, but unlike DRAM cells [14, 15], it does not need to be refreshed. In SRAM architecture, the size of the transistors is most important for the proper operation of the transistors.

2.3 *Sense Amplifiers*

The sensor amplifier amplifies a small analog differential voltage produced on the read-access bit lines. The amplification leads to a complete one-end digital output. Because of the length of the metal and because a lot of transistors take a long time to discharge the bit lines, bit lines have more power. Timing regulation and filling condenser set are hard choices for sensory amplifiers here [16].

2.3.1 Voltage Mode Sense Amplifier.

The power amplifier function is based on the differential voltage produced by the bit lines. The circuit consists of cross-connected inverters that convert the bit-line voltage difference at its entrance to full swing output, as shown in Fig. 6. The cell columns integrate BL and BLB inputs with the cell column bit lines. P_1 binds the memory cell to the P_2 sensory boost and N_3 activates the sensation boost. The inner nodes of the sensory amplifier are separated by output inverters from the external

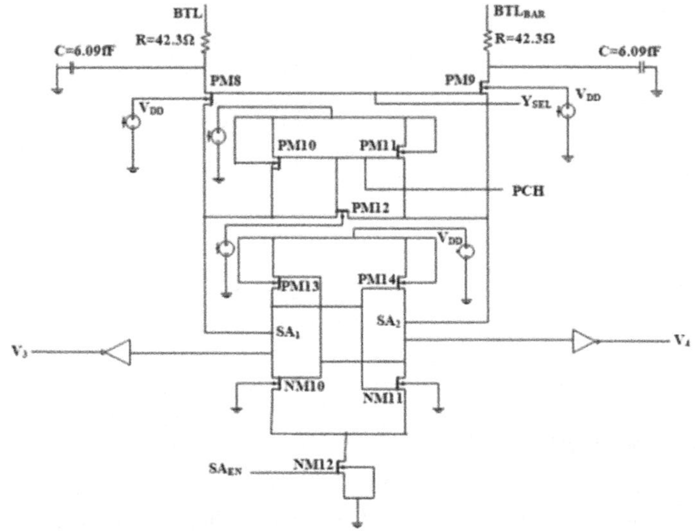

Fig. 6 Schematic of voltage mode sense amplifier

load. The sensor amplification is applied to the memory cell by returning the selected line during the evaluation process (SA_{en}) [17, 18].

2.3.2 Charge-Transfer Sense Amplifier

Figure 7 illustrates the extension of this concept to SRAM sense amplifiers, where the broad data line capacitance of the CDL is related to the limited capacitance of the sensing node, CTSA.

The basic principle behind the amplification of the charging transmission is to generate voltage gains by manipulating the conservation of charge among capacitive devices [19, 20]. The voltage on the first element and its capacity have to equal the voltage on the other element product and its capacity for a set of contacts of two capacitive elements in a device to be loaded.

3 Results and Discussion

Figure 8 describes the output waveform of WDC, for cases arise: (a) when Bit = 0 V and WE = 0 V BL = V_{DD} and BL_{BAR} = V_{DD}, (b) Bit = 0 V WE = V_{DD} so, BL = 0 V and BL_{BAR} = $V_{DD}/2$, c) Bit = V_{DD} WE = 0 V so, BL = 0 V and BL_{BAR} = $V_{DD}/2$ and d) Bit = V_{DD} WE = V_{DD} so, BL = V_{DD} and BL_{BAR} = 0 V.

Figure 9 describes both write operation and hold operation of the SRAMC. There is a pull of the n/w (PM6 and PM7), pull-down network (NM6 and NM7), and access

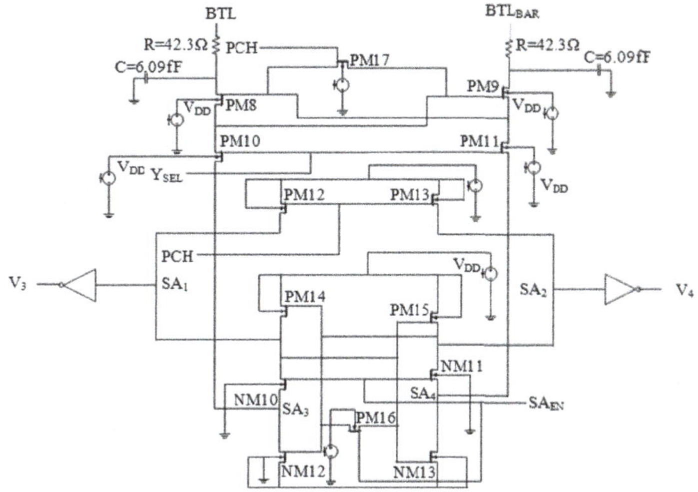

Fig. 7 Schematic of charge-transfer sense amplifier

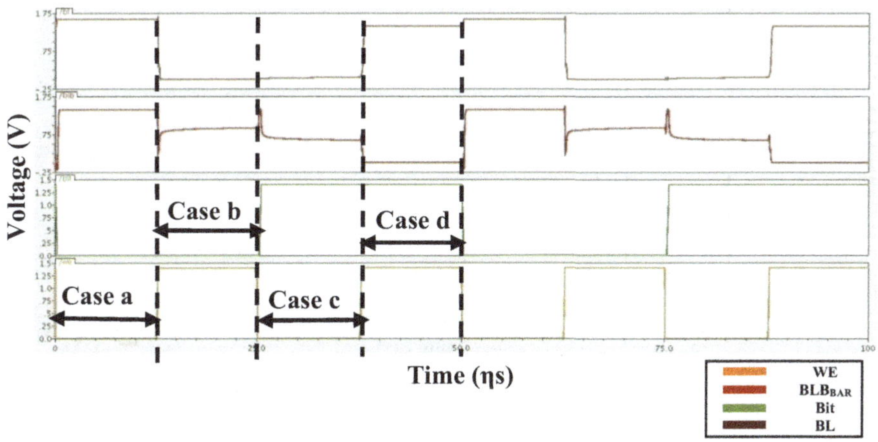

Fig. 8 Output waveform of WDC

transistor (NM8 and NM9) which allows data to store and sense amplifier to read the data.

Figures 10 and 11 describe the read operation of VMSA and CTSA when both SA_{EN} and WL are pulled high; during that time, only the SA senses the data from the SRAMC at bit lines and gives output at V_3 and V_4.

Note: $P = V^2/R$ as this voltage is constant on varying the R and analyzing the power consumption.

Table 1 describes that increase in value of resistance power consumption decreases as because resistance is a path stopper for current in a circuit and no effect on the

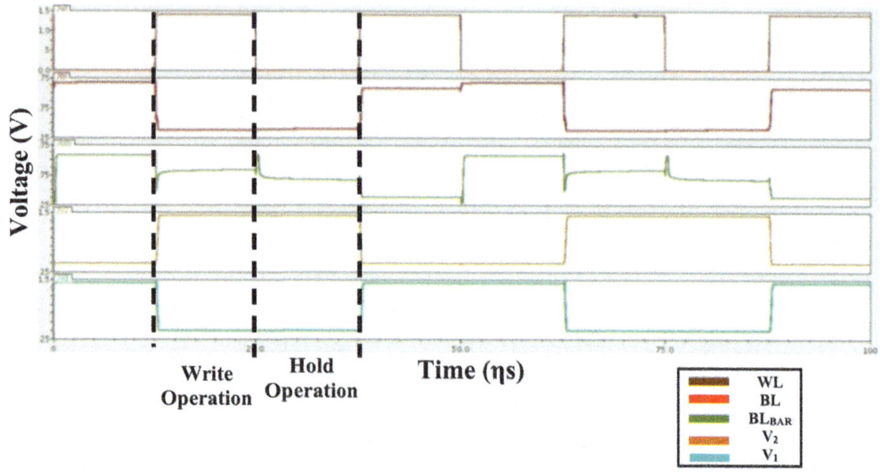

Fig. 9 Output waveform of SRAMC

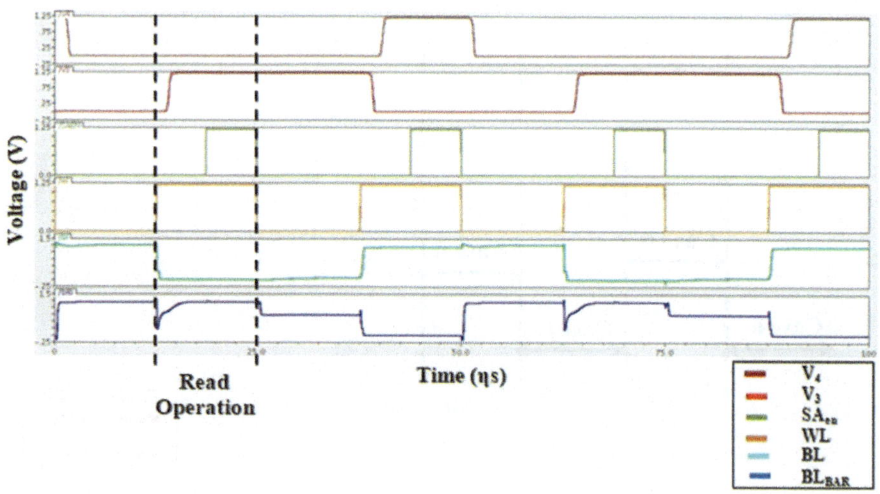

Fig. 10 Output waveform of VMSA

area, performance, and speed, whereas Fig. 12 shows the comparison of the different parameters of a single-bit SRAM VMSA architecture of Table 1 in form of a chart.

Table 2 describes that as increasing in value of resistance power consumption decreases as because resistance is a path stopper for current in a circuit and no effect on the area, performance, and speed, whereas Fig. 13 shows the comparison of the different parameters of a single-bit SRAM CTSA architecture of Table 2 in form of a chart.

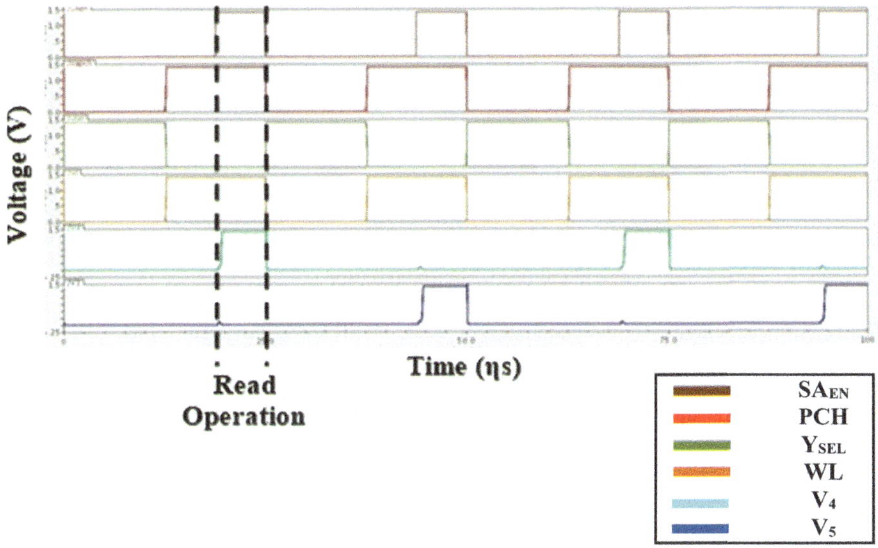

Fig. 11 Output waveform of CTSA

Table 1 Different parameters of single-bit SRAM VMSA architecture

S. No	Parameters	Single-bit SRAM VMSA architecture		
		Power consumption (μW)	No. of transistors	Sensing delay (ηs)
1	$R = 42.3\ \Omega$	13.16	30	13.14
2	$R = 42.3\ \text{K}\Omega$	11.34	30	13.14

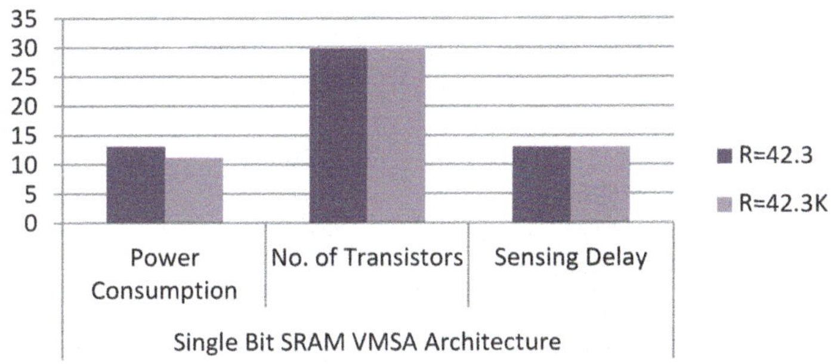

Fig. 12 Comparison of different parameters of single-bit SRAM VMSA architecture

Table 2 Different parameters of single-bit SRAM CTSA architecture

S. No	Parameters	Single-bit SRAM CTSA architecture		
		Power consumption (μW)	No. of transistors	Sensing delay (ηs)
1	$R = 42.3\Omega$	46.35	37	13.51
2	$R = 42.3\text{K}\Omega$	44.32	37	13.51

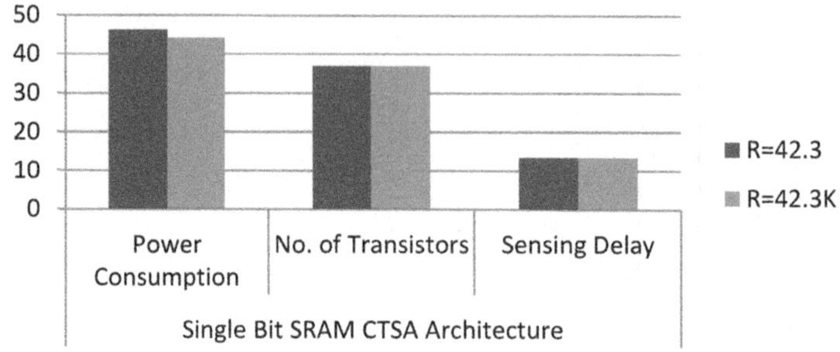

Fig. 13 Comparison of different parameters of single-bit SRAM CTSA architecture

Table 3 Power consumption of single-bit cache memory architecture

S. No		Single-bit SRAM VMSA architecture		Single-bit SRAM CTSA architecture	
		Power consumption (μW)	No. of transistor	Power consumption (μW)	No. of transistor
1	Sleep transistor	11.29	32	20.38	39
2	Forced stack	11.29	32	20.38	39
3	Dual sleep	11.03	34	21.05	41

Table 3 describes that applying power reduction technique over SA forced stack technique in single-bit cache memory architecture reduced power consumption, i.e., 11.03 μW with 34 number of the transistor, whereas Fig. 14 shows the comparison

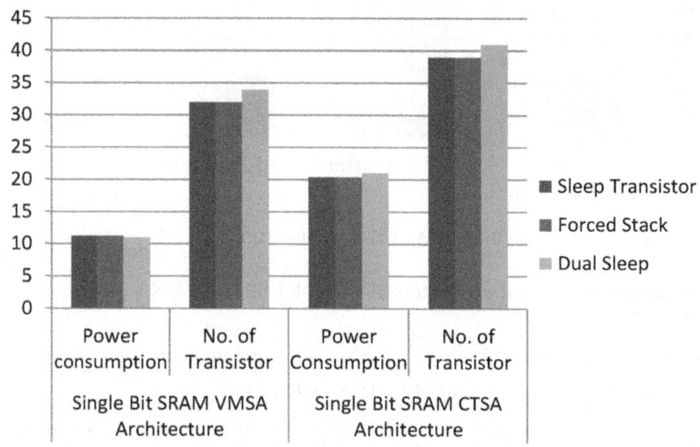

Fig. 14 Comparison of power consumption of single-bit cache memory architecture on applying power reduction techniques over SA

Table 4 Power consumption of single-bit cache memory architecture

S. No		Single-bit SRAM VMSA architecture		Single-bit SRAM CTSA architecture	
		Power consumption (μW)	No. of transistor	Power consumption (μW)	No. of transistor
1	Sleep transistor	9.18	34	18.18	41
2	Forced stack	9.108	34	18.17	41
3	Dual sleep	10.13	38	18.91	45

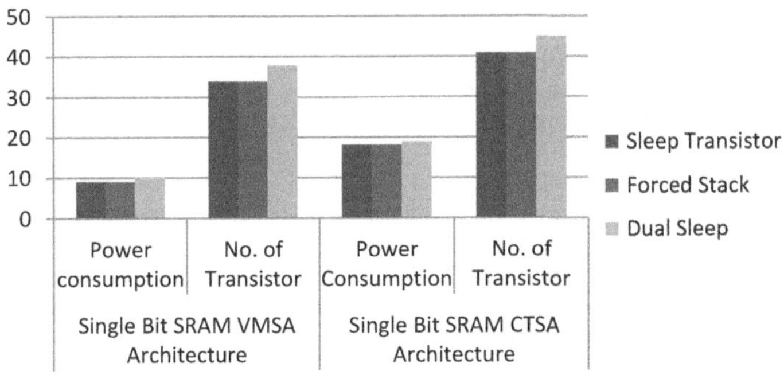

Fig. 15 Comparison of power consumption of single-bit cache memory architecture on applying power reduction techniques over SRAM and SA

of power consumption of single-bit cache memory architecture on applying power reduction techniques over SA of Table 3 in form of a chart.

Table 4 describes that applying power reduction techniques over SRAM and SA consumes up to 9.108 μW power which is lowest as compared to others, but the number of transistors increases transistor, whereas Fig. 15 shows the comparison of power consumption of single-bit cache memory architecture on applying power reduction techniques over SRAM and SA in architecture of Table 4 in form of a chart.

4 Conclusion

In this paper, single-bit cache memory with different sense amplifiers such as voltage differential sense amplifier and charge-transfer sense amplifier has been implemented and compared on different values of resistance (R) with different parameters such as power consumption, sensing delay, and several transistors. Apart from that, power reduction technique has been applied over different blocks of single-bit cache memory architecture and results depicted that the single-bit cache memory architecture having voltage mode differential sense amplifier with forced stacked consumes the lowest power (9.108 μW). In the future scope, this work can be done in form of an array.

References

1. Eslami N, Ebrahimi B, Shakouri E et al (2020) A single-ended low leakage and low voltage 10T SRAM cell with high yield. Analog Integr Circ Sig Process 105:263–274
2. Bazzi, H., Harb, A., Aziza, H. et al. RRAM-based non-volatile SRAM cell architectures for ultra-low-power applications. Analog Integr Circ Sig Process (2020).
3. S. Gupta, K. Gupta, B. H. Calhoun, and N. Pandey, "Low-Power Near-Threshold 10T SRAM Bit Cells With Enhanced Data-Independent Read Port Leakage for Array Augmentation in 32-nm CMOS," in IEEE Transactions on Circuits and Systems I: Regular Papers, vol. 66, no. 3, pp. 978-988, March 2019.
4. H. Dounavi, Y. Sfikas, and Y. Tsiatouhas, "Periodic Aging Monitoring in SRAM Sense Amplifiers," 2018 IEEE 24th International Symposium on On-Line Testing And Robust System Design (IOLTS), Platja d'Aro, 2018, pp. 12-16.
5. Ahmad S, Iqbal B, Alam N, Hasan M (2018) Low leakage fully half-select-free robust SRAM cells with BTI reliability analysis. IEEE Trans Device Mater Reliab 18(3):337–349
6. B. N. K. Reddy, K. Sarangam, T. Veeraiah, and R. Cheruku, "SRAM cell with better read and write stability with Minimum area," TENCON 2019 - 2019 IEEE Region 10 Conference (TENCON), Kochi, India, 2019, pp. 2164-2167.
7. Tripathi Tripti, Chauhan D. S., Singh S. K., and Singh S. V. "Implementation of Low-power 6T SRAM cell using MTCMOS technique", In: Advances in computer and computational sciences, Springer, Singapore, 2017.
8. M.Geetha Priya, Dr.K.Baskaran, D.Krishnaveni "Leakage power reduction techniques in deep submicron technologies for VLSI applications" ELSEVIER, International Conference on Communication Technology and System Design 2011.
9. K Sridhara, G S Biradar, Raju Yanamshetti, "Subthreshold leakage power reduction in VLSI circuits: A survey", Communication and Signal Processing (ICCSP) 2016 International Conference on, pp. 1120–1124, 2016.
10. Gomes Iuri A.C., Meinhardt Cristina, Butzen Paulo F. "Design of 16nm SRAM Architecture" South Symposium on Microelectronics, 2012.
11. Chakka Sri Harsha Kaushik, Rajiv Reddy Vanjarlapati, Varada Murali Krishna, Tadavarthi Gautam, V Elamaran, "VLSI design of low power SRAM architectures for FPGAs", Green Computing Communication and Electrical Engineering (ICGCCEE) 2014 International Conference on, pp. 1–4, 2014.
12. Richa Choudhary, Srinivasa Padhy, Nirmal Kumar Rout, "Enhanced Robust Architecture of Single Bit SRAM Cell using Drowsy Cache and Super cut-off CMOS Concept", International Journal of Industrial Electronics and Electrical Engineering, Volume-3, PP.63–68, July 2011.
13. Jesal P. Gajjar, Aesha S. Zala, Sandeep K. Aggarwal, "Design and analysis of 32 bit SRAM architecture in 90nm CMOS technology" Volume: 03, Issue: 04, Apr-2016, pp:2729–2733.
14. Reeya Agrawal, V. K. Tomar "Analysis of Cache (SRAM) Memory for Core I ™ 7 Processor", 9th International Conference on Computing, Communication and Networking Technologies (ICCCNT), 2018,402.
15. Kundan Vanama, Rithwik Gunnuthula, Govind Prasad, "Design of low power stable SRAM cell", Circuit Power and Computing Technologies (ICCPCT) 2014 International Conference on, pp. 1263–1267, 2014.

16. Rakesh Dayaramji Chandankhede, Debiprasad Priyabrata Acharya, Pradip Kumar Patra, "Design of high-speed sense amplifier for SRAM", IEEE International Conference on Advanced Communication Control and Computing Technologies, pp. 340–343, 2016
17. Zikui Wei, Xiaohong Peng, JinhuiWang, Haibin Yin, Na Gong, "Novel CMOS SRAM voltage latched sense amplifiers design based on 65nm technology" pp.3281–3282, 2016.
18. B. Wicht, T. Nirschl, and D. Schmitt-Landsiedel, "Yield and speed optimization of a latch-type voltage sense amplifier," in the IEEE Journal of Solid-State Circuits, vol. 39, no. 7, pp. 1148–1158, July 2004.
19. T. Kobayashi, K. Nogami, T. Shirotori, Y. Fujimoto and O. Watanabe, "A current-mode latch sense amplifier and a static power-saving input buffer for low-power architecture," *1992 Symposium on VLSI Circuits Digest of Technical Papers*, Seattle, WA, USA, 1992, pp. 28-29.
20. Kobayashi T, Nogami K, Shirotori T, Fujimoto Y (1993) A current-controlled latch sense amplifier and a static power-saving input buffer for low-power architecture. IEEE J Solid-State Circuits 28(4):523–527

The manufacturer's authorised representative in the EU is Springer Nature Customer Service Centre GmbH, Europaplatz 3, 69115 Heidelberg, Germany. If you have any concerns regarding our products, please contact ProductSafety@springernature.com

Printed and bound by CPI Group (UK) Ltd, Croydon, CR0 4YY

30/09/2026

02225332-0015